An Introduction to Synchrotron Radiation

Techniques and Applications

Second Edition

PHILIP WILLMOTT

WILEY

This edition first published 2019
© 2019 John Wiley & Sons Ltd

Edition History
An Introduction to Synchrotron Radiation: Techniques and Applications First edition, Wiley 2011

Registered Offices
John Wiley & Sons, Inc., 111 River Street, Hoboken, NJ 07030, USA
John Wiley & Sons Ltd, The Atrium, Southern Gate, Chichester, West Sussex, PO19 8SQ, UK

Editorial Office
The Atrium, Southern Gate, Chichester, West Sussex, PO19 8SQ, UK

For details of our global editorial offices, customer services, and more information about Wiley products visit us at www.wiley.com.

Wiley also publishes its books in a variety of electronic formats and by print-on-demand. Some content that appears in standard print versions of this book may not be available in other formats.

Library of Congress Cataloging-in-Publication Data

Names: Willmott, Philip (Phil R.), author.
Title: An introduction to synchrotron radiation : techniques and applications/
    Philip Willmott, Swiss Light Source, Villigen, Switzerland.
Description: Second edition. | Hoboken, New Jersey : John Wiley & Sons, Inc.,
    [2019] | Includes bibliographical references and index. |
Identifiers: LCCN 2018054068 (print) | LCCN 2018060648 (ebook) | ISBN
    9781119280378 (Adobe PDF) | ISBN 9781119280385 (ePub) |
    ISBN 9781119280392(pbk.)
Subjects: LCSH: Synchrotron radiation. | X-ray optics.
Classification: LCC QC793.5.E627 (ebook) | LCC QC793.5.E627 W55 2019 (print) |
    DDC 539.7/35–dc23
LC record available at https://lccn.loc.gov/2018054068

Cover Design: Wiley
Cover Images: Image courtesy of the Author using image from the RCSB PDB (www.rcsb.org) of 1PRC (Deisenhofer, J., Epp, O., Sinning, I., Michel, H. (1995) Crystalline refinement at 2.3 angstroms resolution and refined model of the photosynthetic reaction center from rhodopseudomonas viridis J. Mol. Biol. 246 429–457) created with QtMG.

Set in 10/12pt Warnock by SPi Global, Pondicherry, India

10 9 8 7 6 5 4 3 2 1

To Ella, Jo, Nyah, and Hanah.

An Introduction to
Synchrotron Radiation

Contents

Preface

In the seven years since the first edition of this book, there have been critical advances in both x-ray-source technology and experimental methods, leading to previously unimagined vistas of scientific discovery. In particular, two developments in x-ray sources are revolutionizing x-ray science and techniques. First, the emergence of high-gain, hard x-ray free-electron lasers (XFELs) has resulted in, among other things, the birth of new fields in macromolecular crystallography and time-resolved surface and catalytic chemistry. Even more recently, advances in vacuum and computer-numerical-control machining technologies have facilitated the realization of novel magnetic storage-ring components that substantially improve the electron-beam quality, in so-called diffraction-limited storage rings (DLSRs). The greenfield facilities of MAX-IV at Lund, Sweden, and Sirius in Campinas, Brazil, are the first of these fourth-generation synchrotrons, closely followed by several upgrades of third-generation facilities to DLSRs. They promise an increase in brilliance of up to two orders of magnitude compared to the state-of-the-art synchrotrons available at the time of writing the first edition of this book.

This rapid evolutionary phase is reflected in a certain readjustment of the emphasis of some chapters in this second edition. In particular, Chapter 3 has been significantly expanded to include a more thorough description of DLSR science and technology, which in turn has an impact on beamline design (Chapter 5) and experimental techniques, not least in macromolecular crystallography (Section 6.11) and x-ray imaging (Chapter 8). XFELs now command a chapter of their own. All the figures in this second edition are available online at Wiley as PowerPoint slides.

In addition to up-to-date examples of applications of various x-ray techniques, each chapter in this new edition contains problem sets, plus comprehensive solutions in the Appendices. Some of the problems are a straightforward plugging-in of numbers to equations presented in the main text – they are not intended to constitute an intellectual challenge, but rather to provide the student with an opportunity to appreciate the magnitudes of things, and their potential impact on other considerations, both practical and fundamental. Other problems, however, require some thought and a deeper understanding of the background science. It is hoped that, as such, these problems will provide an adjunct to the main text and furnish additional insights into the fascinating, multidisciplinary, and ever-expanding field of x-ray science.

Philip Willmott, Zürich, 2018

Acknowledgements

I am greatly indebted to many colleagues for their patience and tolerance in the face of my repeated questioning (some might say badgering), and for their critical reviewal of the manuscript and suggestions for improvements for this, the second edition.

In particular, I would like to thank Michael Böge, Anne Bonnin, Oliver Bunk, Marco Calvi, Nicola Casati, Ana Diaz, Uwe Flechsig, Ronald Frahm, Daniel Grolimund, Jerry Hastings, Gerhard Ingold, Juraj Krempasky, Federica Marone, Andreas Menzel, Chris Milne, Craig Morrison, Frithjof Nolting, Vincent Olieric, Bruce Patterson, Eduard Prat, Christian Schlepütz, Thomas Schmidt, Marco Stampanoni, Jörg Standfuss, Andreas Streun, Meitian Wang, Ben Watts, Tobias Weinert, Qing Wu, and the editorial team at Wiley, for their insights into all the aspects that hopefully make this edition a more rounded product than its predecessor.

About the Companion Website

To access supplementary materials for this book please use the download links shown below. There you will find valuable material designed to enhance your learning, including:

- PPTs of all the figures

This book is accompanied by a companion website:

http://booksupport.wiley.com

Please enter the book title, author name or ISBN to access this material.

1

Introduction

How does nature create a living thing from simple chemical components? How is a complete human being constructed nine months after the unique mixing of its parents' genetic material contained in a single fertilized egg?

When I was a young child, these questions were only just beginning to be answered. Indeed, just five months before my own birth, the Nobel Committee announced that the 1962 Prize in Physiology or Medicine would be awarded to James Watson, Francis Crick, and Maurice Wilkins[1] for their elucidation of the structure of deoxyribonucleic acid (DNA) [1] 'and its significance for information transfer in living material'.

DNA resides permanently within the nucleus of a living cell. How then is its information conveyed to the rest of the cell and elsewhere in the organism; how is new living material subsequently synthesized; and how do initial ('stem') cells in the blastocyst differentiate into specialized cells as diverse as ganglions, muscle cells, or skin? Major steps towards answering these questions were made in research that would result in Nobel Prizes in Chemistry in 2006 and 2009.

The first step in carrying out the commands encoded in DNA is its transcription into so-called messenger-RNA (mRNA), which then leaves the confines of the nucleus to the protein-producing parts of the cell. How mRNA is synthesized and carries the genetic information with high fidelity, via a very special macromolecule called RNA-polymerase, was reported by Roger Kornberg and co-workers in 2001 [2], for which he received the 2006 Nobel Prize in Chemistry.

The following step, protein synthesis, uses the blueprint of the mRNA, plus raw material in the form of amino acids attached to small transfer-RNA (tRNA) molecules in a chemical factory called the ribosome. The detailed description of the mechanisms by which this highly complex biomolecular process occurs won Ada Yonath, Thomas Steitz, and Venkatraman Ramakrishnan the Nobel Prize in Chemistry in 2009 [3, 4].

As should be apparent from its title, this textbook is not a treatise on molecular biology. It is, however, concerned with diverse aspects of a tool now considered indispensable to molecular biologists – synchrotron radiation. The thematic link between the above trilogy of Nobel Prizes is that of structure, and its inherent importance in determining biomolecular activity. Although the structure of DNA was determined to a large extent by a combination of model building, biochemical know-how (including the application of Chargaff's rules), and some guesswork, it

---

[1] Wilkins' contribution to this work was minor, the crucial crystallographic data being recorded by his King's College colleague, Rosalind Franklin, and her student Raymond Gosling. Controversially, Wilkins showed the now famous 'photograph 51' to Watson and Crick without Franklin's knowledge or permission. Its characteristic cross-motif provided the Cambridge pair with the last key information about the precise helical nature of DNA, from which they swiftly deduced the detailed structure. Sadly, Franklin was to die of ovarian cancer (possibly caused by exposure to x-rays) in 1958, four years before the prize was announced.

An Introduction to Synchrotron Radiation: Techniques and Applications, Second Edition. Philip Willmott.
© 2019 John Wiley & Sons Ltd. Published 2019 by John Wiley & Sons Ltd.

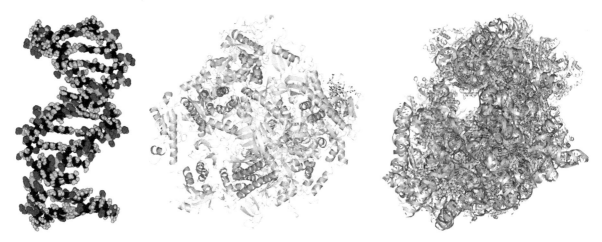

Figure 1.1   The three central biostructures associated with the 'Nobel trilogy of life'. Left: a DNA dodecamer; centre: the yeast RNA polymerase subunit; right: the 80S ribosome from *Tetrahymena cerevisiae*. Rendered from the pdb files 1bna, 1i3q, and (4v5o + 4v8p), respectively.

was x-ray diffraction data which provided the last key information in solving the puzzle[2]. The latter two discoveries of the 'Nobel life trilogy' depended intimately on x-ray diffraction (see Figure 1.1 and Table 1.1).

The relevance of these studies goes well beyond pure academic understanding – the detailed description of DNA transcription and the difference in this process between eukaryotic organisms (with intracellular nucleus) and prokaryotic organisms (without intracellular nucleus), as well as a knowledge of the biochemical function of the ribosome, are already having a huge impact on the design of novel antibiotic substances and other drugs, and provide a deeper understanding of how some cancers arise and how they might be tackled.

The year that Yonath, Ramakrishnan, and Steitz were awarded their Nobel Prize also saw first light at the pioneering hard x-ray free-electron laser (XFEL), the Linac Coherent Light Source (LCLS) at the SLAC facility in California. The first report of a new biological structure (to a resolution of 2.1 Å) determined using the LCLS was reported a little over three years later [5] (see Figure 1.2). XFELs provide x-ray beams with unique properties regarding pulse duration, peak pulse intensity, and beam cross-section. Using a technique coined 'diffraction-before-destruction', crystals with sizes orders of magnitude smaller in volume than hitherto possible can be investigated. As the growth of sufficiently large and well-ordered crystals has to date been *the* major bottleneck in biomolecular-structure studies using x-rays from synchrotrons, the application of this technique at XFELs represents a sea change in this field of research.

The above examples beautifully illustrate that, since their discovery in late 1895 by Wilhelm Röntgen, x-rays have played a pivotal rôle in society, particularly in medicine, pharmacy, physics, and chemistry. Whereas research using x-rays was originally the dominion of physicists, x-rays are now a ubiquitous tool for research in almost all branches of scientific endeavour (see Figure 1.3). They allow us, among many other things, to determine the internal architecture of cells and other biological structures; to identify the chemical composition, fabrication techniques, and provenance of archaeological artefacts; and even to examine the previously hidden earlier artistic efforts of one of the foremost influential post-impressionist painters.

---

[2] In the same year that Watson, Crick, and Wilkins won their Nobel Prize, two of their colleagues at the Cavendish Laboratory, John Kendrew and Max Perutz, won the Nobel Prize in Chemistry for their structural studies of myoglobin and haemoglobin, respectively, also using x-ray diffraction, thereby demonstrating this technique could be used to determine structures of hitherto unimagined complexity.

Table 1.1   Nobel Prizes awarded in the field of x-ray research.

Year	Recipient(s)	Research discipline
1901	W. C. Röntgen	Physics; discovery of x-rays
1914	M. von Laue	Physics; x-ray diffraction from crystals
1915	W. H. Bragg and W. L. Bragg	Physics; crystal structure derived from x-ray diffraction
1917	C. G. Barkla	Physics; characteristic radiation of elements
1924	K. M. G. Siegbahn	Physics; x-ray spectroscopy
1927	A. H. Compton	Physics; scattering of x-rays by electrons
1936	P. Debye	Chemistry; diffraction of x-rays and electrons in gases
1946	H. J. Muller	Medicine; discovery of x-ray-induced mutations
1962	M. Perutz and J. Kendrew	Chemistry; structures of myoglobin and haemoglobin
1962	J. Watson, M. Wilkins, and F. Crick	Medicine; structure of DNA
1964	D. Crowfoot-Hodgkin	Chemistry; structure of penicillin
1976	W. N. Lipscomb	Chemistry; x-ray studies on the structure of boranes
1979	A. McLeod Cormack and G. Newbold Hounsfield	Medicine; computed axial tomography
1981	K. M. Siegbahn	Physics; high-resolution electron spectroscopy
1985	H. Hauptman and J. Karle	Chemistry; direct methods to determine x-ray structures
1988	J. Deisenhofer, R. Huber, and H. Michel	Chemistry; determining the structure of proteins crucial to photosynthesis
1997	P. D. Boyer and J. E. Walker	Chemistry; mechanism of adenosine triphosphate synthesis[†]
2003	R. MacKinnon and P. Agre	Chemistry; structure and operation of ion channels[†]
2006	R. D. Kornberg	Chemistry; atomic description of DNA transcription[†]
2009	V. Ramakrishnan, T. A. Steitz, and A. E. Yonath	Chemistry; structure and function of the ribosome[†]
2012	R. J. Lefkowitz and B. K. Kobilka	Chemistry; studies of G-protein-coupled receptors[†]
2018	F. H. Arnold	Chemistry; the directed evolution of enzymes[†]

[†]Work using synchrotron radiation as a primary tool.

This broad range of applications of x-rays has manifested itself since the turn of this century in the diversity of disciplines served and in the broad palette of techniques now available at synchrotron facilities, which represents one of the principal examples of multidisciplinary research [6]. Today, there are more than seventy facilities in operation, or under construction, worldwide, providing services for well over one hundred thousand users from virtually every discipline of the natural sciences. Each year, several thousand articles are published which explicitly mention synchrotrons (see Figure 1.4). Since 1945, in total almost 90 000 papers relating to synchrotrons and synchrotron radiation have been published and have been cited almost 1 200 000 times.

Why *are* x-rays (and ultraviolet radiation, which bridges the gap between visible light and x-rays) especially important for investigating the properties of materials? There are several reasons. Firstly, x-rays are able to probe deeply and nondestructively into solid materials, with the degree of penetration depending on their energy and on the electron density (and hence the atomic masses and spatial distribution) of the constituent elements of the sample under investigation. Conversely, if the x-ray beam is made to impinge at very shallow angles to the sample

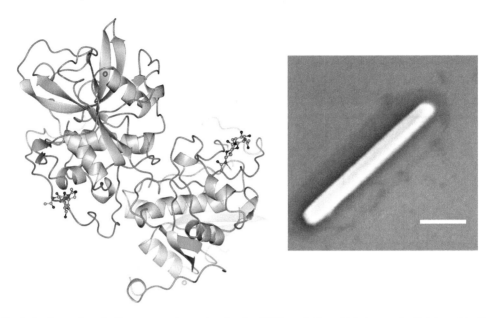

Figure 1.2 Left: the cathepsin B structure determined using XFEL radiation. This is a potential target in the quest to develop new drugs against sleeping sickness, a disease caused by the parasite *Trypanosoma brucei*, which uses cathepsin to break down proteins within the host cell. Rendered from the pdb file 3mor. Right: a typical nanocrystal used in the XFEL 'diffraction-before-destruction' experiment grown *in vivo* in insect cells. The scale bar is 1 µm. Reprinted from [5] with permission from the American Association for the Advancement of Science.

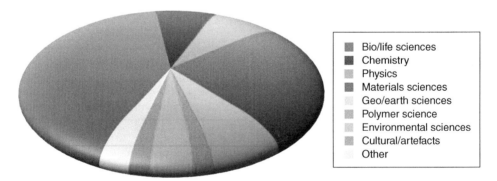

Figure 1.3 Synchrotron users by discipline. Estimated by the author from user data between 2013 and 2016 at the SLS, SSRL, APS, and ESRF.

surface, the beam will be totally reflected and only a so-called evanescent wave will penetrate the surface to a depth of a few nanometres. In this manner, x-rays can also be used as a highly surface-sensitive probe.

Secondly, x-ray photons with energies of several kiloelectronvolts (keV) have wavelengths comparable to those of typical atomic spacings in solid materials, measured in angstroms (1 Å $= 10^{-10}$ m). Under certain conditions, crystalline arrays of atoms can therefore act as interference gratings (that is, ordered arrays of scattering centres) for x-ray light, which can thereby be diffracted. On the other hand, electronic transitions involving valence electrons have energies that are in the range of ultraviolet (UV) and soft-x-ray photons, which are hence useful as probes in the investigation of the chemical nature and electronic properties of materials.

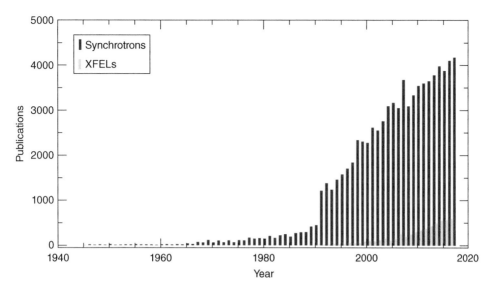

Figure 1.4 Publications which explicitly mention 'synchrotron' or 'synchrotron radiation' (blue data), and 'XFEL', 'x-ray free-electron laser', or 'X-FEL' (yellow data, all entries case insensitive) in the field 'topic' in the Web of Science database, since the first paper by McMillan in 1945 [7]. Papers relating to astronomy, particle fields, and nuclear physics were excluded. Note also that many publications that contain data recorded at synchrotrons and/or XFELs do not include these keywords and that the actual total number of reports relying at least partly on them is substantially higher.

Synchrotron storage rings are very powerful sources of x-rays. Synchrotron facilities are nowadays designed to generate tuneable beams of electromagnetic radiation from the far infrared to the hard x-ray regime. In terms of their 'brilliance', the standard figure of merit for synchrotrons precisely defined in Chapter 3, modern synchrotrons are to laboratory-based x-ray sources what modern computers are to the original Colossus valve-based computer used to crack the Enigma code in 1943! Indeed, in the last six decades, there has been an increase in the brilliance available to researchers by a factor of almost 10^{23} (see Figure 1.5), whereby the most recent advance by nine to ten orders of magnitude of peak brilliance has been obtained through the emergence of XFELs in the first decade of the twenty-first century (see Chapter 4).

Synchrotrons consist of an evacuated storage ring in which high-energy electrons circulate at highly relativistic velocities, and so-called 'beamlines'. These latter utilize the synchrotron light emitted by the electrons tangential to their orbital path, at positions defined by components known as bending magnets and insertion devices (see Figure 1.6).

This text will cover the use of synchrotron radiation for the determination of materials' structures and properties in physics, chemistry, biology, and related disciplines such as archaeology and environmental science. In general, a biologist has a very different training in science compared to that of a physicist. I have therefore tried to pitch the contents of this text in such a manner that it should be accessible to all newcomers to synchrotron radiation. Certain chapters and sections will undoubtedly leave some readers cold – for example, it is unlikely that an archaeologist will see much relevance to his or her research in the arcane joys of surface x-ray diffraction. Accordingly, the tone of the different subjects and the depth to which they are explained has been adapted as much as possible to the demands and scientific profiles of the probable users.

After the Introduction, fundamentals of the interactions of high-energy x-rays with matter are discussed in Chapter 2. Next, the basic physics and the most important technical aspects of synchrotron and XFEL facilities are covered in Chapters 3 and 4, respectively, while details of the optics and instrumentation found at beamlines

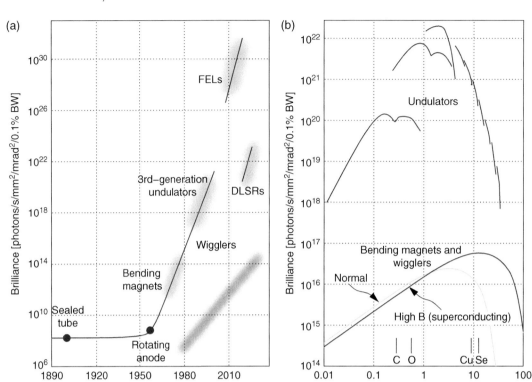

Figure 1.5   The brilliance of x-ray sources. (a) A historical perspective, showing the enormous increase in brilliance, beginning in the second half of the twentieth century. The gradient of the curve (plotted on a logarithmic ordinate axis) since the 1960s equates to, on average, a doubling of brilliance every fourteen months. This fairly comprehensively outstrips the red line showing the approximate doubling every 24 months in computing power, according to the much-vaunted Moore's law. (b) Typical spectral-brilliance curves of devices commonly used in fourth-generation (DLSR) synchrotrons. The positions of the K-absorption edges of carbon, oxygen, copper, and selenium are also shown. Conversion to peak brilliance requires a multiplication of the average brilliance shown here by $\Delta T/\Delta \tau$, the ratio of the pulse separation to the pulse width, and is of the order of 100.

are summarized in Chapter 5. Although many users do not need to understand exactly how the x-rays they are using are generated and manipulated, it is hoped that the information provided in Chapters 3 to 5 will help guide them as to what any particular facility or beamline can realistically provide for potential experiments.

The second part of the book covers synchrotron techniques and applications. The methods described in Chapters 6 to 8 are, as far as possible, accompanied by representative and illustrative examples. The reader is also encouraged to refer to the original sources of these selected examples to obtain still deeper insights and become aware of subtleties peculiar to individual experiments – beamtime is expensive, and a user with an understanding of potential pitfalls and experimental obstacles is a valuable member of any team.

1.1   A Potted History of X-rays

Electromagnetic radiation, across the spectral range from radio waves to gamma rays, has historically been the most ubiquitous probe in the natural sciences for the investigation and understanding of matter. It continues to be so to this day. Before a century or so ago, however, mankind only had the very narrow band of visible light from

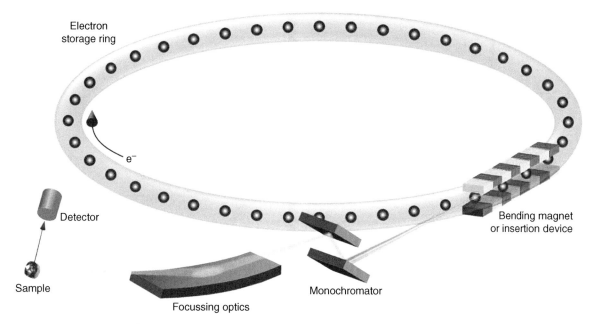

Figure 1.6   Schematic of a modern synchrotron. Electrons moving at highly relativistic velocities in an evacuated storage ring emit electromagnetic (synchrotron) radiation as their direction is changed by bending magnets, used to steer them along a closed path, or by so-called 'insertion devices' placed in straight sections of the storage ring, which 'shake' the electrons to and fro but maintain on average a straight trajectory. At the beamlines tangential to the storage ring (of which there are typically 10 to 50, depending on the size of the facility; only one is shown here) the radiation is normally (but not always) monochromatized and focused using x-ray optics onto a sample.

the electromagnetic spectrum at its disposal. This all changed radically with the discovery of x-rays by Wilhelm Röentgen in 1895.

On the evening of the 8 November 1895, Röntgen first detected x-rays. In his darkened room, he found that, when running a high-voltage discharge tube enclosed in thick black cardboard, which excluded all visible light, a paper plate covered on one side with barium platinocyanide would nevertheless fluoresce, even when it was as far as two metres from the discharge tube. Röntgen concluded that the discharge must be emitting a new form of invisible radiation[3] that could escape the confines of the glass discharge tube and the surrounding covering – their unknown nature led him to name them x-rays [8].

Röntgen immediately discovered that these x-rays were also able to stain photographic plates. Using this phenomenon, he demonstrated in subsequent experiments that objects of different thicknesses and density placed in the path of these rays showed differing degrees of transparency. When Röntgen placed the hand of his colleague at Würzburg, the Swiss anatomist Albert von Kölliker, in the path of the rays over a photographic plate, he observed after development of the plate an image showing the shadows thrown by the finger bones and a

---

[3] Scientific insight and fame is sometimes a fickle mistress – in the decade immediately preceding Röntgen's almost immediate and correct interpretation of the cause of the fluorescing plate, the British scientist Sir William Crookes sought unsuccessfully to explain the repeated fogging of photographic plates stored near his cathode-ray tubes.

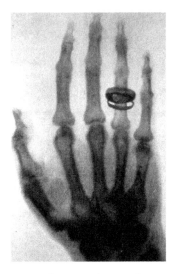

Figure 1.7   Left: this postage stamp, depicting Wilhelm C. Röntgen, was first issued in April 1939. The message translates as 'Fight cancer. Cancer is curable.' – the use of x-rays in oncology was already established soon after their discovery. Reprinted from http://commons.wikimedia.org/wiki/File:DanzigWilhelmKonradR%C3%B6ntgen25Pf1939.jpg. Right: one of the first radiographic images, recorded by Röntgen. Reprinted from Lino Piotto, Toger Gent, and Giovanni Bibbo, Myth busting – in the world of x-rays, *The Radiographer*, **54**, Copyright 2007, with permission of Lino Piotto.

ring, surrounded by the penumbra of the flesh, which absorbed fewer of the rays and therefore threw a fainter shadow[4] [Figure 1.7(right)]. In further experiments, Röntgen showed that the x-rays were produced by the impact of high-energy electrons on a material object. For his discovery of x-rays, Röntgen would receive the first ever Nobel Prize in Physics in 1901.

In 1909, characteristic x-radiation was discovered by the British physicists Barkla and Sadler [9] (see Figure 1.8) – Barkla would also receive the Nobel Prize in 1917. The term 'characteristic' refers to the narrow, intense x-ray emission lines, or x-ray fluorescence, found when a material is bombarded with energetic particles such as electrons or, indeed, x-rays. Each element has a unique set of lines, which, like a fingerprint, characterizes that element alone, and is a manifestation of the discrete energy levels of electronic states in atoms and their dependence on atomic number.

It was Henry G. J. Moseley, however, who would provide the correct interpretation of characteristic radiation, using an x-ray spectrometer of the type invented by the Braggs, and encapsulated in Moseley's law [10] (see Figure 1.9). A brilliant and hugely promising young scientist, tragically, Moseley would die at Gallipoli in the First World War the following year[5].

In 1912, the year he became Professor of Physics at Zürich University, Max von Laue conjectured that periodic, crystalline, structures might diffract waves if their wavelength were of the same order of magnitude as the crystalline periodicity and that x-rays might just fit the bill. He discussed these ideas with his colleagues Arnold

---

[4] There seems to be some dispute as to whether the first x-ray photograph was of von Kölliker's or Röntgen's wife's hand. The image shown in Figure 1.7 was recorded at a demonstration at the Physical Institute on 23 January 1896 and is Kölliker's hand. Another very similar image is undated and was only later attributed to Röntgen's wife.

[5] Isaac Asimov would later write 'In view of what he might still have accomplished (he was only 27 when he died), his death might well have been the most costly single death of a war to mankind generally'. Ironically, this observation is further compounded by the fact that the disastrous Gallipoli Campaign (at least from the perspective of the British) was instigated by arguably one of the other most important characters to influence twentieth century history, Winston Churchill.

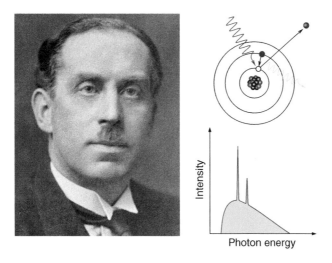

Figure 1.8 Charles G. Barkla, and the characteristic x-radiation he discovered.

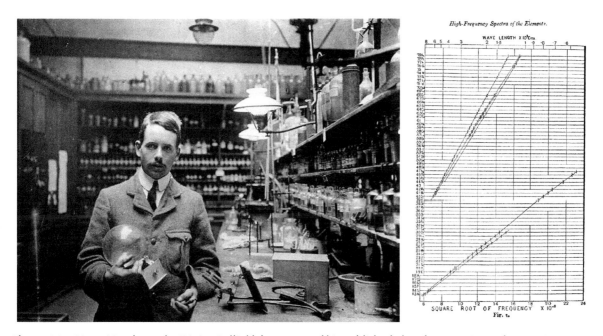

Figure 1.9 Harry Moseley at the Trinity–Balliol laboratory and his published plot of x-ray emission frequencies versus atomic number.

Sommerfeld and Wilhelm Wien during a skiing expedition. Despite Sommerfeld's objection that the supposed amplitude of thermal motion of the atoms within the crystal would be so large (relative to the wavelength of x-rays to be used) that the diffraction pattern would be completely washed out, Laue was able to convince Walter Friedrich, one of Sommerfeld's assistants, and Paul Knipping, a diploma student of Röntgen's, to record the first

Figure 1.10   Max von Laue and the first x-ray diffraction pattern of copper sulfate.

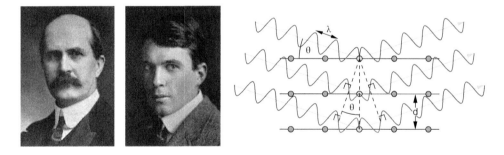

Figure 1.11   W. H. Bragg, his son W. L. Bragg, and the diffraction of x-rays by crystals.

diffraction pattern of a crystal of copper sulfate (Figure 1.10)[6]. Von Laue was awarded the Nobel Prize in Physics in 1914 for this discovery.

This was indeed a hectic and exceedingly productive period. In 1913, the year between the recording of the first diffraction pattern and von Laue receiving his Nobel Prize, the father-and-son team of Sir William Henry Bragg and William Lawrence Bragg were working on the theory and experimental techniques required to obtain detailed atomic structural information about crystals from their diffraction spot intensities and positions (Figure 1.11). The importance of their work is reflected by the fact that they would be awarded the Nobel Prize in Physics only two years later in 1915 'for their services in the analysis of crystal structure by means of x-rays'. Among other

---

[6] This very first diffraction pattern was of a raw, triclinic, copper sulfate pentahydrate crystal (the ubiquitous bright-blue crystal found in school chemistry laboratories) and should not be confused with the much cleaner, and thus more famous, pattern recorded shortly thereafter of a 0.5 mm thick (001)-platelet of cubic zinc sulfide.

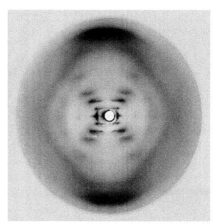

Figure 1.12   The discovery of the structure of DNA. Left: Watson and Crick and their model of DNA. The determination of its structure was much facilitated by the beautiful x-ray diffraction images of crystals of DNA salts recorded by Rosalind Franklin (centre), most notably, the famous 'photograph 51' (right). Centre image reprinted with permission of Image State.

things, they formulated the famous Bragg law, which describes the mechanism by which x-ray diffraction occurs. Crucially, the son, W. L. Bragg, also explained why diffracted spots had the intensities they did, and was able, using these principles, to determine the internal atomic structure of some simple minerals [11, 12]. The foundations of x-ray crystallography had been laid down [13].

This culminated some forty years later in one of the most important scientific discoveries of the twentieth century, namely the determination in 1953 of the structure of deoxyribonucleic acid (DNA) and the mechanism by which it replicates [1], for which James Watson, Francis Crick, and Maurice Wilkins would be awarded the Nobel Prize in Medicine in 1962. Their discovery, which revolutionized biology and provided an atomistic view of the stuff of life, was in no small measure assisted by x-ray diffractograms recorded by Rosalind Franklin at University College, London [14]. The detailed information which was gleaned by these patterns (in particular the central 'X' motif, so characteristic of diffraction from double-helical structures) was only made possible by the high quality of both the crystals of salts of DNA prepared by Franklin, and of the diffractometer that was available (see Figure 1.12).

Although the diffraction patterns recorded by Franklin provided one of the last and most important clues for Watson and Crick in their model-building approach to the structural solution of DNA, the molecular details were primarily deduced using chemical and geometrical arguments. In the same year that Watson, Crick, and Wilkins were awarded the Nobel Prize in Medicine, the Prize in Chemistry was won by their Cavendish Laboratory colleagues, John Kendrew[7] and Max Perutz, for their determination by x-ray diffraction of the atomic structures of myoglobin (Kendrew [16]) and haemoglobin (Perutz [17]) (Figure 1.13). It was Perutz's invention of the technique of multiple isomorphous replacement in x-ray crystallography that provided the breakthrough in what had until then seemed the insurmountable problem of solving the structure of a crystal containing a molecule of a mass of approximately 17 000 Da[8].

---

[7] In his Nobel lecture, Kendrew speculated that, although 'this day will not come soon', a protein's structure could be predicted based solely on its amino acid sequence, and that 'x-ray crystallographers can go out of business, perhaps with a certain sense of relief' [15]. Over sixty years later, we still await this momentous day …

[8] The Dalton, Da, is a unit of atomic mass equal to a twelfth of the mass of a neutral, unbound atom of carbon. The mass of any element expressed in Da is thus simply equal to their atomic mass.

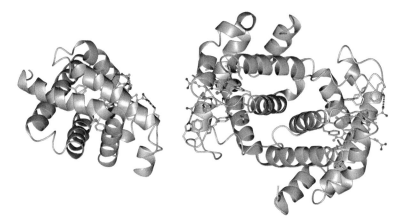

Figure 1.13   The structures of myoglobin (left) and haemoglobin (right), determined by Kendrew and Perutz using x-ray diffraction. Images generated using the pdb files 1mbn and 2dhb, respectively.

Two years later, the widely recognized doyenne of macromolecular crystallography, Dorothy Crowfoot-Hodgkin, would receive the Nobel Prize in Chemistry for her seminal work on crystals of complex molecules, in particular for her determination of the structure of penicillin [18] (see Figure 1.14). Of all the facets of crystallography, protein and macromolecular structure determination have surely had the most impact on society. Hodgkin was the first person to record diffraction patterns of crystals of large molecules, beginning with the digestive enzyme pepsin (molecular weight of approximately 35 kDa) in 1934 [19]. In addition, she went on to develop indexing and diffraction-intensity analysis; she was a pioneer in computer-assisted Fourier analysis; and she solved the structures of cholesterol, ferritin, lactoglobulin, the tobacco mosaic virus, penicillin (for which she would receive the Nobel Prize in Chemistry), vitamin B-12, and insulin [20–22]. This last compound defied structural analysis for over 34 years before Hodgkin cracked the problem in 1969[9].

Most recently, the Nobel Prizes in Chemistry in 2003, 2006, 2009, 2012, and 2018 have all been awarded to biochemistry studies exploiting x-ray crystallography[10] – McKinnon and Agre (2003) described the three-dimensional structure of potassium ion channels across cell membranes [24]; Kornberg was awarded the Nobel Prize in 2006 for his studies of the process by which genetic information from DNA is copied to RNA (this process is common to all life forms containing cell nuclei, that is, all life except the prokaryotae) [2]; Ramakrishnan, Steitz, and Yonath (2009) were able to solve the structure of ribosomes and thereby identify their core function in translating DNA code into living matter [25]; while Lefkowitz and Kobilka (2012) determined the structure and mechanistic pathways of so-called G-protein-coupled receptors (GPCRs), chemicals that allow cells to sense their environments and are responsible, among other things, for our bodies being able to sense light, flavour, odour, and many other stimuli [26–28]. Approximately two thirds of all medications operate through GPCRs. Frances Arnold (2018) used synchrotron radiation in her studies of directed evolution of enzymes [29].

It can be argued, judging by the disproportionate number of Nobel Laureates in the field (see Table 1.1), that x-ray science was one of the most influential areas of pure research in the twentieth century, while its impact has touched many other disciplines. It is interesting to note from Table 1.1 that physics garnered the majority of the early Nobel Prizes concerned with x-rays, from which most of the underlying phenomena were explained,

---

[9] The amino acid sequence of insulin was determined by Frederick Sanger in 1955, but not the detailed arrangement of the atoms in space.

[10] Incidentally, such tours de force of structural biology only became possible with other technological advances, not least in x-ray detectors. The 2009 Nobel Prize in Physics was partly awarded to Willard S. Boyle and George E. Smith for their pioneering work on charged-couple device (CCD) imaging technology [23]. CCDs are nowadays ubiquitous as x-ray detectors. The even more-modern photon-counting hybrid pixel detectors, with their generally superior performance (if not at XFELs, for reasons that will become apparent in Section 5.7.9) are nowadays often the first instrument of choice.

 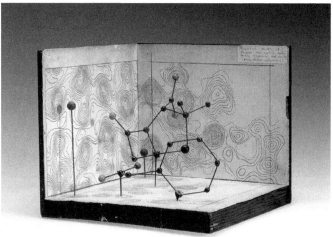

Figure 1.14   Left: Oil painting of Dorothy Hodgkin. Courtesy and copyright, Bryan Organ, 1982. Right: molecular model of penicillin, based on three orthogonal projections of its electron-density distribution, calculated and drawn by Dorothy Hodgkin. Reprinted with permission of the Science Museum, London, UK.

while their application in other scientific disciplines would reap Nobel awards in subsequent years. The use of synchrotron radiation has followed a similar path, and indeed the six most recent Nobel Prizes associated with x-rays used synchrotron radiation as a primary tool.

The importance of x-ray science in physics, chemistry, biology, and medicine can thus hardly be overstated. It seems certain that x-rays will continue to play a hugely important rôle in modern science and technology in the twenty-first century, as ever more materials, compounds, and structures in all their varied guises are discovered or, indeed, designed.

1.2   Synchrotron Sources over the Last Seventy Years

The term 'synchrotron' was coined by Ed McMillan in his seminal 1945 paper [7]. The concept of the synchrotron, however, was also arrived at independently one year earlier, by Vladimir Veksler. The two men began corresponding and developing the idea of a synchrotron and would, in 1963, share the 'Atoms for Peace' Award in recognition of this.

Short-wavelength synchrotron radiation generated by relativistic electrons in man-made circular accelerators dates back to shortly after the Second World War. However, the theoretical basis for synchrotron radiation traces back to the time of Thomson's discovery of the electron in 1897. In the same year, Joseph Larmor derived an expression from classical electrodynamics for the instantaneous total power radiated by an accelerated charged particle (covered in the next chapter), and the following year, the French physicist Alfred Liénard showed the radiated power emitted by electrons moving on a circular path to be proportional to $(\mathscr{E}/m_e c^2)^4/R^2$, where $\mathscr{E}$ is the electrons' kinetic energy, m_e is the electron rest mass, and R is the radius of the trajectory.

The first observation of synchrotron radiation came on 24 April 1947 at the General Electric Research Laboratory in Schenectady, New York (see Figure 1.15), although it was not being sought and was not immediately recognized for what it was. Here is an excerpt from Herb C. Pollock's description of this historical event over two decades later in a letter to Dmitri Ivanenko, dated 25 September 1970:

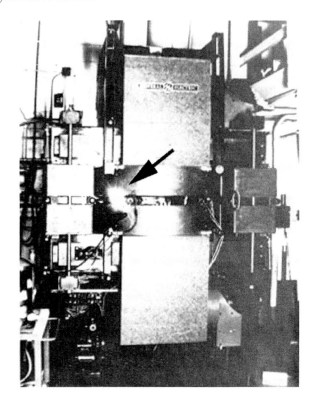

Figure 1.15   Synchrotron light from the 70 MeV electron synchrotron at General Electric.

On April 24, Langmuir and I were running the machine and as usual were trying to push the electron gun and its associated pulse transformer to the limit. Some intermittent sparking had occurred and we asked the technician to observe with a mirror around the protective concrete wall. He immediately signaled to turn off the synchrotron as 'he saw an arc in the tube.' The vacuum was still excellent, so Langmuir and I came to the end of the wall and observed. At first we thought it might be due to Cerenkov radiation, but it soon became clearer that we were seeing Ivanenko and Pomeranchuk radiation.

A more formal description of their findings can be found in [30]. 'Ivanenko and Pomeranchuk radiation' alludes to the two Russian physicists who, in 1944, published their calculations predicting the rate of energy loss due to radiating electrons [31].

Initially, synchrotron radiation was seen as an unwanted but unavoidable loss of energy of charged particles in closed-ring accelerators designed (ironically) to produce intense beams of x-rays by directing accelerated electrons onto a suitable target. Note that this loss is small compared to the particles' kinetic energy, and this remains true today – even the highest photon energies produced by modern synchrotrons are three to four orders of magnitude smaller than the electrons' relativistic kinetic energy (see Figure 1.16).

The potential advantages of synchrotron radiation for its own end were detailed by Diran Tomboulian and Paul Hartman from Cornell University in their seminal paper 'Spectral and Angular Distribution of Ultraviolet Radiation from the 300 MeV Cornell Synchrotron' in the *Physical Review* [32], describing their findings after being granted two weeks access to investigate the possible applications of synchrotron light. The experimental setup and photographic-plate results reported in this paper, showing the broadband synchrotron radiation emitted by monoenergetic electrons, are reproduced in Figure 1.17.

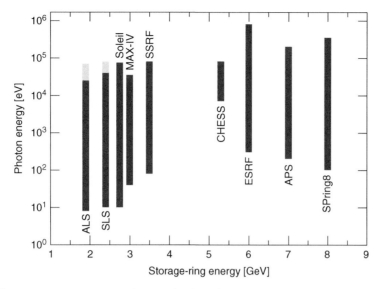

Figure 1.16 Plot of some representative synchrotron-facility photon-energy ranges as a function of the storage-ring (that is, electron) energy. Note the logarithmic ordinate axis. The yellow bars for the ALS and SLS indicate extensions of the accessible photon energies to higher values through the use of so-called 'superbends', described in Section 3.8.1.

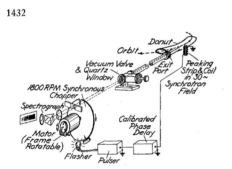

FIG. 11. Diagram showing the arrangement to be used for recording the radiation from essentially monoenergetic electrons. For work in the vacuum ultraviolet the rotating disk is enclosed, the drive shaft coming through a vacuum seal. So far the disk has not been used in the investigation of the far-ultraviolet spectrum. However, the plate of Fig. 12 was obtained with this arrangement and the quartz optical system indicated here.

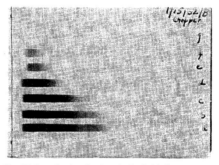

FIG. 12. Reproduction of a plate obtained with the arrangement of Fig. 11, showing spectra of the continuous radiation emitted by essentially monoenergetic electrons. The various exposures correspond to electron energies ranging from 60 Mev at the top to 110 Mev at the bottom. An exposure at 50 Mev is not visible in the reproduction. The exposures were adjusted so that, in each case, approximately the same total number of radiating electrons was involved.

Figure 1.17 Reproduction of Figures 11 and 12 in the *Physical Review* paper by Tomboulian and Hartmann [32], showing the experimental setup and photographic plate of the spectra produced by synchrotron radiation from monoenergetic electrons. Reprinted from [32] with permission from the American Physical Society.

Five years later, in 1961, a pilot experimental program exploiting synchrotron radiation began when the National Bureau of Standards modified its 180 MeV electron synchrotron in Washington, D. C., to allow access to the radiation via a tangent section into the machine's vacuum system. Thus was born the Synchrotron Ultraviolet Radiation Facility (SURF I), the first to cater for regular users of synchrotron radiation.

The first generation of synchrotrons was sometimes referred to as parasitic facilities (reflecting, perhaps, the perception of the primary users of these facilities towards these interlopers), as the synchrotrons were primarily designed for high-energy or nuclear-physics experiments. In addition to SURF I, facilities in Frascati and in Japan soon after began to attract a regular stream of physicists keen to explore the possibilities of synchrotron radiation. More would soon follow suit. Because most of these early facilities had storage-ring energies around or below 1 GeV, experiments were concentrated in the ultraviolet and soft x-ray regimes. 1964 saw the first users of synchrotron radiation from the 6 GeV Deutsches Elektronen-Synchrotron (DESY) in Hamburg – suddenly the range of synchrotron radiation was catapulted to the hard x-ray region down to 0.1 Å (about 125 keV). Many further facilities began to accommodate synchrotron-radiation users. One of the first of these was the 240 MeV machine 'Tantalus' in Wisconsin. Although not originally designed to provide synchrotron radiation, it became the first facility to be exclusively used for synchrotron experiments, providing radiation in the ultraviolet up to a few tens of eV.

After the development of efficient electron-storage rings for long-term operation, the time was ripe to develop the first dedicated greenfield facilities designed specifically for synchrotron radiation. The 2 GeV Synchrotron Radiation Source (SRS) at Daresbury, England, was the first of these so-called 'second-generation' synchrotron sources. Experiments began in 1981. Several more new facilities were soon built and commissioned, while some first-generation sources were upgraded to second-generation status.

At approximately the same time, it was becoming clear that increased x-ray beam brilliance could be achieved by optimizing the property of the electron beam in the storage ring called the 'emittance', described in detail in Chapter 3, through careful design of the array of magnets (the 'magnet lattice') used to manipulate the electrons, and by employing so-called 'insertion devices'. These include 'wigglers' and 'undulators', placed in straight sections in between the curved arcs of large storage rings. They operate by perturbing the path of the electrons in an oscillatory manner, so that, although their *average* direction remains unchanged, synchrotron radiation is produced. Details of undulators and wigglers are covered in Chapter 3.

The inclusion of insertion devices in storage rings defined the third generation of synchrotron sources, which were, at the time, designed for optimal brilliance, that is, the amount of power per unit frequency, surface area, and solid angle. The first third-generation facility to be completed was the European Synchrotron Radiation Facility (ESRF, 6 GeV storage ring) in Grenoble, France, which began user experiments in 1994.

At the time of writing, a fourth generation of storage-ring technology is coming of age, in which the electron beams are significantly narrower and more parallel in the orbital plane than was attainable in third-generation facilities. The first of these so-called diffraction-limited storage rings (DLSRs), MAX IV in Lund, Sweden, was officially inaugurated on the Summer solstice of 2016. Many third-generation facilities are undergoing or planning upgrades to DLSR status, which is therefore fast becoming the new standard.

The projected specifications for these storage rings will redefine the benchmark for synchrotrons – it is planned to provide spatial and energy resolutions of the order of 5 nm and 0.1 meV, respectively, while the brilliance and horizontal emittance, at 10^{22} to 10^{23} photons/mm^2/mrad^2/0.1% bandwidth and approximately 100 pm·rad, respectively, improve on previous performances by one to two orders of magnitude. These significant improvements have been made possible by recent advances in the design and manufacture of storage-ring magnets, and by the fact that, although the preferred storage-ring energies are in general maintained at fairly moderate values (typically 3 GeV), their circumferences are more typical of those previously used in high-energy storage rings. This is discussed in more detail in Section 3.8.5.

Parallel to these developments in synchrotron technology, x-ray free-electron lasers operating in the angstrom regime have emerged in the last decade. The first of these to come online was the LCLS in 2009. It uses one third of the two-mile-long Stanford Linear Accelerator (SLAC), and has pioneered modern femtosecond x-ray science.

The SPring-8 Angstrom Compact Free Electron Laser (SACLA), located at the SPring-8 complex in the Hyōgo Prefecture, Japan, began operation in 2011, offering XFEL radiation up to approximately 20 keV. The European XFEL in Hamburg, Germany, came online in 2017, while the Pohang Accelerator Laboratory XFEL (PAL-XFEL) in South Korea also began pilot experiments in 2017 [33]. The soft x-ray FEL at FLASH (Hamburg) has served external users since 2005; an upgrade, known as FLASH2020, will feature a new flexible undulator scheme, providing coherent radiation for multicolour experiments over a broad energy range [34]. FERMI@Elettra (Trieste,

Italy) provides radiation between approximately 10 and 60 eV at its first source FEL-1, and extends up to 300 eV using its second source, FEL-2. The SwissFEL at the Paul Scherrer Institute saw first lasing at 24 nm (52 eV) in December 2016; 1 Å hard x-rays were produced in late 2018.

Another third of the SLAC LINAC is now being developed to house LCLS-II, the latest example of x-ray light sources. The major upgrade in design is the use of superconducting niobium (also used at the European XFEL), instead of the copper cavities used for the original LCLS, for the accelerating cavities. This will increase the delivery rate of the femtosecond pulses from 120 Hz to 1 MHz, while the accessible photon energy range will be effectively tripled to 25 keV.

Synchrotrons and XFELs are largely complementary, rather than being competing technologies. On the contrary, it appears that both XFELs and cryo electron-microscopy (see Appendix A) are driving an increasing demand for synchrotron beamtime, not least in the life sciences. As more DLSRs come online, this trend looks to continue unabated – the future of x-ray science looks bright indeed!

References

[1] J. D. Watson and F. H. C. Crick, "Molecular structure of nucleic acids – a structure for deoxyribose nucleic acid," *Nature*, vol. 171, pp. 737–738, 1953.

[2] P. Cramer, D. A. Bushnell, and R. D. Kornberg, "Structural basis of transcription: RNA polymerase II at 2.8 Å resolution," *Science*, vol. 292, pp. 1863–1876, 2001.

[3] B. T. Wimberly, D. E. Brodersen, W. M. Clemons Jr., et al., "Structure of the 30S ribosomal subunit," *Nature*, vol. 407, pp. 327–339, 2000.

[4] N. Ban, P. Nissen, J. Hansen, et al., "The complete atomic structure of the large ribosomal subunit at 2.4 Angstrom resolution," *Science*, vol. 289, pp. 905–920, 2000.

[5] L. Redecke, K. Nass, D. P. Deponte, et al., "Natively inhibited Trypanosoma brucei cathepsin B structure determined by using an x-ray laser," *Science*, vol. 339, pp. 227–230, 2013.

[6] M. Mino, E. Borfecchia, J. Segura-Ruiz, et al., "Materials characterization by synchrotron x-ray microprobes and nanoprobes," *Rev. Mod. Phys.*, vol. 90, p. 025007, 2018.

[7] E. M. McMillan, "The synchrotron – a proposed high energy particle accelerator," *Phys. Rev.*, vol. 68, pp. 143–144, 1945.

[8] J. Maddox, "The sensational discovery of X-rays," *Nature*, vol. 375, pp. 183–183, 1995.

[9] C. G. Barkla and C. A. Sadler, "The absorption of Röntgen rays," *Philos. Mag.*, vol. 17, pp. 739–760, 1909.

[10] H. G. J. Moseley, "The high-frequency spectra of the elements. Part II," *Philos. Mag.*, vol. 27, pp. 703–713, 1914.

[11] W. L. Bragg, "The diffraction of short electromagnetic waves by a crystal," *Proc. Cam. Phil. Soc.*, vol. 17, pp. 43–57, 1913.

[12] A. Authier, *Early days of x-ray crystallography*. International Union of Crystallography, 2015.

[13] M. F. Perutz, "How W. L. Bragg invented x-ray analysis," *Acta Crystallogr. A*, vol. 46, pp. 633–643, 1990.

[14] J. D. Watson, *The double helix: a personal account of the discovery of the structure of DNA*. Atheneum, 1968.

[15] J. C. Kendrew, "Myoglobin and the structure of proteins," *Science*, vol. 139, pp. 1259–1266, 1963.

[16] J. C. Kendrew, G. Bodo, H. M. Dintzis, et al., "A three-dimensional model of the myoglobin molecule obtained by x-ray analysis," *Nature*, vol. 181, pp. 662–666, 1958.

[17] M. F. Perutz, M. G. Rossmann, A. F. Cullis, et al., "Structure of haemoglobin: A three-dimensional Fourier synthesis at 5.5 Å resolution, obtained by x-ray analysis," *Nature*, vol. 185, pp. 416–422, 1960.

[18] G. Ferry, *Dorothy Hodgkin: A Life*. Granta, 1998.

[19] J. D. Bernal and D. Crowfoot, "X-ray photographs of crystalline pepsin," *Nature*, vol. 133, pp. 794–795, 1934.

[20]   C. H. Carlisle and D. Crowfoot, "The crystal structure of cholesteryl iodide," *Proc. R. Soc. A*, vol. 184, pp. 64–83, 1945.

[21]   D. Crowfoot, C. W. Bunn, B. W. Rogers-Low, and A. Turner-Jones, "X-ray crystallographic investigation of the structure of penicillin," in *Chemistry of penicillin* (H. T. Clarke, J. R. Johnson, and R. Robinson, eds.), pp. 310–367, Princeton University Press, 1949.

[22]   E. Dodson, M. M. Harding, D. C. Hodgkin, and M. G. Rossmann, "The crystal structure of insulin. III. evidence for a 2-fold axis in rhombohedral zinc insulin," *J. Mol. Biol.*, vol. 16, pp. 227–241, 1966.

[23]   G. E. Smith, "Nobel Lecture: The invention and early history of the CCD," *Rev. Mod. Phys.*, vol. 82, pp. 2307–2312, 2010.

[24]   D. A. Doyle, J. M. Cabral, R. A. Pfuetzner, et al., "The structure of the potassium channel: molecular basis of K^+ conduction and Selectivity," *Science*, vol. 280, pp. 69–77, 1998.

[25]   E. Pennisi, "The race to the ribosome structure," *Science*, vol. 285, pp. 2048–2051, 1999.

[26]   R. A. F. Dixon, B. K. Kobilka, D. J. Strader, et al., "Cloning of the gene and cDNA for mammalian β-adrenergic receptor and homology with rhodopsin," *Nature*, vol. 321, pp. 75–79, 1986.

[27]   S. G. F. Rasmussen, H.-J. Choi, D. M. Rosenbaum, et al., "Crystal structure of the human β_2 adrenergic G-protein-coupled receptor," *Nature*, vol. 450, pp. 383–387, 2007.

[28]   V. Cherezov, D. M. Rosenbaum, M. A. Hanson, et al., "High-resolution crystal structure of an engineered human β_2 adrenergic G protein-coupled receptor," *Science*, vol. 318, pp. 1258–1265, 2007.

[29]   B. Spiller, A. Gershenson, F. H. Arnold, and R. C. Stevens, "A structural view of evolutionary divergence," *Proc. Natl. Acad. Sci. U.S.A.*, vol. 96, pp. 12305–12310, 1999.

[30]   F. R. Elder, R. V. Langmuir, and H. C. Pollock, "Radiation from electrons accelerated in a synchrotron," *Phys. Rev.*, vol. 74, pp. 52–56, 1948.

[31]   D. Iwanenko and I. Pomeranchuk, "On the maximal energy attainable in a betatron," *Phys. Rev.*, vol. 65, p. 343, 1944.

[32]   D. H. Tomboulian and P. L. Hartman, "Spectral and angular distribution of ultraviolet radiation from the 300-MeV Cornell Synchrotron," *Phys. Rev.*, vol. 102, pp. 1423–1447, 1956.

[33]   I. S. Ko et al., "Construction and commissioning of PAL-XFEL facility," *Appl. Sci.*, vol. 7, p. 479, 2017.

[34]   B. Faatz, M. Braune, O. Hensler, et al., "The FLASH facility: advanced options for FLASH2 and future perspectives," *Appl. Sci.*, vol. 7, p. 1114, 2017.

2

The Interaction of X-rays with Matter

2.1 Introduction

How does electromagnetic radiation interact with matter? We are all familiar at some level with phenomena in the visible region (Figure 2.1). Refraction describes the dependence of the velocity of light on the medium in which it is travelling – the higher the refractive index, the slower the light travels. This causes light to be bent when it crosses the boundary between one medium and another. Below a certain incident angle, a beam of light impinging on the surface of a material with a lower refractive index will be completely reflected and only penetrate the boundary in the form of a decaying, or 'evanescent' standing wave to a depth comparable to the light's wavelength.

Conducting materials which contain free or nearly free electrons respond to and reradiate (that is, they reflect) visible light incident at any angle efficiently – the reflectivity of aluminium over the entire visible range and also well into the ultraviolet region is near unity, while gold reflects generally less well and has a yellow (or golden!) appearance, because of an absorption band in the blue.

Visible light, however, spans only a very narrow energy range of about 1.75 to 3.2 eV (or, more familiarly to many, 400 to 700 nm in wavelength), and hence virtually all interactions in this region only involve free electrons or those occupying the outermost (valence) shell[1], as all other electrons are much more strongly bound to atomic nuclei and less affected by such relatively weak perturbations. This begs the question, however, of how material responds to electromagnetic radiation of higher photon energy?

We will see that the above-mentioned phenomena of refraction, reflection, and absorption are all also found for the interaction of ultraviolet rays and x-rays with matter, whereby the main differences compared to the familiar properties of visible light with matter can be explained by the much higher photon energies involved. The most important phenomena are sketched in Figure 2.2.

The binding energy of the only electron in a hydrogen atom is 13.6 eV. To a first approximation, the innermost electrons in an atom of atomic number Z have binding energies which are proportional to Z^2, as shown in Figure 2.3 for the $1s$ electrons. X-rays therefore have energies similar to or larger than those of tightly bound core electrons, and hence the interaction strengths or 'cross-sections' between x-rays and core electrons are stronger than those between x-rays and valence electrons.

In this chapter, we will discuss how x-rays interact with matter and how this manifests itself in the phenomena of refraction, reflection, absorption, and fluorescence. Diffraction phenomena are treated separately in Chapter 6.

---

[1] The electron shell can be thought of classically as the 'orbit' of the electron around the atomic nucleus. It is defined by the principal quantum number, n. For a given atom, the maximum value of n, which is normally associated with the most weakly bound 'valence' electrons, is given by the period (that is, the row) in which the atom resides in the periodic table.

An Introduction to Synchrotron Radiation: Techniques and Applications, Second Edition. Philip Willmott.
© 2019 John Wiley & Sons Ltd. Published 2019 by John Wiley & Sons Ltd.

Figure 2.1 Optical phenomena in the visible regime. Left: reflection off a chrome-plated surface; centre left: refraction of a lime twist in water; centre right: absorption of green laser light by cranberry juice; right: diffraction of laser light off a generic fruit-based smartphone screen.

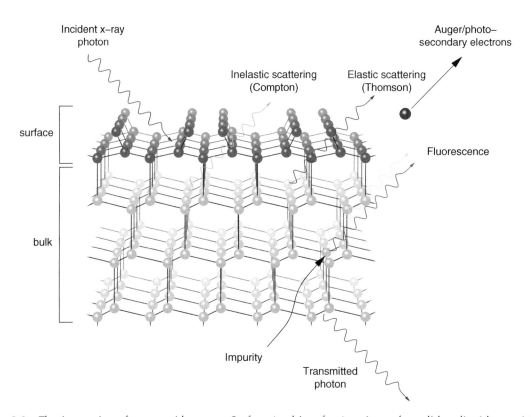

Figure 2.2 The interaction of x-rays with matter. Surface (and interface) regions of a solid or liquid material are characterized by physical properties, structure, and symmetry that may differ significantly from those of the bulk structure. The x-rays impinging on the material may be elastically or inelastically scattered, or absorbed, in which case electrons or lower-energy photons can be subsequently emitted. If none of the above occur, the photon is transmitted through the sample.

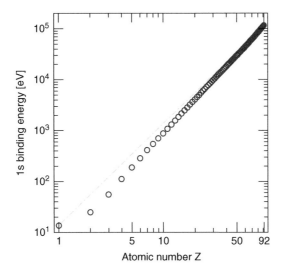

Figure 2.3   The binding energies of the 1*s* electrons in the elements up to uranium. Also shown is the zeroth-order approximation $E_B(1s) = 13.6\,Z^2$ (which ignores any Coulomb interaction between the electrons in the atom) as a comparison. The actual binding energies deviate somewhat from this simple expression, due to the screening effect of the other electrons in the atom, particularly evident for low-*Z* elements, for which the average distances and hence binding energies of the other electrons are more similar to that of the 1*s* electron.

2.2   The Electromagnetic Spectrum

The use of light to investigate the properties and structures of materials goes back to the dawn of science. Before the end of the nineteenth century, mankind only had visible light at its disposal. This covers only a very narrow range of the entire electromagnetic (EM) spectrum, which spans long-wave radio waves, through microwaves, infrared radiation, visible light, ultraviolet light, x-rays, and gamma rays (see Figure 2.4).

The energies in the spectrum shown in Figure 2.4 span some 16 orders of magnitude. In general, the energy of a photon is given by

$$E = h\nu = hc/\lambda, \tag{2.1}$$

where $h = 6.626 \times 10^{-34}$ J s is Planck's constant, ν is the frequency of the radiation in Hz, $c = 2.9979 \times 10^8$ m s^{-1} is the speed of light in vacuum, and λ is the wavelength of light in vacuum, given in metres. A convenient way of re-expressing Equation (2.1) to determine the wavelength of x-ray photons for a given photon energy is

$$\lambda\,[\text{Å}] = \frac{12.3984}{E\,[\text{keV}]}. \tag{2.2}$$

Hence, a 1 Å x-ray photon has an energy of 12 398.4 eV (or about 2 fJ).

In this chapter, we first consider the interaction of x-rays with individual electrons (their primary scatterers in our region of interest), followed by scattering from an ensemble of bound electrons in an atom. From this, we discuss the phenomena of refraction, absorption, and reflection. The phenomena of diffraction from a regular array of atoms, and of small-angle scattering from mesoscopic objects, although logical progressions in complexity, are deferred to Chapter 6.

In general, any interaction of matter with a beam of electromagnetic radiation will result in that beam being attenuated (be that through absorption, elastic scattering, or any other mechanism). The cross-section σ_P for a particular process P is defined as

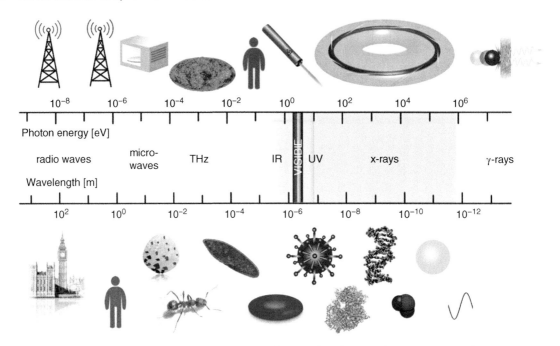

Figure 2.4   The electromagnetic spectrum. Lasers cover the wavelength range from approximately 10 µm (0.124 eV) to 150 nm (8 eV), shown in pink. Worldwide, synchrotrons also cover this range, but, importantly, also extend up to nearly 10^6 eV, shown in yellow. Sources of different types of radiation are shown top: AM radio antenna, FM radio antenna, microwave oven, cosmic background radiation, humans (sources of approximately 10 µm infrared radiation), lasers, synchrotrons, Bremsstrahlung at synchrotrons. Bottom are representative objects at different length scales: Clock tower (100 m), human (2 m), quail egg (3 cm), fire ant (5 mm), Paramecium (200 µm), red blood cell (7 µm diameter), flu virus (100 nm diameter), 80S ribosome (25 nm), DNA (2 nm width), water molecule (96 pm bond length), helium atom (28 pm covalent radius), Compton wavelength $\lambda_C = h/m_e c = 2.4$ pm.

$$\sigma_P = (\Lambda_P N_i)^{-1}, \tag{2.3}$$

where Λ_P is the attenuation length due to process P (that is, the length after which the beam intensity is reduced to $1/e$), and N_i is the atomic number-density (atoms/unit volume) of the atom causing process P.

The cross-sections for the processes involving the interaction of x-rays with matter are shown in Figure 2.5 for the example of elemental barium, given in units of 'barn' (10^{-28} m^2) per atom. For applications using synchrotron radiation, only photoabsorption, elastic Thomson scattering, and (to a lesser extent) inelastic Compton scattering are important [1].

2.3   Compton Scattering

Photoabsorption and elastic (Thomson) scattering are the most important processes in the majority of techniques used to investigate the structure of materials. In the quantum-mechanical description of the excitation of an electron by electromagnetic radiation, however, kinetic energy may be transferred to the electron, resulting in the scattered photon having a lower energy [2]. This inelastic process is called the Compton effect (see Figure 2.6). One should not confuse photoabsorption with Compton scattering. In the former, a photon is entirely removed from the system. This may or may not result in the ejection of a lower-energy photon via fluorescent decay, but these discrete-energy photons do not have the same origin as Compton-scattered photons, which exhibit a continuous spectrum.

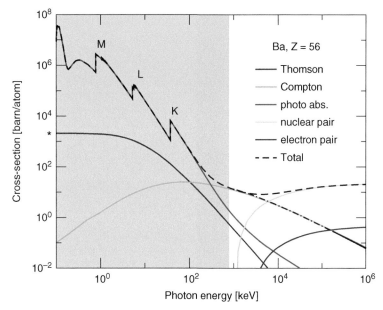

Figure 2.5   The cross-sections for various processes involving the interaction of x-rays with matter, for the example of the element barium (Ba). The unit 'barn' is equal to 10^{-24} cm^2. The yellow region indicates the upper energy range covered by synchrotron sources (which also continues down to a few meV, four orders of magnitude lower still than shown here). In this range, two processes dominate, namely photoelectric absorption (here, the K-, L-, and M-edges of Ba are marked, see later) and elastic (Thomson) scattering, although inelastic (Compton) scattering also becomes significant above approximately 30 keV. The high-energy-approximation for Compton scattering, given by Equation (2.9), is plotted as the dot-dashed curve between 10^3 and 10^6 keV. Note that at low photon energies, the cross-section for Thomson scattering approaches a constant value (highlighted by the asterisk) equal to $8Z^2\pi r_0^2/3$, which for Ba equals 2086 barn.

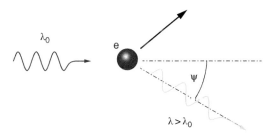

Figure 2.6   The Compton effect, in which a fraction of the energy of a photon is transferred to the kinetic energy of an electron, resulting the scattered photon having a lower energy.

The energy loss in Compton scattering is simple to determine by applying conservation of total energy and momentum, and results in

$$\frac{h\nu_0}{h\nu} = \frac{k_0}{k} = \frac{\lambda}{\lambda_0} = 1 + \frac{\lambda_C}{\lambda_0}(1 - \cos\psi), \tag{2.4}$$

where the incoming photon has the frequency $\nu_0 = c/\lambda_0$, its wavevector is $k_0 = 2\pi/\lambda_0$, ψ is the angle by which the photon is scattered, and λ_C is the Compton scattering length, given by

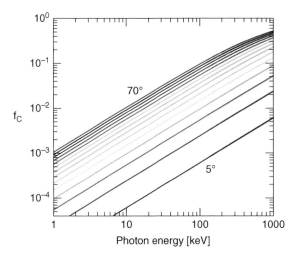

Figure 2.7   The fractional loss f_C in photon energy due to Compton scattering as a function incident photon energy for different scattering angles between 5° and 70°, in 5° steps.

$$\lambda_C = \frac{h}{m_e c} = 2.43 \times 10^{-2} \text{ Å}. \tag{2.5}$$

The *fractional* loss of the photon energy

$$f_C = \frac{h\nu_0 - h\nu}{h\nu_0} = \frac{(E_0/m_e c^2)(1 - \cos\psi)}{1 + (E_0/m_e c^2)(1 - \cos\psi)} \tag{2.6}$$

$$= \frac{1.597 \times 10^{-3} \, E_0[\text{keV}](1 - \cos\psi)}{1 + 1.597 \times 10^{-3} \, E_0[\text{keV}](1 - \cos\psi)} \tag{2.7}$$

increases with increasing scattering angle and incident photon energy (Figure 2.7).

The total cross-section for Compton scattering by an electron bound to an atom has a maximum, typically found at around 100 keV to 1 MeV (see Figure 2.5). The reason for this is that, at low energies, the energy imparted to the bound electron via transfer of momentum from the photon is insufficient to promote it to either an unbound excited state, or completely eject it into vacuum, and hence the atom is left in its ground state and Thomson scattering dominates. As the energy transfer increases, however, the electrons can escape the confines of the atom, and the Compton cross-section for an atom of atomic number Z approaches

$$\sigma_C \approx 2Z\pi r_0^2 \left\{ \frac{1+\zeta}{\zeta^2} \left[\frac{2(1+\zeta)}{1+2\zeta} - \frac{\ln(1+2\zeta)}{\zeta} \right] \right.$$
$$\left. + \frac{\ln(1+2\zeta)}{2\zeta} - \frac{1+3\zeta}{(1+2\zeta)^2} \right\} \tag{2.8}$$

$$\approx \frac{Z\pi r_0^2}{2\zeta}[1 + 2\ln(2\zeta)], \tag{2.9}$$

where $\zeta = h\nu/(m_e c^2) \gg 1$ is the photon energy expressed in units of the electron rest-mass energy (see also Figure 2.5).

For most applications using synchrotron radiation, the cross-section for Compton scattering and its effect are small and are neglected.

2.4   Thomson Scattering

For all but the most energetic gamma rays, the primary scattering particle for x-rays in a medium is the electron. Classically, elastic scattering occurs when an electron is sinusoidally accelerated by the incoming electromagnetic field and reradiates light at exactly the same frequency – the process is therefore elastic. The reason why an accelerated charged particle emits electromagnetic radiation is explained in more detail in Section 3.3.

If the charged particles (here, electrons) are not moving relativistically, they are accelerated mainly by the electric-field component of the incident wave [see also Section 3.4, in particular Equation (3.3)]. As shown in Figure 2.8, the electron will move in an (anti)parallel direction to that of the oscillating electric field, resulting in emission of electromagnetic radiation. The strength of the re-emitted radiation seen at some observation point thus depends on the magnitude of the component of the electric field seen from that point. It should be clear from Figure 2.8 that no scattering occurs along any line parallel to the electric-field vector ($\chi = \pi/2$), as in such directions, the projected component of the oscillation amplitude of the electron is zero. In contrast, an observer anywhere in the plane perpendicular to the electric-field vector sees a maximal scattering amplitude of the electron. The scattered *intensity*, which is proportional to the square of the electric-field amplitude, therefore scales with $\cos^2 \chi$, where χ is the angle between the plane perpendicular to the electric field and the direction of observation. The doughnut (or 'horn-torus') form of this intensity distribution lends it the name electromagnetic *dipole* radiation.

The ability of an electron of rest mass m_e to scatter an x-ray is quantified by a scattering length. For an isolated electron, a convenient unit is the *Thomson scattering length*, which can be thought of as the 'classical' radius of a free electron. This can be physically understood by equating the rest-mass energy of the electron $m_e c^2$ with the electrostatic energy at radius r_0. It is given by

$$r_0 = \left(\frac{e^2}{4\pi\varepsilon_0 m_e c^2} \right) = 2.82 \times 10^{-5} \text{ Å}. \tag{2.10}$$

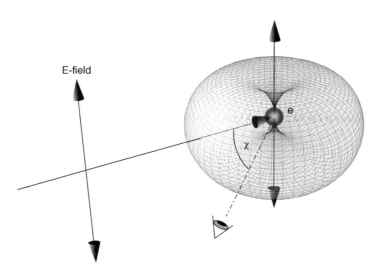

Figure 2.8 Dipole radiation. Electromagnetic radiation drives a free, or unbound, charged particle into oscillatory motion, resulting in reradiation of light at the same frequency. The scattering amplitude varies with the cosine of the polar angle χ, and is also called Thomson scattering.

Elastic scattering from a single electron has a total cross-section

$$\sigma_T = \frac{8\pi r_0^2}{3} = 0.6652 \ \text{barn}, \tag{2.11}$$

which is calculated by averaging over all possible polarization directions.

2.5   Atomic Scattering Factors

2.5.1   Scattering from a Cloud of Free Electrons

The optical properties of materials in the energy range above about 30 eV can be described by the atomic scattering factors (also called the atomic form factors). We begin by assuming the electrons in an atom can move freely and are unaffected by being in bound states, and then introduce the impact of being bound thereafter.

It has already been stated that the electron is the most important scattering particle for the photon energies of interest. A neutral atom, containing Z electrons (where Z is the number of protons in the atom's nucleus), will not act as a point scattering centre, but will scatter x-rays throughout the volume occupied by the electron cloud surrounding the nucleus[2].

The elastic scattering amplitude of an atom is proportional to the Thomson scattering length and Z. The elastic scattering *intensity* therefore increases with Z^2 and is equal to $Z^2 \sigma_T$ for low-energy photons. By 'low energy', it is meant that the wavelength is large compared to the size of the electron cloud surrounding the nucleus. As atoms have typical diameters of the order of 4 Å or smaller, this equates to photon energies less than approximately 3 keV. Below this energy, all the scattering elements within the electron cloud will scatter coherently (that is, in phase) because any differences in their positions in space are very small compared to the wavelength. As there are Z electrons in the cloud, the scattering cross-section is simply $Z^2 \sigma_T$. Note that this is independent of wavelength and, indeed, we see in Figure 2.5 that below approximately 3 keV, the cross-section for Thomson scattering in barium flattens out and asymptotically assumes the value $56^2 \sigma_T = 2086$ barn.

However, as the photon energy increases and the wavelength shortens to distances comparable to and then smaller than the size of the atom, scattering from the different elements in the electron cloud, at some nonzero angle 2θ to the incident beam, begins to interfere partially destructively. This results in a drop-off of the scattering cross-section and intensity, which follows approximately an inverse-square relationship with respect to the photon energy. The precise drop-off is determined by the atomic form factor, or atomic scattering factor, which we discuss in detail in this section.

The total scattering amplitude at 2θ will therefore be equal to the vector sum of the scattering amplitudes in this direction from all volume elements dV of the cloud, taking into account the phases between them, as shown schematically in Figure 2.9(a).

It is a trivial exercise to show that the vectorial difference between an incoming x-ray beam with wavevector **k** and wave **k′** elastically scattered through an angle 2θ, given by

$$\mathbf{Q} = \mathbf{k} - \mathbf{k'}, \tag{2.12}$$

has a magnitude

$$Q = \frac{4\pi}{\lambda} \sin\theta. \tag{2.13}$$

Q is the 'scattering vector', shown in Figure 2.9(b), and $\hbar\mathbf{Q}$ is the momentum transfer imparted on the scattered photon. An important aspect of Thomson scattering is that, at least for hard x-rays, the scattered photon has a

---

[2] One might expect the positively charged nucleus to also scatter x-rays. However, as we have seen in Equation (2.10), the Thomson scattering length is inversely proportional to the mass of the scattering particle, while the scattering cross-section is proportional to the square of the Thomson scattering length, and hence the nucleus contributes less than one part in a million to the scattering amplitude and can be completely ignored.

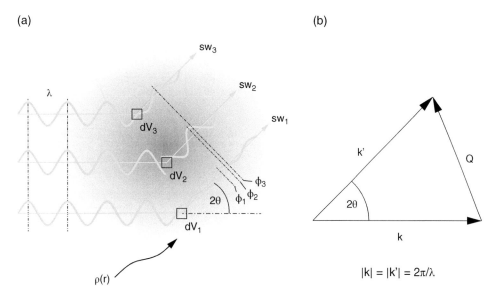

Figure 2.9   Elastic scattering of x-rays by an electron cloud around an atom with an electron density $\rho(\mathbf{r})$. (a) An incident coherent (in-phase) x-ray beam is scattered by all the volume elements constituting the cloud (shown here just three such elements dV_1, dV_2, and dV_3). For a given scattering angle 2θ, each element scatters the incident beam (sw_1, sw_2, and sw_3) with an amplitude proportional to the electron density at the element $\rho(\mathbf{r})$, and a phase, ϕ, determined by its position relative to some arbitrarily chosen origin and the wavelength of the radiation. (b) The scattering vector $\mathbf{Q}$ is the vectorial difference between the incident and elastically scattered wavevectors $\mathbf{k}$ and $\mathbf{k}'$, respectively.

phase shifted by π relative to the incoming photon. This is a consequence of the refractive index of materials in the x-ray regime being less than unity, or in other words, that most of the atom's electrons have binding energies smaller than the x-rays' photon energy.

The atomic scattering factor $f^0(Q)$ describes the scattering amplitude of an atom as a function of $\sin\theta/\lambda$ [or sometimes as a function of Q, the magnitude of $\mathbf{Q}$, which is proportional to $\sin\theta/\lambda$, see Equation (2.13)], and is expressed in units of the scattering amplitude produced by one electron. It is, in fact, the Fourier transform (FT) of $\rho(\mathbf{r})$, the electron density distribution around the atom (see Appendix C for a brief introduction to Fourier theory). As stated above, $f^0(0)$, the scattering amplitude in the forward direction, is nothing more than the integral of the electron density, over the atom's electron cloud, and is thus equal to Z, the atomic number. For nonzero Q, $f^0(Q)$ falls off monotonically with increasing Q. This drop off is steeper for light atoms than for heavy atoms. This might seem counterintuitive at first blush – if the form factor is the FT of the electron cloud, and larger atoms have a larger radius (on average), then should not the FT (which has dimensions inversely proportional to those in real space) be narrower? What this neglects to consider is, however, that although the valence electrons are indeed on average further from the nuclear core, the core electrons are closer, due to the increased Coulomb attraction to the nucleus. The electron density of heavy atoms is therefore concentrated in a smaller central volume than that for lighter atoms. The scattering distribution (the FT of the electron-density distribution) is consequently broader.

It is conventionally assumed that the electron density of an atom is spherically symmetric and hence f^0 depends only on the magnitude of $\mathbf{Q}$ and not on its orientation relative to the scattering atom[3]. Values for $f^0(\sin\theta/\lambda)$

---

[3] This is sometimes a poor approximation. For example, four of the six electrons of carbon in diamond are involved in covalent sp^3-hybridized bonds that are most definitely not spherically symmetric. It is for this reason that some diffraction peaks which, assuming spherical symmetry, would have zero intensity due to their being a so-called 'systematic absence' [for example, the (222)-peak of silicon, see Section 6.4.3], actually do have small but measurable diffraction intensities.

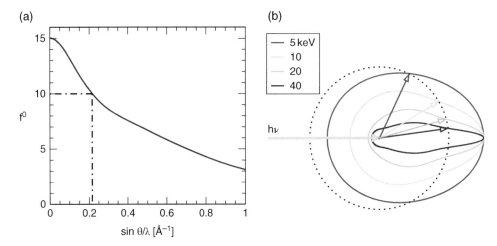

Figure 2.10 The atomic form factor of phosphorus. (a) f^0 as a function of $\sin\theta/\lambda$, according to the tabulated values in the *International Tables for Crystallography*. For $f^0 = 10$, $\sin\theta/\lambda = 0.2152$ Å^{-1} (dashed lines), from which it can be simply calculated that $Q = 2.704$ Å^{-1}. (b) f^0 plotted in a polar diagram as a function of the scattering angle $-\pi \le 2\theta \le +\pi$ for four different photon energies $h\nu$. The origin of the polar figure is at the solid arrow head (representing the incoming beam direction), and the forward-scattering amplitude is $f^0(0) = Z = 15$ for all four energies. The dotted circle represents $f^0 = 10$ ($Q = 2.704$ Å^{-1}). The form factor at any given nonzero scattering angle becomes progressively smaller with increasing photon energy, as destructive interference between scattered waves from different elements of the electron cloud becomes more pronounced, as highlighted by the four scattering arrows (open arrow heads).

can be either directly found in the *International Tables for Crystallography*, or calculated using nine tabulated coefficients and the expression

$$f^0(\sin\,\theta/\lambda) = \sum_{i=1}^{4} a_i \exp(-b_i\sin^2\theta/\lambda^2) + c. \tag{2.14}$$

The atomic scattering factor for phosphorus is shown as a function of $\sin\theta/\lambda$ in Figure 2.10(a), and as a function of scattering angle 2θ in the polar diagram of Figure 2.10(b). Note that for a given Q [which in turn determines $f^0(Q)$], the scattering angle increases with decreasing energy, as shown for $f^0 = 10$ (for which $\sin\theta/\lambda = 0.2152$ Å^{-1} and $Q = 2.704$ Å^{-1}) by the four arrows joining the origin with the points of intersection between the scattering curves with the dotted circle describing $f = 10$. This should be intuitively clear – a lower energy corresponds to a longer wavelength, and the angle required to obtain a given phase difference between two volume elements in the electron cloud around the nucleus [leading to a partial destructive interference and consequent reduction in f^0, see Figure 2.9(a)] will need to be larger.

The form factor for ionic species in the forward direction will differ from Z by an amount equal to the amount of electron transfer to or from neighbouring ions. Hence cations have $f(0)$ values less than Z and anions values greater than Z.

Figure 2.11 pictorially summarizes elastic scattering as described so far, as well as providing a taster of elastic scattering from molecules and crystals, discussed in Chapter 6.

2.5.2 Correction Terms for the Atomic Scattering Factor

In our description of scattering of x-radiation by electrons so far, it has been assumed that the electrons are essentially free; that is, their movement in response to the electromagnetic field of the x-rays is not damped or retarded by the fact that they are bound in an atom. In reality, the response of bound electrons in an atom, molecule, or bulk

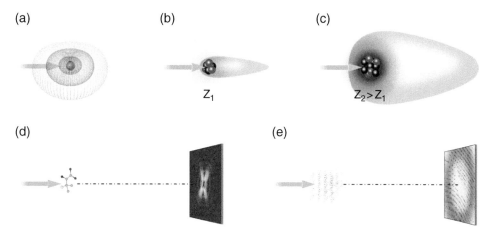

Figure 2.11 Pictorial summary of elastic scattering. (a) Scattering from a single electron results in Hertzian dipole radiation. In this example, the orientation of the dipole torus indicates that the x-rays are vertically polarized. Wavefronts produced by three successive cycles of the driving force are shown. (b) Scattering from a small atom for photon wavelengths shorter than the size of the electron cloud surrounding it (and assuming unpolarized radiation) results in a drop-off of scattering amplitude with scattering angle. (c) Larger atoms scatter more strongly, due to the larger number of electrons in the electron cloud, and the drop-off with scattering angle is less steep because of the higher density of core electrons. (d) and (e) 2D-scattering patterns from a single molecule and a crystal. These are discussed in detail in Chapter 6. They differ from (b) and (c) only in the complexity and size of the electron cloud, which, respectively, now mimic the molecular and crystal structures and can obviously no longer considered to be spherical.

material, to the incoming x-rays is such that additional energy-dependent (i.e. dispersive) terms must be added to more completely describe the scattering factor. We consider these now.

Electrons bound in a material have discrete energy levels and thus defined binding energies E_B. We can model the response of a bound electron to an incident electromagnetic field to that of a damped oscillator responding to an oscillatory driving force. The system has a natural oscillation frequency, given by $\omega_0 = E_B/\hbar$. If the driving frequency ω is much smaller than ω_0, the system can respond without any lag and will thus move in phase with the driving 'handle' or force. This is the case for all those electrons of the atom which have binding energies well in excess of the photon energy. For hard x-rays, however, a good fraction of the electrons may not satisfy this condition – for example, 10 keV photons are approximately 35 times more energetic than the most strongly bound electron in carbon, while even in uranium, which contains 92 electrons, only the ten K- and L-electrons are more tightly bound than 10 keV.

In the other extreme, the electric-field force will oscillate so fast that the electron will respond almost exactly out of phase, that is, lagging by 180°. Note, however, that although the electron moves out of phase with the driving electromagnetic field at high photon energies, the re-emitted radiation is in phase with the incident radiation, as the driving force is $-eE$, that is, it is negative (note that here E is the electric field strength, and should not be confused with the photon energy).

As one tunes the photon energy so that it approaches the binding energy, the amplitude of the electron's motion becomes increasingly damped, because it cannot respond unhindered as a result it being bound to the atom. Exactly when the electron is driven at its natural frequency, ω_0, the system is said to be in resonance and the electron responds with a phase 90° lagging the driving force, and will have a maximum amplitude.

The real part of the elastic-scattering factor, which we henceforth refer to as f_1 (in order to distinguish it from our 'start' definition f^0 for an ensemble of free electrons), will therefore be reduced by an additional energy-dependent term f' describing the damping, i.e.

$$f_1(Q, \hbar\omega) = f^0(Q) + f'(\hbar\omega). \tag{2.15}$$

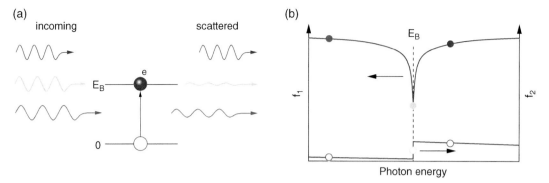

Figure 2.12 Correction terms to the atomic scattering factor. (a) Elastic scattering of an x-ray by an electron bound to an atom with energy E_B. The amplitude of the scattered wave decreases as one approaches a discrete energy level of an electron bound in an atom. Far from resonant edges, the electrons are essentially free and the damping term f' is insignificant. (b) Schematic figure of the qualitative changes in the atomic-scattering-factor components f_1 and f_2 near an absorption edge at energy E_B.

f' is in general negative and hence $f_1 \leq f^0$. At photon energies far above absorption resonances, the electrons are essentially free and $f' \approx 0$ (see Figure 2.12). The effect of f' close to an absorption edge is thus to reduce the apparent number of electrons associated with the absorbing atom.

The absorption and accompanying 90° phase lag is expressed in the description of the atomic form factor by introducing a second additional, energy-dependent, term. Note that the interference between the incoming and emitted radiation causes energy loss (i.e. absorption), and the new second additional term if'' is imaginary. The mathematical justification for this is given in Section 2.6.3.

f'' shows sharp increases at absorption edges [see Figure 2.12(b)], but otherwise falls off as E^{-2} (note the gradient of −2 in the double logarithmic plot of Figure 2.15).

More precisely, f'' is related to the photoabsorption cross-section σ_a at that photon energy by

$$f''(\hbar\omega) = \frac{\sigma_a}{2r_0\lambda}. \tag{2.16}$$

From Figure 2.5, we can see that, for example, at 12.4 keV photon energy (1 Å wavelength), the photoabsorption cross-section of barium (Ba, $Z = 56$) is of the order of 10^{-24} m^2 and hence $f'' \sim 1$. f'' (also called f_2, see below) scales approximately with the fourth power of the atomic number Z, as does σ_a (see Figure 2.13).

In practical units,

$$f'' = 1.431 \times 10^{-5} E \text{ [keV] } \sigma_a \text{ [barn]}. \tag{2.17}$$

We refer to the imaginary component of the total (i.e. complex) atomic scattering factor as $f_2 = f''$, and hence

$$f(Q, \hbar\omega) = f_1(Q, \hbar\omega) + if_2(\hbar\omega). \tag{2.18}$$

This is visually summarized[4] in Figure 2.14, while the atomic form factors for forward scattering for silicon (Si, $Z = 14$) are shown in Figure 2.15. We can qualitatively understand the curves as follows. At the binding

---

[4] See also Appendix D for a description of Argand diagrams, the mathematical constructs that can be used to simply and intuitively describe the addition of vector quantities.

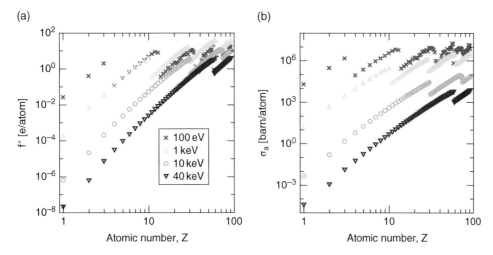

Figure 2.13 The imaginary component of the dispersion correction $f'' = f_2$ and photoabsorption cross-section σ_a as a function of atomic number Z for different photon energies. On this double logarithmic scale, the curves are close to being linear with a gradient of approximately four, except for occasional jumps, due to absorption edges. For example, the drop in absorption between $Z = 10$ (neon) and $Z = 11$ (sodium) for the 1 keV curve is a result of their respective strongest (K) absorption edges lying at 870.2 eV and 1070.8 eV. (a) f_2 increases with decreasing photon energy and can assume several tens of electrons per atom for high Z and low photon energies. This has important consequences for techniques which exploit the anomalous signal associated with f_2, such as SAD (see Section 6.11.4). (b) σ_a as a function of Z.

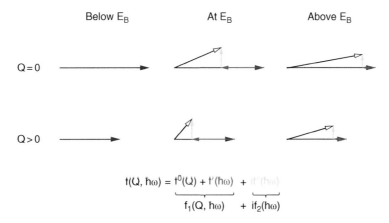

Figure 2.14 Changes in the amplitude and phase (angle) of the total atomic form factor below, at, and above an absorption edge energy E_B, shown for the forward direction ($Q = 0$), and at nonzero scattering angles ($Q > 0$). The components are colour coded according to the given equation.

energies of the K- and L-shells, there are sharp decreases in f_1 and corresponding sharp increases in f_2. Midway between these resonances, coupling of the x-rays to the material is inefficient, and f_1 increases again. At high photon energies, there is only a weak interaction of the x-ray photons with the atoms, and $f_1 \approx Z = 14; f_2$ (which, remember, represents absorption), falls off to very small values.

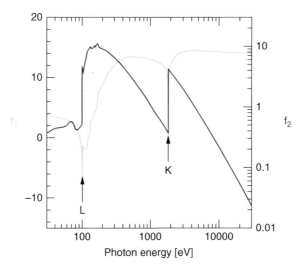

Figure 2.15   The real ($f_1 = f^0(0) + f'$) and imaginary ($f_2 = f''$) components of the atomic form factor for Si in the forward-scattering direction as a function of photon energy. Note that f_1 approaches a constant value at photon energies far above absorption energies equal to the number of electrons in the atom, $Z = 14$. f_2, on the other hand, falls away as E^{-2}.

2.6   The Refractive Index, Reflection, and Photoabsorption

2.6.1   The Refractive Index

The index of refraction n at a wavelength λ describes the response of electrons in matter to electromagnetic radiation. It is a complex quantity related to the atomic form factors of the different atom types in the material, which we express as

$$n = n_R + in_I. \tag{2.19}$$

In the x-ray regime, n is related to the atomic scattering factors of the individual atoms in a material by

$$n = 1 - \frac{r_0}{2\pi}\, \lambda^2 \sum_i N_i f_i(0), \tag{2.20}$$

where N_i is the number of atoms of type i per unit volume and $f_i(0)$ is the complex atomic scattering factor of the ith atom in the forward direction.

Because the phenomena of refraction and absorption both depend on the response of the electrons in a material to the oscillating electric field of an electromagnetic wave propagating through that material, the energy-dependent real and imaginary components f' and f'' are coupled[5], as can be clearly seen in Figure 2.15.

---

[5] The real and imaginary parts of the refractive index are not independent, but are linked by the Kramers–Kronig relations, which describe the interaction between two oscillators.

$$f_1(E) = Z^* + \frac{2}{\pi}P \int_0^\infty \frac{xf_2(x)}{x^2 - E^2}\,dx, \tag{2.21}$$

$$f_2(E) = \frac{2E}{\pi}P \int_0^\infty \frac{f_1(x)}{x^2 - E^2}\,dx. \tag{2.22}$$

Equation (2.21) says that the refraction, f_1, experienced by a photon with energy E is given by the number of electrons present (with a relativistic correction), Z^*, plus an integral over the entire absorption spectrum, f_2, modified by the ratio of the absorption energy, x, to the energy difference between the refracted photon, E,

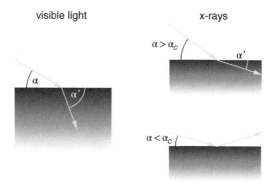

Figure 2.16 The refraction of visible light by a transparent medium is greater than unity (water has a refractive index in the visible of approximately 1.33). In contrast, the refractive index for x-rays is marginally less than one. At grazing angles smaller than the critical angle α_c, total external reflection will occur.

2.6.2 Refraction and Reflection

We have argued that when electromagnetic radiation passes through a solid, it interacts with the electrons within the material. For visible light, this normally means a reduction of the phase velocity by a factor equal to the real part of the refractive index n of that material. This can, for example, be used to bend or focus light with optical lenses. But n is a function of the wavelength of the radiation; in other words, the material disperses the light. This is why rainbows occur – they are manifestations of refraction and reflection (and are *not*, as commonly misconceived, a diffraction phenomenon). Normally, as an absorption maximum of a material is approached from lower frequencies, the refractive index increases. Therefore blue light is refracted more by water or quartz than is red light.

Past the absorption maximum, however, the refractive index drops to values less than unity. The wavelength in the medium thus *increases* to λ/n.

This is the region of so-called *anomalous dispersion*. This means that (a) as radiation enters a material with a refractive index less than unity, it is refracted to *shallower* angles to the interface; and (b) the scattered wave is phase-shifted by π radians with respect to the incident wave (see Figure 2.16). We express the complex refractive index as

$$n = 1 - \delta + i\beta, \qquad (2.23)$$

where the energy-dependent (dispersion) parameters δ and β are called the refractive-index decrement and absorption index, respectively[6]. Comparing this equation with Equation (2.19), we immediately recognize that $n_R = 1 - \delta$, while the absorption index $\beta = n_I$.

The refractive-index decrement δ is equal to the real part of the summed term in Equation (2.20). Far from an absorption edge, we can ignore the correction term f' in Equation (2.15), and the sum $\sum_i N_i f_{1,i}(0) = \rho$ is merely the average density of electrons. Hence

$$\delta = \frac{2\pi\rho r_0}{k^2} = \frac{\rho r_0 \lambda^2}{2\pi}. \qquad (2.24)$$

and the part of the absorption spectrum considered, x. The 'P' in front of the integral sign indicates that this is a so-called 'principal-value' integral, because the integrand is undefined at the point where $E = x$. Equation (2.22) makes the corresponding transform in the opposite direction, calculating the absorption, f_2, from an integral over the refraction spectrum, f_1.

[6] In some texts, n is given as $n = 1 - \delta - i\beta$, which is mathematically more correct, as one subtracts the complex form factors in Equation (2.20). But because the imaginary part (by definition) is at $90°$ to the real part, conventionally it is added in Equation (2.23) rather than subtracted. The magnitude of the complex refractive index remains unchanged.

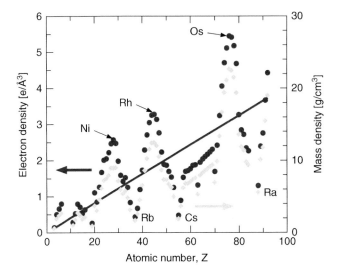

Figure 2.17   The electron density ρ in e/Å^3 (blue circles) and mass density ρ_m in g cm^{-3} (yellow diamonds) as a function of atomic number, Z for those elements which are either liquid or solid at room temperature and pressure. Note the very strong correlation between ρ and ρ_m. To a good approximation, $\rho = \rho_m/4$. The solid blue line is a linear fit relating ρ to Z, and thus also provides a very rough-and-ready rule of thumb that $\rho = Z/25$. Some of the heaviest and lightest elements are highlighted.

A plot of the electron density in those elements that are solid or liquid at room temperature and pressure is shown in Figure 2.17. The values for ρ plotted here can be used to determine δ, as long as the photon energy is far from an absorption edge. Close to edges, however, ρ will be less than its physical value, as already mentioned in our discussion of the atomic form-factor corrections [Equation (2.15)]. Hence, here, ρ represents the 'seen' electron density by the x-rays. A plot of ρ is shown in Figure 2.18 for GaAs.

δ is therefore inversely proportional to the square of the photon energy. If we assume a typical value of one electron per cubic angstrom (for example, silicon has an average electron density of 0.7 e Å^{-3}), we obtain a value of δ for hard x-rays of approximately 5×10^{-6}. Hence, for x-rays, the real part of n is really only *very* slightly less than unity.

Here, we need to distinguish between the group velocity and the phase velocity of a wave packet propagating through a medium of refractive index $n < 1$ (see Figure 2.19). Importantly, in the case of x-rays, the *phase velocity* ($v_p = \omega/k = c/n$) is greater than the velocity of light. But it is $v_g = \partial\omega/\partial k$, the *group velocity*, that carries the energy. The relationship between the group velocity and the phase velocity is

$$v_g = v_p \left(1 - \frac{k}{n} \frac{dn}{dk} \right).$$

$$(2.25)$$

We see from Equation (2.24) that

$$n = 1 - A/k^2,$$

where $A = 2\pi\rho r_0$ is a constant for any given material. From this,

$$\frac{dn}{dk} = \frac{2A}{k^3}.$$

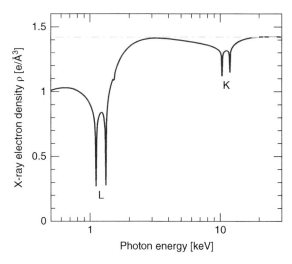

Figure 2.18   The electron density of GaAs as 'seen' by x-rays. The curve was created by taking the known values for δ and calculating ρ using Equation (2.24). The physical electron density is shown as the dot-dashed yellow line. As seen from the perspective of the x-rays, this 'true' value is only assumed for a few keV between the K- and L-edges, and again above the K-edges. This deviation is important to keep in mind when calculating reflectivities and critical angles for, among other things, x-ray mirrors.

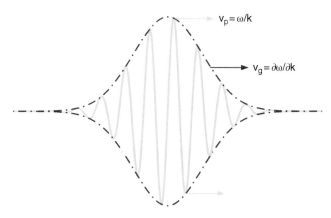

Figure 2.19   Group and phase velocities. A wave packet travelling through a medium of refractive index n will have a phase velocity, v_p, equal to the speed with which the peaks and troughs propagate (yellow arrows). The group velocity, v_g, is the speed of the envelope function (here, the Gaussian envelope described by the dot-dashed curve and blue arrow). For x-rays travelling through matter, $v_g < v_p$.

Inserting this in Equation (2.25) leads to

$$v_g = v_p \left(1 - \frac{2A}{nk^2} \right)$$
$$= v_p \left(\frac{k^2 - 3A}{nk^2} \right).$$

But $v_p = c/n$, hence

$$v_g = c\left(\frac{k^2 - 3A}{(nk)^2}\right).$$

To a high degree of accuracy, $n^2 = (1 - A/k^2)^2 \approx (1 - 2A/k^2)$, and thus

$$v_g = c\left(\frac{k^2 - 3A}{k^2 - 2A}\right) = c\left(\frac{1 - 3\delta}{1 - 2\delta}\right),$$

which can be closely approximated by $c(1 - \delta) = cn$ and always remains below c; the laws of relativity remain inviolate!

2.6.2.1   *Snell's Law and Total External Reflection*

Snell's law relates the incident angle α with the refracted angle α' (see Figure 2.16)

$$\frac{\cos \alpha}{\cos \alpha'} = n_R. \tag{2.26}$$

This means that for a value of n_R of less than unity there will be a minimum incident angle for which the x-rays are refracted into the bulk of the material. This is known as the critical angle, α_c, such that

$$\cos \alpha_c = n_R. \tag{2.27}$$

For x-rays impinging on a surface at an angle below α_c, they will undergo *total external reflection*.

The first two terms of the Taylor expansion of the cosine function $\cos \theta$ are $1 - \theta^2/2$, which is a very accurate approximation for small angles. For example, the fractional difference between the precise value for $\cos(1°)$ and this approximation is smaller than four parts in one billion. So, combining Equations (2.23) and (2.27), we obtain

$$\alpha_c \approx \sqrt{2\delta} \ \text{[rad]}. \tag{2.28}$$

For typical values of δ of around 10^{-5}, α_c is therefore of the order of a few milliradians, or a few tenths of a degree. From Equation (2.24), it can also be deduced that α_c is inversely proportional to the photon energy. For a given incident angle, the photon energy for which $\alpha = \alpha_c$ has a reflectivity of 0.5 and a value of

$$h\nu \ \text{[eV]} = \frac{hc}{\alpha}\sqrt{\frac{\rho r_0}{\pi}} = \frac{37.14}{\alpha\text{[rad]}}\sqrt{\rho[\text{Å}^{-3}]}. \tag{2.29}$$

Hence, at low incident angles, x-rays can be made to reflect from surfaces. Indeed, if the surface of an x-ray mirror is made to be concave (ideally, parabolic), the reflected x-rays can be focussed, as long as the curvature is sufficiently weak that the steepest incident angle is less than α_c. Such x-ray focussing mirrors, as shown in Figure 2.20, are obviously large beasts, as the 'footprint' of the x-ray beam on the mirror surface in the beam direction is $1/\sin\alpha$ times larger than the beam height. X-ray mirrors are described in detail in Section 5.4.

Although x-rays are totally reflected from surfaces if their incident angle is less than α_c, a moment's pause should make it obvious that there must be *some* sort of interaction between the x-rays and the material in order for reflection to occur at all. In actual fact, an *evanescent* standing wave of the x-rays does penetrate exponentially into the refractive medium to a depth of the order of a few nanometres. If the photon energy is close to an absorption edge, the reflection coefficient will be less than unity due to partial absorption of the evanescent wave.

The general expression for the penetration depth Λ into a material in the direction perpendicular to the surface, for any given incident angle α on that surface (as long as $\alpha \approx \sin\alpha$, that is, up to $\alpha \approx 10°$) is given by

$$\Lambda = \frac{\lambda}{4\pi\text{Im}\sqrt{\alpha^2 - \alpha_c^2 - 2i\beta}}, \tag{2.30}$$

Figure 2.20 An x-ray focussing mirror has a radius of curvature such that the variation in the incident beam angle on the mirror surface is much smaller than the critical angle α_c.

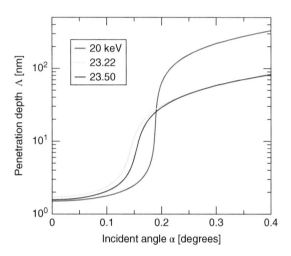

Figure 2.21 The penetration depth of rhodium at different energies close to its K-absorption edge at 23.22 keV. Strong absorption above the edge causes the penetration depth to drop by a factor of four.

where Im refers to the imaginary component of the square-root expression. After some straightforward mathematical manipulation[7], one obtains

$$\Lambda = \frac{\lambda}{4\pi[(-a + \sqrt{a^2 + b^2})/2]^{1/2}}, \tag{2.31}$$

whereby $a = \alpha^2 - 2\delta$ and $b = -2\beta$ (see Figure 2.21). The critical angle and penetration depth therefore both depend on energy. Far from absorption edges, α_c scales inversely with the photon energy [see Equations (2.24) and (2.28)]; Λ is also significantly affected by the changes in δ and β near absorption edges, as shown in Figure 2.21.

Let us consider three limiting cases – large incident angles for which $\alpha \gg \alpha_c$; exactly at the critical angle $\alpha = \alpha_c$; and very shallow angles where α approaches zero.

In the first case, $a \gg b$ and $\alpha \gg \alpha_c$ and so

$$[(-a + \sqrt{a^2 + b^2})/2]^{1/2} \approx [(-a + a(1 + b^2/2a^2))/2]^{1/2}$$

$$= \frac{b}{2\sqrt{a}} \approx -\frac{\beta}{\alpha}.$$

---

[7] $\mathrm{Im}\sqrt{a + ib} = \sqrt{[-a + (a^2 + b^2)^{1/2}]/2}.$

Substituting this into Equation (2.31), we obtain

$$\Lambda = \frac{\lambda \alpha}{4\pi \beta}. \tag{2.32}$$

Λ at high incident angles is thus independent of δ, is proportional to the wavelength and incident angle, and is inversely proportional to β.

Exactly at the critical angle, we obtain

$$\Lambda_c = \frac{\lambda}{4\pi \sqrt{\beta}}. \tag{2.33}$$

In Section 2.6.3, we argue that β scales with the inverse fourth power of the photon energy, hence Λ_c is proportional to $E = h\nu$.

Lastly, at exceedingly shallow angles much smaller than α_c,

$$\Lambda_0 = \frac{1}{4\sqrt{\pi r_0 \rho}}. \tag{2.34}$$

Hence, the minimum penetration depth of x-rays into a material only depends on that material's electron density. For elemental condensed matter at ambient conditions, this ranges from $\rho = 0.139$ e Å^{-3} for lithium to 5.43 e Å^{-3} for osmium (Figure 2.17), for which Λ_0 is therefore 7.1 nm and 1.14 nm, respectively.

Λ for incident angles below α_c is some three orders of magnitude smaller than that well above the critical angle. It is thus possible to use x-rays as a *surface-sensitive probe* down to the atomic level, using grazing-incidence radiation, as will be described in detail in Section 6.12.

Above the critical angle, the reflectivity for an x-ray in vacuum impinging at an angle α on a perfectly flat surface of a material with a real part of the refractive index $n_R = 1 - \delta$ drops off rapidly, and is given by the Fresnel equations. These equations are dealt with in more detail in Section 6.14 describing the technique of x-ray reflectivity.

2.6.3 Photoabsorption

Depending on their energy, photons incident on an atom can be absorbed by that atom to induce excitation of a bound electron either to an unoccupied bound state, or to produce an unbound photoelectron. Details of the absorption and subsequent processes, and how these are used to investigate manifold properties of matter (such as, among others, magnetism, superconductivity, chemical composition, and local structure) are described in Chapter 7. Here, we consider only the fundamentals of photabsorption in the context of the complex refractive index.

2.6.3.1 *The Photoabsorption Coefficient*

From Equations (2.20) and (2.23), it can be seen that the absorption index β is related to f'' by

$$\beta = \frac{r_0}{2\pi} \lambda^2 \sum_i N_i f_i''. \tag{2.35}$$

We have already stated that away from absorption edges, f'' is proportional to the inverse square of E, hence from Equation (2.35) it is apparent that β scales with the inverse fourth power of the energy.

Let us imagine an x-ray beam travelling in vacuum and then penetrating some sort of absorbing medium (Figure 2.22). We begin with the classical description of the electric-field vector of a linearly polarized electromagnetic plane wave propagating through vacuum in the z-direction

$$E(z, t) = E_0 \exp[i(kz - \omega t)], \tag{2.36}$$

where $k = 2\pi / \lambda$ is the wavevector in vacuum and λ is the wavelength in vacuum.

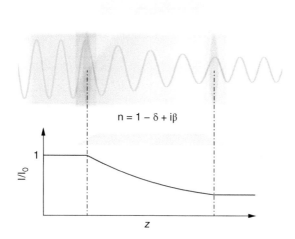

Figure 2.22   The transmission of x-rays through a material falls off in an inverse exponential manner with penetration depth with an absorption attenuation length $\lambda/(4\pi\beta)$. The real part of the refractive index, $1 - \delta$, is smaller than in vacuum, and as a result the wavelength in the medium increases to λ/n.

When the wave travels through the medium, however, the wave function changes to

$$E(z, t) = E_0 \; \exp[i(nkz - \omega t)], \tag{2.37}$$

where we have now included the complex refractive index n in the exponent term. In other words, the wavelength of the x-rays is changed by the medium, although their angular frequency (ω) is, obviously, unaffected. Substituting Equation (2.19) into Equation (2.37), we obtain

$$E(z, t) = E_0 \; \exp(-n_I kz) \; \exp(in_R kz - \omega t). \tag{2.38}$$

The first exponential term describes the attenuation of the amplitude as the wave travels through the medium. One measures, however, the intensity, which is proportional to the square of the amplitude, and therefore drops by an amount $1/e$ over a depth $1/(2n_I k)$ This is the absorption coefficient. The exponential decay behaviour of the transmitted signal

$$\frac{I}{I_0} = \exp(-\mu z) \tag{2.39}$$

is the Beer–Lambert equation for linear absorption. We can now equate the absorption index $n_I = \beta$ to the absorption coefficient μ by

$$\mu = 2n_I k. \tag{2.40}$$

For any given material, therefore, its absorption coefficient μ is defined as the reciprocal of the thickness needed to reduce the intensity of an impinging electromagnetic beam to $1/e$ of its original value. This is the attenuation length.

Clearly, μ is both material specific and a function of the x-ray energy. In general, it is strongly dependent on the atomic number of the element (it varies approximately as Z^4) and decreases with increasing photon energy, for which it approximately scales with E^{-3} [Figure 2.23(a)].

The absorption coefficient is essentially an indication of the electron density in the material and the electron binding energy. Thus if a particular chemical substance can assume different geometric ('allotropic') forms and thereby have different densities, μ will be different. The carbon allotropes, diamond, graphite, C_{60}, and

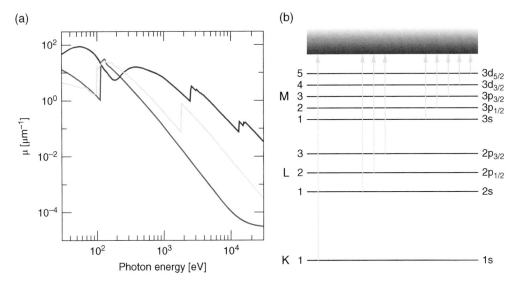

Figure 2.23 (a) The x-ray absorption coefficient μ for beryllium (red curve), silicon (yellow), and lead (blue) as a function of the photon energy. (b) Sharp increases in the absorption occur when the photon energy is just sufficient to eject the electron from the electronic orbital to the continuum. The x-ray absorption (left) and atomic-orbital labellings (right) are shown.

glassy carbon, have densities of 3.53, 2.23, 2.0, and 1.69 g cm^{-3}, respectively. Accordingly, their attenuation lengths (for example, 71.4, 113, 126, and 149 nm, respectively, at 320 eV) are inversely proportional to their densities.

Conversely, compounds that are chemically distinct but contain the same number of electrons per formula unit and have similar mass densities will have similar absorption properties (except close to absorption edges). For example, the attenuation lengths of isoelectronic NaCl, MgS, AlP, and Si (normalized to their mass densities) are shown in Figure 2.24. The average number of electrons per atom is 14. The small differences in absorption at high photon energies are due to differences in the average isotopic ratios of protons to neutrons from element to element. Note also how all the absorption edges are centred around that of Si.

A more fundamental parameter is therefore the *atomic* absorption cross-section of the element, σ_a, briefly introduced in Equation (2.16). It has units of area, such as cm^2, or the barn, where 1 b is equal to 10^{-28} m^2. The relationship between σ_a and μ is

$$\mu = N_i \, \sigma_a = \left(\frac{\rho_m N_A}{A} \right) \sigma_a, \tag{2.41}$$

where N_i, N_A, ρ_m, and A are the atomic number density, Avogadro's number, the mass density, and the atomic mass, respectively[8]. More commonly, however, the absorption cross-section is given in cm^2 g^{-1}, in which case

$$\sigma_a \, [\text{cm}^2\text{g}^{-1}] = \frac{N_A}{A} \, \sigma_a \, [\text{cm}^2] = \frac{\mu}{\rho_m}. \tag{2.42}$$

From Equations (2.40) and (2.41), one obtains

$$\sigma_a = \frac{4\pi A \beta}{N_A \rho_m \lambda}, \tag{2.43}$$

---

[8] The etymology of the unit barn originates from the days of wartime nuclear research, when physicists were so flabbergasted by the enormous neutron scattering strength of the uranium nucleus that they exclaimed its nuclear core to be 'as big as a barn'. One barn is indeed about the size of a uranium nucleus.

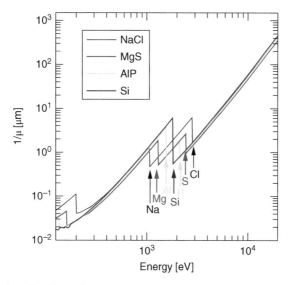

Figure 2.24   The attenuation lengths of NaCl, MgS, AlP, and Si (normalized to their mass density) as a function of x-ray photon energy. Far from absorption edges, the curves have very similar values. Note also how the positions of the K-edges of the component elements converge to that of Si as the number of electrons in those elements approach that of Si ($Z = 14$).

or, in practical units

$$\sigma_a[\text{barn}] = 1.68 \times 10^5 \frac{AE\beta}{\rho_m}, \tag{2.44}$$

whereby A is expressed in g mol^{-1}, E is in eV, and ρ in g cm^{-3}.

Far from an absorption edge, the ratio

$$\frac{\delta}{\beta} = \frac{Z}{f_2}. \tag{2.45}$$

This ratio is important in phase-contrast imaging. It becomes less accurate as the photon energy increases. For the lightest atoms, β begins to drop off less steeply than $(hv)^{-3}$ towards a $1/(hv)$-dependence above approximately 10 keV, while for heavier atoms, this transition lies closer to 100 keV. Nonetheless, it remains a good rule of thumb for most materials at typical synchrotron photon energies.

2.6.3.2   Absorption Edges and Nomenclature

As can be seen in Figures 2.23 and 2.24, the absorption spectra show occasional sudden sharp increases, known as absorption edges. These are produced by resonances between the x-ray photon energy and the binding energy of an electron in the atom. The electronic state of a bound electron is defined by the principal, orbital angular momentum, and spin quantum numbers n, l, and s, respectively. In atomic physics, these states are labelled as nl'_j. The angular momentum $l = 0, 1, 2$, etc., is denoted by the letters $l' = s, p, d, f, \ldots$ and $\mathbf{j}$ is the total angular momentum given by the vector sum of the orbital and spin momenta, that is

$$\mathbf{j} = \mathbf{l} + \mathbf{s}. \tag{2.46}$$

Note also that $\mathbf{s} = 1/2$ for an electron.

Table 2.1 Correspondence between x-ray absorption edges, quantum numbers (n, l, and j), and their electronic configurations.

Edge	n l j	Configuration	Edge	n l j	Configuration
K	1 0 1/2	$1s$	N_1	4 0 1/2	$4s$
L_1	2 0 1/2	$2s$	N_2	4 1 1/2	$4p_{1/2}$
L_2	2 1 1/2	$2p_{1/2}$	N_3	4 1 3/2	$4p_{3/2}$
L_3	2 1 3/2	$2p_{3/2}$	N_4	4 2 3/2	$4d_{3/2}$
M_1	3 0 1/2	$3s$	N_5	4 2 5/2	$4d_{5/2}$
M_2	3 1 1/2	$3p_{1/2}$	N_6	4 3 5/2	$4f_{5/2}$
M_3	3 1 3/2	$3p_{3/2}$	N_7	4 3 7/2	$4f_{7/2}$
M_4	3 2 3/2	$3d_{3/2}$	O_1	4 0 1/2	$5s$
M_5	3 2 5/2	$3d_{5/2}$	O_2	4 1 1/2	$5p_{1/2}$

The nomenclature for absorption edges in x-ray spectroscopy is more pragmatic. The first 18 and their 'translation' to atomic-physics electron-configuration notation are given in Table 2.1.

2.7 X-ray Fluorescence and Auger Emission

The two dominant processes occurring subsequent to absorption of an x-ray and the ejection of an x-ray photoelectron are fluorescence and Auger electron emission (see Figure 2.25).

In fluorescence, the excess energy remaining in the atom after relaxation of an electron from a shell further out to the hole left behind by the photoelectron is removed by emission of an x-ray photon. In contrast, Auger emission is a nonradiative process, in which the excess energy is carried away by the ejection of a second electron. We discuss each process in more detail now.

2.7.1 X-ray Fluorescence

Any material can be made to fluoresce by bombarding it with x-rays or high-energy electrons. If electrons are used, the energy spectrum of the emitted electromagnetic radiation is composed of two parts – the broad and continuous Bremsstrahlung background signal produced as electrons collide with the atoms and are strongly decelerated, described in more detail in Section 3.5.2, and intense, sharp, and discrete fluorescence lines on top of this, called characteristic radiation. The photon energies of this fluorescence are 'characteristic' (that is, a fingerprint) for the type of atom from which they originate, as shown in Figure 1.8. In the case of excitation by x-rays, the Bremsstrahlung is absent.

These characteristic x-ray lines result from the transition of an outer-shell electron relaxing to the hole left behind by the ejection of the photoelectron from the atom. This occurs on a time-scale of the order of 10 to 100 fs. As the energy difference between the two involved levels is in general well-defined, these lines are exceedingly sharp. Indeed, from Heisenberg's uncertainty principle, $\Delta E \Delta t \sim \hbar$, the natural linewidth can be calculated to be of the order of 0.01 eV, although this depends on the element and the transition. Commonly, values for the energies of characteristic radiation are given to an accuracy of 0.1 eV.

Not all transitions are allowed, as the relaxation of an electron from one bound electronic state to another within an atom (or ion) follows the selection rules for electric dipole radiation (or, in other words, for the emission of an x-ray photon). Photons have an angular momentum of $1\hbar$, hence emission of an x-ray photon in fluorescence must be accompanied by a change in the orbital angular momentum l by one, i.e.

$$\Delta l = \pm 1. \tag{2.47}$$

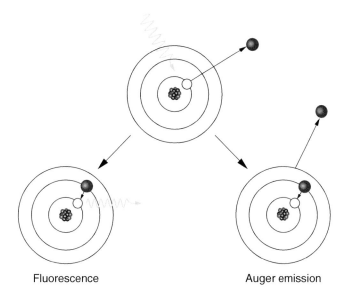

Fluorescence   Auger emission

Figure 2.25 The absorption of an x-ray photon results in the ejection of a core electron. This can then be filled by an electron from a shell further out. The excess energy can either be emitted in the form of characteristic x-ray radiation or by the ejection of an Auger electron.

Hence, for example, dipole transitions from the $2s$ to the $1s$, or from the $3p$ to the $2p$ level are forbidden.

The spin of the electron has no effect on the spatial distribution of the electron in the atom, and therefore cannot affect the orbital angular momentum. Using this somewhat reverse-logic argument, we arrive at the second selection rule, namely

$$\Delta s = 0. \tag{2.48}$$

From our definition of **j** [Equation (2.46)], we immediately obtain the third selection rule

$$\Delta j = 0, \pm 1; \text{ but not } 0 \rightarrow 0. \tag{2.49}$$

In general, fluorescence lines have lower energies (or longer wavelengths) than the absorbed radiation, due to nonradiative relaxation (normally down a cascade of closely lying excited vibrational states) in the time between absorption and emission. This is known as the Stokes shift. For example, the difference in energy between the K-absorption in copper (8980.5 eV) and the $K\alpha_1$ emission (8047.8 eV) is 932.7 eV.

2.7.1.1 Nomenclature and Emission Energies

The nomenclature in x-ray absorption lines of K, L, M, etc., described above, actually originates in the early studies of Barkla concerning x-ray emission spectra. In 1911, he described his findings that narrow emission lines characteristic for each element increased in their energy (and thereby, also their penetrating power) with atomic number. He wrote in *The Philosophical Magazine* [3]

It is seen that the radiations fall into two distinct series, here denoted by the letters K and L*.

He justified the use of this nomenclature in the footnote indicated by the asterisk:

Previously denoted by letters B and A. The letters K and L are, however, preferable as it is highly probable that series of radiations both more absorbable and more penetrating exist.

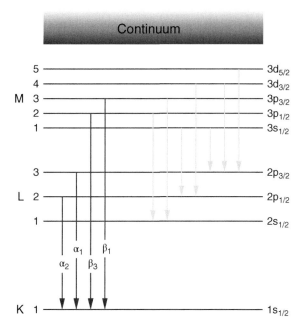

Figure 2.26   The Siegbahn nomenclature of x-ray fluorescence lines. The blue set represents K emission, the yellow set L emission.

Barkla speculated that there might be more energetic series than the K-series, hence his decision to label the observed series in the middle of the alphabet, reserving the letters A to J for up to ten higher-energy series of transitions. In this one instance, he was wrong, although the nomenclature has remained.

Two years later, Henry Moseley recorded the energies of characteristic radiation by dispersing the spectra using an x-ray spectrometer based on diffraction crystals developed by the Braggs. He discovered a systematic relationship between the energies of the most intense characteristic radiation (which turned out to be the Kα line, see below) and the atomic number, which he defined in what is now known as Moseley's law:

$$\nu = K(Z - 1)^2, \tag{2.50}$$

where $K = 2.47 \times 10^{15}$ Hz, and ν is also given in Hz. He was thus able to show that the characteristic radiation followed the Bohr model of the atom, and that the lines had energies corresponding to the difference in energy of the discrete orbitals [4].

The nomenclature for x-ray fluorescence lines is more complex than that for absorption lines, as they involve two bound orbitals, not one (see Figure 2.26). The Siegbahn K-series involves transitions in which the level to which the electron relaxes is the K (or $1s$) state. Similarly, L lines have L states ($n = 2$) as the lower energy level, and so on. The longest wavelength (smallest energy-difference) transition within a series is labelled α, with successive lines denoted β, γ, etc. These lines are actually further split due to the fact that the energy depends subtly on the orientation of the spin momentum vector with respect to the orbital angular momentum (so-called 'spin-orbit' interactions), labelled by the subscripts 1, 2, etc.

The IUPAC notation is more logical and requires no rote-learning. The format is simply (filled state)-(source state); so for example, emission caused by relaxation of the L_2 to K state is the K-L_2 line; that from the M_4 state to L_2 state is the L_2-M_4 line. Note that if the filled state is the K state, no subscript is used, as only one K state exists.

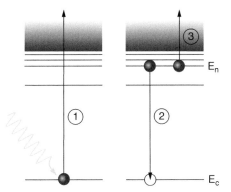

Figure 2.27 The three-electron process in Auger emission. First, a core-electron is ejected from an atom by absorption of a photon. Next, the system relaxes by an outer electron falling into the core-level, releasing an energy $|E_n - E_c|$. In the Auger process, this energy is channelled into the ejection of a second, or Auger, electron.

2.7.2 Auger Emission

Auger emission is a three-electron process first discovered by Lise Meitner in 1923, but was named after the French physicist Pierre Victor Auger, who independently discovered and explained the Auger process three years later.

Auger electrons are produced when an outer shell electron relaxes to the core-hole produced by the ejection of a photoelectron. The excess energy produced in this process is $|E_n - E_c|$, whereby E_c and E_n are the core- (filled) and outer- (source) shell binding energies, respectively. Instead of being manifested as a fluorescence x-ray photon, the energy can also be channeled into the ejection of another electron if its binding energy is smaller than the excess energy. The process is shown in Figure 2.27 for ejection of an Auger electron from the same shell as that of the electron which relaxed to the core-level hole.

In this case, (ignoring relaxation effects) the electron energy is $|E_c - 2E_n|$. More generally, the kinetic energy is $|E_c - E_n - E'_m|$, where E'_m is the binding energy of the Auger electron. The prime shows that the binding energy of this level has been changed (normally increased) because the electron ejected from this level originates from an already ionized atom.

The notation of Auger electrons is *ABC*, where *A*, *B*, and *C* relate to the three electronic processes involved in Auger emission, and assume Barkla's notation of K, L, M, etc. An example would be a KLL Auger electron. *A* is the core level from which the photoelectron originates (with energy E_c), *B* is the level from which the electron that fills the core-hole originates (energy E_n), and *C* is the level from which the Auger electron is ejected (energy E'_m).

Typical Auger electron energies are in the range of 100 to 500 eV, which, as we shall see in the Chapter 7, have escape depths of only a few nanometres, hence Auger spectroscopy is very surface sensitive.

Of particular importance is the fact that, in contrast to photoelectrons, the energies of Auger electrons ($|E_c - E_n - E'_m|$) are independent of the incident photon energy, although the amount of Auger electrons emitted is directly proportional to the absorption cross-section in the surface region.

2.7.3 Fluorescence or Auger?

Auger-electron emission and x-ray fluorescence are competitive processes [5]. The rate of spontaneous fluorescence, given by the Einstein A-coefficient, is proportional to the third power of the energy difference between the upper and lower state. Hence, for a given atom, K emission lines are more probable (and therefore more intense) than L emission (assuming L radiation is allowed at all), while fluorescence is stronger for heavier atoms, which have a more attractive positive nuclear charge and thus a larger energy difference separating adjacent shells (see Figure 2.3).

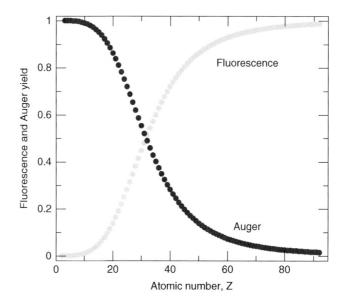

Figure 2.28   The fluorescence and Auger yields of the elements as a function of atomic number, Z.

The empirical expression for K-shell fluorescence is

$$Y_K = \frac{Z^4}{10^6 + Z^4}. \tag{2.51}$$

Figure 2.28 shows the normalized yields for Auger and fluorescence processes as a function of the atomic number, Z, defined as $Y = n/N$, whereby n is the number of emitted electrons (Auger) or photons (fluorescence), and N is the number of core-shell ionizations.

In contrast, the probability of an Auger electron being emitted increases with decreasing energy difference between the excited atom and the atom after Auger emission. Hence, LMM events are more likely than KLL events. Also, low atomic-number atoms have higher Auger yields than do heavier atoms. Similarly, high atomic-number elements have a large positive charge at the nucleus, which binds electrons more tightly, reducing the probability of Auger emission.

The relatively high yield for Auger emission in light atoms has deleterious consequences in macromolecular crystallography. The dominant mechanism resulting in radiation damage of protein crystals is the secondary ionization of atoms by Auger electrons, which have a significantly larger cross-section than the initial photoelectrons. The number of secondary electrons generated can exceed the number of photoionization events by over an order of magnitude. The rate of Auger decay is of the order of a femtosecond, hence any one atom can in principle generate many Auger electrons within the typical pulse durations of synchrotron radiation of a few tens of picoseconds. Sub-femtosecond x-ray pulses, now possible to generate using x-ray free-electrons lasers (Chapter 4), allow this problem to be ameliorated, as a maximum of just one Auger electron per atom can be generated per x-ray pulse [6].

2.8   Concluding Remarks

In this chapter, the basic concepts of the interaction of x-rays with matter have been introduced. The most important scatterer of x-rays is the electron. Because x-rays have such high energies compared to the binding energies of all

Table 2.2 Summary of the dependence of some basic material properties as a function of photon energy E and atom type Z.

	f''	δ	β	σ_a	σ_T	α_c	μ
E	E^{-2}	E^{-2}	$E^{-4\dagger}$	$E^{-3\ddagger}$	E^0 ($E \lesssim 1$ keV), E^{-2} ($E \gtrsim 5$ keV)	E^{-1}	E^{-3}
Z	Z^4	Z^{1*}	Z^{4*}	Z^4	Z^2	$Z^{-1/2*}$	Z^4

*: Approximate.
$\dagger$ Changes to E^{-2} at high photon energies.
$\ddagger$ Changes to E^{-1} at high photon energies.

but the most tightly bound core electrons, their interaction with matter is relatively weak – that is why medical radiographic images through the torso of a human can be recorded, not something that could be achieved using visible light!

We have discussed how the phenomena of refraction, reflection, and absorption are all intimately related to one another through the complex property of the atomic scattering factor and how these depend on the atom type (Z) and photon energy E. The most important relationships in this context are summarized in Table 2.2.

The knowledge we have gained here will provide the basis for understanding both x-ray optics hardware, discussed in Chapter 5, and also in fathoming the various experimental methods described in the second half of the book.

Problems

1. Order the following electromagnetic radiation from low to high energy:
   (a) An 8 keV photon
   (b) A photon with $\lambda = 1.5895$ Å
   (c) A photon with wavevector $k = 4.0036$ Å^{-1}
   (d) A photon with energy 1.2177 fJ
   (e) A photon with frequency $\nu = 1.8619 \times 10^{18}$ Hz
   (f) A photon with angular frequency $\omega = 1.1395 \times 10^{19}$ rad. (3P)
2. The cosmic background radiation, a microwave background relic of the big bang, has a blackbody spectrum at a temperature of 2.7255 K. Using Wien's displacement law, determine the wavelength of the maximum of the spectrum, and also express this in electronvolts. (2P)
3. The earth has a radius of 6400 km. Calculate its surface area to the nearest order of magnitude, expressed in barns. (1P)
4. From Equations (2.4) and (2.5), prove Equation (2.6). (2P)
5. Beginning with Equation (2.8), derive Equation (2.9). (2P)
6. The nine coefficients a_i, b_i, and c ($i = 1 \ldots 4$) used to calculate the atomic form factor f^0 of phosphorus as a function of $\sin\theta/\lambda$ are:
   $a_1 = 6.4345$, $b_1 = 1.9067$, $a_2 = 4.1791$, $b_2 = 27.157$, $a_3 = 1.78$, $b_3 = 0.526$, $a_4 = 1.4908$, $b_4 = 68.1645$, $c = 1.1149$.
   Using your favourite plotting program (mine is xmgrace), plot f^0 as a function of $\sin\theta/\lambda$ up to $\sin\theta/\lambda = 1$ Å^{-1}. At which value of $\sin\theta/\lambda$ does $f^0 = f^0(0)/3$? What value does Q assume here and what is the minimum photon energy required to access this wavevector? (4P)
7. Calculate the average electron density ρ of GaP in e Å^{-3}. The atomic numbers of Ga and P are 31 and 15, respectively. The (zincblende) face-centred unit cell of GaP is shown in Figure 2.29 and has a lattice constant of 5.451 Å. (1P)
8. Calculate the elastic scattering vector Q for 20 keV electrons scattered through an angle of 40°. (1P)

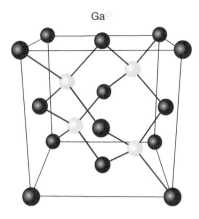

Figure 2.29   The fcc unit cell of GaP.

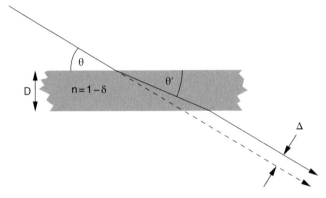

Figure 2.30   The displacement Δ of an x-ray beam due to refraction after it passes through a parallel plate of a material of refractive index $n = 1 - \delta$ and thickness D at an angle θ.

9. If the refractive-index decrement δ of a given material has a value of 8×10^{-6} at 1 Å, what value does it assume at 20 keV?   (1P)

10. A certain element has an atomic form factor at a certain angle $f^0 = 14$. Close to an absorption edge, this is modified by the dispersive terms $f' = -6.1$ and $f'' = 1.9$. Calculate the ratio in magnitudes f/f^0.   (1P)

11. The imaginary component f_2 of the atomic form factor of sulfur is 1.295 e at 5 keV. Using this information, what value does f_2 assume for calcium at 20 keV?   (2P)

12. Starting with Snell's law [Equation (2.26)], demonstrate that, for x-rays in vacuum incident on the interface to a medium of refractive index $1 - \delta$ at a grazing angle $\alpha \geq \alpha_c$ (but which nonetheless remains small),

$$\alpha^2 - \alpha'^2 = 2\delta,$$

    where α' is the angle of the transmitted beam relative to the interface.   (2P)

13. Consider Figure 2.30. A beam is incident at a grazing angle θ on a parallel plate of material of refractive index $n = 1 - \delta$ and thickness D. Prove that, upon emerging from the far side of the plate, the x-rays are displaced by an amount

$$\Delta = D \cos \theta \left[\frac{\sin \theta}{(n^2 - \cos^2 \theta)^{1/2}} - 1 \right],$$

and that, in the regime $2\delta \ll \theta^2 \ll 1$, this can be accurately approximated by

$$\Delta = D \left[\frac{1}{\theta^2/\delta - 1} \right].$$

Plot Δ and the beam transmission for 25 keV photons and a 0.5 mm Be plate as a function of θ between $\theta = 0.3°$ and $5°$. The refractive-index decrement for Be at 25 keV is $\delta = 5.45 \times 10^{-7}$ and the attenuation coefficient is $\mu = 0.0315$ mm^{-1}. (5P)

14. Using Equation (2.31), describing the penetration depth of x-rays in a material as a function of the incident angle, derive Equations (2.33) and (2.34). (4P)

15. Which of the following dipole transitions, using the IUPAC notation, are allowed: K-L$_2$; L$_1$-M$_4$; K-M$_2$; L$_2$-M$_4$; L$_2$-M$_5$; L$_3$-M$_3$? Explain your reasoning. (2P)

16. Why does one never observe helium Auger electrons? (1P)

References

[1]   J. Als-Nielsen, "X-rays and matter – the basic interactions," *C. R. Physique*, vol. 9, pp. 479–486, 2008.

[2]   A. H. Compton, "A quantum theory of the scattering of x-rays by light elements," *Phys. Rev.*, vol. 21, pp. 483–502, 1923.

[3]   C. G. Barkla, "The spectra of the fluorescent Röntgen radiations," *Phil. Mag.*, vol. 22, pp. 396–412, 1911.

[4]   H. G. J. Moseley, "The high-frequency spectra of the elements. Part II," *Phil. Mag.*, vol. 27, pp. 703–713, 1914.

[5]   W. Bambynek, C. D. Swift, B. Crasemann, et al., "X-ray-fluorescence yields, Auger, and Coster-Kronig transition probabilities," *Rev. Mod. Phys.*, vol. 44, pp. 716–813, 1972.

[6]   L. Young, E. P. Kanter, B. Krässig, et al., "Femtosecond electronic response of atoms to ultra-intense x-rays," *Nature*, vol. 466, pp. 56–61, 2010.

3

Synchrotron Physics

3.1 Introduction

For many synchrotron-facility users, a detailed knowledge of the working principles behind the generation of synchrotron radiation may seem of secondary importance. In general, however, both users trying to optimize their experimental setup, and prospective users trying to decide which facility or beamline might be appropriate for a planned experiment, will benefit greatly from having a working knowledge of how synchrotrons perform. Importantly, the demand for access to synchrotron radiation in modern facilities far outstrips availability, hence a submitted proposal which might otherwise be ranked positively by a review committee set up to decide which users will obtain beamtime, will be in danger of summary rejection if the foreseen experiment demands operational parameters outside the capabilities of the facility in question. Understanding something of synchrotron physics is therefore advantageous.

In this chapter, I will convey the most important aspects of synchrotron physics, without bogging down the reader with unnecessary mathematical derivations. This should provide essential rules of thumb and simple equations to determine what any given facility has to offer a potential user. For a deeper understanding of the myriad details of 'machine physics', as it is also known, the reader is referred to excellent primers by Herman Winick [1], Helmut Wiedemann [2], and Philip Duke [3].

3.2 Overview

Third- and, even more so, fourth-generation synchrotrons[1] are characterized by their ability to produce highly collimated and narrow beams of high-intensity x-rays [4]. These properties can be expressed in a single quantity called 'brilliance', described in detail below. Another important property is the energy of the electrons within the storage ring that generate the x-radiation, which influences the range of photon energies that any one facility can practically cover (see also Figure 1.16). The most relevant parameters of a selection of third-generation synchrotron facilities are listed in Table 3.1; the performance of the emerging class of fourth-generation synchrotrons is detailed in Table 3.2.

A synchrotron consists of five main components (see Figure 3.1). These are

---

[1] On perhaps a pedantic note, the term 'synchrotron' to describe electron-storage rings has been somewhat hijacked from its original meaning. Synchrotron accelerators were first so named to indicate that the magnetic fields used to maintain the charged particles on a closed path needed to be increased and synchronized with the particles' increasing relativistic mass (i.e. energy). A storage ring is a special case of a synchrotron, in which the particle energy remains constant. Booster rings at modern 'synchrotron' facilities are, however, true synchrotrons.

Table 3.1 Important properties of selected third-generation synchrotrons, listed in order of their storage-ring energy.

Facility, Country	Storage-ring energy [GeV]	Current [mA]	Circumference [m]	Emittance (x, y) [nm rad × pm rad]	Brilliance [ph/s/mm^2/mrad^2/0.1% BW]
BESSY II, Germany	1.7	100	240	6 × 100	5×10^{18}
ALS, US	1.9	400	198	6.8 × 8	3×10^{18}
Elettra, Italy	2 – 2.4	320	260	7 × 70	10^{19}
SLS, Switzerland	2.4	400	288	5 × 0.9	4×10^{19}
Soleil, France	2.75	500	354	3.7 × 11	10^{20}
Diamond, England	3.0	300	562	2.7 × 27	3×10^{20}
NSLS-II, USA	3.0	500	792	0.6 × 8	3×10^{21}
Petra-III, Germany	6.0	100	2304	1.0 × 10	2×10^{21}
ESRF, France	6.0	300	846	3.8 × 10	8×10^{20}
APS, US	7.0	100	1104	3.0 × 25	8×10^{19}
SPring8, Japan	8.0	100	1436	2.8 × 6	2×10^{21}

Table 3.2 Selected parameters and performance metrics of DLSRs in the design phase (D), under construction (C), or in operation (O) at the time of writing.

Facility, Country	$\mathscr{E}$ [GeV]	Circumf. [m]	Current [mA]	Lattice	ϵ_x^e [pm rad]	λ_{DL}* [nm]	$h\nu_{DL}$* [eV]	$\mathscr{M}$**	Status
ALS-U, USA	2.0	197	500	9-BA	50#	0.62	2000	0.095	D
APS-U, USA	6.0	1104	200	7-BA	67	0.84	1476	2.5	D
DLS-II, England	3.0	562	300	4, 5, or 6-BA?	270–100	3.39–1.26	366–984	5.33–1.97	D†
Elettra2.0, Italy	2.0	260	400	6-BA?	230–280?	2.89–3.52?	429–352	1.01–1.23?	D†
ESRF-EBS, France	6.0	844	200	7-HBA‡	140	1.76	704	2.34	O
HEPS, China	6.0	1360.4	200	7-HBA‡	35	0.44	2818	2.45	D
MAX-IV, Sweden	3.0	528	500	7-BA	200–330††	2.51–4.15††	494–299	3.27–5.40††	O
PETRA-IV	6.0	2304	100	7-HBA‡	≈ 15	≈ 0.21	≈ 6020	≈ 5.1	D
SLS-2.0, Switzerland	2.4	290	400	7-BA	120	1.65	752	0.56	D
Soleil-II, France	2.75	354	N.A.	6/7BA?	∼ 200?	∼ 2.5?	∼ 496?	∼ 1.2?	D†
Sirius, Brazil	3.0	518	350	5-BA	250	3.14	395	3.86	O
SPring8-II, Japan	6.0	1436	100	5-BA	190	2.39	519	15.6	D

* The diffraction-limited photon wavelength $\lambda_{DL} = 4\pi\epsilon_x^e$ matches the horizontal electron emittance. The corresponding diffraction-limited photon energy $E_{DL}[\text{eV}] = 1239.8/\lambda_{DL}[\text{nm}]$.
** Units: ×10^6 [nm rad m^3 GeV^{-2}].
# For fully coupled beam. Natural lattice emittance ≈100 pm rad.
† Early design phase at the time of writing.
‡ HBA: hybrid bend achromat.
†† With/without insertion devices.

1. A source of electrons (normally generated by thermionic emission from a hot filament) in an electron gun. The electrons are accelerated using a linear accelerator (LINAC) to about 100 MeV. A regular supply of electrons is required, as they are always being lost in the machine, due to collisions with residual gas particles in the storage ring.

2. A booster ring into which the electrons are injected from the LINAC and further accelerated. They may either be accelerated to the energy of the electrons in the main storage ring or (less commonly, especially for modern facilities) to a somewhat lower energy. They are then periodically injected into the storage ring, so that the specified storage-ring current is maintained. Historically, this was performed when the storage-ring current

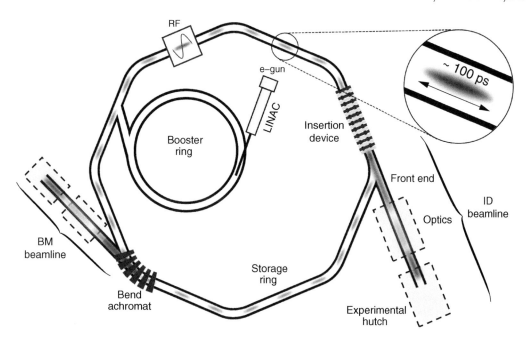

Figure 3.1 A schematic of the most important components of a modern synchrotron facility. Electrons from a source (a heated filament in an electron gun) are accelerated in a linear accelerator (LINAC) into an evacuated booster ring, where they undergo further acceleration. They are then injected into the storage ring. There, they are maintained in a closed path using bending-magnet achromats at arc sections. The beamlines use the radiation emitted from insertion devices (IDs, either wigglers or undulators) placed at the straight sections between the arcs, and from the bending magnets, on the axes of emission. The energy lost by the electrons through radiation of synchrotron light is replenished by a radio frequency (RF) supply.

dropped to about $1 - 1/e \approx 70\%$ of the initial current. Such injections of a large amount of current required 'downtime' at the beamlines, while the insertion devices were opened (because their magnetic fields must be minimized), as otherwise the perturbation to the electrons' orbit was too strong during injection, resulting in a spontaneous 'beam dump'.

Modern facilities, however, nowadays operate in a so-called 'top-up' mode, in which the booster ring accelerates the electrons to the same energy as the storage ring (nothing new there) and, importantly, the stored current in the ring is quasi-continuously 'topped up' by small and frequent injections of electrons from the booster, allowing uninterrupted user operation. The storage-ring current is allowed to drop by approximately 1 mA, or less than a percent, before it is topped up. For typical machine losses, this means injection approximately every two minutes. This greatly enhances stable operation at the beamlines, because it provides a constant heat load on x-ray optical components, which, once optimized, are not subsequently distorted by a change in the intensity of the incident synchrotron radiation.

The electrons from the booster are injected into the storage ring in a parallel orbit to the electrons already there, as injection exactly into the orbit would eject the already-stored beam, causing a beam dump. The new electrons then merge with the main electron beam over thousands of cycles, requiring some milliseconds to reach equilibrium. This process limits the smallest lateral extent of, and oscillations in, the electron beam, which in turn sets a lower limit to the brilliance. Novel so-called 'swap-out' injection modes which circumvent this merging process of the freshly introduced electrons are, at the time of writing, being implemented in some upcoming fourth-generation facilities for which the electron-beam quality is at a premium. This is described in more detail in Section 3.8.5.

Figure 3.2   View in the downstream direction of the tunnel of the SLS, which accommodates both the booster ring (close to the wall on the right) and the storage ring (on the left). Both rings contain bending magnets, quadrupoles, and sextupoles. The storage ring also accommodates insertion devices; one is just visible in this image before the storage ring turns behind the inner wall.

A view within the storage-ring tunnel of the Swiss Light Source is shown in Figure 3.2 – note that the large booster ring is accommodated in the same tunnel as the storage ring.

3. The storage ring contains the electrons and maintains them on a closed path by the use of an array of magnets, commonly referred to as the 'magnet lattice' of the ring. The magnets are most commonly of three types: dipole (or bending) magnets cause the electrons to change their path and thereby follow a closed orbit; quadrupole magnets are used to focus the divergent electron beam; and sextupole magnets correct for chromatic aberrations that arise from the focussing by the quadrupoles (see Section 3.8).

   The electrons have kinetic energies $\mathscr{E}$ measured in GeV, and their velocities are highly relativistic[2], i.e. only *very* marginally smaller than the velocity of light. The relative spread in electron energy $\Delta\mathscr{E}/\mathscr{E}$ is of the order of 0.1%.

   The ring is a structure consisting of arced sections containing bending, quadrupole, and sextupole magnets (i.e. the magnet lattice) and straight sections used for insertion devices (IDs), which generate the highest-brilliance synchrotron radiation. The bending magnets (BMs) used to deflect the electrons round the arced sections that connect the straight sections are also often used to provide bending-magnet radiation – although their brilliance is significantly lower than that produced by IDs, even monochromatized BM-radiation is still orders of magnitude more intense than that which can be provided by laboratory-based sources.

4. Energy is lost by the electrons, due to emission of synchrotron radiation. This must be replenished if the electrons are not to spiral into the inner walls of the storage ring and be lost, and is achieved by a radio frequency (RF) cavity, which supplies the electrons just the right amount of extra energy every time they pass through it.

---

[2] Among other parameters, storage rings are characterized by the electrons' kinetic energy, as in 'the 6 GeV storage ring at the ESRF'. This should not be confused with the energy of the photons produced by deflecting these electrons in a magnetic field, i.e. the synchrotron radiation, which is several orders of magnitude smaller. The highest-energy synchrotron storage ring presently in existence is at SPring8, which stores electrons at 8 GeV, hence its name.

5. The beamlines run off tangentially to the storage ring, along the axes of the insertion devices and tangentially at bending magnets. The first section of a beamline, referred to as the 'front end', has several functions and safety features – it isolates the beamline vacuum from the storage ring vacuum; it monitors the position of the photon beam; it defines the angular acceptance of the synchrotron radiation via an aperture; it blocks, when required, the x-ray and Bremsstrahlung radiation to the next sections containing the optics and experimental hutches; and it filters out, if necessary, the low-energy tail of the synchrotron-radiation spectrum, which is strongly absorbed by matter and can thus damage optical components.

The beam is then normally focussed and/or monochromatized in the optics hutch, before it enters the experimental hutch. For those beamlines generating high-energy x-rays, the hutches are shielded using lead-lined, thick concrete walls, to protect users not only from x-rays, but also from gamma rays and high-energy neutrons, which can be produced in the storage ring when the relativistic electrons collide with stray gas particles. The maximum energy of such radiation is equal to the storage-ring energy, of the order of a few GeV. Such gamma rays penetrate very deeply in matter, hence effective radiation shields may consist of tens of centimetres of lead blocks. Experiments in the hutch are therefore performed remotely, from outside the radiation area.

In this chapter, we will consider the generation of synchrotron radiation in more detail. Beamline components are dealt with separately in Chapter 5. To begin with, we familiarize ourselves with the generation of electromagnetic radiation through the acceleration of charged particles and, conversely, quantify the forces acting on these particles by electromagnetic radiation. In addition, we consider the most relevant features of electrons moving at relativistic velocities.

3.3 Production of Light by Acceleration of Charged Particles

We begin this section by stating that, although the physics presented here is not essential to understand subsequent material in this book, it does provide an elegant insight into many of the phenomena described in both this chapter and Chapter 2. The more physics-phobic reader may, however, skip both this section and Section 3.4.

In the absence of other nearby charges, the electric-field lines of a charged particle at rest or in uniform motion propagate radially outwards from the particle at the speed of light. Any observer looking at this particle therefore sees no component of the field in any transverse direction. But electromagnetic radiation is a transverse wave, and hence the stationary particle does not emit light [Figure 3.3(a)].

Now consider a situation in which the charged particle is abruptly accelerated, such as in Figure 3.3(b), where it bounces off a wall. Because of the finite velocity of light, c, an observer at a distance R from this event will only observe this event after a time R/c, at which moment, the field lines will appear to shift rapidly in a lateral direction. This sudden distortion of the electric-field lines and the momentary generation of a lateral component to them is what we call electromagnetic radiation. Note that in Figure 3.3(b), the observer at A detects no distortions and hence in this direction, that is, the direction of the acceleration, no light is emitted. Conversely, at 90° (observer at B), the distortion is maximized and the electromagnetic-field amplitude is strongest. More precisely, the field strength in a given direction is proportional to the acceleration observed perpendicular to that direction. The angular distribution in the amplitude of the electromagnetic radiation is thus proportional to $\cos\theta$, while the intensity follows a $\cos^2\theta$ relationship, where θ is the angle between the observer and the axis normal to the acceleration.

Next, consider Figure 3.3(c), in which the charged particle bounces back and forth between two walls separated by a distance L. The electric-field lines shift rapidly at regular intervals given by L/v, where v is the magnitude of the particle's velocity. This pattern is 'rounded off' if the particle is forced instead to execute simple harmonic motion, as sketched in Figure 3.3(d). Situations in which this occurs include, for example, conduction electrons in the metal of a mirror, which re-emit at the same frequency as the incident light (the mirror reflects); electrons driven by an external source in a radio antenna; or electrons travelling along an undulator, as will be discussed

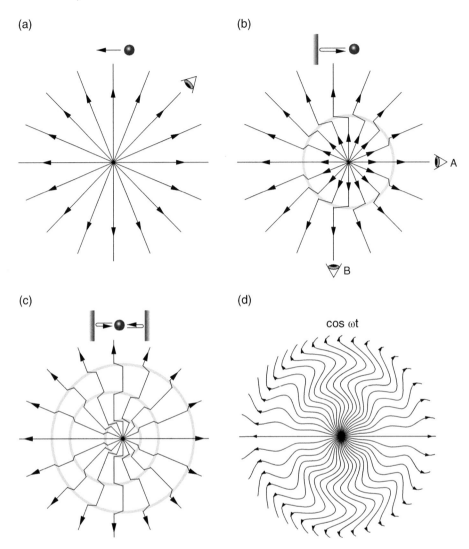

Figure 3.3 Generation of electromagnetic radiation by acceleration of a charged particle. (a) A charged particle at rest or moving at uniform speed will not emit light, as any observer of the particle detects no lateral component of the electric-field lines. (b) If, however, the particle undergoes acceleration, an observer positioned anywhere other than along the axis of that acceleration (position A), will see a shift in the position and direction of the electric-field lines as the event horizon washes over them at the speed of light (for example, at position B). (c) A charged particle bouncing between two boundaries will generate a corresponding set of pulses of electromagnetic radiation at regular intervals. (d) A simple-harmonic driving force will generate radiation at the same frequency.

in Section 3.9.3. This angular distribution is referred to as 'dipole radiation', which has already been sketched in Figure 2.8.

What sort of spectra can one expect as a result of accelerating charged particles? In cases (b) and (c) of Figure 3.3, the acceleration is described by a short-lived pulse during the 'bounce' off the wall, whereas in case (d), the acceleration is simple harmonic, described by a single frequency. This latter acceleration therefore results in the generation of radiation of the same single frequency as the driving force. In contrast, no single frequency

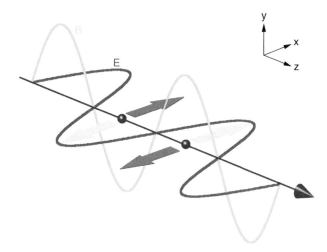

Figure 3.4 The fields and associated forces acting on electrons by an electromagnetic plane wave co-propagating in the +z-direction. The electric force, $F_E = -eE$, shown here as red arrows, acts in the opposite direction to that of the oscillatory electric-field vector. The magnetic Lorentz force (yellow arrows) acts in the opposite direction to that of the electric force and has a magnitude $F_L = eBv$, whereby v is the electron's axial velocity.

can describe the pulsed acceleration of cases (b) and (c), and thus here, the resulting spectra are broad, containing a continuous range of frequencies required to describe this sudden change in direction. As we will see later in this chapter, cases (b), (c), and (d) correspond closely to light generated by bending magnets, wigglers, and undulators, respectively.

3.4 Forces Acting on a Charged Particle by Electromagnetic Radiation

By convention, a linearly polarized electromagnetic plane wave propagating in the positive z-direction will, for an observer looking in the forward z-direction, have its magnetic-field vector always rotated $+\pi/2$ (90°) clockwise relative to the electric-field vector (see Figure 3.4). So, for example, if the magnetic field points in the positive y-direction, the electric field will point in the positive x-direction.

In which directions do the electric- and magnetic-field components of the electromagnetic radiation act upon an electron? The force imparted by an electric field on an electron (with its negative charge) always acts perpendicular to the propagation direction of the electromagnetic wave and in the opposite direction of the electric-field vector[3] **E**, that is

$$\mathbf{F}_E = -e\mathbf{E}. \tag{3.1}$$

The direction of the magnetic Lorentz force acts perpendicularly to the plane defined by the electron's direction of motion and the magnetic-field vector (it is proportional to the cross-product of **v** and **B**, Figure 3.5). If, as the case in Figure 3.4, the electrons co-propagate in the same positive z-direction as the electromagnetic field, then the Lorentz force will have its maximum magnitude

---

[3] Both the electric-field amplitude and the photon energy are denoted by E in this text. Where possible, the former, which is a vector, is written in bold face to distinguish it. In those instances where this is not the case, the context should remove any ambiguities.

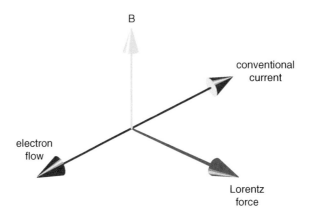

Figure 3.5 The direction of the Lorentz force on a moving electron. Note that the conventional current is opposite in direction to the electron movement, and that the direction of the magnetic field **B** is defined as being from north to south. Hence, bending magnets in synchrotron facilities in which the electrons circulate in a clockwise sense (viewed from above) will have the north poles above the south poles, and vice versa for 'anticlockwise' storage rings.

$$\mathbf{F}_L = e|\mathbf{B}||\mathbf{v}| \tag{3.2}$$

and in the direction opposite to that of the electric force.

From classical electromagnetism, it emerges that the relationship between the magnetic-field amplitude B and the electric-field amplitude E of an electromagnetic wave is

$$|\mathbf{B}| = \frac{|\mathbf{E}|}{c}. \tag{3.3}$$

Therefore the ratio $F_L/F_E = -v/c$. As a consequence, the Lorentz force induced by an electromagnetic wave only becomes nonnegligible compared to the force from the electric-field component for electrons travelling at relativistic velocities.

3.5   Radiation from Relativistic Electrons

3.5.1   Synchrotron Radiation

According to special relativity, the total energy $\mathscr{E}$ of any particle of rest mass m_e travelling at a velocity v is given by

$$\begin{aligned}
\mathscr{E} &= \frac{m_e c^2}{\sqrt{1 - (v/c)^2}} \\
&= \frac{m_e c^2}{\sqrt{1 - \beta^2}},
\end{aligned} \tag{3.4}$$

where $\beta = v/c$ [not to be confused with the same symbol used for the absorption index, first introduced in Equation (2.23)]. The kinetic energy, $\mathscr{E}_{KE}$, of the same particle is equal to the total energy minus the rest-mass energy, that is

$$\mathscr{E}_{KE} = \mathscr{E} - m_e c^2 = m_e c^2 \left(\frac{1}{\sqrt{1 - \beta^2}} - 1 \right). \tag{3.5}$$

Note that for $\beta \ll 1$ (i.e. nonrelativistic particles, here electrons), Equation (3.5) reduces to $m_e v^2/2$, the Newtonian expression for kinetic energy. In contrast, for highly relativistic electrons, the fractional difference between $\mathscr{E}_{KE}$ and $\mathscr{E}$ is negligibly small.

We define the so-called Lorentz factor γ as

$$\gamma = \frac{\mathscr{E}}{m_e c^2}. \tag{3.6}$$

$m_e c^2$ is the rest-mass energy of the particle, which for an electron of mass $m_e = 9.11 \times 10^{-31}$ kg is equal to 511 keV. γ can thus be re-expressed simply in terms of the storage-ring energy in GeV as

$$\gamma = 1957\ \mathscr{E}[\text{GeV}]. \tag{3.7}$$

From Equation (3.4),

$$\gamma = \frac{1}{\sqrt{1 - \beta^2}}. \tag{3.8}$$

We see, therefore, that as the particle approaches the velocity of light, it gains mass by a factor of γ. As it would have infinite mass exactly at $v = c$, it can only asymptotically approach but never reach this limit. But from Equation (3.6), we see that the relativistic mass γm_e of an electron in the storage ring is simply

$$\gamma m_e = \frac{\mathscr{E}}{c^2}. \tag{3.9}$$

We manipulate Equation (3.8) to obtain

$$\beta = \left(1 - \frac{1}{\gamma^2}\right)^{1/2}. \tag{3.10}$$

For typical γ values of 10^4, $1/\gamma^2$ is very small, so we can ignore all but the first two terms in the Taylor expansion of Equation (3.10) and thereby obtain

$$\beta \approx 1 - \frac{1}{2\gamma^2}. \tag{3.11}$$

The radiation emitted by accelerated electrons moving at a small fraction of the speed of light has a dipole distribution, whereby the nodal axis lies in the direction of acceleration, as shown earlier in Figure 2.8. In contrast, radiation from a relativistic electron moving along a circular arc is like a sweeping searchlight – the highly directed nature of the photon beam is produced by the dipole distribution being greatly suppressed in the backward direction (i.e. opposite to the electrons' motion), and enhanced in the forward direction. The radiation power from a relativistic particle of charge e executing a circular trajectory of radius ρ is given by Liénard's equation

$$P = \frac{2}{3}\frac{e^2}{c^3}\gamma^4 a^2, \tag{3.12}$$

where a is the acceleration experienced by the charged particle. The centrifugal acceleration of electrons in a synchrotron is in general given by v^2/ρ, where v is the particle velocity (which can be accurately approximated by c, the speed of light). In Section 3.5.3 below, ρ is shown to equal $\gamma m_e c/eB$ [Equation (3.23), where γm_e is the relativistic mass of the electrons in the storage ring and B is the magnitude of the magnetic field used to achieve the acceleration through the Lorentz force]. From this, we thus obtain

$$a = \frac{Bec}{\gamma m_e} = 5.273 \times 10^{19}\frac{B[\text{T}]}{\gamma}\ \text{m s}^{-2}. \tag{3.13}$$

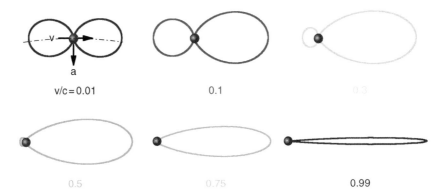

Figure 3.6   The angular distribution of electromagnetic radiation emitted by acceleration of a charged particle travelling at different velocities. For small v/c, the pattern is that of dipole radiation, shown earlier in Figure 2.8. As v/c approaches unity, the distribution becomes increasingly pencil-like in the forward direction and diffuse and suppressed in the backward direction.

Substituting Equations (3.6) and (3.13) into Equation (3.12), we obtain

$$P = \frac{2}{3} \frac{e^4 c^3}{(m_e c^2)^4} \mathscr{E}^2 B^2 = 6.763 \times 10^{-18} \mathscr{E}^2 B^2, \tag{3.14}$$

whereby P is in watts, $\mathscr{E}$ has been expressed in GeV, and B in tesla. The typical current of synchrotrons is measured in hundreds of mA. For example, a 3 GeV synchrotron storing a current of $I = 250$ mA and using bending magnets with field strengths of 1.4 T equates to a radiative power output due to circulation of the electrons alone of 186 W.

Note that Equation (3.14) scales very strongly with the particle mass and is the reason why electrons instead of protons are used at synchrotron facilities (in contrast to CERN, where energy loss by radiation should be kept to a minimum).

The angular power distribution of a charged particle moving with a velocity v is given by

$$\frac{dP}{d\Omega} = \kappa a^2 \frac{[(v/c) - \cos\theta]^2}{[1 - (v/c)\cos\theta]^5}, \tag{3.15}$$

where $\kappa = e^2/(16\pi^2\epsilon_0 c^3) = 6.814 \times 10^{-55}$ kg s is a constant (assuming a is expressed in m s^{-2}), a is the centrifugal acceleration of the electrons directed at right angles to the electron velocity v, and θ is the angle between v and the point of observation (see Figure 3.6). Note that in the nonrelativistic limit of $v \ll c$ ($\beta \ll 1$), Equation (3.15) reduces to the simple $\cos^2\theta$ dependence of dipole radiation, as was asserted above.

In the forward direction ($\theta = 0$)

$$\frac{dP}{d\Omega}(\theta = 0) = \frac{\kappa a^2}{(1-\beta)^3}. \tag{3.16}$$

Substituting Equation (3.11) in to Equation (3.16), we immediately obtain

$$\frac{dP}{d\Omega}(\theta = 0) = 8\kappa a^2 \gamma^6. \tag{3.17}$$

Substituting Equation (3.13) into Equation (3.17), we obtain

$$\frac{dP}{d\Omega}(\theta = 0) = \kappa' B^2 \gamma^4, \tag{3.18}$$

whereby $\kappa' = e^4/(2\pi^2 m_e^2 \epsilon_0 c) = 1.5156 \times 10^{-14}$ m^2 C^2(kg s)$^{-1}$ (or W T^{-2}).

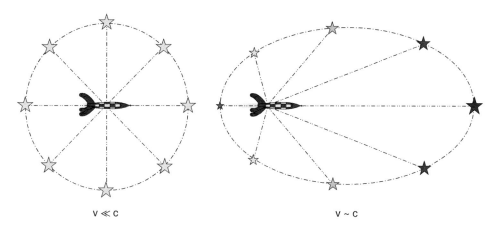

Figure 3.7 Distortion of perspectives at relativistic velocities. Imagine a spacecraft moving at small velocities relative to the speed of light. Identical stars at the same distance in every direction from the ship will be indistinguishable to an observer aboard the ship. At relativistic velocities, however, this 'celestial sphere' will, from the perspective of the ship, be stretched in the direction of motion by the Lorentz factor γ to an ellipsoid, with the observer placed at the 'upstream' focus. This causes the field of view to become increasingly wrapped around the observer and compressed in the forward direction. In addition, light arriving at larger angles from the forward direction becomes increasingly dim and red-shifted. The ratio of the ellipse's major to minor axis is γ. Here, $\gamma = 1.689$ ($v/c = 0.806$).

The maximum intensity thus scales with the fourth power of the electron-beam energy. At highly relativistic velocities, we can substitute $\cos\theta$ with $1 - \theta^2/2$, from which it emerges that

$$\frac{dP}{d\Omega} = \kappa' B^2 \gamma^4 \frac{[1 - (\gamma\theta)^2]^2}{[1 + (\gamma\theta)^2]^5}. \tag{3.19}$$

The entire beam lies within $\theta = \pm 1/\gamma$, and has a full-width at half-maximum of approximately $1/\gamma$. This 'natural opening angle' (or divergence) of the narrow radiation cone is, for typical storage-ring energies of 1 to 8 GeV, equal to 0.5 to 0.06 milliradians (mrad), respectively (1 mrad is approximately equal to $0.057°$).

Why should the angular distributions described by Equation (3.15) and shown in Figure 3.6 become increasingly elongated in the direction of motion as v approaches the speed of light, c? This is a consequence of Einstein's theory of special relativity, namely time dilation and Lorentz contraction. Although this topic lies outside the scope of this book, it is qualitatively summarized in Figure 3.7.

A second intuitive way to understand why the production of x-radiation is concentrated in the forwards direction is shown in Figure 3.8. The velocity of the electrons in a storage ring is very close to c, so an observer looking at an electron that is turning into her line of sight sees an enormous relativistic Doppler shift, as the distance between radiation wavefronts is hugely relativistically compressed (because the electrons are moving almost as fast as the wavefronts).

At other positions on the orbit, for which the angle θ between the observer's line of sight and the electron-velocity vector is larger than $1/\gamma$, the relativistic compression of the wavefront quickly falls off. The on-axis rapid and large Doppler shift is synonymous with a large acceleration. This means that in the observer's frame of reference, radiation is only significantly observed close to the observation axis, as emission of radiation requires acceleration of the electron. Hence, the less pronounced Doppler effect observed slightly off axis means that here, synchrotron radiation has its maximum intensity at lower energies compared to that of on-axis radiation (shown in Figure 3.11).

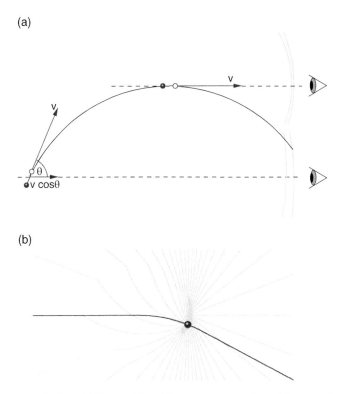

Figure 3.8 Bending-magnet radiation. (a) The rapid and intense compression of the wavefronts of the radiation emitted from electrons moving along the line of sight due to a blue Doppler shift means the radiation appears to the observer to be far more intense here than at other positions on the orbit. (b) Simulation of the radiation field emitted by an electron executing an arc of 30° with a velocity of 0.925 c and a 1 m bending radius.

3.5.2 Bremsstrahlung

Bremsstrahlung (German for braking radiation) can be thought of as a 'shock wave' of electromagnetic radiation produced by the sudden deceleration of the electrons, as shown in Figure 3.3(b). It has a maximum cut-off energy corresponding to the highest kinetic energy of the electrons impinging on the target. No Bremsstrahlung will be observed exactly in the direction of the deceleration, as we have already argued in Section 3.3, in particular Figure 3.3.

The power dP emitted in a solid angle $d\Omega$ by Bremsstrahlung for the case of deceleration of a particle with an initial velocity v along the same axis as the velocity is given by

$$\frac{dP}{d\Omega} = \kappa a^2 \frac{\sin^2\theta}{[1 - (v/c)\cos\theta]^5}, \tag{3.20}$$

where κ is the same constant as that in Equation (3.15), a is the (negative) acceleration, and θ is the angle between the deceleration and the point of observation. Polar plots of the Bremsstrahlung power distribution for different values of v/c are shown in Figure 3.9(a), while the dependence of the angular separation of the lobes $2\alpha_{\mathrm{BS}}$ on v/c is plotted in Figure 3.9(b). Note the similarity of Equations (3.15) and (3.20). The essential difference between the two is due to the fact that in the former, the acceleration a acts perpendicularly to the electrons' motion, while in the latter it is (anti)parallel.

At the sort of relativistic velocities typically used in synchrotron storage rings, Bremsstrahlung produced by collisions between the electrons and residual gas molecules in the imperfect vacuum is highly forward directed.

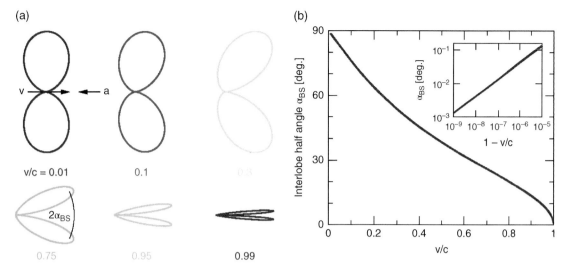

Figure 3.9 The polar distribution of Bremsstrahlung radiation. (a) Normalized polar plots for a selection of values of v/c [Equation (3.20)]. The acceleration a is so oriented that it slows the electron (deceleration). (b) Plot of the interlobe half-angle α_{BS} as a function of v/c. The double-logarithmic inset shows α_{BS} for values of $1 - v/c$ typically found at synchrotron facilities.

The angular lobe separation is measured in microradians and the central node is entirely smeared out by variations in the direction of deceleration resulting from the electron–molecule collisions.

3.5.3 Magnetic Deflection Fields

How exactly *are* the electrons in a storage ring deflected into a closed orbit? A magnetic field[4] induces a Lorentz force on a charged particle perpendicular to both the motion of that particle and the magnetic-field vector (it is proportional to the cross-product of **v** and **B**, see Section 3.4 and Figure 3.5).

Classically, we equate the Lorentz force to a centripetal force such that

$$e\,\mathbf{v} \times \mathbf{B} = \frac{m_e v^2}{\rho},\tag{3.21}$$

where ρ is the orbital radius of the arc. Let us assume that the magnetic field is perpendicular to the electron path, so that the cross-product $\mathbf{v} \times \mathbf{B}$ is merely $|v|\,|B|$. Here, we are dealing with highly relativistic electrons, and so we must replace m_e with the relativistic mass γm_e. We can also, to a high degree of accuracy, replace v with c. Therefore,

$$ecB = \frac{\gamma m_e c^2}{\rho}$$

$$\Rightarrow \rho = \frac{\gamma m_e c}{eB}.\tag{3.22}$$

$$= \frac{\mathscr{E}}{ceB}.\tag{3.23}$$

---

[4] Note that the magnitude of a radial electric field required to induce a comparable centripetal force as the magnetic-field-induced Lorentz force would be an impractically hazardous 10^9 V m^{-1}!

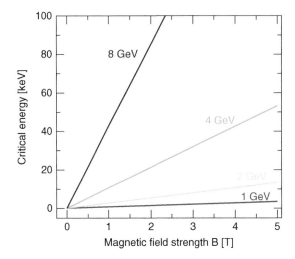

Figure 3.10   The critical energy $\hbar\omega_c$ as a function of magnetic-field strength for different storage-ring energies.

If we express $\mathscr{E}$ in GeV, we can rewrite this equation using practical units, such that

$$\rho[\mathrm{m}] = 3.3\ \frac{\mathscr{E}[\mathrm{GeV}]}{B[\mathrm{T}]}. \tag{3.24}$$

As typical magnetic-field strengths of bending magnets and insertion devices are of the order of 1 tesla[5] and storage-ring electron energies are normally of the order of a few GeV, the bending radius is typically a few metres.

Another important parameter is the angular frequency of the orbiting electrons $\omega_0 = c/\rho$ (that is, the number of radians turned by the electrons per second). In calculating the angular frequency, it is important to remember that only a part of the storage ring path is curved. Hence, for example, a storage ring with a 300 m circumference, of which 200 m are straight sections[6], has an angular frequency of $2\pi \times c/(300 - 200) = 18.8$ Mrad s^{-1}.

The characteristic, or critical, angular frequency ω_c of the emitted radiation from a given synchrotron source is given by

$$\omega_c = \frac{3}{2}\gamma^3\,\omega_0 = \frac{3}{2}\gamma^3\,\frac{c}{\rho} = \frac{3}{2}\gamma^2\,\frac{eB}{m_e}, \tag{3.25}$$

whereby the term c/ρ was substituted with $eB/\gamma m_e$, derived from Equation (3.22). The critical energy, $E_c = \hbar\omega_c$, exactly divides the total emitted radiative power from a bending magnet in half. As ρ is proportional to $\mathscr{E}$ [Equation (3.23)], we can re-express the above equation in practical units (see also Figure 3.10)

$$\hbar\omega_c[\mathrm{keV}] = 0.665\ \mathscr{E}^2[\mathrm{GeV}]\ B[\mathrm{T}]. \tag{3.26}$$

The emitted bending-magnet spectrum from the orbiting electrons is very broad[7] [see Figures 1.5 and 3.8(b)]. However, for photon frequencies higher than about $\gamma^3\omega_0$, the spectral intensity falls off sharply. The

---

[5] As a comparison, the earth's magnetic field is strongest around the magnetic poles in northern Canada and southern Australia, and in parts of Siberia, at a little above 60 μT (0.6 gauss), while a common-or-garden kitchen magnet may have a field strength at its surface of 0.03 T.

[6] There are also short straight sections between dipoles. A typical storage ring might have in total only 25 to 100 m of bending arcs.

[7] The reason for this is that, in contrast to the quasi-monochromatic emission induced by a sinusoidal driving force of the same frequency, as shown in Figure 3.3(d), the observed acceleration viewed tangentially to the bending magnet is described by a single transverse 'pulse' of acceleration, which can only be described by a continuous spectrum of frequencies.

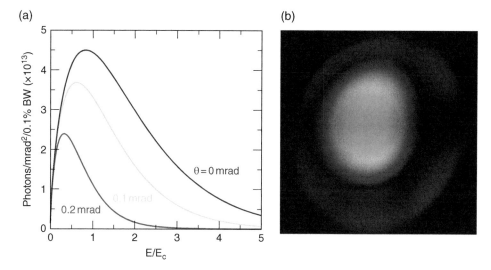

Figure 3.11 The spatial distribution of synchrotron radiation. (a) The spectrum from a 1.41 T bending magnet in a 2.4 GeV storage ring. The abscissa is in dimensionless units of the photon energy E divided by the characteristic, or critical, energy $E_c = \hbar\omega_c$, defined in the text. The weight of the spectrum changes according to the angular position relative to the central axis. On axis, there is a larger contribution from high-energy photons compared to that in the spectra off axis. (b) This is vividly demonstrated in the image taken on axis of the visible/UV spectrum of the wiggler at the Femto Project of the Swiss Light Source. Courtesy Gerhard Ingold, Paul Scherrer Institute.

bending-magnet spectrum for a 2.4 GeV storage ring running at 400 mA, containing bending magnets with a magnetic-field strength of 1.41 T, is shown in Figure 3.11.

3.5.4 Radiated Power Loss in Synchrotrons

The radiated power integrated over all angles from a circular arc of length L is stated without derivation[8] as

$$P[\text{kW}] = 1.266 \ \mathscr{E}^2[\text{GeV}] \ B^2[\text{T}] \ L[\text{m}] \ I[\text{A}]. \qquad (3.27)$$

For a storage ring with arc sections of equal radius, the power loss of the entire storage ring due to bending-magnet radiation only is calculated by setting $I = 2\pi\rho$. The total loss in third-generation synchrotrons is more, because of the additional loss generated by the insertion devices installed in the straight sections of the storage ring.

The power per unit solid angle per 0.1% bandwidth can be calculated by weighting the curves of Figure 3.11, which show the brilliance in terms of the rate of emission of photons, by the photon energy, as shown in Figure 3.12. It is immediately obvious that, on axis, the area under the curve is most certainly not divided equally left and right of E_c. However, the maximum of radiation off axis shifts more and more to lower energies with off-axis angle. Importantly, the weighted contribution to the total power for a spectrum at a given off-axis angle is proportional to the circumference of the associated cone, and hence to the subtended angle. Once these factors are taken into account, it can be shown that E_c indeed does divide the total emitted power from a bending magnet in half.

---

[8] The quadratic dependence of Equation (3.27) on the electron-beam energy $\mathscr{E}$ is consistent with the fourth-power dependence of Equation (3.19), as the latter describes the angular power density (i.e. not the integrated power). The integrated power, however, is proportional to the natural opening angles of the radiation both in the in-orbital-plane and out-of-orbital-plane, each of which are themselves proportional to $1/\gamma$.

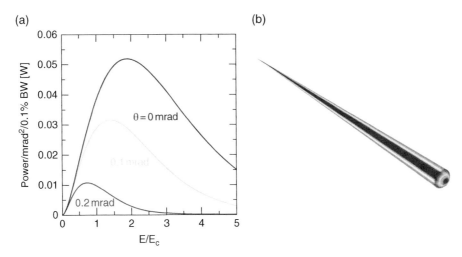

Figure 3.12 Bending-magnet power spectra. (a) The power spectra per mrad² and 0.1% bandwidth of bending-magnet radiation as a function of E/E_c, the ratio of the photon energy to the critical energy $\hbar\omega_c$, for different angular positions out of the plane of the electron orbit. The power spectrum falls off and the positions of the spectrum maxima move to lower energies with increasing angle from the axis. (b) The contribution to the integrated power spectrum of a cone associated with a certain off-axis angle is proportional to that angle. The spectrum associated with the central (indigo) cone has a higher power density and is shifted to higher photon energies than that of the blue cone, and still more than that from the green cone, and so on. However, the weighted contribution of any given cone to the integrated power is proportional to the angle it subtends with the central axis.

3.6   Radio-frequency Power Supply and Bunching

The dissipated kinetic energy of the electrons due to emission of radiation at the bending magnets and insertion devices must be replenished before they spiral into the inner wall of the storage ring. This is achieved by giving them a small boost at every turn as they pass through a radio-frequency (RF) cavity [normally powered by a so-called klystron, or similar device, Figure 3.13(a)]. How does one ensure that the electrons enter the cavity at the correct phase of the RF voltage cycle to obtain the correct boost in energy?

Well, one doesn't. Either the electron is in the right environment and it survives, or it isn't and it doesn't. In other words, the manner in which this is done is in fact an automatic, self-correcting process and, as we shall see, also results in the electrons settling into a series of small discrete packets, or 'bunches'.

First, what is the drop in energy of the electrons for each cycle around the ring? This is simply the total power loss of the storage ring divided by the storage-ring current, and is equal to approximately 0.2 to 1 MeV, or of the order of 0.05% of the nominal electron energy. The RF cavity must therefore have a peak voltage in excess of this. Typical klystrons at synchrotrons have peak voltages of about 3 MV.

Consider Figure 3.13(b). On average, the electrons require a certain boost in order to maintain them on a stable path, given by an amount eV_{ref}. If an electron dissipates more than this amount of energy in the following turn around the storage ring, it will enter the RF cavity somewhat earlier at point A [because it takes a shorter path in the bends, Equation (3.24)], and will experience a larger acceleration than if it were at the reference voltage. Likewise, if the electron is too fast (i.e. it has dissipated less energy than eV_{ref} in the last turn round the storage ring), it will receive less of a boost.

Electrons outside this range above and below the reference voltage will not gain the correct energy and will be lost to the system. The electrons therefore quickly bunch into packets associated with each cycle of the RF cavity.

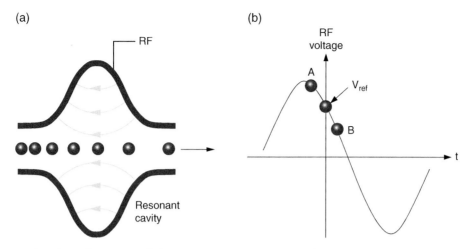

Figure 3.13   Replenishing the energy of electrons in a storage ring. (a) Electrons entering the resonant cavity at the right moment in its voltage cycle are accelerated by the electric field within the cavity generated by the klystron source. Note that the field lines point in the opposite direction to the acceleration [Equation (3.1)]. (b) 'Slow' electrons entering the RF cavity at *A* will be given an extra boost, while 'fast' electrons (at *B*) receive less energy.

For example, electrons in a storage ring with a 300 m circumference complete one cycle every microsecond. A 500 MHz RF supply will therefore generate 500 bunches in the storage ring, separated from one another by 60 cm (or, in time, by 2 ns). Typical bunch lengths are measured in mm (or of the order of 20 to 100 ps). This ratio of bunch length to bunch separation is of the order of 0.005.

The short length of the bunches allows users to exploit the time structure of synchrotron radiation down to well below the nanosecond time scale. These time-resolved experiments are becoming increasingly important in modern synchrotron facilities; examples will be given in later chapters.

To perform time-resolved studies, storage rings can be run in different modes (see Figure 3.14). In the 'single-bunch' mode, only one bunch of electrons (or sometimes two) circulates in the ring; the time-averaged flux is concomitantly reduced, making this operational mode unpopular with those beamlines not performing time-resolved experiments. To overcome this, a hybrid 'camshaft' mode can be employed, in which one bunch (often containing more charge than that stored in the bunches in the main bunch train) is temporally isolated by approximately ±100 ns [Figure 3.14(c)]. This mode has the advantage that 'normal' users can operate at reasonable average photon fluxes, while those beamlines wanting to perform time-resolved experiments can receive a trigger from the storage-ring control centre, in order to synchronize their experiments with the isolated bunch. This does require, however, that those beamlines performing time-resolved experiments using this camshaft bunch must gate their detectors and/or install expensive and fabricationally challenging chopper wheels, which must absorb the photon flux originating from the main bunch train.

A novel approach to this technological problem is to displace the camshaft bunch in the direction perpendicular to the plane of the storage ring, by a short-pulse 'kicker' magnet, that induces this bunch to oscillate up and down [Figure 3.14(d)]. This 'pseudo single-bunch' (PSB) mode thus spatially decouples the camshaft pulse from the main pulse train, so that the former can be selected by a simple, stationary aperture. In addition, the temporal separation of the arrival of successive camshaft bunches can be adjusted from the minimum of one for every full cycle (i.e. the storage-ring circumference divided by the speed of the electrons ≈ *c*) to several milliseconds, by adjusting the camshaft-bunch oscillation period [5].

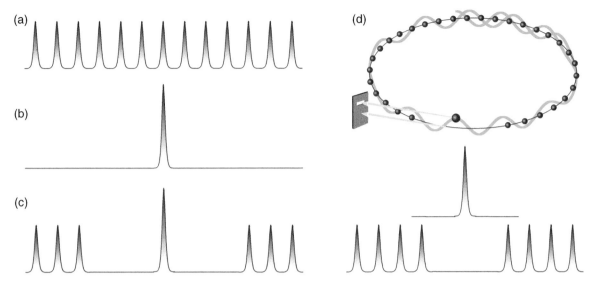

Figure 3.14   Different filling modes in a synchrotron storage ring. (a) Normally, the ring is filled with bunches of electrons equally spaced from one another by a few nanoseconds. (b) Some time-resolved experiments may require that only a single bunch of electrons is in the storage ring. (c) In the hybrid 'camshaft' mode, a bunch (often containing more charge) is isolated by approximately ±100 ns by dropping bunches on either side. (d) Problems associated with the limited time in which detectors can be gated and/or the high precision required by mechanical choppers to block the main bunch train in the camshaft mode are overcome in the pseudo single-bunch mode, in which the camshaft bunch is spatially separated from the main bunch train using a short-pulse kicker magnet, which induces it to execute an oscillatory motion (shown in green).

Coulomb repulsion between electrons in a given bunch, which will defocus and energetically disperse the electron beam in the ring, is ameliorated by relativistic effects. In the direction of movement, relativistic length contraction means that the distances 'seen' by electrons moving together in the same frame of reference are γ times larger than in the laboratory frame of reference, and so the Coulomb forces are $1/\gamma^2$ weaker. The so-called 'beam stiffness', or 'magnetic rigidity', is a parameter which increases with energy and is defined as

$$\frac{p}{e} = B\rho, \tag{3.28}$$

where p is the magnitude of the electron momentum, e is the elementary charge, B is the magnetic-field strength, and ρ is the bending radius associated with the electron immersed in the magnetic field B. We substitute for ρ using Equation (3.22) to obtain

$$\frac{p}{e} = \frac{\gamma m_e c}{e}, \tag{3.29}$$

which, in practical units, can be rewritten as

$$\frac{p}{e} = 3.336 \, \mathscr{E} \; [\text{T m}], \tag{3.30}$$

where $\mathscr{E}$ is given in GeV. A 3 GeV storage ring thus has a beam stiffness of 10.0 T m.

Even for high-energy synchrotron facilities, the finite beam emittance must still be corrected for using the magnet lattice. The precision with which this is achieved fundamentally influences the emittance of the facility, described in the following section.

3.7   Photon-beam Properties

3.7.1   Flux and Brilliance

Flux and brilliance are indicators of the quality of a synchrotron facility. The spectral flux is defined as the number of photons per second per unit bandwidth (normally 0.1%) passing through a defined area, and is the appropriate measure for experiments that use the entire, unfocussed x-ray beam. Brilliance essentially states how the flux is distributed in space and angular range (or, for the *cognoscenti*, in 'phase-space'). It is defined as

$$\text{Brilliance} = \frac{\text{photons/second}}{(\text{mm}^2 \ \text{source area}) \ (\text{mrad})^2 \ (0.1\% \ \text{bandwidth})} \tag{3.31}$$

and is therefore the flux per unit source area and unit solid angle, or, as we will shortly see, flux per total emittance. If the x-ray optics of a beamline are less than perfect (which is always the case to a greater or lesser extent), the brilliance at the experimental station will fall short of theory. It is the goal of any self-respecting beamline engineer to approach as best as possible this upper limit.

Third-generation storage rings optimize both the photon flux and the brilliance, required for modern experimentation. Fourth-generation 'diffraction-limited storage rings' (DLSRs), which, at the time of writing, are on the cusp of becoming the new norm, further improve (i.e. reduce) the total emittance by up to two orders of magnitude, by lowering the contribution from the electron beam, which, in third-generation facilities, dominates, at least in the orbital (x-) plane (see Section 3.8.5).

The highest brilliance of a fourth-generation DLSR undulator is approximately 10^{22} photon/s/mm²/mrad²/0.1% bandwidth (see Figure 1.5), some eleven orders of magnitude higher than that of a rotating anode Cu Kα line and only one or two orders of magnitude lower than the highest-quality visible laser sources. This is, above all, the reason why synchrotrons have become such important research tools[9].

Why do synchrotrons have such high brilliances? Firstly, the size of the radiation source is the convolution of the size of the photon source [determined below for undulators, Equation (3.66)] and the transverse size of the electron beam. This is of the order of a hundred microns at third-generation facilities, and is an order of magnitude smaller still for fourth-generation DLSRs. Secondly, synchrotrons emit an enormous amount of light. The flux emitted by an electron can be shown to be proportional to the square of the electron's acceleration. The centripetal acceleration exerted on electrons in a storage ring is proportional to γ^2, hence the flux is proportional to γ^4. High-energy synchrotron storage rings therefore have correspondingly high brilliances.

3.7.2   Emittance, Radiation Equilibrium, and Quantum Excitation

From Equation (3.31), it can be seen that the brilliance is inversely proportional to both the source size and the x-ray beam divergence. The product of the linear source size σ and the beam divergence σ' in the same plane is known as ϵ, the emittance in that plane, that is

$$\epsilon_x = \sigma_x \, \sigma'_x, \tag{3.32}$$

$$\epsilon_y = \sigma_y \, \sigma'_y. \tag{3.33}$$

One would therefore like to obtain as low an emittance as possible, in which an exceedingly small source size emits x-rays that are almost perfectly collimated (Figure 3.15). For a given synchrotron storage ring, the electron emittance in a given transverse direction is a constant, though this constant is different for different facilities,

---

[9] Note that the *peak* flux and brilliance are two orders of magnitude larger than the average values, the exact factor given by the ratio of the electron-bunch separation to the electron-bunch duration.

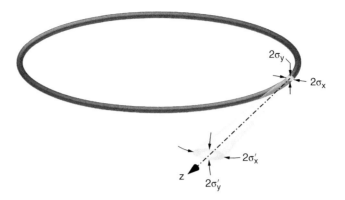

Figure 3.15   The four parameters defining the emittances ϵ_x and ϵ_y of a storage ring.

being determined primarily by the degree of sophistication and perfection of the magnet lattice. The goal then is to minimize this constant.

The total source size $\sigma_{x,y}$ and divergence $\sigma'_{x,y}$ in the x- and y-planes perpendicular to propagation of a given storage ring are convolutions of contributions from the electron and photon beams, that is

$$\sigma_{x,y} = [(\sigma^e_{x,y})^2 + (\sigma^p)^2]^{1/2}, \tag{3.34}$$

$$\sigma'_{x,y} = [(\sigma'^e_{x,y})^2 + (\sigma'^p)^2]^{1/2}. \tag{3.35}$$

The intrinsic photon-beam sizes and divergences in the x- and y-planes produced by an undulator (normally the 'x-ray source of choice' at synchrotron facilities, though not always, see Section 3.9.3) of length L and emitting photons at a wavelength λ, are given by

$$\sigma^p = \frac{1}{4\pi}\sqrt{\lambda L}, \tag{3.36}$$

$$\sigma'^p = \sqrt{\frac{\lambda}{L}}, \tag{3.37}$$

$$\Rightarrow \epsilon^p = \sigma^p\,\sigma'^p = \frac{\lambda}{4\pi} = \frac{98.66}{E[\text{keV}]}\ [\text{pm rad}]. \tag{3.38}$$

ϵ^p is the fundamental lower limit to the total emittance, given by Heisenberg's uncertainty principle; in other words, by the fact that the electron recoils as a result of emitting a photon. Hence, for 1 Å radiation emitted from a 2 m long undulator, the photon source size, divergence, and diffraction-limited photon-beam emittance are 1.6 μm, 5 μrad, and 8 pm rad, respectively.

In third-generation storage rings, however, the contributions σ^e and σ'^e from the electron beam dominate and the total emittance is thus considerably larger than the Heisenberg limit in the horizontal (x-) plane. In fourth-generation DLSRs, now defining modern storage-ring designs, the electron emittance has been reduced to values close to or below (in the case of soft x-rays) the intrinsic photon emittance, described in detail in Section 3.8.5.

In our definitions above, $\sigma_{x,y}$ and $\sigma'_{x,y}$ are the *standard deviations* of Gaussians describing the beam profile in the x- and y-directions, that is

$$I = I_0\,\exp(-x^2/2\sigma_x^2)\,\exp(-y^2/2\sigma_y^2), \tag{3.39}$$

and similarly for the divergence σ'. Hence the relationship between $\sigma_{x,y}$ and the full-widths at half-maximum (FWHM$_{x,y}$) is

$$\text{FWHM}_{x,y} = \sqrt{8 \ln 2} \, \sigma_{x,y} = 2.355 \, \sigma_{x,y}. \tag{3.40}$$

σ'_x and σ'_y are the corresponding angular divergences.

If an electron deviates from the ideal reference orbit in the storage ring and possesses a nonzero transverse momentum, it cannot be maintained in a closed orbit using dipole bending magnets alone. In order to refocus the electrons and bring them back towards the ideal orbit, pairs of alternating vertical and horizontal focussing quadrupole magnets are employed. Hence, the beam shape, given by σ_x and σ_y, varies along the magnet lattice, although the emittances do not.

A convenient measure of these variations around the storage ring is the so-called beta function, given by

$$\beta_x = \sigma_x/\sigma'_x, \tag{3.41}$$

$$\beta_y = \sigma_y/\sigma'_y. \tag{3.42}$$

Inserting Equations (3.41) and (3.42) into Equations (3.32) and (3.33), respectively, we obtain

$$\sigma_x = \sqrt{\epsilon_x \, \beta_x}, \tag{3.43}$$

$$\sigma_y = \sqrt{\epsilon_y \, \beta_y}. \tag{3.44}$$

Since the emittance is a constant for any given storage ring, focussing the electron beam results in a small beam size with a large divergence (low β), while a broader beam will be more collimated (high β). The preferred combination of β_x and β_y changes around the ring according to the elements of the magnet lattice that the electron beam is passing through. An example of how the beta functions typically change along a section of a magnet lattice in a fourth-generation facility is shown in Figure 3.16.

Coupling of the horizontal emittance to the vertical emittance can occur if the magnetic-dipole fields at the position of the electron beam contain some horizontal component. This induces a vertical dispersion of the electrons. A particularly common source of this sort of coupling is a misalignment of the focussing quadrupole magnets. This coupling should be minimized. At the third-generation Swiss Light Source, this cross-coupling of the electron-beam emittances is exceedingly low, at 5×10^{-4}. The emittances are $\epsilon^e_y = 0.9$ pm rad and $\epsilon^e_x \leq 5$ nm rad. Note that this vertical electron emittance is considerably smaller than the theoretical Heisenberg limit for the photon emittance of $\lambda/4\pi$ [Equation (3.38)] for all photon energies below approximately 80 keV, well above the highest accessible photon energy at the facility. In the vertical plane, therefore, most third-generation facilities are already 'diffraction limited'. As already mentioned, it is in the horizontal (orbital) plane where substantial improvements in electron emittances have been made for DLSRs.

The emittance of a storage ring is determined by the equilibrium between two opposing phenomena – radiation damping and quantum excitation. Radiation damping describes how axial acceleration by the RF cavity reduces the angular deviation of the electron from the ideal orbit [see Figure 3.17(a)]. Radiation damping thus minimizes the transverse momenta of the electrons. By introducing high-field bending magnets and/or damping wigglers (see Section 3.9.2), radiation damping can be maximized.

The opposite effect of quantum excitation sends the electrons into oscillatory paths inside the ideal orbital, due to the emission of a photon and consequent loss in the electron energy [Figure 3.17(b)]. This results in a stochastic distribution of the electrons' transverse momenta, significantly increasing the emittance. Quantum excitation can be reduced by designing the magnet lattice so that the electrons' energy dispersion is minimized at the main locations of radiation, namely the bending magnets. This can be achieved by horizontal focussing at the bends and

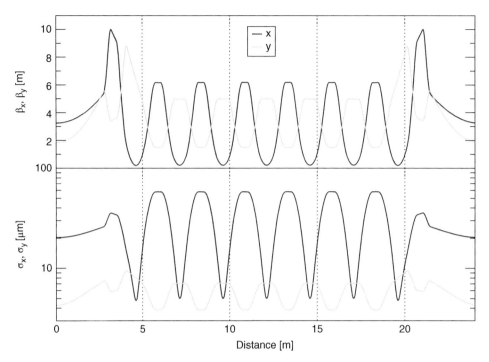

Figure 3.16 Variations in the beta function and the electron-beam profile around one of the twelve identical sectors (each subtending 30°) in the storage ring of the SLS-2 DLSR. Top: the beta functions in the x- and y-directions. Bottom: the corresponding beam dimensions, calculated from $\sigma = \sqrt{\epsilon\beta}$. Note the seven-bend achromat (7-BA) in the centre and the space left free for insertion devices (left and right extremes). Courtesy Andreas Streun, Paul Scherrer Institute.

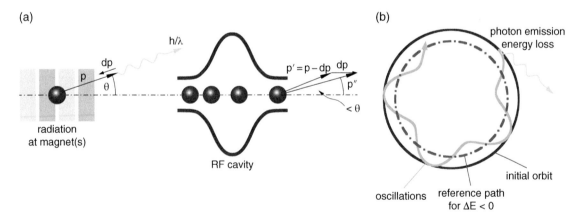

Figure 3.17 Radiation equilibrium between radiation damping and quantum excitation. (a) Radiation damping. An electron traversing a magnet deviates from the central axis by an angle θ. Emission of a photon with momentum $h/\lambda = \hbar k$ will be in the direction of the electron at that instant in time. From the conservation of momentum, the electron's momentum will be reduced to $p' = p - h/\lambda$. The same electron will regain this momentum dp after travelling through the RF cavity, but importantly, this will now be parallel to the central axis, thus reducing the angle of the electron's momentum p'' to that axis and hence also the electron beam's emittance. (b) Quantum excitation. Due to emission of a photon, an electron loses energy and begins to oscillate around a new reference path with smaller radius. These oscillations produce a stochastic distribution of transverse momenta, thereby increasing the emittance.

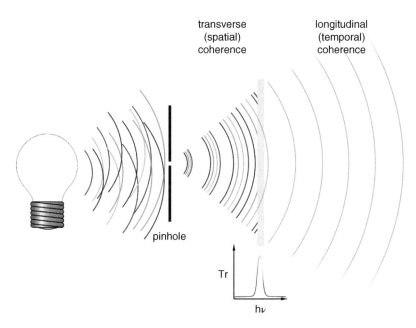

Figure 3.18   Coherent radiation can be extracted from a broadband, spatially extended, source as follows. First, a pinhole acts as a secondary, quasi-point-like, source by only letting through a small, spatially constrained, fraction of the radiation. This is said to have an improved transverse (or spatial) coherence, as the product of the radiation's divergence and source size is now much smaller – in other words, it has a smaller emittance. Secondly, a filter (which might, for example, be a monochromator) suppresses all radiation apart from a narrow bandwidth, much narrower than the original source. Now, the radiation is spatially *and* longitudinally (or temporally) coherent. Both the emittance and relative spectral bandwidth are included in the definition of brilliance.

the use of many small-deflection-angle bends in multibend-achromat lattices (see Section 3.8.5) to limit dispersion growth.

A final word of caution. A high brilliance is not always advantageous; it depends how it was achieved. Brilliance can be improved either by reducing the emittance or by increasing the flux. A beamline with a high brilliance may therefore offer an exceptionally parallel beam and tight focus (and, consequentially high transverse coherent fraction, see Sections 3.7.3 and 3.8.5) but not necessarily a high number of photons per second within a given bandwidth (i.e. flux). So, although experiments requiring high spatial resolution and/or coherence would welcome a low emittance, there are many experimental setups which are photon-hungry but do not require the smallest emittance – most powder-diffraction and total-scattering experiments are good examples. How the cited brilliance of a facility was achieved – through the numerator or denominator – is therefore a very important piece of information to know …

3.7.3   Coherence

Bound up in the figure of merit of brilliance are the parameters that quantitatively define coherence – the emittance and the relative spectral bandwidth. Figure 3.18 summarizes coherence schematically.

Because no x-ray source has an infinitely narrow bandwidth, the different frequencies within the bandwidth of the beam will sooner or later drift out of phase with one another. The time required for the phase between two sinusoidal waves differing in frequency by an amount $\Delta\nu$ to change by π radians (e.g. from fully constructive to fully destructive) is simply $1/2\Delta\nu$, and is called the longitudinal coherence time, $\Delta\tau_c^{(l)}$. During that time, the

(a)

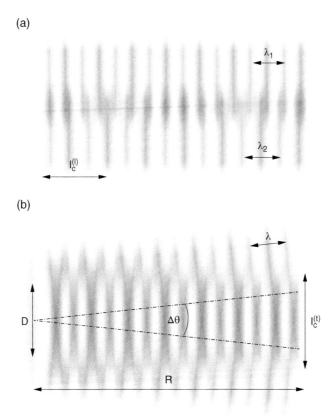

(b)

Figure 3.19   Beam coherence. (a) The longitudinal, or temporal, coherence length is determined by the monochromacity of the source, while (b) the transverse (or spatial) coherence length depends on the beam divergence and source size.

waves have travelled in vacuum a distance $l_c^{(l)} = c\Delta\tau_c^{(l)}/2$, referred to as the longitudinal (or temporal) coherence length of the light source [see Figure 3.19(a)], given by

$$l_c^{(l)} = \frac{\lambda^2}{2\Delta\lambda}. \tag{3.45}$$

The longitudinal coherence after a monochromator is usually determined by the quality of the crystal or grating used in the monochromator, which defines $\lambda/\Delta\lambda$. This can exceed easily 10^4 and the longitudinal coherence length can be several microns, even in the hard x-ray regime.

We define the *transverse coherence length* $l_c^{(t)}$ (also called the spatial coherence length) resulting from the interference of waves of the same wavelength but with slightly different directions of propagation. This arises because all sources have a finite size D (for example, the size of the pinhole in Figure 3.18) and a nonzero divergence $\Delta\theta$, as shown in Figure 3.19(b). In this case

$$l_c^{(t)} = \lambda/2\Delta\theta = \lambda R/2D, \tag{3.46}$$

where D is the lateral extent of the finite source and R is the distance from the source to the observation point. If we assume the source has a Gaussian profile, determination of D requires integration of interference contributions across the entire source's intensity distribution. It emerges that $l_c^{(t)}$ is related to the standard deviation of the beam $\sigma_{x,y}$ by

$$l_c^{(t)} = \lambda R/(2\pi^{1/2}\sigma_{x,y}), \tag{3.47}$$

or, in practical units

$$l_c^{(t)}[\mu m] = 28.21\frac{\lambda\,[\text{Å}]\,R\,[m]}{\sigma_{x,y}\,[\mu m]}. \tag{3.48}$$

In the vertical direction, the source size at an undulator of, say, 2 m length, is of the order of $\sigma_y = 2\,\mu m$. For 1 Å radiation, this yields a vertical spatial coherence for an observer at 40 m of $l_c^{(t,y)} = 564\,\mu m$. The horizontal spatial coherence has traditionally been two orders of magnitude smaller than this, due to the very much larger extent of the electron beam in the orbital plane. In the case of DLSRs, however, the source size in the horizontal direction may be as small as $\sigma_x = 10\,\mu m$ and the corresponding coherence length of the order of $l_c^{(t,x)} \sim 100\,\mu m$.

Interference/diffraction phenomena such as x-ray diffraction or small-angle x-ray scattering can only occur between scattering elements within the coherence volume. As the largest unit-cell sizes found in x-ray diffraction are over an order of magnitude smaller than $l_c^{(l)}$ and approximately three orders of magnitude smaller than $l_c^{(t)}$, sharp diffraction maxima originating from 10^7 or more repeat units are typically obtained – although synchrotron light is often said to be 'incoherent', it *is* sufficiently parallel and monochromatic (usually after some encouragement from beamline optics) to be an excellent source for interference and diffraction experiments.

As will become apparent in Section 5.3.3, the tightest possible focus of x-radiation corresponds closely to the transverse coherence length. On the other hand, the longitudinal coherence length, of the order of a micron, is over four orders of magnitude smaller than the pulse lengths ($c\Delta\tau \sim 3 \times 10^8$ m s^{-1} $\times 5 \times 10^{-11}$ s = 15 mm). As each pulse only contains of the order of 10^4 photons, this means that the coherence volume contains on average about one photon[10].

Imperfections in optical components (such as the slope errors of x-ray mirrors, see Section 5.4.1) can further contribute significantly to the effective spatial coherence and increase this far beyond the theoretical limit given by the divergence of the source.

The coherent fraction of a synchrotron source has contributions from the electron beam and the natural emittance of the emitted radiation, and is given by

$$f_{\text{coh}} = \left\{\left[1 + \left(\frac{\sigma_x^e}{\sigma^p}\right)^2\right]\left[1 + \left(\frac{\sigma_x^{\prime e}}{\sigma^{\prime p}}\right)^2\right]\right.$$
$$\left.\left[1 + \left(\frac{\sigma_y^e}{\sigma^p}\right)^2\right]\left[1 + \left(\frac{\sigma_y^{\prime e}}{\sigma^{\prime p}}\right)^2\right]\right\}^{-1/2} \tag{3.49}$$

$$= \frac{(\epsilon^p)^2}{\epsilon_x \cdot \epsilon_y}. \tag{3.50}$$

Thus, if the electron-beam emittance in the horizontal direction is reduced to that of the photon beam, and the vertical electron emittance is significantly smaller than ϵ^p, a coherent fraction of 50% is obtained, two orders of

---

[10] Photon 'Bose–Einstein' statistics are well outside the remit of this book. It is briefly mentioned, however, that it is often stated that 'a photon can only interfere with itself'. This is not strictly true, however, as has been experimentally demonstrated many times. Indeed, the use of the word 'itself' is ill-advised, because all photons of the same wavelength are identical in quantum mechanics. Because they are bosons, the wavefunction of all photons is symmetric, in other words, it is invariant under all permutations of the individual photons.

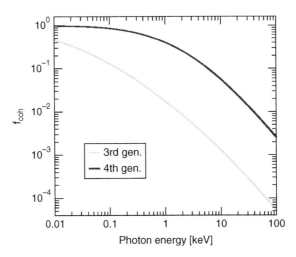

Figure 3.20   The calculated coherent fraction f_{coh} as a function of photon energy for a 2 m undulator at a third-generation facility, for which $\epsilon_x^e = 5$ nm rad, $\epsilon_y^e = 1$ pm rad, $\sigma_x^e = 100$ μm, and $\sigma_y^e = 2$ μm; and for a fourth-generation DLSR facility ($\epsilon_x^e = 100$ pm rad, $\epsilon_y^e = 1$ pm rad, $\sigma_x^e = 10$ μm, and $\sigma_y^e = 2$ μm).

magnitude larger than possible at third-generation facilities. DLSRs achieve this for photon energies of the order of a few hundred eV to a keV (see Figure 3.20).

For a given beam brilliance $\mathscr{B}$[11], the spatially coherent flux F_{coh} scales with the square of the wavelength, while for a given absolute bandwidth $\Delta\lambda$, the coherence is linearly proportional to the wavelength. Specifically,

$$F_{coh} = \mathscr{B}\,\lambda^2\left(\frac{\Delta\lambda}{\lambda}\right). \tag{3.51}$$

It is therefore easier to produce coherent soft x-ray sources, although the ultimate resolution is poorer than that when using hard x-rays[12].

3.7.4   Polarization of Synchrotron Radiation

The polarization of the x-rays emerging from a storage ring depends on the line of sight, illustrated in Figure 3.21. Viewed in the electrons' orbital plane, the polarization of the radiation is linear in that plane (assuming one has not installed an APPLE undulator, see below), as the electrons appear to oscillate in the horizontal plane. If the observer is above the plane of the storage ring, the electrons will appear to execute an elliptical orbit in a clockwise direction. This angular momentum from the observer's frame of reference is transferred to the emitted photons, which are left-circularly polarized (LCP). The electric-field vector of the x-rays also rotates clockwise around the direction of propagation as viewed back down the beam from the observer. The angular momentum of the rotating electric field of the LCP photons is $-\hbar$, and is described by a vector of that magnitude pointing opposite to the direction of propagation. The opposite is true for an observer below the electrons' orbital plane – the polarization is right-circular (RCP) and the photons carry an angular momentum $+\hbar$. Note that the angular momentum of a photon is transferred to systems which absorb that photon. This is an important fact for understanding (among other things) dichroism in spectroscopy, as we shall see in Section 7.8. Generation of different polarizations using undulators is described in detail in Section 3.9.6, while the polarization relative to a sample (for example, one of its crystallographic planes or physical surface) is explained in Section 7.3.2.

---

[11] The calligraphic font $\mathscr{B}$ is used here to distinguish brilliance from the magnetic field B.

[12] Beware when using Equation (3.51) to calculate the brilliance, that it is defined in the appropriate units, particularly with regards to the fractional bandwidth $\Delta\lambda/\lambda$.

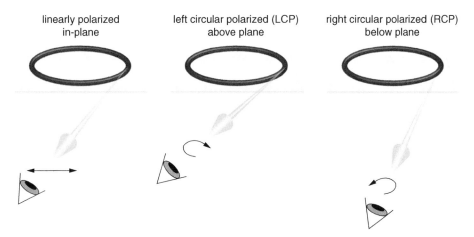

linearly polarized
in-plane

left circular polarized (LCP)
above plane

right circular polarized (RCP)
below plane

Figure 3.21   The polarization of synchrotron light depends on the position of the observer relative to the plane of the electron orbit.

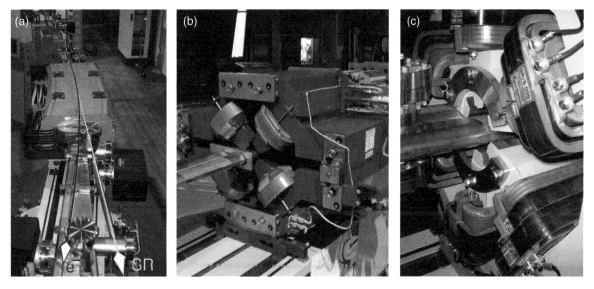

Figure 3.22   Elements of the magnet lattice. (a) Bending or dipole magnets force the electrons to execute a curved path, thereby emitting synchrotron radiation. (b) A focussing quadrupole magnet, and (c) a correcting sextupole magnet.

3.8   The Magnet Lattice

The magnet lattice of a storage ring consists of three main magnet types – bending magnets for diverting the electrons' path, and quadrupole and sextupole magnets, which focus and correct for chromatic aberrations (see Figure 3.22). Their design and the accuracy of their construction critically influence the performance of the storage ring with regards to brilliance.

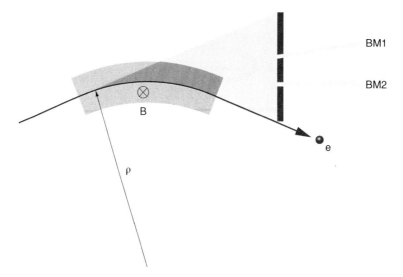

Figure 3.23 The fan of radiation produced by a beam of relativistic electrons passing through a bending magnet of arc radius ρ, determined by Equation (3.23). The magnetic field is into the paper. The large fan angle allows more than one bending-magnet beamline to be served by a single bend.

Magnet lattices in storage rings are generally periodic, that is, there is a regular repetition of an array of bends, quadrupoles, and sextupoles. Such periodic lattices are, moreover, usually symmetric – a set of magnet cells make up a so-called 'super-period'. A number of super-periods constitute the complete ring.

3.8.1 Bending Magnets and Superbends

The primary purpose of bending magnets is to circulate the electron beam in the storage ring on a closed path. Bending magnets have typical magnetic-field strengths of around a tesla. They produce bending-magnet radiation in a flattened cone with a fan angle equal to that swept out by the path of the electrons due to the Lorentz forces they are subjected to (plus the small extra amount due to the photon-beam divergence, equal to $1/\gamma$). Because of the relatively large subtended angle of bending-magnet radiation, measured in degrees (see Figure 3.23), it is possible to position more than one so-called 'bending-magnet beamline' at a single bending magnet, using two or more apertures.

The critical energy and emitted power, defined by Equations (3.25) and (3.27), respectively, are determined by the storage-ring energy and the magnetic-field strength. The bending-magnet spectra for three different combinations are shown in Figure 3.24. Particularly at low- and medium-energy facilities, the maximum values of both these quantities may not be normally sufficient to extend the spectrum far into the hard x-ray regime. By 'normally', the use of conventional magnetic-field sources is meant. However, if superconducting magnets are employed, one can begin to access harder x-radiation as the critical energy is proportional to the magnetic-field strength. In addition, from Equation (3.27), one can see that there is much to be gained in radiative power by increasing the magnetic-field strength B.

As mentioned above, the relative spread in electron energy $\Delta\mathscr{E}/\mathscr{E}$ is of the order of 10^{-3}. But the bending radius ρ is directly proportional to the electron energy [Equation (3.23)], hence the path of those electrons with less (more) than the central energy will have a tighter (larger) radius, resulting in an unwanted increase in emittance. This problem is solved by using a particular arrangement of magnets known as a double-bend achromat (DBA, also called a Chasman–Green lattice, after its inventors [6]), shown in Figure 3.25.

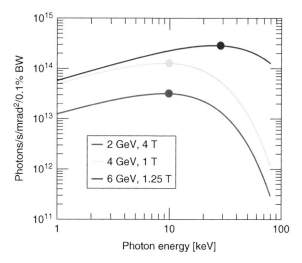

Figure 3.24 The bending-magnet spectra for horizontally polarized radiation on axis for three different combinations of the storage-ring energy and magnetic-field strength. Also shown as circles for each curve are the critical energies $\hbar\omega_c$, given by Equation (3.26).

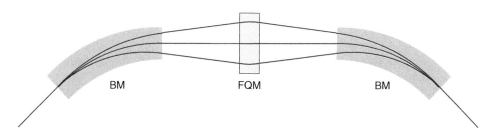

Figure 3.25 Angular dispersion of electrons of different energies as they pass through a bending magnet and the consequent increase in the electron-beam emittance can be corrected in a double bend achromat by placing a focussing quadrupole magnet (FQM) symmetrically in between two identical dipole bends (BM). Lower-energy electrons are bent through larger angles by the dipoles than are those of higher energy.

The natural (i.e. minimum) horizontal emittance of a DBA with bending angle 2θ (that is, θ for each dipole pair) is given by

$$\epsilon_{x,\mathrm{DBA}} = C_{\mathrm{DBA}}\gamma^2\theta^3, \tag{3.52}$$

whereby $C_{\mathrm{DBA}} = 11\sqrt{5}\hbar/(384\,m_ec) = 2.474 \times 10^{-5}$ nm [7]. So, for example, the lower-limit emittance of a 3 GeV storage ring containing 20 DBAs would be 3.3 nm rad, larger than the diffraction-limited value of 8 pm rad calculated above for 1 Å radiation by nearly three orders of magnitude. Efforts to approach the diffraction limit by using multiple-bend achromats (MBAs) are discussed in Section 3.8.5.

Present technology limits the maximum field strength of permanent magnets to a little more than a tesla – superconducting electromagnets, operating at cryogenic temperatures, can routinely provide field strengths in excess of 5 tesla. Typical superconducting materials for these applications are based on liquid-helium-cooled niobium alloys. In this manner, the radiated power output (proportional to the square of B) compared to

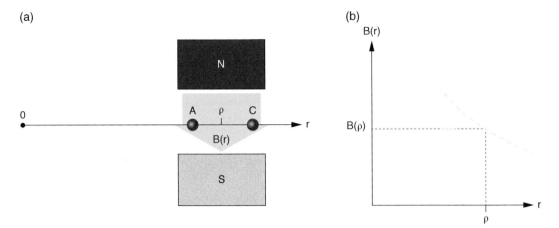

Figure 3.26   Betatron oscillations. (a) View of the electrons looking in the upstream direction, i.e. opposite to their motion ($-z$). An electron at A, for which $r < \rho$, must experience a weaker Lorentz force $evB(r)$ than $\gamma m_e v^2/r$ in order for it to drift back towards the ideal orbit radius at $r = \rho$ and not spiral into the inner wall of the storage ring. Conversely, an electron at C, outside the ideal orbit, must be pulled more strongly than the nominal centripetal force in order to maintain it in a stable orbit. (b) These conditions for stability are met if the magnetic field $B(r)$ lies within the shaded areas bounded by the horizontal line passing through $B(\rho) = \gamma m_e v/e\rho$ representing a uniform magnetic field, and the dashed yellow curve $B(r) = \gamma m_e v/er$, for which the Lorentz force exactly equals the centripetal force.

conventional bending magnets can be increased by approximately a factor of 20 or more, while the critical energy, which is proportional to B [Equation (3.26)], can be increased by a factor of four to five.

For certain experiments, a high brilliance and high degree of monochromacity are not needed – for example, in many computed-tomography experiments (see Section 8.2), the sample might have to be illuminated by a uniform field of several square millimetres, while little is gained by highly monochromatizing the beam. In such instances, flux rather than brilliance is at a premium, and bending-magnet beamlines are very competitive. Yet more photon intensity[13] can be obtained by relaxing the bandwidth of wavelengths that are selected by the monochromator. This can be simply achieved by using multilayer monochromators (see Section 5.4.2), which have bandwidths approximately ten to a hundred times larger than those of crystal monochromators.

3.8.2   Betatron Oscillations and the Dynamic Aperture

What happens to an electron if its path deviates slightly from the precise closed orbit of radius ρ? If the variation of the vertical magnetic field $B(r)$ with the electron's radius r from the origin of the perfect orbit satisfies certain conditions, then $B(r)$ can provide a stable restoring Lorentz force. The electrons' trajectories can thus oscillate periodically about the ideal orbit – these are called betatron oscillations [8].

Consider Figure 3.26. The magnetic field will produce stable oscillations if the Lorentz force increases more than the centripetal force $\gamma m_e v^2/r$ for electrons outside the ideal orbit, and vice versa for those electrons inside the orbit, that is

---

[13] I have used here the phrase 'photon intensity' to distinguish it from the formal definition of flux. As flux is measured in photons per second per unit bandwidth, two beamlines with the same source, but for which one transmits a 0.1% and the other a 1% bandwidth, will have the same flux – the increased arrival rate of photons is exactly cancelled by the increased relative bandwidth. Although they have the same flux, some techniques, such as tomography, would gain greatly from using the beamline with the larger bandwidth.

$$evB(r) < \frac{\gamma m_e v^2}{r}; \quad (r < \rho),$$

$$evB(r) > \frac{\gamma m_e v^2}{r}; \quad (r > \rho).$$

This turns out to be satisfied when

$$-\frac{\rho}{B(\rho)} \left.\frac{\partial B}{\partial r}\right|_{r=\rho} \leq 1. \tag{3.53}$$

So, for a bending magnet of radius 10 m and magnetic-field strength at ρ equal to 1 tesla, the radial magnetic-field gradient should be $\partial B/\partial r \leq -0.1\ \mathrm{T\,m^{-1}}$. Even in a perfectly uniform field for which $\partial B/\partial r = 0$, the orbits are stable because all orbits are circles of equal radius. Actually, since the late 1950s, synchrotrons have been designed to rapidly alternate the value of Equation (3.53) between large positive and negative values, thereby greatly increasing the stability of vertical and radial oscillations [9].

The dynamic aperture is defined as being the maximum betatron oscillation that can be tolerated for many turns, equating to a few milliseconds 'damping time'. It is normally limited by the perturbing effects of nonlinear magnetic fields generated (primarily) by the sextupole magnets (see Section 3.8.3) and ideally it should be larger than the physical lateral size of the storage-ring vacuum vessel. Storage rings with high 'periodicities' (i.e. the repeat unit of the magnet lattice) normally have high dynamic apertures. For most instances, the dynamic aperture is of the order of a few millimetres, up to ten or twenty mm.

3.8.3   Quadrupole and Sextupole Magnets

The geometries and transverse magnetic fields of quadrupole and sextupole magnets are shown in Figure 3.27. Quadrupoles consist of four magnets with alternating inward pointing pole ends. In a perfectly aligned quadrupole, the magnetic field at the centre is zero, but grows rapidly with radial distance from the central axis. From Figure 3.27(a), it is evident that if the Lorentz forces act on the electron beam to focus it in one plane, they also defocus it in the orthogonal plane. Quadrupoles installed so that focussing is in the horizontal plane are referred to as 'F-quadrupoles', while those positioned so that the beam is defocussed horizontally are called 'D-quadrupoles'. If a D-quadrupole and an F-quadrupole are placed directly together with no gap, their fields cancel and the beam will not be focussed. If, however, they are separated by a carefully chosen distance, the combined effect is an overall focussing in both the horizontal *and* vertical planes.

For a given lateral position off axis within a quadrupole, the bending radius is proportional to the electron energy $\mathscr{E}$ [see Equation (3.23)]. The effect of this is that those electrons with marginally higher/lower kinetic energies will be less/more tightly focussed, leading to a longitudinal chromatic dispersion of the beam. The magnetic field of sextupole magnets is such that all electrons that have reasonably small deviations from the nominal storage-ring energy are focussed to the same point, thereby correcting for the chromatic dispersion induced by the focussing quadrupole pair.

3.8.4   Orbit Control and Feedbacks

The electrons within the storage ring of a synchrotron have an ideal reference orbit, from which they will deviate to a certain extent. There are several sources for drifts (on a longer time-scale) and vibrations (normally categorized as having characteristic frequencies above 1 Hz). There are some fascinating causes for both drift and vibrations: Earth tides cause the Earth's surface to move up and down twice daily by approximately 25 cm. Although this is only about 4×10^{-8} of the Earth's radius it does mean that the size of a storage ring, if anchored firmly to the ground, will vary by a similar fraction – a 200 m diameter facility will therefore fluctuate by 8 μm. A second consequence of tidal action is a peak at 0.14 Hz, due to the pounding of surf at coastlines worldwide.

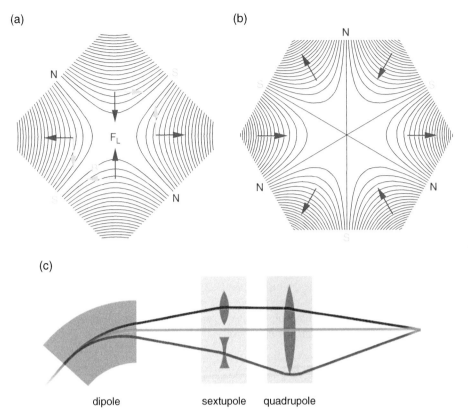

Figure 3.27   Magnetic fields in quadrupole and sextupole magnets. (a) In a quadrupole the Lorentz forces (F_L, shown as red arrows) are such that in one plane (here, vertical) they focus the electron beam, while in the orthogonal plane (here, horizontal), the electrons are defocussed. The action in the horizontal plane defines the nomenclature, hence this configuration is called a defocussing quadrupole, or D-quadrupole. The electron-beam direction is out of the image plane (towards the reader). (b) Sextupoles have focal lengths that are inversely proportional to the distance of the electrons from the central axis and, if they are focussing at **r**, they are defocussing by the same extent at −**r**. (c) Schematic of the combined chromatic correction of a sextupole and focussing of a quadrupole. The dispersion of the electrons' energies are sorted for their momentum by the sextupole in the plane perpendicular to its axis, which corrects for the quadrupole chromatic focussing error.

In addition to naturally self-correcting measures such as careful selection of the magnetic-field gradient of the bending magnets, discussed in Section 3.8.2, the position of the electron beam is monitored all around the ring using so-called beam-position monitors (BPMs), which provide feedback to correcting magnets. In this manner, the electron-beam position is stabilized to significantly better than one micron, an important feature for those experiments that also require photon-beam stability at this level.

3.8.5   Multiple-bend Achromats and DLSRs

The main limit to the electron-beam emittance in storage rings is due to the spread induced at bending-magnet achromats. In our discussion of double-bend achromats (DBAs) in Section 3.8.1, the lower limit to the electron emittance, its so-called 'natural emittance', was given by Equation (3.52). If we extend an arc section by increasing the number of dipoles from 2 to M in a so-called multibend achromat (MBA), keeping the swept angle per dipole constant, the ratio between the natural emittances of the DBA and MBA is

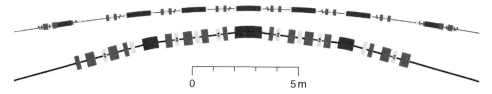

Figure 3.28 Multibend achromats. The fourth-generation of synchrotron facilities are defined by the use of multibend achromats. This has been made practicable by miniaturization of the elements of the magnet lattice (blue: dipole bends; yellow and red: chromatic and focussing elements) with micron-accuracy machining. Above: the MAX-IV 7-bend achromat; below: the SLS triple-bend achromat.

$$\frac{\epsilon_{x,\text{DBA}}}{\epsilon_{x,\text{MBA}}} = 3\frac{M-1}{M+1}. \tag{3.54}$$

Hence, a 10° DBA will have a natural emittance 2.25 times that of a 35° 7BA ($M = 7$). So, simply increasing the number of dipoles in the achromats (thereby proportionally increasing the swept angle) will improve the emittance by up to almost a factor of three (for large M).

A far greater improvement in the natural emittance of an MBA can be realized by reducing the swept angle θ per dipole, due to the cubic dependence of the former on the latter [Equation (3.52)]. So, if in the example above, the 7BA is designed to sweep the same total angle as the DBA, the reduction in emittance will now be $(4/9) \times (2/7)^3 = 32/3087 \approx 1/100$, thus producing emittances which assume values of the order of 100 pm rad (Figure 3.28). This is only an order of magnitude larger than the diffraction-limited photon emittance for hard x-rays of approximately 10 pm rad. Indeed, soft x-ray beamlines for which $\lambda/4\pi \sim \epsilon_{x,\text{MBA}}$ are diffraction limited.

Note also, that the natural emittance scales with the square of the storage-ring energy $\mathscr{E}$. The two highest-energy third-generation storage rings, namely the APS (7 GeV) and SPring8 (8 GeV) are both planning to decrease $\mathscr{E}$ to 6 GeV in their upgrades, despite the associated decrease in the critical energy [see Equation (3.26)].

In most instances, the circumference C of a storage ring is approximately inversely proportional to θ. Thus, for a fixed cell design, a convenient figure-of-merit

$$\mathscr{M} = \frac{\epsilon_x C^3}{\mathscr{E}^2} \tag{3.55}$$

describes how well optimized the magnet lattice is [10] (see Figure 3.29 and Table 3.2).

The newest (fourth) generation synchrotron facilities are thus dubbed 'diffraction-limited storage-rings' (DLSRs) [10–13]. Strictly speaking, a light source is diffraction limited if the horizontal electron-beam emittance is smaller than that of the radiated photon beam ($\lambda/4\pi$). In practical terms, the diffraction-limited photon energy E_{DL} with a wavelength equal to $4\pi\epsilon_x$, is given by

$$E_{\text{DL}}[\text{keV}] = \frac{98.66}{\epsilon_x[\text{pm rad}]},$$

which is essentially a re-expression of Equation (3.38). By this metric, even second-generation facilities would be DLSRs for infrared radiation, while most third-generation facilities are diffraction limited in the very soft x-ray regime at around 25 eV. Note that beamlines that use photon energies that lie near or below the diffraction limit for the facility where they are installed gain little or nothing from the emittance of the storage ring being further improved (i.e. reduced). This is the true meaning of the diffraction limit – the fundamental lower value set by the photon beam rather than the electron beam (see Figure 3.30).

DLSRs are thus characterized by their use of MBAs, in large-circumference storage rings, at medium storage-ring energies. This seemingly obvious and long-understood approach to improve the horizontal emittance

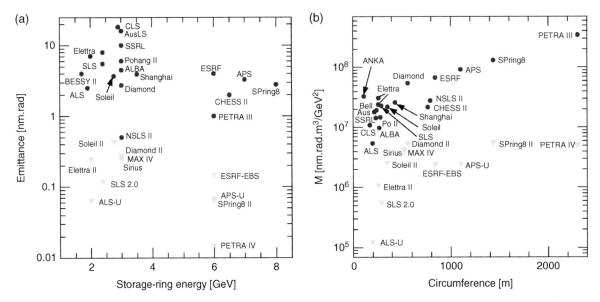

Figure 3.29   Performance of some synchrotron facilities worldwide. (a) Plot of horizontal emittance as a function of storage-ring energy. Data points in blue circles represent third-generation facilities employing either double-bend or triple-bend achromats, those given as yellow triangles are for operational and planned fourth-generation facilities installed with multiple-bend achromats. (b) Plot of $\mathcal{M}$, given by Equation (3.55), for the same facilities. In general, the lower the facility storage-ring energy, the better the value of $\mathcal{M}$, although it should be remembered that this also adversely impacts the highest accessible photon energies.

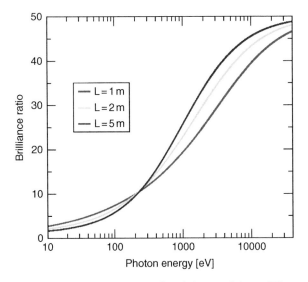

Figure 3.30   The calculated brilliance ratios $\mathcal{B}_{\text{DLSR}}/\mathcal{B}_{\text{3rd}}$ of undulators of three different lengths installed at a DLSR with $\epsilon_x^e = 100$ pm rad and a horizontal electron-beam width $\sigma_x^e = 10$ μm, and in a third-generation facility with $\epsilon_x^e = 5000$ pm rad and $\sigma_x^e = 100$ μm. The vertical emittance is the same for both facilities.

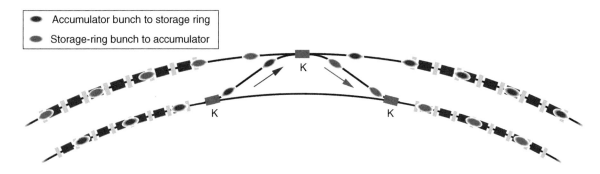

Figure 3.31 Swap-out injection. In swap-out injection, two electron rings are required; an outer storage ring and an inner accumulator ring. Electron bunches are replaced by quasi-simultaneous injection of one or more pulses from the accumulator ring into the storage ring and the same number of pulses out of the storage ring into the accumulator, both actions using fast kicker magnets, labelled here K. The advantages of swap-out operation include the on-axis injection of electrons, which produces smaller betatron oscillations and improved emittance compared to conventional off-axis top-up injection.

and thereby the brilliance has only recently been pursued because, until now, the costs and the introduction of mechanical misalignments associated with increasing the number of elements in the magnet lattice, were considered unacceptable. Miniaturization of these magnet-lattice components and development of multifunctional magnets machined from a single yoke block have decreased both costs and the necessary circumference of the ring to accommodate the MBAs and thus achieve these goals. Another obstacle to reducing the size of the magnets is that the dimensions of the storage-ring vacuum vessels containing the electron beam become so small that pumping them with traditional pumping equipment, in particular, ion-getter pumps, becomes increasingly difficult.

In recent years, however, a novel approach to achieving ultra-high vacuum conditions has been developed, namely the use of nonevaporable getter (NEG) coating of the inner walls of the storage-ring vacuum vessels [14, 15]. NEGs are porous alloys of Al, Ti, Fe, V, and Zr sintered onto the inner walls of the vacuum vessel to a thickness of the order of a micron. After installation, the vacuum vessel is pumped to a moderately high vacuum using traditional pumps, and then heated out to temperatures below 200 °C. This activates the NEG material, allowing pressures to drop to approximately 10^{-10} mbar. Importantly even very narrow spaces can be readily coated – recent developments in computer numerical control (CNC) of machining storage-ring components allow micron accuracy such that geometrically near-perfect miniature vacuum vessels with cross-sections measured in a few square millimetres can be constructed.

The much reduced vacuum-pipe cross-sections associated with DLSRs set stringent specifications on the chamber aperture. DLSRs tend to be designed with high periodicities (repeat units of the magnet lattice), for which the dynamic aperture is on average larger, due to the need to use more sextupole magnets. Thus, any innovation reducing the betatron oscillations might be advantageous. One approach is so-called swap-out injection [16, 17], which enables on-axis injection of electrons into the storage ring, thus avoiding the perturbations and betatron oscillations associated with standard top-up injection (see Figure 3.31).

The improved brilliance of DLSRs has several fundamental and technological facets, in particular:

Small source size This allows a smaller focus for a given optical magnification. This is beneficial for all scanning methods, including scanning x-ray fluorescence, absorption, micro-x-ray diffraction, scanning x-ray transmission microscopy; and also opens opportunities to investigate heterogeneous systems, *in operando* electronic devices, or systems with domain structure, using, for instance, resonant inelastic x-ray scattering (RIXS, see Section 7.6.3).

Alternatively, for a given desired spot size, DLSRs accommodate a larger working distance compared to third-generation sources between focussing optics and sample, allowing the implementation of more sophisticated sample environments, such as used in high-pressure, or ambient-pressure photoelectron spectroscopy, and in high-pressure x-ray diffraction (see Section 7.9.4).

A small illuminated spot on a sample, for which the emission spectrum is of interest, such as in x-ray fluorescence using a von Hamos arrangement, or in '$h\nu^2$-RIXS' [18], directly translates to a higher spectral resolution.

Because the reduction in emittance is in the electrons' orbital plane, DLSR beams are significantly rounder, resulting in a less angularly anisotropic distribution of the coherent flux, thereby reducing astigmatic artefacts in techniques such as phase-contrast tomography.

Parallel beam A parallel beam is essential for high-resolution studies using x-ray diffraction, including macromolecular crystallography (MX). In MX, there are conflicting desires – one wants a small spot size for the increasingly small crystals produced for MX, but also parallelism. The approximately two orders of magnitude reduction in divergence of DLSRs can thus be invested in either or both of these properties.

Scanning tomographies, that is, those which generate the tomograms by scanning a pencil beam perpendicular to the rotation axis, also gain from the ability to produce both small-cross-section and parallel x-ray beams.

Coherence Lensless-imaging techniques such as ptychography are becoming increasingly popular to investigate systems in the 'resolution gap' between scanning x-ray methods (normally limited to a few tens of nm) and electron microscopy (few-angstrom resolution). These methods depend intimately on the coherent flux of the x-ray beam, which increases by the same factor promised in brilliance, up to a coherent fraction of 10% in the hard x-ray regime [see Equations (3.46) and (3.49) in Section 3.7.3].

Novel source technologies As detailed towards the end of Section 3.9.4 (see in particular Figures 3.42 and 3.43), the spectra produced by small-period undulators at DLSRs are much cleaner than those at third-generation facilities, due primarily to the fact that the electron-beam width σ_x^e is an order of magnitude smaller. As a result, entire undulator harmonics can be used for those experiments that do not demand the narrowest spectral bandwidths, thereby providing an additional factor of 100 or more photons on the sample. It is expected that some methods, particularly fast tomography and ptychography, will be able to record data at DLSRs at rates three to four orders of magnitude faster or more, revolutionizing the breadth and complexity of experiments that can be performed.

The reduction of the horizontal emittance in DLSRs thus lends many potential opportunities and is driving exciting innovations along the full technology chain, from the sources (in particular, undulators, see also Section 3.9.5), through improved x-ray optics (regarding imperfections and aberrations), detector technology, to handling of large data sets. The interplay of machine physics, source technologies, and experimental techniques is summarized in Figure 3.32.

3.9   Insertion Devices

Third-generation synchrotrons were defined by the use of insertion devices (IDs), placed in the straight sections between the bending-magnet arc segments. IDs produce significantly higher fluxes and brilliances than do bending magnets. They operate by forcing the electrons to execute an oscillatory path in the plane of the storage ring, employing a set of dipoles that produce magnetic fields that point alternately up and down (see Figure 3.33).

There are two types of ID, distinguished from each other by the degree to which the electrons are forced to deviate from a straight path. This at first seemingly subtle distinction has a fundamental effect on the nature of the radiation, however. For angular excursions substantially larger than the natural opening angle γ^{-1}, the radiation cones from each 'wiggle' do not overlap. Under these conditions, the intensities produced from each dipole are added and the ID is referred to as a *wiggler*.

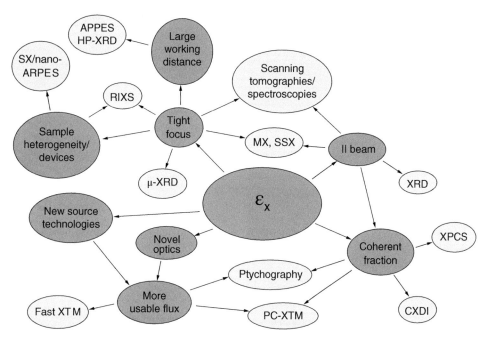

Figure 3.32 Interplay of technology and experimental methods resulting from the advent of DLSRs. From the reduction of the horizontal emittance ϵ_x, certain parameters and properties (blue ellipses) directly (or indirectly, through enabling of other technological developments) impact many different experimental methods (yellow ellipses). As DLSRs mature and become more commonplace, this (far from exhaustive) flow diagram will undoubtedly increase in complexity and interdependence.

For gentler excursions of the order of γ^{-1}, the ID is called an *undulator*. Here, the radiation cones emitted by the electrons overlap as they execute the slalom motion. Consequently, radiation from the dipoles interfere with one another. In this case, the field *amplitudes* are added vectorially (i.e. including the phase difference from each contribution) and the sum of this is squared to produce the intensity, which peaks at those wavelengths where interference is constructive (see Figure 3.34).

The maximum angular deviation $\phi_{\max}$ of the electron oscillations in an ID is defined by the dimensionless 'magnetic-deflection parameter' K, given by

$$\phi_{\max} = K/\gamma. \tag{3.56}$$

K can be expressed in terms of the maximum magnetic field B_0 as

$$K = \frac{eB_0}{m_e c k_{u,w}} = 0.934 \, \lambda_{u,w}[\text{cm}] \, B_0[\text{T}], \tag{3.57}$$

where λ_u and λ_w are the periods of the oscillations in the undulator or wiggler, respectively, and $k_{u,w} = 2\pi/\lambda_{u,w}$. For a wiggler, K is typically between 10 and 50, while for undulators, K is close to unity. The horizontal spread in the electron-beam divergence is

$$\theta_x = 2K/\gamma. \tag{3.58}$$

So, for example, a wiggler having $K = 20$, operating in a 4 GeV storage ring would have a horizontal divergence of 5.2 mrad (0.30°).

(a)

(b)

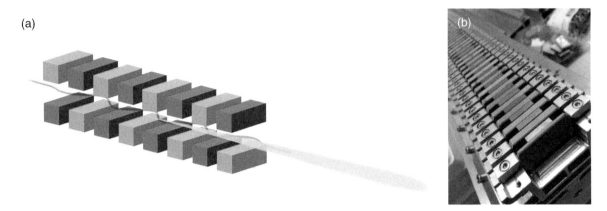

Figure 3.33   Insertion devices. (a) A schematic diagram of an insertion device. (b) The lower magnet array of a U15 (15 mm period) undulator.

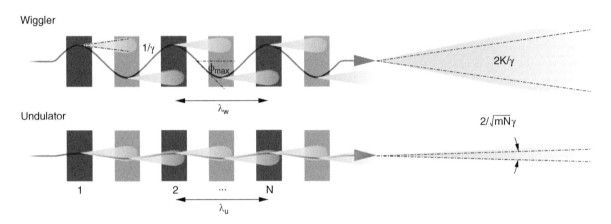

Figure 3.34   Wigglers and undulators in principle differ only in the size of the angular excursions from a straight path that they force the electrons to execute. The maximum angular deviation ϕ_{max} in wigglers is substantially larger by a factor K than the natural opening angle $1/\gamma$, whereas in undulators, $\phi_{max} \sim 1/\gamma$. The output radiation of wigglers has a standard deviation divergence σ_x' in the horizontal plane of K/γ, dominated by the electron path; and in the vertical plane $1/\gamma$, the natural synchrotron-radiation opening angle, which is in general significantly larger than the electron-beam vertical divergence. The corresponding radiative divergences of undulator radiation in both the x- and y-planes is $\sigma'^p \approx 1/\sqrt{mN}\gamma$, though in the x-direction, its convolution with the electron divergence of approximately $\pm 1/\gamma$ means that in this plane, the latter dominates.

3.9.1   Wigglers

A wiggler can be thought of as a series of bending magnets that turn the electrons alternately to the left and to the right, as shown in Figure 3.34. The maximum angular excursion from the wiggler axis is larger than the natural opening angle of the radiation, γ^{-1}. For each oscillation, the electrons are twice moving parallel to (and in reality also very close to) the wiggler axis. The radiation is therefore enhanced by a factor of $2N$, where N is the total number of wiggler periods. The spectrum from a wiggler has the same form as that from a bending magnet and thus the equation describing the emitted power is similar to Equation (3.27) for the radiative power output of a bending magnet. Because in a bending magnet the magnetic field is a constant, while in a wiggler the field strength oscillates along the ID axis (B drops to zero exactly between magnet pairs), the average of the square of the field in a wiggler is $B_0^2/2$, where B_0 is the maximum magnetic-field strength. Hence Equation (3.27) is modified to

$$P_w[\text{kW}] = 0.633\ \mathscr{E}^2[\text{GeV}]\ B_0^2[\text{T}]\ L[\text{m}]\ I[\text{A}]. \tag{3.59}$$

The excursions in Figure 3.34 are much exaggerated, and in fact L is to a very good approximation the linear length of the wiggler, which is typically of the order of 1 to 5 m.

The amount of power radiated by a wiggler increases as the gap between the upper and lower sets of magnets is reduced. At large gap distances, a wiggler will eventually become an undulator, as will become clear below. Wigglers are not designed for this purpose, however, and the radiative power (which is proportional to the square of the magnetic field, which drops off with increasing gap size) becomes too low to be of any practical use.

In the case of a wiggler, for which there is no overlap of the radiation emitted by successive dipoles, $\phi_{\text{max}} = K/\gamma$ is approximately equal to the ID's horizontal standard deviation in divergence σ_x'. In the vertical direction, $\sigma_y' = 1/\gamma$ remains unchanged compared to a bending magnet.

The heat load on optical equipment downstream from a wiggler, such as monochromators, beryllium windows, and x-ray mirrors, can be very high, as matter generally absorbs efficiently for x-ray energies below about 5 keV and wiggler spectra, in contrast to that of undulators (see Section 3.9.3), are broadband. It is therefore often necessary in the case of beamlines using hard x-rays to install a low-energy absorbing filter in the front end to protect components downstream. This filter should be transparent in the energy region of interest above approximately 5 keV, but absorb the longer wavelength radiation, without being damaged itself. This limits the choice of possible material to one only – carbon. Carbon has a low Z-value, is very robust, and has a very high vaporization temperature. It can be heated in vacuum to 1800 K before its vapour pressure becomes problematic. As graphite or glassy carbon, it has a high emissivity, allowing efficient radiative-heat dissipation. In the form of diamond it has an excellent thermal conductivity. High-pass filters of this nature are discussed in more detail in Chapter 5.

3.9.2   Damping Wigglers

A relatively straightforward approach to improve (i.e. reduce) the horizontal emittance of a storage ring is to insert so-called 'damping wigglers' into one or more of the straight sections. A given ring's emittance is determined by different contributions from different elements in the magnet lattice – this is normally dominated by the quantum excitations induced by bending magnets (see Section 3.7.2). However, incorporation of damping wigglers can lower the natural emittance of the facility by a factor of four to five compared to the same lattice without these components through the phenomenon of radiation damping. The details of the mechanism responsible for this lie outside the remit of this book [19]. However, we can explain this heuristically as follows. Consider again Figure 3.17(a). An electron in the damping wiggler, moving at some nonzero angle relative to the wiggler axis (approximately the maximum deviation angle K/γ of the wiggler) emits a photon of wavelength λ and momentum h/λ ($\sim 5 \times 10^{-24}$ kg m s^{-1} for hard x-ray emission) in the direction it was travelling at that moment. Conservation of momentum means that the electron recoils as a result of the photon emission, reducing p by an amount $dp = h/\lambda$. In order for the electron to remain in a stable orbit, it must regain this momentum, which it does by being accelerated by the longitudinal electric field inside the RF cavity. Importantly, dp is now directed parallel to the cavity axis, which reduces the angle of motion to lower values, thereby reducing σ' and the emittance.

The momentum of the highly relativistic electron is $\mathscr{E}/c \sim 2 \times 10^{-18}$ kg m s^{-1}, some five orders of magnitude larger than the recoil momentum dp. At first blush, one might therefore assume that such recoil effects are negligible. However, the equilibrium time for electrons in a storage ring is of the order of milliseconds, in which this recoil/acceleration process will occur thousands of times, inducing a cumulative effect that can dominate the magnet lattice and reduce the natural emittance significantly. This approach is employed in the NSLS-II and Petra-III facilities.

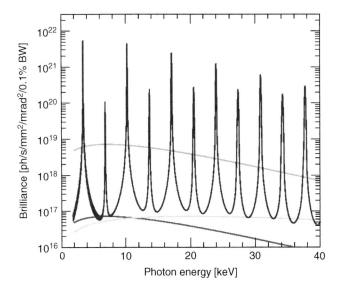

Figure 3.35 Comparison of brilliances at a 3 GeV DLSR running at 400 mA between a U14 undulator with $K = 1.6$ (blue), a bending magnet with $B = 1.41$ T (red), a superbend with $B = 4$ T (yellow), and a wiggler with the same field strength as the bending magnet and 100 periods (green).

The reduced emittance ϵ_W induced by a damping wiggler is determined by the increased radiated-energy loss per turn (U_W) compared to the loss U_0 without the wiggler. To a first approximation this is given by

$$\epsilon_W = \epsilon_0 \frac{U_0}{U_0 + U_W}, \tag{3.60}$$

where ϵ_0 is the emittance in the absence of the damping wiggler. The emittance can easily be reduced by a factor of four or more. Therefore, such a device produces photon-beam powers typically of several tens of kilowatts.

The most important design parameters of damping wigglers are the K-value [and, through Equation (3.57), the magnetic-field strength B_0 and wiggler periodicity λ_w], and the number of periods N. Although a large K will increase the impact of each recoil/acceleration process, increasing it too much will cause the electrons to drift out of phase with the wiggler array, inducing self-dispersion and nonlinear effects, which in turn will blow up the emittance again. A better strategy is to use a 'moderate' K-value, and increase the number of periods N [19].

The advantages of this approach to reduce the emittance are its relative ease of implementation and the possibility of using the generated photon flux at high-energy-photon beamlines. On the other hand, the high power output is not trivial to manage and damping wigglers significantly increase the power consumption of the facility. Also, in many cases, the space occupied in the straight section(s) by the damping wiggler(s) could be better utilized by installing undulators.

3.9.3   Undulators

Undulators differ from bending magnets and wigglers in that their spectral flux is concentrated in evenly separated, narrow bands of radiation (Figure 3.35). This is a result of the fact that the radiation fields, shown schematically in Figure 3.34, overlap and interfere with each other. Only certain wavelengths interfere constructively, and the undulator spectrum therefore consists of a fundamental frequency plus a series of regularly spaced higher harmonics.

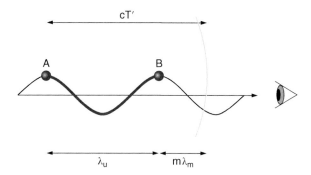

Figure 3.36 Constructive interference between electromagnetic wavefronts emanating from equivalent points on the undulations occurs when their separation is an integer multiple of the radiation wavelength. The curved path S taken by the electron travelling at the relativistic velocity v between A and B is highlighted in blue.

Although the concept of undulators harks back to the 1940s, and the first successful demonstration of undulator radiation was in the early 1950s (albeit producing microwave radiation), the first practical device to operate in the x-ray regime was constructed by Klaus Halbach and co-workers at the Lawrence-Berkeley National Laboratory and tested at the SSRL synchrotron at Stanford in 1981. This breakthrough was largely thanks to the development of novel magnetic alloys such as $SmCo_5$ [20], which allowed the construction of magnet arrays with the required small periodicity and high magnetic-field strength [21].

The basic parameters for undulator radiation are the relativistic Lorentz factor γ, the undulator spatial period λ_u, and K. For an undulator, K is about unity (i.e. the maximum angular deviation is about the same as the natural opening angle $1/\gamma$). K can be varied by changing the gap size between the upper and lower array of magnets, in order to tune the spectrum so that a spectral maximum sits at the desired photon energy.

Both the horizontal and vertical divergences in radiation (as against the electron divergence) are given by

$$\sigma_x'^p = \sigma_y'^p = \frac{1}{\gamma}\left[\frac{(1 + K^2/2)}{2mN}\right]^{1/2} \approx 1/\sqrt{mN}\gamma, \tag{3.61}$$

where $K \sim 1$, N is the number of periods in the undulator ($2N$ being the number of magnet poles), and m is the harmonic number [see Equation (3.62) below]. For a typical undulator consisting of one hundred poles or more, $\theta_{x,y}^p \sim 10$ µrad.

It has already been noted in Section 3.7.1, however, that, invariably, the divergence measured at an undulator beamline is larger in the horizontal direction than that given by Equation (3.61). This is because the observed emittance is a convolution of the photon beam emittance with that of the electron beam. This is why the photon divergences in Equation (3.61) have been labelled with the superscript p. The electron emittance is generally significantly larger for third-generation facilities. For DLSRs, the electron emittance only begins to dominate above photon energies typically of the order of a keV.

The transformation from wiggler to undulator radiation is achieved in practice not by reducing the lateral excursions through reduction of the magnetic-field strength between the magnetic pole-pairs – this would result in an unacceptable drop in flux – but instead by reducing the magnetic-pole spatial periodicity λ_u [Equation (3.57)].

We now derive an expression for the set of discrete wavelengths which results in constructive interference in an undulator in terms of the undulator period λ_u. Consider Figure 3.36. Radiation is emitted at time $t' = 0$ at point A. After a time $t' = T'$ the electron has moved one undulation downstream to point B, and the radiation that originated at A has propagated a distance cT'. The difference in these two distances is $cT' - \lambda_u$, and only that radiation with a wavelength λ_m equal to an integral fraction $1/m$ of this will interfere constructively, i.e.

$$m\lambda_m = cT' - \lambda_u. \tag{3.62}$$

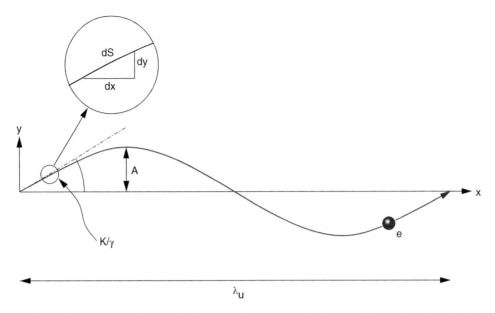

Figure 3.37   The path taken by an electron of velocity v during one cycle in an undulator insertion device. The y-scale is much exaggerated.

The fundamental wavelength λ_1 satisfies Equation (3.62) for $m = 1$. What we need to do now is determine T'. For low-amplitude oscillations of the electron beam path, it is marginally greater than λ_u/v. However, we cannot ignore the oscillatory nature of the path and simply assume it is straight, as the electron velocity v is anyway so close to c, so this small deviation has a significant impact on the condition for constructive interference. Consider Figure 3.37.

The path $S = vT'$ taken by the electron in one cycle in the undulator is

$$S = \int_0^{\lambda_u} dS. \tag{3.63}$$

But by Pythagoras' theorem,

$$
\begin{aligned}
dS &= \sqrt{dx^2 + dy^2} \\
&= \left[1 + \left(\frac{dy}{dx} \right)^2 \right]^{1/2} dx \\
&\approx \left[1 + \frac{1}{2} \left(\frac{dy}{dx} \right)^2 \right] dx.
\end{aligned}
\tag{3.64}
$$

Also,

$$y = A \sin \left(\frac{2\pi x}{\lambda_u} \right),$$

$$\frac{dy}{dx} = \frac{2\pi A}{\lambda_u} \cos \left(\frac{2\pi x}{\lambda_u} \right) \tag{3.65}$$

and $dy/dx \ll 1$. What is the value of the oscillation amplitude A? For small angles, $\tan \theta \approx \theta$, and hence

$$\tan \left(\frac{K}{\gamma} \right) \approx \frac{K}{\gamma} = \frac{dy}{dx}\bigg|_{x=0}$$

$$= \frac{2\pi A}{\lambda_u} \cos \left(\frac{2\pi x}{\lambda_u} \right)\bigg|_{x=0}$$

$$\Rightarrow \frac{K}{\gamma} = \frac{2\pi A}{\lambda_u}$$

$$\Rightarrow A = \frac{K\lambda_u}{2\pi\gamma}. \tag{3.66}$$

Note that for typical values for K, λ_u, and γ, A is of the order of a micron. Also, the lateral extent of the radiation cone with a natural opening angle $1/\gamma$ after a pole separation of, say, 25 mm, is of the order of a few microns – in other words, the radiation emitted by an undulator does indeed overlap if K is kept to around unity.

Inserting Equations (3.64), (3.65), and (3.66) into Equation (3.63), we obtain

$$S = \int_0^{\lambda_u} \left[1 + \frac{K^2}{2\gamma^2} \cos^2 \left(\frac{2\pi x}{\lambda_u} \right) \right] dx. \tag{3.67}$$

We now use the geometrical identity

$$\cos^2 A = \frac{1 + \cos 2A}{2}$$

to obtain

$$S = \lambda_u + \frac{K^2}{4\gamma^2} \int_0^{\lambda_u} \left[1 + \cos \left(\frac{4\pi x}{\lambda_u} \right) \right] dx$$

$$= \lambda_u + \left(\frac{K}{2\gamma} \right)^2 \left[x + \frac{\lambda_u}{4\pi} \sin \left(\frac{4\pi x}{\lambda_u} \right) \right]_0^{\lambda_u}$$

$$- \lambda_u \left[1 + \left(\frac{K}{2\gamma} \right)^2 \right]. \tag{3.68}$$

The condition for constructive interference is therefore

$$m\lambda_m = \frac{\lambda_u}{\beta} \left[1 + \left(\frac{K}{2\gamma} \right)^2 \right] - \lambda_u. \tag{3.69}$$

But we know from Equation (3.11) that $\beta \approx 1 - 1/2\gamma^2$. Inserting this into Equation (3.69), we obtain

$$m\lambda_m = \lambda_u/2 \, \frac{(4\gamma^2 + K^2)}{(2\gamma^2 - 1)} - \lambda_u$$

$$= \frac{2\gamma^2 \, \lambda_u + \lambda_u K^2/2 - 2\gamma^2 \, \lambda_u + \lambda_u}{2\gamma^2 - 1},$$

which, once we recognize that $\gamma^2 \gg 1$, reduces to

$$m\lambda_m = \frac{\lambda_u}{2\gamma^2} \left(1 + \frac{K^2}{2} \right), \tag{3.70}$$

or in practical units

$$m\lambda_m[\text{Å}] = \frac{13.056\,\lambda_u\,[\text{cm}]}{\mathscr{E}^2[\text{GeV}]}\left(1 + \frac{K^2}{2}\right). \tag{3.71}$$

Inserting this into Equation (2.2), we obtain

$$E_m[\text{keV}] = 0.95\,\frac{m\mathscr{E}^2[\text{GeV}]}{(1 + K^2/2)\,\lambda_u\,[\text{cm}]}. \tag{3.72}$$

The interference spectrum at an angle θ away from the central axis of the undulator is shifted towards lower energies, and is given here without derivation as

$$m\lambda_m(\theta) = \frac{\lambda_u}{2\gamma^2}\left(1 + \frac{K^2}{2} + \gamma^2\theta^2\right). \tag{3.73}$$

Thus, an observer positioned off axis by half the natural opening angle $\theta = 1/2\gamma$ (approximately 3.5 mm at a distance of 40 m for a facility with a 3 GeV storage-ring energy) would, for $K = 1$, see a spectrum shifted by a factor 7/6 to the red.

The more observant reader might have objected to the implicit geometrical assumption in Figure 3.36 – the condition for constructive interference seems to imply that the distances between adjacent pole pairs (positions A and B in Figure 3.36) must be adjusted to an accuracy better than a wavelength, that is, of the order of an angstrom, which is, needless to say, an unreasonable engineering demand. Fortunately, in arriving at the above condition for interference, we needed to exploit Equations (3.10) and (3.11), which derive from special relativity. The consequence of this is that, from the frame of reference of the electrons, the magnetic lattice is strongly Doppler- and Lorentz-contracted (to be precise, by an amount equal to $\gamma^2/2$) and therefore the undulator period λ_u appears to be shrunk by this amount (of the order of 10^{-7}). The distortion of the electric-field lines through Doppler shifts compared to the electron oscillating around a fixed point shown in Figure 3.3(d) is illustrated in Figure 3.38 [22].

The term $\lambda_u/2\gamma^2$ in Equation (3.70) shows that in order to access a particular photon wavelength, we require either a longer-period undulator for high-energy facilities, or shorter undulator periods at lower-energy storage rings. So, for example, photon energies as high as 40 keV are accessible using an in-vacuum, 14 mm period undulator at the Materials Science beamline at the medium-energy SLS facility [23, 24]. This was made possible by the development of novel radiation-hard and high magnetic-field-strength magnets that permit undulator-gap sizes as small as 3.5 mm.

The undulator spectrum consists of a set of narrow lines equally spaced in energy, ΔE, which from Equation (3.70) can be shown to be given by

$$\Delta E = \frac{2hc\gamma^2}{\lambda_u(1 + K^2/2)}. \tag{3.74}$$

The spectrum can therefore be tuned by varying K. This is achieved by changing the gap between the two sets of magnetic poles and thereby the magnetic-field strength B_0 (see Figure 3.39).

Note that a high-K undulator gives more intense higher-energy harmonics than a low-K device. The detailed appearance of the undulator spectrum depends on the K-value and the point of observation. We now discuss the form of undulator spectra as a function of K.

For an observer on axis and for low K-values, the even harmonics are largely suppressed, due to symmetry considerations. The spectrum is dominated by $m = 1$. As one increases K (for example, by closing the undulator gap), the spectrum acquires an increasing content of higher harmonics (see Figure 3.40). In addition, the increased angular spread means that even harmonics also begin to appear, as the tight symmetry defining the undulator axis becomes relaxed to an extent.

The spectral width of the undulator harmonics depends on the number of periods, N. As in any interference or diffraction setup, the condition for constructive interference becomes increasingly strict the larger the number

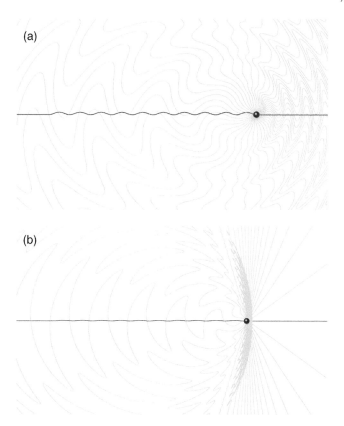

Figure 3.38 Electric-field lines of undulator radiation. These differ from Figure 3.3(d) in that the electron is moving perpendicular to the sinusoidal driving force at velocities that are significant fractions of c, the speed of light, resulting in a Doppler blue-shift in the direction of propagation and a red-shift in the opposite direction. (a) At relatively low velocities (here, $\mathcal{E} = 0.1$ MeV, $v/c = 0.548$), the Doppler compression in the forward direction is less pronounced. (b) As one approaches c (here, $\mathcal{E} = 2$ MeV, $v/c = 0.979$) the Doppler compression in the forward direction forms a clear 'wake' of closely packed electric-field lines (in other words, short wavelength) and higher oscillation amplitude (that is, the electric-field amplitude). Other parameters common to both graphs: $\lambda_u = 100$ mm, 10 periods (20 poles), $K = 1$. Note that the electron energies chosen here, even in (b), are far smaller than typically used in synchrotrons, in order to highlight the salient features. Choosing more representative values would, on this scale, show almost no discernible features other than a single highly flattened arc ever so slightly downstream of the electron, and part of a greatly stretched, red-shifted oscillation in the counterpropagating direction.

of 'scatterers' or 'sources' (in this instance, the $2N$ magnet pairs). In other words, a small relative deviation in the frequency from resonance will still produce a more or less strong positive interference if there is only a small number of participating waves. In contrast, for high N, constructive interference is only satisfied for a very narrow range of frequencies[14]. It turns out that the inverse of the relative bandwidth, called the monochromaticity, $\lambda_m/\Delta\lambda_m = \nu_m/\Delta\nu_m$, is equal to the number of periods N multiplied by the harmonic number m. As an example, the tenth harmonic of an undulator consisting of $N = 70$ periods will have a relative bandwidth $\Delta\lambda_m/\lambda_m$ of 1.4×10^{-3}.

---

[14] Exactly the same phenomenon is described in Section 5.4.2 to explain energy selection in grating monochromators. See particularly Figure 5.23.

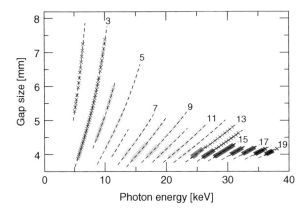

Figure 3.39 Measurements of optimized gap positions for different photon energies and utilized harmonics of the U14 undulator of the Materials Science beamline, SLS, from the second to nineteenth harmonic. In normal operation, scanning to higher energies within any given harmonic is achieved by opening the undulator gap. One moves from a given harmonic to the next higher when the desired photon energy can be accessed at the higher harmonic with a gap size no smaller than the minimum allowed value, here 3.8 mm. In most instances, a 'swap-gap size' of 4 mm is preferred, in order to minimize background radiation doses caused by the electron-beam halo interacting with the undulator's magnet array. Adapted from [24], with permission from IUCr.

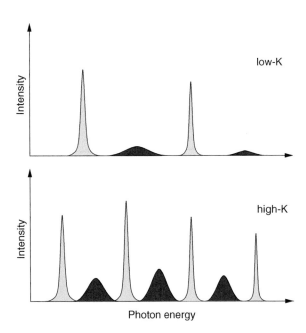

Figure 3.40 Schematic spectra of low-*K* and high-*K* undulators. Note that the harmonics of the low-*K* undulator have larger photon energy separations and smaller contributions of the even harmonics (blue) compared to the odd harmonics (yellow) than those of the high-*K* device.

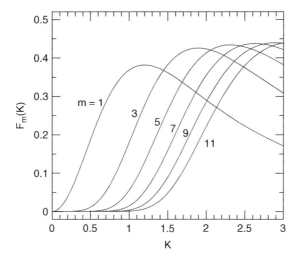

Figure 3.41 The tuning function $F_m(K)$ for the first six odd harmonics in undulator radiation used to calculate the on-axis peak intensities.

The on-axis peak intensity I_m of the mth harmonic of undulator radiation is given by

$$I_m = \alpha N^2 \gamma^2 \frac{\Delta v}{v} \frac{I}{e} F_m(K), \tag{3.75}$$

where $\alpha = 1/137.04$ is the fine-structure constant, $\Delta v/v = 1/(mN)$ is the relative spectral bandwidth of the harmonic peak, I is the current, and

$$F_m(K) = \chi^2 K^2 [J_{(m-1)/2}(K^2\chi/4) - J_{(m+1)/2}(K^2\chi/4)]^2, \tag{3.76}$$

whereby

$$\chi = \frac{m}{1 + K^2/2} \tag{3.77}$$

and the J's are Bessel functions. These tuning functions $F_m(K)$ are plotted for the first six odd harmonics in Figure 3.41. In units of [ph/s/mrad^2/0.1% BW], Equation (3.75) simplifies to

$$I_m = 1.74 \times 10^{14} N^2 \mathscr{E}^2 [\text{GeV}] I[\text{A}] F_m(K). \tag{3.78}$$

So, for example, an undulator with $K = 1.5$ containing 70 periods in a storage ring running at 3 GeV and 200 mA would provide an on-axis peak intensity for the 5th harmonic of 4.05×10^{17} ph/s/mrad^2/0.1% BW. For a typical undulator source size of $150 \times 20~\mu\text{m}^2$, this equates to a brilliance of 1.35×10^{20} ph/s/mm^2/mrad2/0.1% BW.

To scan the photon energy produced by an undulator, one tracks the resonant energy E_m of the mth harmonic [Equation (3.72)] by opening the undulator gap (decreasing K). Different harmonics are used in different energy ranges: above a certain photon energy, the intensity of a given (normally odd) harmonic, given by the function $F_m(K)$, drops below that of the next harmonic at the highest accessible K-value, given by the minimum permitted undulator gap. The undulator is closed accordingly and this next harmonic is used.

3.9.4   Undulators at DLSRs

With the emergence of DLSRs and improvements in magnetic materials, undulator spectra have undergone important transformations. Figure 3.42 compares the brilliance of the same undulator at a third-generation source and at

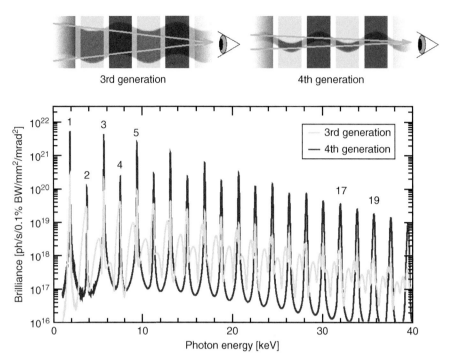

Figure 3.42 Comparison of the brilliance of undulator spectra at third- and fourth-generation facilities. Top: the lateral extent of the electron beam passing through an undulator at third-generation facilities is approximately two orders of magnitude larger than the oscillation amplitude *A*, given by Equation (3.66), while at DLSRs, it might only be approximately 10*A*. Bottom: as a consequence, an observer on axis will 'see' less off-axis radiation, given by Equation (3.73), thereby suppressing the lobes on the low-energy flanks of the main spectral maxima. Note also the enhanced brilliance at the spectral peaks for the DLSR. Both simulated spectra were generated for a U12 undulator (that is, $\lambda_u = 12$ mm) containing 120 magnet periods, for $K = 1.6$, 400 mA, and a storage-ring energy of 2.4 GeV. Courtesy Marco Calvi, Paul Scherrer Institute.

a DLSR having a 40 times smaller total horizontal emittance. In addition to the expected 40 times increase in peak brilliance, the DLSR spectrum is substantially cleaner – the lobes seen on the low-energy flanks of the spectral peaks for the third-generation source are completely absent in the DLSR spectrum. This is because the horizontal width of the electron beam is much smaller (typically by an order of magnitude). In third-generation facilities, the horizontal electron-beam width is, at approximately 100 μm, two orders of magnitude larger than the oscillation amplitude *A*, which we determined earlier in this section to be of the order of a micron. This means that an observer on axis at the DLSR undulator will see a much smaller contribution from emission from off-axis electrons, and it is these that produce the low-energy lobes in the spectra, according to Equation (3.73). Many experiments do not require the relative bandwidths of the order of 10^{-4} provided by crystal monochromators, and would profit from using the entire flux from any given harmonic. The relative bandwidth of the *m*th undulator harmonic is $1/mN$, whereby *N* is the number of undulator periods. For the lower harmonics, this is of the order of 0.005 to 0.01; for higher harmonics, the relative bandwidth does not drop as steeply as $1/m$, as gradually the off-axis contributions do begin to leak in to the peaks, causing them to broaden marginally. Nonetheless, the spectral quality remains sufficiently high to use the entire flux of any given harmonic for small-period undulators at DLSRs. Note, however, spectral filtering is still required to remove the other harmonics. This can be effectively achieved using multilayer monochromators (see Section 5.4.2) or, in some instances, refractive optics such as CRLs or prisms [25] (see Section 5.4.3).

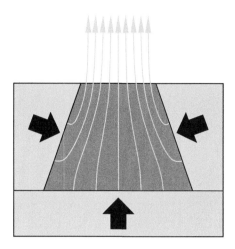

Figure 3.43   Magnetic funnels. A high magnetic field can be generated by sandwiching a soft yoke material (grey) between novel permanent-magnet elements (yellow).

Improvements don't stop here, however. A critical aspect of undulator design is that the magnetic field must be exceedingly homogeneous in the x-direction (that is, in the horizontal plane, perpendicular to the undulator axis) in the region where the electron beam propagates. This has meant that, for third-generation facilities, the magnet yokes need to be a few centimetres wide or more in the x-direction. The approximately five to ten times smaller extent of electron beams passing through undulators installed in DLSRs means that the yokes can be of the order of one centimetre. This allows the design of compact magnetic 'funnels', as shown schematically in Figure 3.43, concentrating magnetic-field lines and thus increasing the magnetic-field strength, which, in turn, allows for a more compact design and shorter undulator periods. Moreover, this means that more periods (N) can be fit into a given undulator length L.

As a result of the discreteness of their spectra, the efficiency of undulators, that is, the ratio of the output x-ray power after monochromatization to the total electromagnetic power generated by the device, is, at approximately 10^{-4} or even larger for DLSRs, superior by well over an order of magnitude compared to that for wigglers.

3.9.5   Echo-enabled Harmonic Generation at DLSRs

Echo-enabled harmonic generation (EEHG) is a promising method for producing high-brilliance beams of soft x-rays in XFELs. Recent studies indicate that it could also be implemented at DLSRs, for which the electron-density of the circulating beam is sufficiently high to interact with modulating lasers.

EEHG depends on an energy-modulated electron beam undergoing so-called 'slippage' in a dispersive element. Consider Figure 3.44. Slippage describes the advancement/retardation of electrons with higher/lower energy than the nominal storage ring energy of $\mathscr{E}_0 \pm \sigma_{\mathscr{E}}$ after passing through a magnetic dispersive element, where $\sigma_{\mathscr{E}}$ is the root-mean square energy spread of the unmodulated beam. The physics behind this is the same as that responsible for bunch compression in XFELs described in Section 4.2 and shown schematically in Figure 4.3.

The setup for EEHG at DLSRs is shown in Figure 3.45. It consists of two modulator undulators and high-power lasers separated by a dispersing element (in the case of DLSRs, a bending magnet, BM), followed by a four-pole dispersing chicane and a radiator undulator. The electron beam is modulated twice in the modulator arrays by

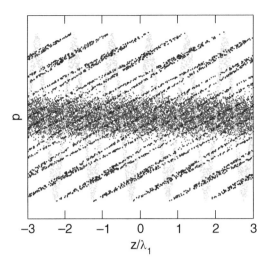

Figure 3.44   Electron-beam slippage. An electron beam with an rms energy spread of $\sigma_{\mathscr{E}}$ but otherwise of constant energy $\mathscr{E}_0$ (shown in red) has its energy modulated by interaction with the electromagnetic field of a high-power laser of wavelength λ_1 (yellow). The increased spread in energy is insufficient to noticeably advance or retard the electrons because they are anyway moving so close to the speed of light. However, by passing them through a magnetic dispersive element, electrons of different energy execute different paths under the influence of the magnetic field, the higher-energy electrons executing a shorter path than those of lower energy, resulting in the modulation to be distorted along z, the axis of propagation, in a phenomenon known as slippage (shown in blue).

the lasers of wavelengths λ_1 and λ_2 – the electric field of the laser light modulates the electrons' kinetic energy, described by the dimensionless quantity

$$p = \frac{\mathscr{E} - \mathscr{E}_0}{\sigma_{\mathscr{E}}}. \tag{3.79}$$

The electron beam after the first modulation and dispersion undergoes severe 'slippage' which is itself then more modestly modified by the second modulator/chicane system. The result is that, as the beam enters a third, 'radiator', undulator, the beam has charge oscillations with a wavelength typically measured in a few nm, resulting in undulator radiation also at this wavelength, that is, in the soft x-ray regime [26]. This EEHG scheme causes coherent radiation through this microbunching of the electrons at the resonant wavelength $m\omega_1 + n\omega_2$, where $\omega_{1,2} = 2\pi c/\lambda_{1,2}$ and m and n are integers. The brilliance can be boosted by up to four orders of magnitude to essentially 100% longitudinal and transverse coherent radiation. Moreover, as the bandwidth is so small, the monochromator could become obsolete. Lastly, a key advantage of EEHG is that the high-harmonic intensities decay only slowly with increasing m and n, possibly allowing EEHG to be implemented in the tender (2 to 4 keV) or even hard (above 4 keV) x-ray regime [27]. EEHG is possible at DLSRs only because the fine structure of the current would not be blurred by emittance-related path-length effects, thanks to their small emittance.

3.9.6   Control of Polarization using Undulators

The polarization of undulator radiation can be controlled and varied in so-called APPLE (Advanced Planar Polarized Light Emitter) undulators[15]. Instead of the two dipole arrays found in 'normal' undulators discussed thus far,

---

[15] APPLE undulators have undergone successive refinements (reflected by their generational names of APPLE-I, II, III, and X), based on the flexibility of movements of their magnet arrays, not detailed here.

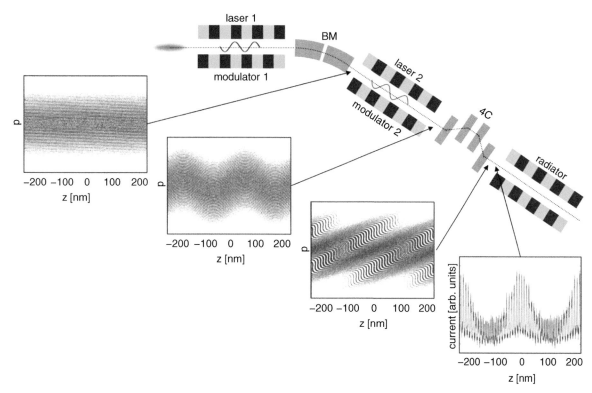

Figure 3.45   EEHG at DLSRs. The first modulator causes extreme slippage of the electron beam after it passes through the bending magnet (BM). This modified beam is further modulated a second time by a second laser/modulator array, which, after passage through a four-dipole chicane (4C) produces a complex energy modulation. The current profile provided by the electron density as a function of *z* (essentially given by integrating along *p* after modulator 2) exhibits oscillations with a wavelength of a few nm, producing photon energies emitted in the radiator undulator that lie in the soft x-ray regime. Adapted from [26] with permission from the American Physical Society.

APPLE undulators consist of four arrays (two above, two below) that can be longitudinally shifted relative to each other [A1 to A4, Figure 3.46(a)] [28]. Each array consists of a repetition of four magnet orientations [down (blue ⊗); reverse (green arrow); up (yellow ⊙), and forward (red arrow), see Figure 3.46(b)]. With all array-components aligned longitudinally, normal linear horizontal (LH) radiation is produced. Antisymmetric shifts (that is, one array moving in the positive *z*-direction, the other in the negative) of A1 and A4 between $-\lambda_u/2$ and $+\lambda_u/2$ will cause the radiation to remain linearly polarized but to vary the tilt angle from linear vertical (LV at $-\lambda_u/2$), through LH (zero shift) back to LV ($+\lambda_u/2$). Conversely, symmetric movements of A1 and A4 by approximately $\pm\lambda_u/4$ will produce right-circularly polarized light (at approximately $-\lambda_u/4$), or left-circularly polarized light (at $+\lambda_u/4$). The exact shift in the arrays depends on the gap size and details of the magnet strengths. Symmetric shifts smaller than this result in elliptical polarization, that is, radiation with both a linear and circularly polarized component.

3.10   Concluding Remarks

With the development of third-generation synchrotron facilities, characterized by their use of insertion devices, particularly undulators, the brilliance available for researchers in the field of x-ray science jumped by some four orders of magnitude. This enormous increase in the number of x-ray photons in tightly focussed bundles

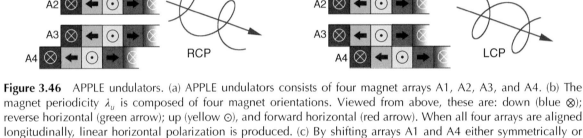

Figure 3.46   APPLE undulators. (a) APPLE undulators consists of four magnet arrays A1, A2, A3, and A4. (b) The magnet periodicity λ_u is composed of four magnet orientations. Viewed from above, these are: down (blue $\otimes$); reverse horizontal (green arrow); up (yellow $\odot$), and forward horizontal (red arrow). When all four arrays are aligned longitudinally, linear horizontal polarization is produced. (c) By shifting arrays A1 and A4 either symmetrically or antisymmetrically by $\pm\lambda_u/2$, linear vertically polarized radiation is produced. (d) Linear polarization of any desired tilt angle can be selected by moving A1 and A4 antisymmetrically relative to A2 and A3. (e) and (f) Circularly polarized radiation is generated by symmetric shifts of A1 and A4 by approximately $\lambda_u/4$, the exact shift depending on the gap size and details of the magnetic-field strengths.

has revolutionized x-ray techniques, enabling the study of increasingly small structures over shorter and shorter timescales. This has driven a parallel thrust in the sophistication of x-ray optical components and detectors, which we discuss in some detail in Chapter 5.

The number of dedicated, third-generation sources has burgeoned over the last two decades and has opened the field of synchrotron physics, transforming it into a multidisciplinary enterprise, nowadays attracting some hundred-thousand scientists across the broadest spectrum of the natural sciences.

Third-generation storage rings are now a mature technology, providing stable, high-brilliance, photon beams over a broad range of photon energies from the infrared to ultrahard x-rays measured in hundreds of keV. In 2015, MAX-IV, the first of the next-generation DLSR facilities had first light. Sirius, in Campinas, Brazil, followed in early 2019. Several more third-generation facilities are either planning or are in the construction phase of upgrades to DLSR status, the latest benchmark in machine-physics technology, promising up to another two orders of magnitude increase in brilliance.

Several orders of magnitude higher peak brilliance are provided by high-gain x-ray free-electron lasers (XFELs). The machine science and technologies for this paradigm shift in x-ray sources are based on fundamentally different principles, and consequently XFELs are discussed separately in the next chapter.

Problems

1. Beginning with Equation (3.4), demonstrate that, for relativistic particles, their velocity v is smaller than the speed of light c by an amount equal to $m_e^2 c^5 / 2\mathscr{E}^2$, whereby $\mathscr{E}$ is the total particle energy, equal to the storage-ring energy. Express this in practical units for relativistic electrons.   (3P)

2. Equation (3.20) describes the angular distribution of Bremsstrahlung for charged particles decelerated from an initial velocity v along the same axis as v. Demonstrate that the general equation for the interlobe half-angle α_{BS} shown in Figure 3.9(a) and plotted in Figure 3.9(b) is given by

$$\alpha_{\mathrm{BS}} = \arccos\left[\frac{\sqrt{1 + 15\beta^2} - 1}{3\beta} \right],$$

where $\beta = v/c$. Demonstrate also that, for small angles and highly relativistic charged particles, Equation (3.20) reduces to

$$\frac{dP}{d\Omega} = \frac{32\kappa a^2 \theta^2}{(\theta^2 + 1/\gamma^2)^5},$$

where γ is the Lorentz factor, accurately approximated by Equation (3.11) relating it to β, and θ is expressed in radians. In addition, prove that the maxima of the Bremsstrahlung lobes will be observed at $\theta = \alpha_{\mathrm{BS}} = \pm 1/2\gamma$.   (8P)

3. Calculate the light power density in W cm^{-2} of a 60-watt tungsten bulb at a distance of 2 m. The bulb's light-conversion efficiency is 2%.   (2P)

4. Calculate the incident solar-power density in W cm^{-2} in London at 13:00 British Summer Time at the September equinox. The total absorbed solar power of the earth is 1.2×10^{17} W. The earth's diameter is 12 800 km. London lies at latitude 51.52° N.   (2P)

5. Calculate the brilliance of an LED consuming 40 mW electrical power in ph/s/mm^2/mrad^2/0.1% BW. The LED emits at 700 nm from an active surface area of 4 mm^2, and has a bandwidth of 25 nm; it has a beam divergence in both transverse directions of 10°, and a light-conversion efficiency of 70%.   (2P)

6. Calculate the brilliance of a He-Ne laser emitting 20 mW light power from a cavity with a partially silvered exit mirror of 1 mm diameter. The laser's bandwidth is 0.1 nm, and its divergence in both directions is 1 mrad. He-Ne lasers emit at 633 nm.   (2P)

7. Calculate the brilliance of a naked human body at the maximum of its blackbody radiation.   (4P)

8. An undulator provides a brilliance of 2×10^{19} ph/s/mm^2/mrad^2/0.1% BW at 15 keV, with an effective source size of 50×5 μm^2 and a divergence of 75×15 μrad^2. What is the on-axis flux in photons/s after a Si(111) monochromator ($\Delta E/E = 1.4 \times 10^{-4}$)? Calculate also the optical power in watts.   (2P)

9. The 6 GeV storage ring of the ESRF has an 844 m circumference and is powered by a 352 MHz klystron. The storage-ring current is 200 mA. Calculate the number of bunches in a full filling pattern, the bunch separation, and the number of electrons in each bunch. Calculate the total kinetic energy of the electrons stored within the ring.

   The average photon flux at an undulator beamline is 4×10^{13} photons s^{-1}. If the bunch duration is 50 ps, calculate the peak flux and the average separation in time and space (in the direction of propagation) between photons within a bunch.   (5P)

10. The 2.4 GeV storage ring of the SLS operates in top-up mode, whereby the beam current is allowed to drop from 401 to 399 mA between top-ups. The typical interval between injections is 3 minutes. Calculate the decay time of the storage ring (i.e. the time for the ring current to drop by a factor $1/e$) if top-up operation is halted. Calculate also the Bremsstrahlung power output, assuming all the electrons lost to the storage current contribute exclusively towards this. The SLS storage-ring circumference is $L = 288$ m.   (3P)

11. Calculate the Lorentz factor γ and the velocity of the electrons circulating in the 8 GeV storage ring of SPring8, plus the natural opening angle $1/\gamma$ (in degrees) of the emitted synchrotron radiation. If an electron

within the SPring8 storage ring emits a photon at $t = 0$, how much further will this photon have travelled than the electron after one hour?   (4P)

12. Usain Bolt's top speed was clocked at the World Championships in 2009 as being 44.72 km hr^{-1}. Calculate his relativistic mass increase in picograms from the perspective of a stationary observer, assuming his stationary mass is 94 kg.   (2P)

13. The circumference of the SPring8 storage ring is 1436 m. The bending magnets used to keep the electrons on a closed orbit have a magnetic-field strength of 0.68 tesla. What fraction of the circumference is taken up by bending magnets? Calculate the critical energy of the photon spectrum produced by the bending magnets.   (3P)

14. The Large Hadron Collider (LHC) at CERN differs from electron storage rings in several ways. One qualitative difference is that most of the 27 km long ring consists of the bending magnets, and only a small fraction is reserved for the focussing quadrupoles and sextupoles. There are no 'straight sections' reserved for insertion devices. Given a kinetic energy of the protons of 6.5 TeV, calculate γ and the fraction of the ring taken up by the 5.5 T superconducting bending magnets. In addition, calculate the critical photon energy under these conditions.   (4P)

15. Beginning with Equation (3.49), derive Equation (3.50).   (2P)

16. In the text, the intrinsic photon-beam divergence of undulators was provided twice, and differently, in Equations (3.37) and (3.61). Demonstrate that these are equivalent to each other. Hint: use also Equation (3.70).   (2P)

17. MAX-IV has a horizontal emittance of 250 pm rad. Although this value for the first of the fourth-generation facilities is smaller by an order of magnitude compared to third-generation facilities, it still lies well above the theoretical limit. Given that the storage-ring energy is 3 GeV, and the lattice consists of 20 identical straights separated by 7-bend achromats, determine this theoretical limit.   (3P).

18. Given the equation describing the condition for constructive interference in an undulator

$$m\lambda_m = \frac{\lambda_u}{2\gamma^2}\left(1 + \frac{K^2}{2}\right),$$

prove that the separation in energy between harmonics is given by

$$\Delta E = \frac{2hc\gamma^2}{\lambda_u(1 + K^2/2)}.$$ (2P)

19. The Materials Science beamline at the Swiss Light Source (storage energy = 2.4 GeV) has an undulator source with a magnet periodicity $\lambda_u = 14$ mm (and is thus called the 'U14' undulator). The U14 deviation parameter can be varied between $0.65 \leq K \leq 1.6$ by adjusting the undulator gap. Determine which harmonics (m) can in principle access photons with a wavelength of 1.6 Å. Of these possibilities, and based on Figure 3.41, which harmonic would you use? You can assume that the curves for the even harmonics lie evenly in between those shown for the odd harmonics.

    Within the above-given limits of K, determine the lowest photon energy which can in principle be accessed using the U14 undulator. Suggest reasons why such a low energy is never used at this beamline?   (5P)

20. The magnetic-field variation $B_y(z)$ along the central axis of an insertion device perpendicular to the plane of the electrons' trajectory (i.e. in the y-direction) is given by

$$B_y(z) = B_0 \sin\left(\frac{2\pi z}{\lambda_u}\right),$$

whereby λ_u is the magnetic period of the insertion device. Derive from this the maximum deflection angle

$$\phi_{max} = \frac{ecB_0\lambda_u}{2\pi\gamma\, m_e c^2} = \frac{K}{\gamma},$$

whereby $K = B_0 \lambda_u (e/2\pi m_e c)$ is the deviation parameter. The term in parentheses is a constant. Calculate its value and relate this to Equation (3.57). (5P)

21. A U12 undulator at a third-generation, 2.4 GeV storage ring has its on-axis fifth harmonic at 9.399 keV and an off-axis lobe on its lower flank at 8.968 keV for $K = 1.6$ (see Figure 3.42). Determine the angle θ from which the off-axis lobe originates. Is this reasonable? Justify your reasoning. (3P)

References

[1] H. Winick, *Synchrotron radiation sources – a primer*. World Scientific, 1995.

[2] H. Wiedemann, *Synchrotron radiation*. Springer, 2003.

[3] P. Duke, *Synchrotron radiation – production and properties*. Oxford Science Publications, 2009.

[4] R. Clarke, "Incoherent sources: Synchrotrons," in *Encyclopedia of Modern Optics*, pp. 217–224, Elsevier, 2005.

[5] C. Sun, G. Portmann, M. Hertlein, et al., "Pseudo-single-bunch with adjustable frequency: a new operation mode for synchrotron light sources," *Phys. Rev. Lett.*, vol. 109, p. 264801, 2012.

[6] R. Chasman, G. K. Green, and E. M. Rowe, "Preliminary design of a dedicated synchrotron radiation facility," *IEEE Trans. Nucl. Sci.*, vol. 22, pp. 1765–1767, 1975.

[7] A. Wolski, "Low-emittance storage rings," in *CERN Accelerator School Series – Advanced Accelerator Physics*, pp. 245–294, 2014.

[8] D. W. Kerst and R. Serber, "Electronic orbits in the induction accelerator," *Phys. Rev.*, vol. 60, pp. 53–58, 1941.

[9] E. D. Courant and H. S. Snyder, "Theory of alternating-gradient synchrotron," *Annals Phys.*, vol. 281, pp. 360–408, 2000.

[10] M. Borland, "Progress toward an ultimate storage ring light source," *J. Phys. Conf. Ser.*, vol. 425, p. 042016, 2013.

[11] M. Eriksson, Å. Andersson, S. Biedron, et al., "MAX 4, a 3 GeV light source with a flexible injector," in *Proc. EPAC 2002*, pp. 686–687, 2002.

[12] M. Eriksson, L.-J. Lindgren, M. Sjostrom, et al., "Some small-emittance light-source lattices with multi-bend achromats," *Nucl. Instrum. Methods A*, vol. 587, pp. 221–226, 2008.

[13] E. Weckert, "The potential of future light sources to explore the structure and function of matter," *IUCrJ*, vol. 2, pp. 230–245, 2015.

[14] C. Benvenuti, J. M. Cazeneuve, P. Chiggiato, et al., "A novel route to extreme vacua: the non-evaporable getter thin film coatings," *Vacuum*, vol. 53, pp. 219–225, 1999.

[15] C. Benvenuti, P. Chiggiato, P. Costa Pinto, et al., "Vacuum properties of TiZrV non-evaporable getter films," *Vacuum*, vol. 60, pp. 57–65, 2001.

[16] A. Xiao, M. Borland, and C. Yao, "On-axis injection scheme for ultra-low-emittance light sources," in *Proceedings of PAC2013*, pp. 1076–1078, 2013.

[17] M. Borland, G. Decker, L. Emery, et al., "Lattice design challenges for fourth-generation storage-ring light sources," *J. Synchrotron Rad.*, vol. 21, pp. 912–936, 2014.

[18] V. N. Strocov, "Concept of a spectrometer for resonant inelastic x-ray scattering with parallel detection in incoming and outgoing photon energies," *J. Synchrotron Rad.*, vol. 17, pp. 103–106, 2010.

[19] M. Borland and L. Emery, "Evaluation of the possibility of using damping wigglers in the Advanced Photon Source," in *2007 IEEE Particle Accelerator Conference*, pp. 1124–1126, 2007.

[20] K. Halbach, "Physical and optical properties of rare-earth cobalt magnets," *Nucl. Instrum. Methods Phys. Res.*, vol. 187, pp. 109–117, 1981.

[21] K. Halbach, J. Chin, E. Hoyer, et al., "A permanent-magnet undulator for SPEAR," *IEEE Trans. Nucl. Sci.*, vol. 28, pp. 3136–3138, 1981.

[22]   T. Shintake, "Real-time animation of synchrotron radiation," *Nucl. Instrum. Methods A*, vol. 507, pp. 89–92, 2003.

[23]   M. Calvi, T. Schmidt, A. Anghel, et al., "Commissioning results of the U14 cryogenic undulator at SLS," *J. Phys. Conf. Ser.*, vol. 425, p. 032017, 2013.

[24]   P. R. Willmott, D. Meister, S. J. Leake, et al., "The Materials Science beamline upgrade at the Swiss Light Source," *J. Synchrotron Rad.*, vol. 20, pp. 667–682, 2013.

[25]   I. Inoue, T. Osaka, K. Tamasaku, et al., "An X-ray harmonic separator for next-generation synchrotron X-ray sources and X-ray free-electron lasers," *J. Synchrotron Rad.*, vol. 25, pp. 346–353, 2018.

[26]   D. Xiang and G. Stupakov, "Echo-enabled harmonic generation free electron laser," *Phys. Rev. ST Accel. Beams*, vol. 12, p. 030702, 2009.

[27]   E. Hemsing, M. Dunning, B. Garcia, et al., "Echo-enabled harmonics up to the 75th order from precisely tailored electron beams," *Nature Phot.*, vol. 10, pp. 512–515, 2016.

[28]   M. Calvi, C. Camenzuli, E. Prat, and T. Schmidt, "Transverse gradient in Apple-type undulators," *J. Synchrotron Rad.*, vol. 24, pp. 600–608, 2017.

4

Free-electron Lasers

4.1 Introduction

Radiation from fourth-generation DLSRs has much in common with laser radiation – it is highly intense, parallel, and, in the case of undulator sources, partially monochromatic. However, the degree of transverse (spatial) coherence, although much improved compared to that produced by third-generation facilities, is, in the hard x-ray regime, still only of the order of a few percent (see Figure 3.20). This is because, although radiation from any single electron travelling along an undulator is coherent, there is no spatial correlation between different electrons, and hence their combined radiation remains largely incoherent. In contrast, visible lasers are normally close to being 100% coherent.

In addition, the shortest pulse duration of x-rays from synchrotrons (with the exception of femto-slicing [1, 2], a 'bridging' technique that has become largely redundant since the advent of XFELs, the subject of this chapter) is a few tens of picoseconds, and may be a few hundred picoseconds in the case of DLSRs. Lasers in the visible and near-visible regimes can have pulse lengths as small as 70 attoseconds in exceptional cases[1], while pulse lengths of a few femtoseconds are routinely achieved.

The production of femtosecond pulses has been motivated by studies of the dynamics of chemical reactions – this is one of the most fundamental and important goals in science, as a biomolecular understanding of the processes of life, plus vast swathes of industrial chemical processes, depend intimately on this[2] (see Figure 4.1).

The speed of atomic motion is approximately 1000 m s^{-1}. If, therefore, one would like to follow a chemical reaction stroboscopically with a resolution of 1 angstrom, 100 fs long flashes are required. Similarly, vibrational frequencies are typically measured in tens or hundreds of terahertz (corresponding to the infrared regime), meaning that dynamics – the breaking and making of bonds and reconfiguration of the atomic structure – happens on the time-scale of a few femtoseconds to tens of femtoseconds or longer. Lasers are thus used to probe vibrational and vibronic (vibrationally and electronically excited) states. The temporal evolution of these states is followed using many different probes, such as the arrival times and orientations of photofragments, fluorescence, and electron spectroscopy, to name just three [4]. What lasers *cannot* do, however, is directly probe the atomic structure, as their wavelengths exceed typical interatomic distances by three orders of magnitude or more.

---

[1] The hard UV pulses here were so short that they contained only three or four oscillations, resulting in a Fourier-transform-limited relative bandwidth of the order of 25% [3].

[2] As an example, nitrogen fixing through the highly energy-hungry Haber–Bosch process feeds nearly half the world's population and accounts for almost 2% of the worldwide energy consumption – the Aswan Dam in Egypt was constructed over a century ago specifically to fuel this. On an atomic level, however, this catalytically driven reaction is still incompletely understood. A detailed picture of the chemical dynamics might provide insights into making this process more efficient and less costly.

An Introduction to Synchrotron Radiation: Techniques and Applications, Second Edition. Philip Willmott.
© 2019 John Wiley & Sons Ltd. Published 2019 by John Wiley & Sons Ltd.

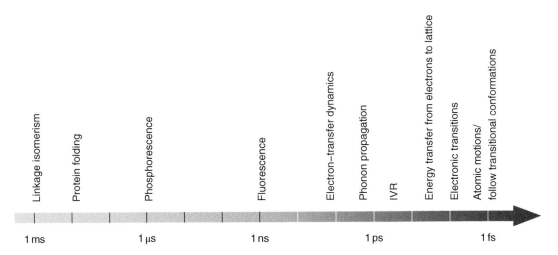

Figure 4.1 Typical time-scales for different dynamical processes in matter. IVR = intramolecular vibrational energy redistribution. The position of any one process is only meant to be representative, and may in reality span a broader range, depending on the details of the system; for example, protein folding is known to cover approximately eight decades, from minutes to microseconds.

X-ray free-electron lasers (XFELs) were developed to fill this need [5]. They provide extremely intense and short pulses of x-radiation. The difference between XFELs and synchrotron facilities, ignoring for the present their engineering architecture (that is, the manner in which the radiation is generated), is one of degree (see Table 4.1). While the total average number of photons per second delivered to a third- or fourth-generation synchrotron beam-line after monochromatization (typically between 10^{13} and 10^{14} s^{-1}) is comparable to that of an XFEL beamline (though the latter can vary enormously, depending on the facility and whether the beam is monochromatized or not), their time structures are very different: synchrotrons deliver several hundred million pulses per second ($1/\Delta T$), each a few tens of picosecond long ($\Delta\tau$) and containing of the order of 10^4 to 10^5 photons; XFELs deliver anything between approximately 100 and several hundreds of thousands of pulses per second. Crucially, each pulse is only a few tens of femtoseconds long, or even shorter, and contains of the order of 10^{12} photons[3]. The peak arrival rate of photons at XFELs is therefore well over a billion times higher than at synchrotron beamlines.

XFELs have thus laid the foundations for studies of the dynamics, properties, and structure of materials with unsurpassed spatial detail on a time-scale three to four orders of magnitude shorter than that available from synchrotron facilities.

Conventional lasers cannot generate photons more energetic than about 8 eV (i.e. ultraviolet light of around 150 nm), far less energetic than x-rays. This is primarily because, at higher photon energies, first absorption, then reflectivity, conspire to thwart laser operation. Lasers rely on mirrors, which not only become less efficient as they reflect higher-energy light, but also absorb too heavily, increasing the cavity losses, thereby forcing the cavity below the lasing threshold. They will also begin to burn!

This problem was resolved by developing the XFEL, first proposed in 1971 by John Madey at Stanford University [7]; he also coined the term XFEL in 1976. XFELs operate in the far ultraviolet and x-ray regimes.

XFELs are not lasers in the strictest sense, as they do not produce an avalanche-type increase in the flux of photons of the same wavelength and phase through stimulated emission[4], but instead use the phenomenon known

---

[3] As an amusing side note, after 10 years and assuming 5000 hours of operation per year, LCLS-I will have delivered photons for a total of 600 µs. Admittedly, in this time, it also provided 100 MJ of photonic energy ... [6]

[4] Laser is an acronym for light-amplification through stimulated emission of radiation.

Table 4.1 Comparison of orders-of-magnitude synchrotron- and XFEL-radiation[†] properties.

Property	Synchrotron	XFEL
$\Delta\tau$	50–400 ps	5–50 fs
ΔT	5 ns	10^{-2} to 2×10^{-7} s*
Average flux**	2×10^{14}	10^{14}
Peak flux**	6×10^{15}	2×10^{25}
#$h\nu$/pulse	4×10^{4}[‡]	4×10^{12}[¶]
Peak power	1 W[‡]	10^{11} W[¶]
Average power	25 mW[‡]	600 mW[¶] to 140 W[¶]*[§]

[†]XFEL values derive from LCLS unless otherwise stated. 8 keV photons assumed.
*EuroXFEL time structure: 2700 pulses at 4.5 MHz, 10 such bursts per second.
**Photons/s/0.1% BW.
[‡]After Si(111) monochromator, $\Delta\nu/\nu = 1.4 \times 10^{-4}$.
[¶]Unmonochromatized, full SASE spectrum.
[§]23 kW during pulse burst.

as 'self-amplified spontaneous emission' (SASE, described in Section 4.3). They are sometimes referred to as 'fourth-generation light sources', an unfortunate name in this author's opinion, as DLSRs are the natural evolutionary successors to third-generation synchrotrons (see Section 3.8.5). XFELs, on the other hand, are based on LINAC technology, not storage rings, hence a better definition would be 'first-generation coherent x-ray source', eschewing the word 'laser' in favour of 'coherent x-ray source', as SASE is a stochastic, spontaneous process, and not stimulated (the 's' in 'laser').

Crucially, SASE does not require the mirrors used in a conventional laser cavity. Instead, a SASE free-electron laser uses a monoenergetic electron beam and a very long, high-quality undulator, typically of the order of a few hundred metres in length.

The beam quality in storage rings is limited by stochastic radiation emission and perturbations of the electron beam by the magnet lattice. High-gain XFELs are made possible by the SASE runaway process, first proposed by Anatoli Kondratenko and Evgeni Saldin in 1980 [8] and worked out in detail by Rodolfo Bonifacio, Claudio Pellegrini, and Lorenzo Narducci in 1984 [9]. Because the electrons in XFELs are not stored and require only a few microseconds to traverse the length of the LINAC and undulator, they are far less perturbed than the equilibriated electrons in a storage ring, and can thereby maintain a very low emittance. The linear architecture of XFELs allows only a limited number of beamlines, in 'multiplexed' operation, to take beam at any one time. This is in contrast to synchrotron facilities, which can accommodate of the order of 20 or more experiments in parallel.

Let us make a back-of-the-envelope calculation of how much each photon costs at a synchrotron and XFEL, taken over the facilities' entire operational lifetime and considering only the initial costs of construction and commissioning. For the sake of simplicity, we assume that both types of source cost 1 billion dollars to build and commission, and that they both provide 15 years (5×10^8 s) of uninterrupted operation. A synchrotron with 20 beamlines will provide, on average, of the order of 5×10^{14} ph/s (photons per second), while the few beamlines served by an XFEL[5] may also deliver of the order of 5×10^{14} ph/s (but with a significantly larger spectral bandwidth than that from the synchrotron). The cost of each photon at a synchrotron and XFEL is thus about 4 femtodollars. Because in the case of the synchrotron, the photons can be distributed across many experiments and beamlines, whereas at an XFEL at most two or three multiplexed experiments can be performed in parallel, one might therefore deem XFELs to be a poor investment of precious scientific funding.

---

[5] In this somewhat whimsical back-of-the-envelope calculation, I consider hard x-ray photons. The LCLS-II will be capable of producing soft x-ray pulses at rates up to 1 MHz, which changes the numbers significantly.

Table 4.2 Parameters of important worldwide XFEL facilities operating between the vacuum-ultraviolet and the x-ray regimes and exploiting SASE. Not all parameters at a given facility can be achieved simultaneously.

Facility	LINAC type S/C*	$\mathscr{E}$ [GeV]	Wavelength range [nm]	Max. rep. rate [Hz]	Max. pulse energy [mJ]
Euro XFEL, Germany	S	17.5	0.05–4.7	$4.5 \times 10^{6\dagger}$	100
FERMI, Italy	C	1.2	4–90	50	0.2
FLASH, Germany	S	1	4.2–45	$10^{6\dagger}$	0.5
LCLS, USA	C	13.6	0.1–5	120	5
LCLS-II, USA	S/C	4/15	0.05–5	$10^6/120$	8
PAL-XFEL, Sth. Korea	C	10	0.06–10	60	2
SACLA, Japan	C	8.45	0.06–3	60	0.5
SwissFEL, Switzerland	C	5.8	0.08–5	100	2

* Superconducting/conducting LINAC.
† During burst in burst mode.

However, let us now consider the cost for the peak (i.e. pulsed) photon rate and the number of pulses provided over the same lifetimes of 15 years. We assume a peak synchrotron photon arrival rate from 20 beamlines of 2×10^{16} ph/s, and that from an XFEL to be 10^{26} photons per second; while the total number of pulses provided over a lifetime would be approximately 2.5×10^{17} and 6×10^{10}, respectively (the latter number corresponding to the modest pulse rate of the LCLS of 120 pulse/s), resulting in costs of 2×10^{-25} dollars/[(ph/s) pulse] and 1.7×10^{-28} dollars/[(ph/s) pulse], respectively. Using this metric, therefore, XFEL photons are 1200 times 'cheaper' than those provided by synchrotrons.

Both of the above calculations, interesting as they might be, somewhat miss the point, as the comparisons drawn from these numbers are predicated on the assumption that synchrotrons and XFELs are used for the same type of experiment. The second calculation does suggest, however, that the high intensity and temporal structure of XFEL radiation are *the* key attributes that justify their existence. In other words, XFELs are used for those experiments that absolutely require the high peak flux and short pulse lengths of XFELs to provide invaluable scientific insights otherwise unobtainable using conventional storage-ring facilities.

Table 4.2 shows a list of the XFELs worldwide at the time of publication, including some of their most important parameters.

4.2 XFEL Architecture

The progression of electrons in a high-gain XFEL is as follows. First, a bunch of electrons is generated by photoemission from a conducting surface radiated by a laser pulse of typically 10 ps duration, resulting in a photoemitted charge of the order of 200 pC (or 1.25×10^9 electrons, see Figure 4.2). Shorter pulse durations are excluded due to Coulomb repulsion, which would blow up the beam. Only once the electrons become relativistic after acceleration in a short LINAC can the bunch be compressed without spoiling the emittance (due to the 'beam stiffening' effect mentioned in Section 3.6). This is achieved in one or more bunch compressors, then the electrons are further accelerated in a longer LINAC, before entering the undulator array. Here, the SASE process takes place, resulting in the emission of femtosecond-duration pulses of x-rays.

The initial electron bunch produced by the picosecond laser is also in the picosecond regime; the transient peak current is about 50 A. Bunch compression is therefore needed in order to reduce the pulse duration to less than 100 fs and produce the several thousand amps peak current required to induce SASE (Figure 4.3). The necessary two orders of magnitude increase in n_e, the electron density, is realized by adjusting the RF phase in the first accelerator module (LINAC 1 in Figure 4.2) to allow the electrons to 'surf' down the slope of the sinusoidal

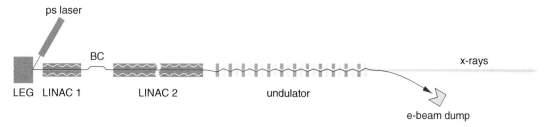

Figure 4.2 Schematic of XFEL facilities. Electron bunches are emitted from a low-emittance gun (LEG) irradiated by picosecond laser pulses. They are then accelerated in a short LINAC (LINAC 1), compressed longitudinally using one or more bunch-compressor magnet chicanes (BC), then further accelerated using a much longer LINAC (LINAC 2) before entering a long undulator, typically a few hundred metres in length. The SASE process along the undulator produces highly intense x-ray pulses with durations of the order of 50 fs. The electrons are deflected after the undulator using a bending magnet and subsequently dumped.

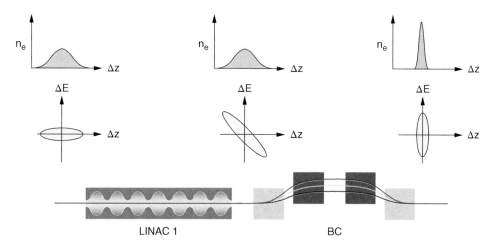

Figure 4.3 Bunch compression in high-gain XFELs. Before entering the first LINAC, the electron density n_e and spread in kinetic energy ΔE are both low. The phase of the LINAC RF-field relative to the passage of the bunch is so adjusted to induce a larger spread of the electrons' kinetic energies, whereby the trailing edge is made to be more energetic than the leading edge. Compression is then achieved by allowing the bunch to pass through a four-dipole chicane, whereby the faster electrons at the back can catch up with the slower electrons at the front, thanks to the shorter path they execute.

RF field – the leading electrons are thus accelerated less than those electrons towards the back of the bunch (this phenomenon is referred to as 'chirping'). Nonetheless, the electrons' velocities are so close to the speed of light that their differences are far too small to allow the faster electrons at the back of the bunch to catch up with the slower electrons positioned further forward and thereby squeeze the bunch length. Instead, the bunch passes through a magnetic four-dipole chicane. The trailing (high-energy) electrons execute a shorter path on account of their being less deviated by the magnetic field. In this manner, they are able to catch up with the less energetic leading electrons, which are more deviated by the magnet chicane. This compression thus shortens the bunch duration to approximately 300 fs (equating to a bunch length of 100 μm) and increases n_e by a factor of approximately 100. The increased energy spread of the chirped electron beam after bunch compression must be removed through 'dechirping', which can be achieved in one of several ways [10, 11], such as off-crest acceleration in the LINAC section downstream of the last bunch-compressor chicane.

4.3   The SASE Process

As electrons propagate down an undulator, they emit electromagnetic radiation (i.e. x-rays) independently and stochastically and are thus bathed in this radiation. Classical electromagnetism tells us that they will interact with this electromagnetic field. We begin our discussion of SASE by calculating the magnitude of these forces in conventional undulators at third- and fourth-generation synchrotrons.

The power transmitted per unit area by an electromagnetic plane wave is given by

$$\frac{P}{A} = \frac{E_0 H_0}{2}, \tag{4.1}$$

whereby E_0 is the amplitude of the electric-field component of the electromagnetic wave and H_0 is the amplitude of the magnetic intensity, which, in vacuum, is related to the magnetic-field-strength amplitude B_0 by

$$H_0 = \frac{B_0}{\mu_0}, \tag{4.2}$$

whereby $\mu_0 = 4\pi \times 10^{-7}$ m kg C^{-2} is the permeability of free space (given here in SI units). Note also from Equation (3.3) that

$$B_0 = \frac{E_0}{c}. \tag{4.3}$$

From Equations (4.1) to (4.3) we thus obtain

$$\frac{P}{A} = \frac{E_0^2}{2\mu_0 c}. \tag{4.4}$$

A typical synchrotron undulator may have a source size of 100×10 μm^2, and generate bunches of 50 ps duration containing 5×10^6 1 Å photons (within the full width of a harmonic). This equates to an areal power density of the order of 1.6×10^{11} W m^{-2}, and, from Equations (4.3) and (4.4), an electric-field amplitude $E_0 \approx 10^7$ V m^{-1}, and $B_0 \approx 40$ mT, nearly two orders of magnitude smaller than that imposed by the undulator's magnet array. We can thus conclude that, *in a conventional third-generation facility, the impact of the electromagnetic field generated by undulator radiation on the electrons' trajectory is completely negligible.*

In the case of XFELs, however, the beam is tailored to have as low an emittance and as high an electron density as possible, through the low-emittance gun and bunch compression, respectively. Although at the upstream end of the XFEL undulator, the interaction between the emitted radiation and the electrons is still very weak, it is sufficient to seed the runaway process of SASE, which is now described in detail.

The magnitudes and directions of the forces imparted upon electrons by the electric-field and magnetic-field components of electromagnetic radiation have been discussed in Section 3.4. We also know from Equation (3.3) that $B = E/c$, and therefore

$$F_L = eE\frac{v}{c} = -F_E \beta. \tag{4.5}$$

The electrons within XFELs (and storage rings) are highly relativistic; to a high degree of accuracy, $\beta \approx 1$ and $F_L \approx -F_E$ (the fractional imbalance is in fact equal to $1/2\gamma^2 \sim 2 \times 10^{-9}$). Thus, for an electron moving in exactly the same direction as the electromagnetic radiation (that is, ignoring for the time being the oscillations induced by the magnet array of an undulator), the forces are equal and opposite and lie in the direction perpendicular to the beam propagation (that is, the x-direction in Figure 3.4).

In the simplification made above that the electrons move in a straight line parallel to the electromagnetic plane wave, both the electric and magnetic forces (which anyway cancel each other out) always act at right angles to the electrons' direction of propagation and thus cannot transfer energy to or from the electrons. What happens if we now remove this simplification? Let us consider those positions in between the magnet poles where the electrons have their maximal transverse velocity induced by the undulator magnet array.

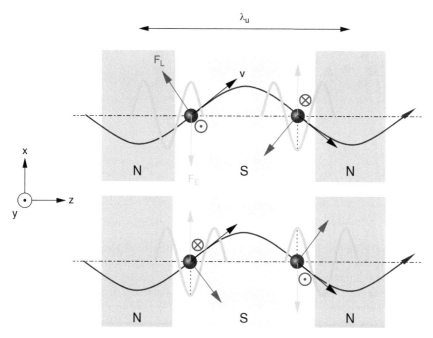

Figure 4.4 Different possible phase relationships between the electromagnetic field and the electron path in an undulator. In the upper case, when the electron has a transverse velocity component v_x in the positive x-direction, the electric field (yellow sine wave) is also positive and the magnetic field points in the positive y-direction (red $\odot$). As a consequence, the electric force points in the negative x-direction (yellow arrow) and will decelerate the electron; the Lorentz force (red arrow) points at 90° to the electron's trajectory in the x-z-plane so cannot transfer energy to or from the electron. Conversely, half an undulator period downstream, the electron's transverse velocity is now in the negative x-direction and the electromagnetic field has advanced relative to the electron by half a cycle (π radians). These two reversals cancel each other out, resulting in the electric force continuing to have a decelerating effect on the electron. In the lower panel, the phase relationship between the electron's transverse velocity component and the electromagnetic field is shifted by π radians, so that the electric field now accelerates the electron. These two opposing situations result in a spatial redistribution of the electrons within the wavelength of the emitted radiation, causing microbunching.

The first observation to make is that, because the magnetic field of the emitted electromagnetic radiation always lies perpendicular to the electrons' trajectory, it cannot be responsible for transfer of energy, but only in varying the direction of the electrons' motion[6]. Next, we should recognize that, in deriving Equation (3.70) for the resonant fundamental x-ray wavelength ($m = 1$) generated by undulators, we started off by assuming that the difference in length between the path executed by the electron as it oscillates through a single undulator period λ_u and that of the electromagnetic field emitted by the electron was a single wavelength (see Figure 3.36). Hence, from the perspective of the electron, after travelling a single period of the undulator, the electromagnetic field has gained on it by a single x-ray wavelength and hence it (the electron) has experienced a single 2π cycle change of that electromagnetic field. Importantly, during the same period, the electron has itself executed one cycle of transverse motion induced by a single period of the undulator's magnet array. This is shown in Figure 4.4.

---

[6] From Newtonian mechanics, the energy transfer $d\mathscr{E}$ of a particle experiencing a force $\mathbf{F}$ over a distance $d\mathbf{x}$ is $d\mathscr{E} = \mathbf{F} \cdot d\mathbf{x}$, and the energy-transfer rate is thus $d\mathscr{E}/dt = \mathbf{F} \cdot \mathbf{v}$. As $\mathbf{F}_L$, the Lorentz force, is always perpendicular to $\mathbf{v}$, their scalar product is zero and there can be no energy transfer between the magnetic field and electron.

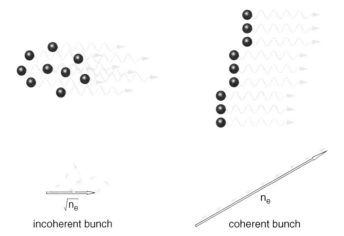

Figure 4.5 Incoherent and coherent radiation from an ensemble (bunch) of electrons. If the positions of the electrons bear no correlation to the radiation they emit, the phases of that radiation are likewise random. Their summed amplitude is equal to the amplitude of the radiation emitted from one electron multiplied by the square root of the number of electrons in the bunch. The intensity is therefore proportional to n_e. In coherent radiation, such as induced by SASE, emission from all electrons are in phase. The amplitudes of the radiation from each of the n_e electrons thus add linearly, resulting in an increase in intensity by n_e^2.

At the four positions of the electrons shown in Figure 4.4, the electric-field component of the emitted radiation has a component parallel to the electrons' trajectory and can either decelerate them (upper panel), or accelerate them (lower panel). So, within the wavelength of the emitted radiation, some electrons are accelerated while others are decelerated. The consequence of these two opposing interactions is that the electrons begin to form microbunches within the 100 μm bunch, separated by a distance equal to the wavelength of the light they both generate and are bathed in. Although, for hard x-rays, the bunch length equates to approximately 10^6 microbunches, SASE only occurs in the most intense portion of the bunch. The duration of the SASE radiation therefore depends on the degree of bunch compression and the integrated charge of the initial bunch as it enters the undulator array. Low-charge bunches will thus produce XFEL pulses with lower peak brilliance but with durations that can be shorter than 1 fs. More commonly, 'standard' bunch durations are a few tens of femtoseconds.

Another way, intuitively pleasing and simple, to visualize the microbunching process is as follows: in the laboratory frame of reference, the electron interacts with two fields, namely its own emitted radiation and the static undulator magnetic field. In the rest frame of the electron, however, relativistic effects transform the static undulator field into a counterpropagating wave with a wavelength equal to the relativistically (Lorentz) contracted undulator period, which is exactly equal to the radiation wavelength. This interferes with the copropagating emitted field, resulting in a standing wave which exhibits the spatial potential responsible for the microbunching.

In addition to breaking up a 'normal' electron bunch into microbunches, what other purpose could the SASE process serve? If, as is the case for normal synchrotron radiation, the emission of radiation from n_e electrons within a bunch is stochastic and uncorrelated, the vector addition of the individual amplitude contributions will, on average, be equal to $A_{\text{tot}} = \sqrt{n_e}\,A$, whereby A is the amplitude of the contribution to the radiation of any one of the electrons. The intensity, proportional to the square of the amplitude, is thus proportional to n_e.

On the other hand, if the bunch is shorter than the wavelength of radiation being emitted, the individual contributions to the radiation amplitude will all be more or less in phase (that is, the electrons all emit coherently). Thus $A_{\text{tot}} = n_e A$ and the intensity will be proportional to n_e^2 (see Figure 4.5). This is the mechanism behind the 'runaway' amplification in SASE – the more the electrons bunch and radiate in phase (that is, coherently), the stronger is the electromagnetic-field interaction with them, which in turn further enhances the microbunching phenomenon associated with coherent emission, thus increasing the degree of coherent emission ...

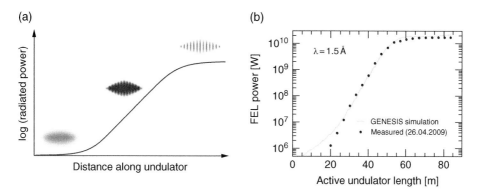

Figure 4.6 The self-amplification spontaneous-emission (SASE) process in an XFEL undulator. (a) Because of the interaction of the electrons with the synchrotron radiation they emit, a longitudinal density modulation (microbunching) develops together with a resulting exponential growth of the radiation power along the undulator. Note that in reality the number of microbunches within a conventional electron bunch is much larger than shown here ($\geq 10^5$ for x-ray FELs). (b) The gain/saturation curve measured at the LCLS in April 2009, along with the theoretically expected curve, calculated using the GENESIS simulation package. Reprinted from [12] courtesy Paul Emma and John Galayda, SLAC National Accelerator Laboratory, with permission of Macmillan Publishers Ltd.

This favourable runaway instability means that the light intensity grows exponentially along the undulator until the process saturates, bringing the beam to its highest possible intensity [12] (see Figure 4.6). This amplification operates only within a narrow bandwidth around the undulator resonant wavelength, given by Equation (3.70).

The microbunches produced by the SASE process each contain approximately $n_e = 10^9$ electrons, resulting in a peak brilliance for XFELs approximately one billion times more intense than that from fourth-generation synchrotron sources!

The quality of a high-gain XFEL is encapsulated in the dimensionless so-called FEL parameter, ρ_{FEL} (also called the Pierce parameter), given approximately by

$$\rho_{FEL} = \left[\frac{\lambda_u^2 r_0 n_e K^2}{32 \pi \gamma^3} \right]^{1/3}, \qquad (4.6)$$

whereby λ_u is the undulator period length, r_0 is the Thomson scattering length [Equation (2.10)], n_e is the electron density of the pulse bunch entering the undulator, K is the undulator parameter, and $\gamma = \mathscr{E}/m_e c^2$ is the Lorentz factor. Equation (4.6) can be re-expressed in practical units as

$$\rho_{FEL} = \frac{1.55 \times 10^{-5}}{\mathscr{E}[\text{GeV}]} [(\lambda_u[\text{mm}])^2 n_e[\mu\text{m}^{-3}] K^2]^{1/3}. \qquad (4.7)$$

Typical electron densities in the compressed electron bunch entering the undulator are of the order of 10^4 μm^{-3} (that is, 10^9 electrons in a volume of the order of 10^5 μm^3); the undulators have periods of 15 to 30 mm and K values of approximately 2 to 3; and the LINACs produce electron energies of the order of 10 GeV. Because of the cube-root dependency of ρ_{FEL} on all the design variables except $\mathscr{E}$, its spread is fairly narrow across all facilities designed or commissioned to date, and lies around 5×10^{-4}.

ρ_{FEL} determines three important characteristics of high-gain XFEL radiation. First, it describes the fraction of the electrons' power converted at SASE saturation to photon power, that is

$$\rho_{FEL} = \frac{P_{ph}}{P_e}. \qquad (4.8)$$

The larger is ρ_{FEL}, the more efficiently will the FEL convert electron energy into photon energy. ρ_{FEL} also describes the relative spectral bandwidth at saturation,

$$\rho_{FEL} = \frac{\Delta \omega}{\omega}, \qquad (4.9)$$

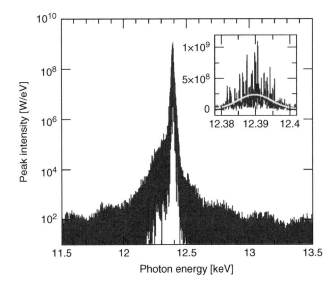

Figure 4.7 Simulated SASE power spectrum of the XFEL U15 undulator at the SwissFEL. The inset shows the central 20 eV of the spectrum on a linear scale, highlighting the spectrally noisy character of SASE pulses. The yellow curve is a best-fit Gaussian with a FWHM of 11.5 eV, or approximately 0.1%. Courtesy Marco Calvi and Sven Reiche, Paul Scherrer Institute.

which means that at photon energies of the order of 10 keV, the instantaneous bandwidth (that is, the bandwidth of any one FEL pulse) is of the order of 5 eV, well over an order of magnitude larger than the transmitted bandwidth after monochromatization with a diamond (004) single crystal. In addition, the position of the centre of the SASE spectrum jitters by a few tens of eV from pulse to pulse. It is the stochastic nature of the initial spontaneous generation of SASE radiation which is responsible for this fairly large bandwidth (see Figure 4.7). In contrast to the transverse coherence length, the longitudinal coherence length of SASE pulses [which depends inversely on the bandwidth, see Equation (3.45)] is thus small, of the order of 200 nm in the hard x-ray regime.

Lastly, the distance within an XFEL undulator, L_G, required to obtain a gain in SASE radiation by a factor $e \approx 2.72$ is inversely proportional to ρ_{FEL} and given by

$$L_G = \frac{\lambda_u}{4\pi\sqrt{3}\rho_{FEL}}. \qquad (4.10)$$

This is of the order of a few metres. Because saturation occurs after a SASE gain of approximately five orders of magnitude, the total undulator should exceed L_G by a factor of 11 or more (because $10^5 \approx e^{11}$). XFEL undulators are therefore measured in several tens or even hundreds of metres.

Why cannot one produce XFEL radiation by installing a long undulator in a storage ring? The reason is that the SASE runaway gain in XFELs requires that the radiation field interacting with the electrons is sufficiently strong to induce microbunching via the initial stochastic variations in the beam intensity. Dilation of the equilibriated electron bunches in synchrotrons, caused by Coulomb repulsion, is far too pronounced to allow SASE to begin. Importantly, the field strength depends linearly on the electron-gas density, hence extremely short and narrow-diameter pulses are required. Because the LINACs used in XFELs do not store electrons, it is possible to produce much higher ('pre-equilibrium') electron densities within the bunches, due on the one hand to the much lower electron-beam emittance, and on the other, to the bunch compressor(s) in the upstream region of the facility (see Figure 4.2), which squeeze the electron-bunch length to a few tens of microns, approximately two to three orders of magnitude shorter than in synchrotron facilities.

The very short pulse duration of light emitted by all the microbunches means that atomic motions in condensed matter, that is, vibrations which have periods of the order of picoseconds down to tens of femtoseconds, can be tracked using x-radiation, a hitherto impossible feat. Hence, in summary, FELs provide peak brilliances many orders of magnitude greater than radiation from third- and fourth-generation sources; 100% transverse coherence; and pulse durations typically of a few tens of femtoseconds, but which can also be tailored to be less than a femtosecond [13, 14].

4.4 Properties of XFEL Beams

Radiation from XFELs differs substantially from synchrotron radiation. The most pertinent aspects are covered in this section.

4.4.1 Tuning the Photon Energy

In contrast to synchrotron storage rings, the electron energy in LINAC-based sources such as FELs can more easily be changed, as the electrons are not required to execute a closed orbit, and therefore do not require bending magnets, whose bending radius depends on the electron energy. Hence, according to Equation (3.70), the photon energy can be adjusted by changing the electrons' energy. This has been the approach at the LCLS.

A different approach was adopted at the SACLA facility in Japan. By employing sophisticated feed-forward control, the undulator gap [or K, given by Equation (3.57)] can be controlled with an accuracy along the entire array of undulators, that phase errors (which induce partial destructive interference and a suppression of SASE) could be sufficiently minimized. All subsequent XFELs have LINACs optimized for a set of fixed electron energies, and for each of these, the photon energy is tuneable across a certain range.

4.4.2 Source Fluctuations

Users of third- and fourth-generation synchrotrons are used to x-ray beams that are stable in energy, flux, and position to better than one part in 10^4, 0.5%, and one micron, respectively. In contrast, the stochastic nature of SASE means that XFEL beams fluctuate significantly in these metrics, which therefore need to be determined shot-for-shot using sophisticated diagnostics [15, 16]. In addition, the temporal jitter associated with arrival of individual XFEL pulses can be over 100 fs, up to an order of magnitude larger than the pulse duration itself. This needs to be accurately tracked if the full potential of time-resolved experiments using XFELs is to be exploited.

4.4.3 Degree of Monochromacity

The unfiltered output from a SASE FEL, although highly spatially coherent in the transverse plane, has a very spiky energy spectrum, which seriously compromises the temporal (or longitudinal) coherence [see Equation (3.45)]. This limits the longitudinal coherence length to approximately 100 nm. The reason for the relatively broad spectrum is the fact that the amplification process starts from noise and the output therefore consists of a number of independent wave packets originating from different parts of the electron bunch. A typical SASE spectrum may have a relative bandwidth for any given pulse of the order of 0.1%, and the centre of the spectrum has a jitter of around 0.25%. This should be compared to monochromatized radiation at a synchrotron, which may have a bandwidth of approximately 10^{-4} and a long- and short-term stability better than one part in 10^5 (see Figure 4.8).

The impact of the variation in the SASE spectrum on the experimental setup at any given XFEL endstation depends intimately on that station's function. So, for example, serial femtosecond macromolecular crystallography (described in more detail in Section 6.11.5) involves recording crystallographic features (Bragg peaks) that in general are larger in extent (in reciprocal space) than the width of the 'Ewald shell' associated with SASE radiation. Because the protein crystals are destroyed within the duration of a single XFEL pulse, their diffraction patterns

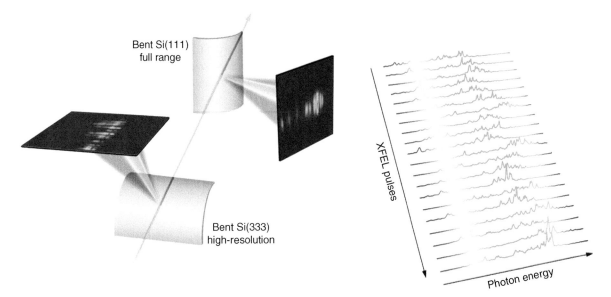

Figure 4.8 In-line spectral diagnostics for hard x-ray FEL pulses. Left: by bending thin single-crystal wafers of Si(111), the SASE spectral features can be spatially dispersed onto an area detector. The bent Si(111) crystal used for the full-range diagnostics provides a resolution of $\Delta E/E \approx 1.3 \times 10^{-4}$, or $\Delta E = 1$ eV at 8 keV. The Si(333) crystal, on the other hand, has a resolution of 9×10^{-6}, or 0.07 eV at 8 keV, sufficiently narrow to resolve individual SASE spikes, which are typically separated at 8 keV by approximately 0.2 eV. Right: the SASE spectra can fluctuate significantly from shot to shot, not least regarding the photon energy for which the maximum intensity is found. Adapted from [16] with permission of the American Institute of Physics.

are of stationary crystals, precluding the rotation method described in Section 6.8.2, the traditional approach to obtaining reliable diffraction intensities. The larger 'thickness' of the SASE-Ewald shell can be seen, in this respect, to be advantageous.

On the other hand, spectroscopic measurements, which may require resolutions well below an electronvolt, cannot use unfiltered SASE radiation. In this case, either the XFEL must be seeded in order to promote a narrow and well-defined band of the SASE spectrum, or the beam must be subsequently monochromatized.

4.5 Seeding

It can be easily demonstrated from the properties of so-called conjugate variables (here, time and energy) that the ultimate limit (the so-called 'Fourier-transform limit') to the temporal coherence $\Delta \nu$ of a Gaussian packet of radiation of a pulse length FWHM $\Delta \tau$ is given by

$$\Delta \tau \Delta \nu = \frac{2}{\pi} \ln 2 = 0.44, \tag{4.11}$$

whereby $\Delta \nu$ is the FWHM spread of frequencies given in Hz and $\Delta \tau$ is the pulse duration in seconds. Therefore the spectral purity is

$$\frac{\Delta \nu}{\nu} = \frac{2\lambda \ln 2}{\pi c \, \Delta \tau} \tag{4.12}$$

$$= 1.472 \times 10^{-4} \, \frac{\lambda[\text{Å}]}{\Delta \tau[\text{fs}]}. \tag{4.13}$$

A Fourier-transform-limited 50 fs pulse of 1 Å radiation would have a relative bandwidth of 3×10^{-6}, almost two orders of magnitude smaller than the best monochromatization possible using a Si(111) crystal. Therefore, any process which could clean up the spectral purity of FEL radiation would have an enormous technological benefit insofar that one could dispense with the need for a downstream monochromator. Seeding has other added advantages: the transition from incoherent radiation at the beginning of the undulator to fully saturated radiation occurs earlier, allowing the use of shorter (and cheaper) devices; the stability of the resonant wavelength and the output power is much improved; and synchronization from the seed radiation with external signals for pump-probe experiments is simplified.

Several seeding strategies for XFELs are currently being explored; some are more suited to soft x-ray generation, others to the production of angstrom-scale radiation. High-harmonic generation (HHG), high-gain harmonic generation (HGHG), and echo-enabled harmonic generation (EEHG) are just three notable examples; the reader is directed to [17], and references therein, for more detailed descriptions.

Seeding can thus be accomplished in one of several possible ways; at this relatively youthful phase of XFEL developments, it is difficult to predict whether one or two approaches will dominate in the future, or, if that indeed is the case, which these approaches will be. Nonetheless, we briefly cover two strategies, that, at the time of writing, seem to be particularly promising.

4.5.1 High-brilliance SASE using an Array of Short Undulators and Chicanes

The inherently stochastic process of SASE leads to the amplification of many regions of microbunching within the original electron bunch (see Section 4.4.3). Although each region has a more or less well-defined phase, there is no correlation from region to region. The longitudinal distance of any one in-phase region is called the 'cooperation length'. The goal of the configuration used in high-brilliance SASE (HB-SASE) is to significantly increase the cooperation length and thereby improve the longitudinal coherence (that is, reduce the bandwidth).

In HB-SASE, short undulator sections, each typically 2 m long, are separated by weak chicanes consisting of four dipole pairs (see Figure 4.9) [18]. The XFEL field generated in the first undulator thus induces SASE in a new region of the original electron bunch further downstream in the second undulator. Importantly, the separated SASE regions are in phase with each other. Both of these can now induce new SASE regions in the next undulator section. In the configuration shown schematically in Figure 4.9, each successive delay induced by the chicane is half as long as the previous, causing the SASE regions to 'fill the gaps'. A simulation of HB-SASE for an array of 1.5 m undulators is shown in Figure 4.10. HB-SASE can be employed for both soft and hard x-rays.

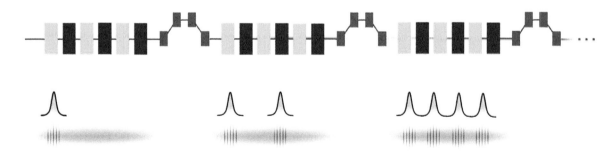

Figure 4.9 High-brilliance SASE. The cooperation length of the SASE process is increased by introducing tailored delays for the electrons with respect to the generated XFEL radiation using dipole chicanes interspersed between short undulator sections. Each new microbunch region maintains a well-defined phase relationship with microbunches produced in the previous undulator sections, thus increasing the cooperation length. In the configuration shown here, each successive delay is half as long as the previous.

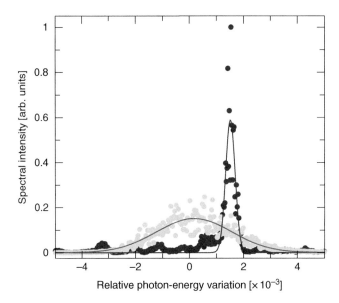

Figure 4.10   Comparison of spectra produced by 'normal' SASE (yellow circles and red best-fit Gaussian) and HB-SASE (blue circles and solid line best-fit Gaussian), averaged over 10 shots, and relative to a photon energy of 1.2375 keV. The bandwidth of the HB-SASE is almost ten times narrower than that for normal SASE. Adapted from [18] with permission from Elsevier.

Not only does HB-SASE improve XFEL brilliance by over an order of magnitude, but also the total length of undulators required to reach saturation is reduced to approximately two-thirds that of 'normal' SASE. Finally, it should be noted that HB-SASE is not strictly a seeding mechanism, as no 'seed' radiation upstream of the HB-SASE array is used.

4.5.2   Self-seeding of Hard XFEL-radiation using Diamond Monochromatization

 Self-seeding of hard x-rays is now routine at the LCLS, using an intra-undulator diamond single crystal [19, 20]. This works best for short pulse lengths below approximately 20 fs, for reasons that will become immediately apparent. In this scheme (see Figure 4.11), a diamond single crystal acts as a Bragg-transmission monochromator. At the Bragg diffraction condition, the temporal form of the transmitted beam has a delayed long monochromatic 'wake pulse' with a width inversely proportional to the bandwidth of the extinction line in the transmission spectrum (the feature highlighted by an asterisk in Figure 4.11). This radiation is made to physically overlap with the electron pulse in the second undulator array and thus provide the monochromatic seeding. The required overlap is achieved by the weak four-magnet chicane that is also used to provide space for the diamond crystal and wash out the incipient SASE microbunching produced by the SASE undulator.

It should be noted, however, that the Darwin width of the diamond (004) diffraction peak (that is, the range of photon energies that satisfy the diffraction condition, see Section 6.4.4) is narrower than typical gaps in the SASE spectra, and hence a significant fraction of shots using self-seeding in this manner result in no amplification, due to the relatively high probability that the stochastic spectrum will have zero intensity across the Darwin width.

4.6   Radiation Damage and Heat Loads

The high transient intensities associated with XFEL pulses place stringent specifications on any x-ray optics used, and more often than not (though not always) also result in the destruction of the probed sample.

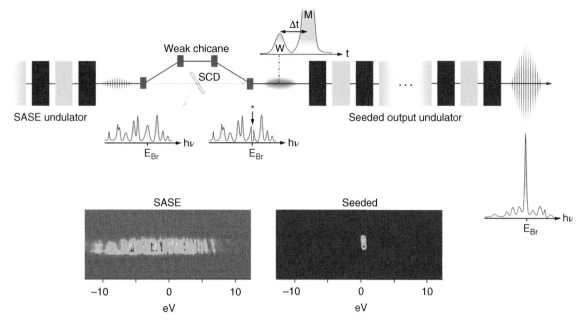

Figure 4.11 Self-seeding using a diamond crystal. Upper panel: the first part of the undulator induces a modest SASE signal which is nevertheless sufficiently intense to seed the second section. Both this and the electron bunch exit the SASE undulator. The photon pulse passes through a single-crystal diamond wafer positioned accurately at the (004) Bragg condition. The main transmitted beam ('M') has both a notch of missing radiation at the energy E_{Br} that satisfies the Bragg condition (given by the asterisk), and, importantly, a delayed monochromatic tail, or 'wake pulse' ('W'). Meanwhile, the electrons travel through a weak chicane. This not only provides the necessary space for the diamond wafer, but also destroys the microbunching, and retards the electrons by an amount $\Delta t \sim 20$ fs, so that they overlap after the chicane with the monochromatic wake pulse. It is this overlap that then seeds the radiation in the second undulator array to produce a significantly improved monochromatic output x-ray pulse. Lower panels: Comparison of spectra recorded for a 'normal' SASE beam and a self-seeded beam, using the diagnostic tool shown in Figure 4.8. Adapted from [21] with permission from the Institute of Physics.

4.6.1 Thermal Loads on Optics

Let us begin by determining the typical heat loads experienced by different optical elements, in particular focussing optics such as K-B mirrors, compound refractive lenses, and Fresnel zone plates (see Chapter 5).

The FWHM divergence of XFEL radiation in both transverse planes is of the order of 3 μrad. At a distance of approximately 200 m downstream of the XFEL undulator, the radiation will thus have a FWHM[7] of 600 μm. The areal energy and power densities are therefore 0.88 J cm^{-2} and 17.7 TW cm^{-2} (for a 50 fs pulse duration), respectively.

Consider this beam incident at $\alpha = 0.1°$ on a silicon mirror. The footprint is increased by $1/\sin \alpha \approx 570$ compared to the beam's cross-section, and the energy density on the silicon surface drops to 1.5 mJ cm^{-2}. The ablation thresholds for hard x-rays on silicon and silica (SiO_2) have been shown to approximately equate to the thermal melting dose, of the order of 80 and 400 J cm^{-2} [22], while the damage thresholds, although an order of magnitude smaller, are still significantly larger than that experienced by the mirrors. Although the performance of silicon is

---

[7] Note that half the total power of a two-dimensional Gaussian profile lies within the FWHM.

inferior to that of silica with regards to its damage and ablation thresholds, the former's good thermal conductivity means that it remains the material of choice for those XFELs, such as the European XFEL and LCLS-II, that provide tens of thousands of pulses per second and thus have a significant average thermal load in addition to the instantaneous values.

Two common materials used for x-ray focussing elements are diamond (for Fresnel zone plates, FZPs [23]) and beryllium (for compound refractive lenses, CRLs [24]). Their attenuation lengths at 8 keV are 630 μm and 5280 μm, respectively. Diamond FZPs have a thickness of the order of 5 μm, hence the absorbed energy within the FWHM of a 5 mJ pulse is approximately 20 μJ, whereby the irradiated area is 2.8×10^{-3} cm^2. For a repetition rate of 120 Hz, this equates to an absorbed areal power density of 0.85 W cm^{-2}. Even if we ignore the excellent thermal dissipation properties of diamond (with a thermal conductivity of the order of 2000 W m^{-1} K^{-1}), and consider pure blackbody radiation, the FZP will only assume a temperature of perhaps 200 °C. If one considers also thermal conductivity, the FZP will heat only by a few kelvin.

Beryllium has an order of magnitude lower thermal conductivity, but on the other hand has an order of magnitude larger attenuation length than diamond. Be-based CRLs thus also remain robust when irradiated with unfocussed XFEL pulses. It is noted, however, that the much higher repetition rates generated by the European XFEL and LCLS-II may pose more serious problems for x-ray optical components.

4.6.2   Sample Irradiation

From Equation (4.4) it can be calculated that the electric-field amplitude of XFEL radiation focussed down to 1 μm^2 is of the order of 10^{13} V m^{-1}, or 10^5 V across 100 Å. This distance (100 Å) is the approximate distance from an atom that an electron has to move to be considered to be unbound. Thus, it might be assumed that all bound electrons other than the K electrons for the very heaviest elements would be ripped from their respective atoms within the attosecond time-scale associated with a single x-ray period. However, a detailed quantum-mechanical description of the response of the electrons to high-frequency- and high-field electromagnetic pulses shows that atoms are in fact *stabilized* against ionization [25]. Instead, classical photoabsorption dominates, with the exception of hydrogen, for which Compton scattering is the dominant process (Section 2.3).

Subsequent to absorption, fluorescence is most likely for the heavier elements, though for the lighter elements associated with protein structures, Auger emission will dominate, resulting in the loss of a second electron. In either case, irradiation by a micron-sized focussed XFEL pulse having an energy of the order of millijoules, will result in extensive damage, or, in the case of micron-sized samples such as found in serial femtosecond macromolecular crystallography, a Coulomb explosion within a few tens of femtoseconds [26]. Indeed, the medium in which such crystals are transported into the beam path may be liquid jets, which can be locally obliterated and require a significant recovery time until they once more become continuous [27] (Figure 4.12). Very recent results indicate that if the jets are kept to diameters below approximately 10 μm and travel at about 100 m s^{-1} (corresponding to a consumption rate of nearly 10 μl s^{-1}), repetition rates for the XFEL pulses as high as 1.1 MHz can be tolerated without collateral damage from one pulse impacting neighbouring irradiated material [28].

Focussing down to a micron, or even tighter, is a prerequisite in serial femtosecond crystallography because of the generally small crystallite sizes and weak elastic scattering cross-sections. Expanding the focus would, at first, simply reduce the signal intensity while still destroying the sample. By the time the focus would be sufficiently large that the crystal would survive more than one XFEL pulse, the signal would be too weak to detect.

On the other hand, other experiments use larger samples that scatter x-radiation more strongly, and which are, relatively speaking, fairly radiation hard. In these cases, an unfocussed, or only weakly focussed, beam is adopted and it is thus possible to irradiate the same sample many times. So, for example, a large single crystal of the multiferroic material TbMnO$_3$ was investigated in a THz-pump–XFEL-probe experiment using unfocussed

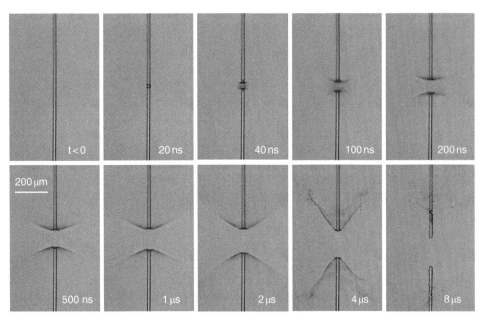

Figure 4.12   Explosion of delivery jet by XFEL pulses. The pulse energy was 0.75 mJ; the water jet has a diameter of 20 μm and travels at a speed of 10 m s^{-1}. Adapted from [27] with permission from Macmillan Publishers Ltd.

100 fs, 1 mJ x-ray pulses, resulting in a spot size of approximately 300 μm and an areal energy density of only approximately 40 mJ cm^{-2} [29].

4.7   XFELs and THz Radiation

That part of the electromagnetic spectrum between approximately 0.1 and 10 THz (corresponding to 0.4 and 40 meV photon energy, respectively) is especially interesting, as it can excite low-energy states in matter, such as those associated with rotation, vibration, and excitation of electron-hole bound states ('excitons'). Until recently, there has been a lack of sources of intense THz radiation[8], but as its relevance in the study of both condensed and gas-phase matter has become more apparent, there has been a concerted effort to rectify this, primarily through the optical rectification of visible and infrared pulses [30].

In addition to resonant excitation via THz absorption, these sources can provide sufficient intensities in 'half-cycle' rectified pulses to liberate electrons through field ionization – state-of-the-art THz sources, if focussed, are capable of producing field strengths in excess of 100 MV cm^{-1}. This corresponds to a potential difference of 100 eV across 100 Å, enough to remove valence electrons, or core electrons of the lighter elements.

Transient states induced either by resonant absorption of THz radiation, or field-induced modifications, lend themselves to be investigated by XFELs in pump-probe experiments, as the XFEL pulse lengths are typically two orders of magnitude shorter than the THz perturbation (see Figure 4.13). Phenomena such as phonon excitations,

---

[8] Ironically, the cosmic microwave background, having a peak wavelength of approximately 1 mm (3×10^{11} Hz), constitutes most of the radiation energy in the universe.

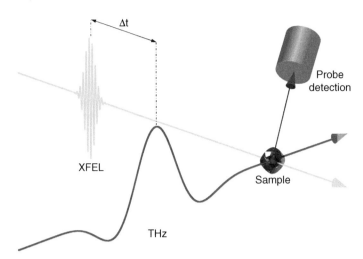

Figure 4.13 Schematic of a THz-XFEL pump-probe experiment. The THz radiation excites the sample either resonantly or nonresonantly. The dynamics of the response of the sample both during the excitation and thereafter can be probed using XFEL radiation in a stroboscopic manner by tuning Δt, the delay between the THz and XFEL pulse.

spin manipulation, field ionization of excitons, and electron acceleration can all be probed using a combination of THz and XFEL radiation, in a burgeoning field that remained, until recently, largely unexplored [29, 31].

4.8 Concluding Remarks

The primary goal of this chapter was to describe the machine physics of XFELs at a level that is accessible to a wide spectrum of potential users. XFEL science is liable to further expand and morph in the next decades, as XFELs are still very much in their infancy. They are, however, maturing rapidly, not least thanks to the knowledge base already established for synchrotron facilities.

The radiation produced by XFELs differs from synchrotron radiation primarily in the peak brilliance and associated very short pulse durations. These unprecedented sea changes are so extreme that experimental procedures commonly used at synchrotrons need to be entirely rethought in order to be carried out at XFELs; indeed, some experimental methods at synchrotrons are entirely excluded at XFELs. On the other hand, whole new vistas of experimental disciplines, particularly in the fields of time-resolved x-ray science and ultra-intense (nonlinear) light–matter interactions, have been opened up using XFELs and are still rapidly evolving; this is the reason, in fact, why they have been developed in the first place. Examples of experiments performed using XFELs will be included in the relevant sections in Chapters 6 to 8. It remains an exciting prospect to see how these flourish in the forthcoming years and decades!

Problems

1. The LCLS provides 30 fs, (8 ± 0.025) keV pulses, each of total energy equal to 5 mJ. The Materials-Science beamline at the Swiss Light Source provides an average flux of 10^{14} ph/s/0.1% BW after monochromatization with a Si(111) monochromator, comprising of 5×10^8 pulses per second, each pulse being 40 ps long. Assuming that the emittance of the LCLS is 1000 times smaller than at the SLS, calculate the ratio of their peak brilliances. (2P)

2. In Section 4.3, the electric-field and magnetic-field strength amplitudes, E_0 and B_0, of the radiation produced by a synchrotron-based undulator were calculated using Equations (4.1) to (4.4) in order to demonstrate that the interaction of the generated radiation with the electrons was negligible compared to the Lorentz force imposed by the static magnetic field of the undulator magnets.

   Consider now a 'typical' 70 fs XFEL pulse emerging from the LCLS facility in Stanford, California. It contains 8.4×10^{11} 1.5 Å photons with a cross-section as it emerges from the undulator of 15×15 μm^2. Calculate P/A (the peak power density), E_0, and B_0. (3P)

3. The LCLS runs east to west, while the earth's magnetic field at Menlo Park runs north–south and has a magnitude of approximately $B_{\oplus} = 47\,000$ nT. If no measures were taken to compensate for this, calculate the lateral distance the electrons would drift after the length of the facility of 1.5 km due to the Lorentz force generated by $B_{\oplus}$. Make the simplification that the electrons immediately reach their terminal energy of 15 GeV at the upstream end of the LINAC. The mass of an electron at rest is $m_e = 9.11 \times 10^{-31}$ kg. (3P)

4. Given that the Fourier transform of a temporal Gaussian pulse $\exp(-t^2/2\sigma_t^2)$ is $C \exp(-\omega^2\sigma_t^2/2)$, whereby σ_t is the standard deviation of the temporal Gaussian (and hence the FWHM $\Delta\tau$ equals $\sqrt{8\ln 2}\,\sigma_t$), $\omega = 2\pi\nu$, and C is a constant, prove Equation (4.11). (3P)

5. The SwissFEL free-electron laser is being designed to provide as short as 1 fs pulses at 1 Å. Determine the narrowest possible bandwidth $\Delta\nu/\nu$ that could be achieved under these conditions. Describe qualitatively the consequences of attempting to improve the spectral purity of such pulses by monochromatization through, for example, a crystal monochromator. (2P)

6. A water jet with a diameter ϕ_w less than 20 μm is irradiated with 8 keV XFEL pulses. The attenuation length of x-rays in water at this photon energy is $1/\mu = 1$ mm. Assuming a beam diameter of $\phi_x = 1$ μm, show that the average absorbed energy per unit volume by a single XFEL pulse is accurately approximated by

$$E/V = \frac{4E_p\mu}{\pi\phi_x^2},$$

   where E_p is the energy per XFEL pulse. Assuming $E_p = 5$ mJ, determine E/V and compare this with the energy per unit volume required to boil water from room temperature.

   The LCLS-II upgrade promises photon energies up to 25 keV. Determine E/V for the same setup, but using 24 keV photons. (4P)

7. The molecule I_2 has a bond length of $R = 2.666$ Å. The rotational energy levels assume values of

$$E_J = \frac{J(J+1)\hbar^2}{2I},$$

   whereby $I = [m_1 m_2/(m_1 + m_2)]R^2 = \mu_r R^2$ is the moment of inertia of a diatomic molecule containing atoms of mass m_1 and m_2 and J is the total angular-momentum quantum number. Transitions are only allowed for $\Delta J = \pm 1$.

   Determine I in kg m^2, and the maximum value of J which can be excited to $J+1$ using 0.1 THz radiation. (3P)

8. In the harmonic approximation, vibrational-energy levels of a given vibrational mode are separated by ΔE_v, whereby

$$\Delta E_v = \hbar\sqrt{\frac{k}{\mu_r}},$$

   $\mu_r = m_1 m_2/(m_1 + m_2)$ is the reduced mass, and k is Hooke's force constant, which for I_2 is equal to 172 J m^{-2}. Determine ν for iodine and thereby demonstrate that this is in the THz regime. (2P)

9. A state-of-the-art THz source is able to generate a 1 mJ THz radiation pulse with a duration of 0.2 ps and focus this down to 0.1 mm^2. Determine the associated electric- and magnetic-field strengths. (2P)

References

[1]   A. A. Zholents and M. S. Zolotorev, "Femtosecond x-ray pulses of synchrotron radiation," *Phys. Rev. Lett.*, vol. 76, pp. 912–915, 1996.

[2]   P. Beaud, S. L. Johnson, A. Streun, et al., "Spatiotemporal stability of a femtosecond hard-x-ray undulator source studied by control of coherent optical phonons," *Phys. Rev. Lett.*, vol. 99, p. 174801, 2007.

[3]   K. Zhao, Q. Zhang, M. Chini, et al., "Tailoring a 67 attosecond pulse through advantageous phase-mismatch," *Optics Lett.*, vol. 37, pp. 3891–3893, 2012.

[4]   A. H. Zewail, "Femtochemistry: Atomic-scale dynamics of the chemical bond," *J. Chem. Phys. A*, vol. 104, pp. 5660–5694, 2000.

[5]   P. Schmüser, M. Dohlus, J. Rossbach, and J. Behrens, *Free-electron lasers in the ultraviolet and x-ray regime*. Springer, Second edn., 2014.

[6]   Christoph Quitmann, private communication.

[7]   J. M. J. Madey, "Stimulated emission of Bremsstrahlung in a periodic magnetic field," *J. Appl. Phys.*, vol. 42, pp. 1906–1913, 1971.

[8]   A. M. Kondratenko and E. L. Saldin, "Generation of coherent radiation by a relativistic electron beam in an undulator," *Part. Accelerators*, vol. 10, pp. 207–216, 1980.

[9]   R. Bonifacio, C. Pellegrini, and L. M. Narducci, "Collective instabilities and high-gain regime in a free-electron laser," *Opt. Commun.*, vol. 50, pp. 373–378, 1984.

[10]   S. Antipov, S. Baturin, C. Jing, et al., "Experimental demonstration of energy-chirp compensation by a tunable dielectric-based structure," *Phys. Rev. Lett.*, vol. 112, p. 114801, 2014.

[11]   P. Emma, M. Venturini, K. L. F. Bane, et al., "Experimental demonstration of energy-chirp control in relativistic electron bunches using a corrugated pipe," *Phys. Rev. Lett.*, vol. 112, p. 034801, 2014.

[12]   P. Emma, R. Akre, J. Arthur, *et al.*, "First lasing and operation of an ångstrom-wavelength free-electron laser," *Nature Phot.*, vol. 4, pp. 641–647, 2010.

[13]   J. Feldhaus, J. Arthur, and J. B. Hastings, "X-ray free-electron lasers," *J. Phys. B. At. Mol. Opt. Phys.*, vol. 38, pp. S799–S819, 2005.

[14]   G. Margaritondo and P. R. Ribic, "A simplified description of x-ray free-electron lasers," *J. Synchrotron Rad.*, vol. 18, pp. 101–108, 2011.

[15]   S. Moeller, G. Brown, G. Dakovski, et al., "Pulse energy measurements at the SXR instrument," *J. Synchrotron Rad.*, vol. 22, pp. 606–611, 2015.

[16]   D. Zhu, M. Cammarata, J. M. Feldkamp, et al., "A single-shot transmissive spectrometer for hard x-ray free electron lasers," *Appl. Phys. Lett.*, vol. 101, p. 034103, 2012.

[17]   S. Reiche, "Overview of seeding methods for FELs," in *Proceedings of IPAC2013*, pp. 2063–2067, 2013.

[18]   E. Prat, M. Calvi, R. Ganter, et al., "Undulator beamline optimization with integrated chicanes for X-ray free-electron-laser facilities," *J. Synchrotron Rad.*, vol. 23, pp. 861–868, 2016.

[19]   G. Geloni, V. Kocharyan, and E. Saldin, "A novel self-seeding scheme for hard X-ray FELs," *J. Mod. Opt.*, vol. 58, pp. 1391–1403, 2011.

[20]   J. Amann, W. Berg, V. Blank, et al., "Demonstration of self-seeding in a hard-X-ray free-electron laser," *Nature Phot.*, vol. 6, pp. 693–698, 2012.

[21]   C. Bostedt, J. D. Bozek, P. H. Bucksbaum, et al., "Ultra-fast and ultra-intense x-ray sciences: first results from the Linac Coherent Light Source free-electron laser," *J. Phys. B*, vol. 46, p. 164003, 2013.

[22]   T. Koyama, H. Yumoto, Y. Senba, et al., "Damage study of optical substrates using 1-μm-focusing beam of hard x-ray free-electron laser," *J. Phys. Conf. Ser.*, vol. 463, p. 012043, 2013.

[23]   C. David, S. Gorelick, S. Rutishauer, et al., "Nanofocusing of hard x-ray free electron laser pulses using diamond based Fresnel zone plates," *Sci. Reports*, vol. 1, p. 57, 2011.

[24]   F. Seiboth, A. Schropp, R. Hoppe, et al., "Focusing XFEL SASE pulses by rotationally parabolic refractive x-ray lenses," *J. Phys. Conf. Ser.*, vol. 499, p. 012004, 2014.

[25] H. R. Reiss, "Physical basis for strong-field stabilization of atoms against ionization," *Laser Phys.*, vol. 7, pp. 543–550, 1997.

[26] R. Neutze, R. Wouts, D. van der Spoel, E. Weckert, and J. Hajdu, "Potential for biomolecular imaging with femtosecond x-ray pulses," *Nature*, vol. 406, pp. 752–757, 2000.

[27] C. A. Stan, D. Milathianaki, H. Laksmono, et al., "Liquid explosions induced by X-ray laser pulses," *Nature Phys.*, vol. 12, pp. 966–971, 2016.

[28] M. O. Wiedorn, D. Oberthür, R. Bean, et al., "Megahertz serial crystallography," *Nature Comms.*, vol. 9, p. 4025, 2018.

[29] T. Kubacka, J. A. Johnson, M. C. Hoffmann, et al., "Large-amplitude spin dynamics driven by a THz pulse in resonance with an electromagnon," *Science*, vol. 343, pp. 1333–1336, 2014.

[30] D. M. Mittleman, "Frontiers in terahertz sources and plasmonics," *Nature Phot.*, vol. 7, pp. 666–669, 2013.

[31] T. Kampfrath, K. Tanaka, and K. A. Nelson, "Resonant and nonresonant control over matter and light by intense terahertz transients," *Nature Phot.*, vol. 7, pp. 680–690, 2013.

5

Beamlines

5.1 Introduction

For users, the beamline is the part of a synchrotron facility of primary interest. Although for some, a beamtime at a synchrotron consists of bringing along a sample, placing it in the beam and starting an experiment (or getting the beamline scientist to do this), and then leaving again, clutching a CD-ROM, USB stick, or hard drive replete with experimental data[1], many experiments require a deeper level of understanding of the beamline components in order to optimize the measurements. This can range from being able to change the x-ray photon energy, to preparing sophisticated time-resolved experiments that use triggering from the time structure of the electron bunches in the storage ring.

In this chapter, we will look at the main beamline components in the same order as they are encountered by the photons emerging from the storage ring. Not all beamlines contain all the components described below, although the general features in Figure 5.1 are fairly standard.

5.2 Front End

The 'front end' of a beamline consists of the components in the ring tunnel, after the insertion device or bending-magnet source. Its functions are to (a) monitor the position of the photon beam; (b) define the angular acceptance using a beam-defining aperture; (c) block, when required, some part of the x-rays and the Bremsstrahlung radiation; (d) filter out, if necessary, the portion of the source spectrum in the soft x-ray region, which would otherwise interact very strongly with the beamline components; and (e) when possible, isolate the beamline vacuum from the storage-ring vacuum.

5.2.1 X-ray Beam-position Monitors

Beam-position monitors (BPMs) measure the lateral position of the x-ray beam at different positions along the beamline from the front end to the experimental hutch. There are several possible designs, with typical accuracies of the order of a micron. Wire monitors, shown in Figure 5.2(a), consist of a thin wire of, for example, tungsten, and can be scanned across the beam to record its profile in one direction, by measuring the photocurrent induced by the x-rays. Their resolution is, of course, limited by the diameter of the wire, and is typically of the order

---

[1] Indeed, it is becoming increasingly common in macromolecular crystallography for users to simply send samples to the facility from their home institute for investigation, and await the results via the internet.

An Introduction to Synchrotron Radiation: Techniques and Applications, Second Edition. Philip Willmott.
© 2019 John Wiley & Sons Ltd. Published 2019 by John Wiley & Sons Ltd.

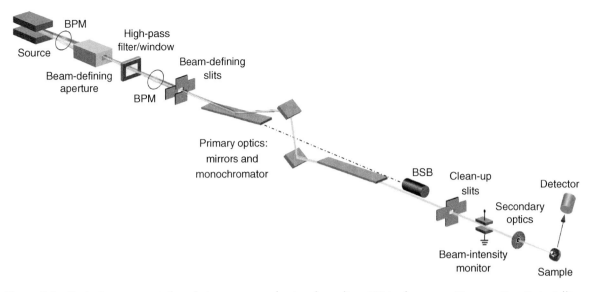

Figure 5.1   Typical components found at an x-ray synchrotron beamline. BPM = beam-position monitor. By installing two or more BPMs, both the beam height and the tilt angle can be determined. BSB = Bremsstrahlung blocker. The precise nature of each component can vary significantly from beamline to beamline, while not all components shown here are necessarily used.

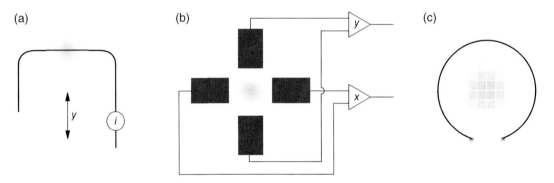

Figure 5.2   Different designs for beam-position monitors: (a) a simple 1D wire monitor, (b) a 2D blade monitor, and (c) a diamond CVD profile monitor.

of 5 µm. Other common designs include blade BPMs [Figure 5.2(b)], in which four metallic blades intersect with the outer halo of the x-ray beam, thereby producing four independent photoelectron currents, from which the position in two dimensions can be ascertained using comparators; and CVD diamond BPMs [Figure 5.2(c)], in which a set of photocurrents from an array of ultrathin metallic electrodes deposited on thin and transparent synthetic diamond discs can provide not only the beam position, but also a two-dimensional profile of the beam intensity [1].

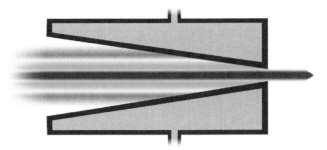

Figure 5.3 Beam-defining primary apertures. Normally the first component downstream of the source, these water–cooled apertures define the angular range to the beamline and often have a rectangular cone shape, in order to distribute the thermal load over a larger area. Because the outer parts of the synchrotron radiation contains lower-energy photons than the central cone, these apertures also act as high-pass filters.

5.2.2 Primary Aperture and Front-end Slits

The size of storage rings normally means that the nearest position to the x-ray source where components can be placed in a beamline branch is of the order of $l_{min} = 10$ m. Because the synchrotron radiation emitted from the insertion device or bending magnet has a finite size and divergence, as already discussed in Chapter 3, it will have a lateral size at this minimum distance of

$$l_{h,v} = l_{min}\,\theta_{h,v}, \tag{5.1}$$

where $\theta_{h,v}$ is the full-width at half-maximum (FWHM) opening angle of the source in the horizontal (vertical) direction, given in radians. We have seen in Chapter 3 that $\theta_{h,v} = 2.355\sigma'_{h,v}$ is typically of the order of 20 μrad (horizontal) and 5 μrad (vertical) for undulators at DLSRs. Hence, the typical size of undulator radiation at a distance of 10 m is 2 mm (H) by 0.5 mm (V).

The spectrum of synchrotron radiation, outside the central cone, peaks at lower energies (see, for example, Figure 3.11), hence the outer regions of the radiation are generally more strongly absorbed by matter (see Figures 2.23 and 3.12), a phenomenon known as 'beam hardening.' It is therefore important in hard x-ray beamlines to define the beam using a fixed aperture at the front end, not just for purposes of collimation, but also to suppress the lower-energy part of the photon spectrum incident on the beamline components, which could otherwise be destroyed. Because the power that needs to be absorbed can be considerable (measured in several kW cm^{-2}), the aperture is often in the form of a water-cooled rectangular funnel, in order to dissipate the power over a large area (see Figure 5.3).

The opening angle of bending-magnet sources in the horizontal direction is very large and in fact is given by the angle subtended by the bending magnet itself (see Figure 3.23). Here, the need to restrict the beam to define the divergence is even more important.

In addition to the primary aperture, the divergence of the beam can be further reduced in the front end by two pairs of adjustable (and usually water-cooled) slits. These are normally protected from the high-flux heat-load experienced by the beam-defining aperture by one or more filters used to further remove soft x-rays. We describe these now.

5.2.3 Low-energy Filters

The sources at high-photon-energy beamlines nonetheless produce a large flux of soft x-rays, which, below approximately 5 keV, interacts strongly with matter and can thereby damage beamline components. In order to filter out these soft x-rays, one needs to use a low-Z material, which is robust, able to withstand high temperatures, have

(a)

(b)

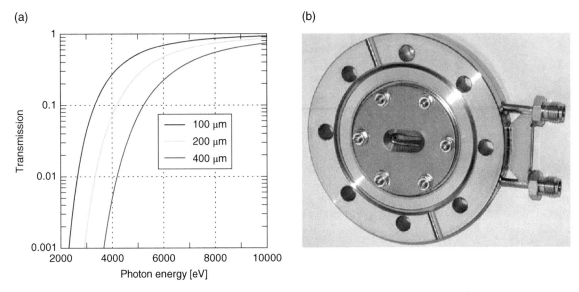

Figure 5.4 High-pass diamond front-end filters used to remove the low-energy component of the x-ray spectrum. (a) The transmission curves for various thicknesses of diamond filters, assuming a density of $3.5\,\mathrm{g\,cm^{-3}}$. (b) A 100 µm-thick, ultra-high-vacuum-compatible, water-cooled diamond window, used to remove soft x-rays and provide isolation between the storage-ring vacuum and the beamline vacuum. Courtesy of Max Kleeb, Paul Scherrer Institute.

good thermal conductivity, and preferably be highly crystalline. The only realistic candidate is carbon, either in the form of graphite or glassy carbon (especially if the filter has to be physically large), or synthetically grown diamond, in the form of thin sheets [see Figure 5.4(a)].

Diamond has the highest known thermal conductivity of any material[2], and is hence ideally suited to conduct away the absorbed power – commonly, such diamond filters are brazed onto a water-cooled copper block for this purpose [Figure 5.4(b)].

After the removal of the soft x-ray portion of the spectrum, other materials can be placed in the beam. So, for example, it is common to use thin beryllium windows[3] to isolate the vacuum of the storage ring from that of the beamline. It also further reduces mechanical distortions (or 'thermal bumps') due to localized thermal loads on the optical components used for focussing and monochromatizing. Caution must be exercised when using beryllium. Firstly, the oxide of beryllium is exceedingly toxic if it enters the bloodstream. Secondly, beryllium components are invariably polycrystalline, producing diffraction rings, while microscopic inhomogeneities and inclusions produce significant and undesired diffuse scattering signal in the forward direction.

5.3   Basics of X-ray Optics

Before we discuss actual beamline components, it is incumbent upon us to review basic elements of optics theory. We begin with the simpler case of ray optics, which suffices for many applications. However, particularly

---

[2] In contrast to other good thermal conductors such as silver, copper, and aluminium, which conduct heat primarily via the kinetic energy of the conducting electrons, diamond conducts heat via quantized vibrations of the lattice, known as phonons. This is particularly efficient in diamond, because of the very high strength of its covalent bonds. The vibrational energy levels lie in a deep parabolic (i.e. symmetric) potential well, where scattering of the phonons, given by *odd* terms in the potential-well description, is exceedingly weak. The phonons therefore move ballistically, resulting in very efficient transport of heat. The thermal conductivity of diamond is, at approximately $20\,\mathrm{W\,cm^{-1}\,K^{-1}}$, some four to five times larger than that of copper.

[3] A 0.5 mm-thick sheet of Be has a transmission of 70% at 5 keV, which rises to 95% at 10 keV.

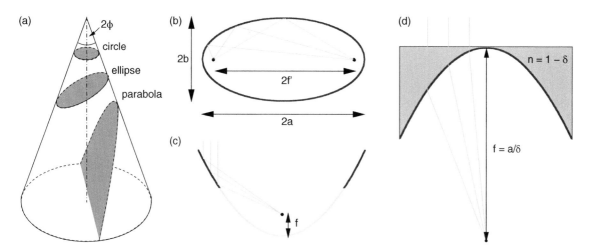

Figure 5.5 Elliptical and parabolic focussing. (a) Three sections through a cylindrical cone. Sections using planes at angles to the cone's symmetry axis greater than the half opening angle ϕ result in closed ellipses. The circle is a special case when the subtended angle is 90°. Parabolas are open curves produced by section planes tilted exactly at ϕ. (b) Light emitted from one of the foci of an ellipsoid (that is, the cylindrically symmetric three-dimensional object formed by rotation of an ellipse around its long axis), given by Equation (5.3), will be refocussed after reflection off the inner walls of the ellipsoid at the other focus. (c) A paraboloid of the form $y = r^2/4f$ aligned coaxially with a parallel beam of light will focus it to a point at f through reflection, assuming, in the case of x-rays, that the incident angles are smaller than α_c, the critical angle for total external reflection. (d) Refractive focussing by transmission through a paraboloid of form $y = r^2/2a$ has a focal length $f \approx a/\delta$, whereby δ is the refractive-index decrement.

with the advent of DLSRs and ever more instances of experiments in which x-rays are focussed down to well below the micron scale, and even down to a few nanometres, diffraction effects must be considered. The description of such phenomena in wave optics and consequent limitations in focussing of x-rays is also briefly described.

5.3.1 Ray Optics

5.3.1.1 Ellipsoids and Paraboloids

The mathematical description of a planar section through a circular cone depends on the angle of the sectioning plane relative to the cone's surface. If the angle between the plane and the cone's symmetry axis is larger than the half opening angle of the cone ϕ [see Figure 5.5(a)], the section will be a closed ellipse. In the special case that the subtended angle is $\pi/2$, the section is circular. If the subtended angle is exactly ϕ, in other words, when the section plane is parallel to a plane tangential to the conic surface, the intersection is described by a parabola. Intersections produced by planes with shallower angles to the symmetry axis than ϕ are hyperbolas [not shown in Figure 5.5(a)].

Ellipsoids and paraboloids (the cylindrically symmetric three-dimensional objects formed by rotation of ellipses and parabolas around their symmetry axis, respectively) are important in ray optics, as they are capable of focussing light to a single point, or focus (ignoring diffraction effects). We begin with the ellipsoid.

The general equation describing an ellipse with a major axis $2a$ in the x-direction and minor axis $2b$ in the y-direction [Figure 5.5(b)] is

$$\frac{x^2}{a^2} + \frac{y^2}{b^2} = 1. \tag{5.2}$$

The two *foci* of an ellipsoid are positioned at $\pm f'$ along the major axis relative to the ellipsoid's centre[4], whereby

$$f' = (a^2 - b^2)^{1/2}. \tag{5.3}$$

These positions have the property that light emanating from one of these foci will be reflected from the ellipsoid walls and focussed perfectly at the other (again, ignoring diffraction effects). The shallowest approach to the ellipsoidal surface is when the rays strike the surface in the central plane between the foci. Here, the incident angle is approximately b/a, assuming $b \ll a$. This is required to be smaller than the critical angle of the material used to fabricate the inner surface of the ellipsoid, that is, $b/a < \alpha_c \sim 10^{-3}$. When the rays strike the surface elsewhere on the ellipsoid, the incident angle increases, and this condition becomes still stricter.

A reflecting paraboloid of the form

$$y = r^2/4f \tag{5.4}$$

will focus any parallel beam of light that is also parallel to the paraboloid's symmetry axis [Figure 5.5(c)] at f, as long as the beam's incident angle on the paraboloidal surface is smaller than the critical angle α_c.

Low-Z materials such as beryllium can also be shaped into planar-paraboloids [see Figure 5.5(d)] to act as refractive lenses. The focal length of a single paraboloid lens with the form $y = r^2/2a$ is

$$f = \frac{a}{\delta}, \tag{5.5}$$

where δ is the refractive-index decrement. Thus, a single Be lens with $a = 100$ μm would have a focal length for 10 keV photons of 29.35 m.

Note that although paraboloids focus rays that are parallel to the symmetry axis perfectly (within the limits set by diffraction), they do a poor job of focussing nonparallel rays.

5.3.1.2   *Source Demagnification*

Demagnification is simply achieved by placing the primary beamline optics closer to the sample than to the source. In general the magnification factor is equal to

$$M = \frac{q}{p} = \frac{1}{D}, \tag{5.6}$$

where p is the source–optics distance, q is the optics–focus distance, and D is the *de*magnification factor (Figure 5.6).

5.3.2   **Spherical Surfaces and Aberrations**

Despite many recent advances in fabrication technology and resulting reductions in costs, ellipsoidal and paraboloidal surfaces remain a challenge to manufacture. Spherical and cylindrical surfaces are far easier to grind and polish, although they focus less precisely. Many experiments can tolerate these aberrations and consequently use x-ray optics with spherical or cylindrical symmetry. In general, the smaller the footprint of the x-rays on the spherical or cylindrical surface, the smaller will be the aberration.

The position of focussing mirrors with respect to the x-ray source and sample decides the magnification, the divergence, and the degree of spherical aberration.

---

[4] The position of the foci, $\pm f'$ is labelled with a prime to distinguish it from the focal length f.

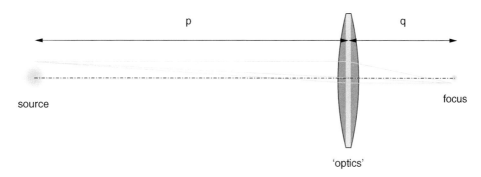

Figure 5.6   The magnification factor of the x-ray source size is equal to q/p, which can also be expressed as a *de*magnification factor of p/q. Note that the lens sketched here generically representing 'optics' would in fact be a *concave* structure if used to converge x-rays.

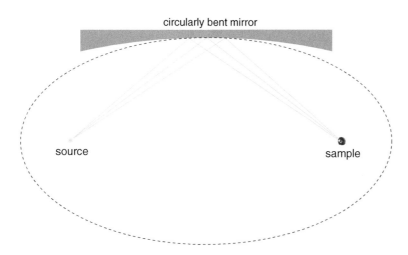

Figure 5.7   By placing the beamline optics at a distance approximately in between the source and sample, any cylindrically or toroidally bent optical element most closely approximates an ideal elliptical surface, and also has the longest radius of curvature.

Wiggler and bending-magnet radiation normally has a relatively large horizontal divergence and therefore a large cross-section/footprint at the primary optics. In order to minimize the radius of curvature of any optical components that need to be bent in order to collimate or focus, they should be as far as possible be positioned centrally between the source and sample (see Figure 5.7). This has the added advantage that this so-called 'one-to-one' configuration is optimal for approximating ideal elliptical focussing surfaces by cylindrical ones, thereby minimizing spherical aberrations. Assuming near-perfect optics, the size of the focus should then be equal to that of the source.

Focussing undulator radiation is in general more forgiving, due to its intrinsic low divergence and small cross-section. Here, one can more comfortably place the optics away from the central position between source and sample without significantly compromising the focus. The main reason for moving away from the one-to-one configuration is to demagnify the source, and thereby obtain a tighter focus at the sample. This is important if the sample size is smaller than the one-to-one image of the source.

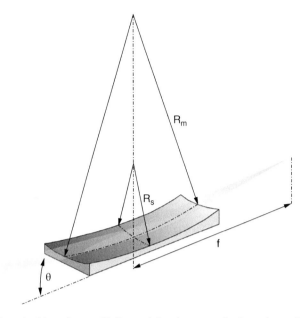

Figure 5.8   The meridional and sagittal bending radii R_m and R_s of a curved mirror focussing a parallel x-ray beam both vertically and horizontally to a focal point f.

5.3.2.1   *Mirror Focal Lengths – the Coddington Equations*

For a glancing incidence angle θ, the 'meridional' (vertical) and 'sagittal' (horizontal) focal lengths of a toroidally curved mirror of large radius R_m and small radius R_s (shown schematically in Figure 5.8) are given by

$$f_m = \frac{R_m \sin\theta}{2}, \tag{5.7}$$

$$f_s = \frac{R_s}{2\sin\theta}. \tag{5.8}$$

For a source–mirror distance p and a mirror–image distance q, the standard 'lensmaker' equation

$$\frac{1}{f} = \frac{1}{p} + \frac{1}{q} = \frac{p+q}{pq} \tag{5.9}$$

yields the Coddington equations for meridional and sagittal focussing

$$R_m = \frac{2}{\sin\theta}\left(\frac{pq}{p+q}\right) = \frac{2f}{\sin\theta}, \tag{5.10}$$

$$R_s = 2\sin\theta\left(\frac{pq}{p+q}\right) = 2f\sin\theta. \tag{5.11}$$

Remember that θ is, in general, very small for x-rays, as it is limited by α_c, the critical angle for total external reflection. For $\theta = 0.15°$ (≈ 2.5 mrad), the meridional radius R_m is θ^2 times larger ($\sim 10^5$) than R_s and may be of the order of several kilometres, while R_s is typically only a few centimetres. Note also that if either p or q is infinite (corresponding, respectively, to a parallel incoming beam and a parallel outgoing beam), then the fraction $pq/(p + q)$ reduces to q and p, respectively, and we regain the mirror focal-length equations, given above. Finally, the subtended meridional half-angle of the mirror is $l_{\mathrm{mirror}}/2R_m$, where $l_{\mathrm{mirror}} \sim 50$ cm is the length of the mirror. This is therefore of the order of 50 μrad, some two orders of magnitude smaller than typical critical angles for total external reflection. Hence, one can ignore any variations in reflectivity over the length of a curved mirror on which a parallel beam is incident.

Optical elements that exploit refraction or diffraction will suffer from *chromatic* aberrations, that is, a dependence of their focal length on photon energy. These are discussed quantitatively in Section 5.5.

5.3.3 Wave Optics

5.3.3.1 The Airy Pattern and the Abbe Limit

We have seen in Section 5.3.1 that when assuming geometrical, or ray, optics, it is possible to mathematically construct an optical element that perfectly refocusses a point source to a point image, such as in the case of the source and image lying at the foci of a reflecting ellipsoid surface. This description ignores the wave nature of light and its interaction (diffraction, or scattering) with the optical elements it passes through or is reflected off. There is, therefore, a so-called 'diffraction limit' which arises because of this interaction, which results, for a given wavelength, in a lower limit to the resolution with which any object can be imaged.

Consider a perfectly monochromatic, parallel, and uniform plane wave of light passing through a circular aperture of diameter $D = 2a$. The transmitted pattern will appear to be a well-defined disc with the same dimensions on a detector, if this is placed immediately behind the aperture. However, as the detector is moved further downstream, the edges of the disc become modulated, until, at large distances, a pattern is recorded that is in fact the square of the Fourier transform of the aperture, called an 'Airy pattern', after George Airy, the first person to explain the phenomenon mathematically [see Figure 5.9(a)]. A 'large distance' is one defined as being larger than a^2/λ, for which one is said to be in the 'Fraunhofer', or 'far-field-diffraction' regime (see also Figure 8.14).

The radial profile of an Airy pattern follows the form

$$I(\theta) = I_0 \left[\frac{2J_1(ka \sin \theta)}{ka \sin \theta} \right]^2, \tag{5.12}$$

whereby $k = 2\pi/\lambda$, θ is the angle with respect to the symmetry axis, and J_1 is the Bessel function of the first kind, and is plotted in Figure 5.9(b). This has its first minimum at $x = ka \sin \theta = 3.8317$, or, in other words, when

$$\sin \theta = (3.8317/\pi)\frac{\lambda}{D} = 1.22\frac{\lambda}{D}. \tag{5.13}$$

The circular area bounded by this first minimum is referred to as the 'Airy disk'. As an example, 2 nm (620 eV) radiation passing through a 100 μm diameter hole would generate an Airy pattern that could be recorded at distances greater than 1.25 m from the aperture, for which the central Airy disk would subtend an angle of $\pm 0.0014°$.

Now consider a situation in which a perfect, aberration-free, converging lens is placed directly behind the aperture [Figure 5.9(c)]. Before we discover what the consequences of this are, let us ask the question 'What is the function of a converging lens?' In everyday parlance, a lens collects the scattered light off an object and refocuses it to produce an image of that object (as long as the object is further away from the lens than the lens' focal length). However, without the lens, remember, the scattered light from the aperture generates (at sufficiently large distances) the Airy pattern, the squared Fourier transform of the aperture. We therefore conclude that *the function of a converging lens is to perform an inverse-Fourier transform of the scattered light from an object to regain an image of that object*. What a clever device a lens is!

The action of inserting the lens directly behind the circular aperture is, therefore, to produce the Airy pattern, but now not at large distances, but instead at the focal plane of the lens [Figure 5.9(c)]. There is no angular

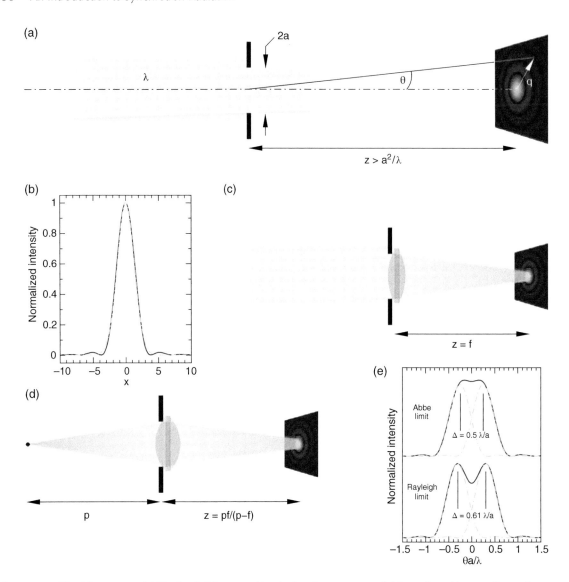

Figure 5.9   The Airy pattern. (a) The far-field pattern (that is, for distances $z > a^2/\lambda$) produced by a uniform plane wave passing through an aperture of diameter $D = 2a$ is the Airy pattern, which is the square of the Fourier transform of the aperture, and given by Equation (5.12). This is plotted as the solid blue curve in (b), whereby the abscissa $x = ka \sin \theta$. The first minimum corresponds to $x = \pm 3.8317$, for which $\sin \theta \approx \theta = 0.61\lambda/a$. This minimum defines the boundary of the so-called 'Airy disk'. The best-fit Gaussian to the Airy function, given by Equation (5.16), is also plotted as a dot-dashed curve. (c) The same Airy pattern can be generated at the focal plane of a converging lens placed directly behind the aperture. The observation angle θ remains the same as in the lensless case. (d) As a result of diffraction, a point source cannot be imaged precisely but will have a point-spread function equal to the Airy function in the focal plane. (e) The ultimate spatial resolution of a lens is thus determined by the Airy pattern. Abbe defined this as being equal to the angular separation between two points to be $\Delta = \lambda/2a$, resulting in two Airy patterns being separated by their FWHM; while Rayleigh defined it as being at $\Delta = 0.61\lambda/a$, that is, the Airy patterns being separated by the Airy-disk radius.

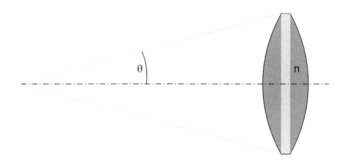

Figure 5.10   The numerical aperture of an optical element is given by $n \sin\theta$, where n is the refractive index.

dependence of the Airy pattern [Equation (5.12)] on the lens's focal length, although its lateral extent in the focal plane is proportional to the focal length.

The numerical aperture, or NA, is a dimensionless number describing the acceptance angle of a lens or other optical system. It is given by (Figure 5.10)

$$NA = n \sin\theta, \tag{5.14}$$

where n is the refractive index, which for x-rays can be approximated to be unity, and θ is the half-angle of the maximum cone of radiation the optical element can capture. If, for example, the optical element is a converging lens, then the maximum value of θ is determined by the lens's radius a and the focal length f. Assuming, as is always the case when dealing with x-rays, that $f \gg a$, then $\tan\theta = a/f \approx \sin\theta \approx \theta = NA$.

Returning to the condition for the first minimum in the Airy pattern, the corresponding radius on the image plane is $q = f \sin\theta \approx f\theta$. Substituting this and Equation (5.14) into Equation (5.13) leads to

$$q = \frac{0.61\lambda}{NA}. \tag{5.15}$$

Finally, consider a lens imaging a point source [Figure 5.9(d)]. Once again, the image will be the Airy pattern, which by definition is necessarily larger than the source. Any two objects, no matter how small, that subtend a half-angle with a focussing lens smaller than $\approx 0.5\lambda/a$ cannot be resolved, because of diffraction effects [Figure 5.9(e)] This is called the Abbe limit [2].

Synchrotron radiation can often be approximated as having a Gaussian profile rather than one of a flat-top uniform intensity. The Fourier transform of a Gaussian (and thus its far-field scattering pattern) is another Gaussian. The Gaussian profile that best approximates an Airy pattern is given by

$$I(\theta) = I_0 \exp[-(ka \sin\theta)^2/3.482], \tag{5.16}$$

for which the FWHM is equal to $2(3.482 \ln 2)^{1/2} = 3.107$, or 41% of the Airy-disk diameter [Figure 5.9(c)].

5.3.3.2   The Numerical Aperture and Rayleigh Length

From the above, we can conclude that any focus has a finite extent. A Gaussian beam will therefore have a minimum 'waist' diameter $2w_0$ equal in size to that of the Airy disk (Rayleigh) or approximately 82% of this (Abbe), depending on your definition (see Figure 5.11). The Rayleigh length z_R is the distance along the propagation direction (z) between this waist and where the beam's linear extent is larger by a factor of $\sqrt{2}$ (i.e. where the beam cross-section increases by a factor of two), and is given by

$$z_R = \frac{\pi w_0^2}{\lambda}. \tag{5.17}$$

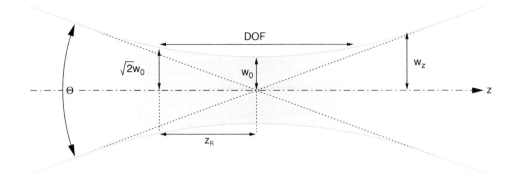

Figure 5.11   The relationship between the FWHM beam divergence $\Theta = (8\ln 2)^{1/2}\sigma'$ the depth of field (DOF $= 2z_R$), and the minimum waist radius w_0 of a focussed Gaussian beam.

The validity of this equation assumes w_0 remains significantly larger than λ, which certainly holds true for hard x-radiation of wavelengths of the order of an angstrom to a nanometre.

The beam radius w_z at a distance z from the beam waist is given by

$$w_z = w_0\sqrt{1 + \left(\frac{z}{z_R}\right)^2},\tag{5.18}$$

from which it can be shown (see Problems) that the full-width half-maximum divergence $\Theta = 2\mathrm{NA}$ is related to w_0 and z_R by

$$\Theta = \frac{2w_0}{z_R}.\tag{5.19}$$

Using Equation (5.17) to substitute for either z_R or w_0, we can also express the beam divergence as

$$\Theta = \left(\frac{2\lambda}{\pi w_0}\right) = \left(\frac{4\lambda}{\pi z_R}\right)^{1/2}.\tag{5.20}$$

Thus, the more focussed is the beam (large Θ), the smaller is the focal spot, and the shorter is the DOF. As an example, a Gaussian beam of 1 Å radiation, focussed using parabolic optics with $\Theta = 10$ mrad, will produce a spot size of 6.4 nm with a DOF of 2.55 μm, whereas a ten times more gentle focus of $\Theta = 1$ mrad can only be focussed to 64 nm, although the DOF will increase a hundredfold to 255 μm. The cost of strong demagnification is thus an accompanying rapid (inverse square) reduction in the depth of field (see Figure 5.12).

Finally, note that we can rearrange Equation (5.20) to obtain $w_0 = \lambda/(\pi\mathrm{NA})$. If we compare this to the FWHM intensity of the Airy disk, and remembering that this is 41% the disk width, we obtain $q = 0.31\lambda/\mathrm{NA}$, which is essentially identical to w_0, the very small difference being due to the fact that the Gaussian fit to the Airy pattern is not perfect.

5.3.3.3   Limits to Magnification

It should be apparent from the preceding section that the geometric relationship describing the magnification factor of an optics setup given by Equation (5.6) breaks down for large demagnification approaching the diffraction limit.

In our description of the Airy pattern, a perfectly monochromatic incident plane wave was assumed. But due to the finite size and nonzero divergence of any source, the transverse coherence length is not infinite, as already discussed in Section 3.7.3. If the size of any x-ray optics used to focus the source onto a sample is larger than the

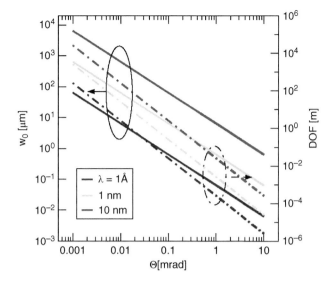

Figure 5.12   Dependence of the beam waist w_0 (left ordinate, solid lines) and depth of field (DOF, right ordinate, dot-dashed lines) on the FWHM beam divergence Θ for different photon wavelengths.

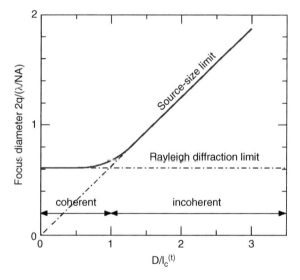

Figure 5.13   Transition of focussing the source from normal demagnification when the focussing element is larger than the transverse coherence length and thus collects incoherent radiation, to the diffraction limit, when the optics are smaller than $l_c^{(t)}$.

transverse coherence length, the normal magnification condition given by Equation (5.6) applies. If, however, the lens is entirely bathed in coherent light, that is, if $D < l_c^{(t)}$, the focus will be limited by diffraction effects to the size of the Airy disc (Figure 5.13). Importantly, DLSR insertion-device sources have transverse coherent lengths in the horizontal planes an order of magnitude larger than those associated with third-generation facilities, which thus allow proportionally more coherent photons within the diffraction-limited spot size.

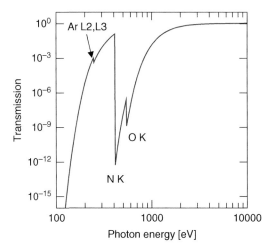

Figure 5.14　The transmission of x-rays through 1 cm of air at room temperature and pressure. The composition of air was taken to be 78.5% N_2, 21% O_2, 0.03% CO_2, and 0.94% Ar.

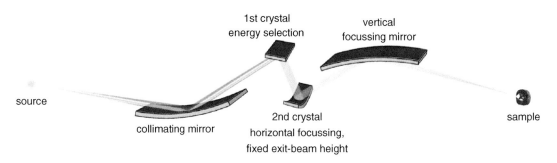

Figure 5.15　Schematic of an x-ray optical system, consisting of a double-crystal monochromator and two mirrors. Note that the second crystal of the DCM can also be bent to provide horizontal focussing.

5.4　Primary Optics

Optical components for x-rays are quite different than those used for visible light. The primary reason is that, as we have seen in Chapter 2, the refractive index of x-rays is very close to *and slightly below* unity. In addition, soft x-rays are very strongly absorbed by condensed matter, and can even be strongly attenuated by gas (e.g. air at one atmosphere, see Figure 5.14).

This makes it difficult to bend or redirect x-rays. The phenomena of total external reflection, refraction, and diffraction are all used in x-ray optics to achieve these ends.

An x-ray optical system consists normally of a monochromator and one or two x-ray mirrors, plus sundry slits and filters. It performs up to three tasks; (a) photon energy selection using the monochromator; (b) optimization of the energy resolution using a collimating mirror upstream of the monochromator; (c) refocussing using either curved mirrors or gratings, or via bending one of the dispersing elements within the monochromator. An example of one possible configuration is shown schematically in Figure 5.15.

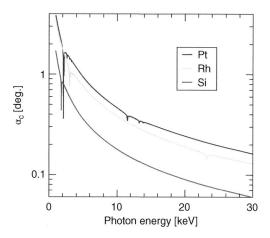

Figure 5.16 A plot of the critical angle for total external reflection as a function of energy for silicon, rhodium, and platinum. Note the dips at absorption edges. At these energies, the effective electron density drops due to f' making $f^0(Q)$ smaller through destructive interference.

5.4.1 X-ray Mirrors

As we have already hinted, x-ray mirrors are in general very different beasts than the piece of aluminium-coated glass we peer into in the bathroom in the morning. Perhaps the most obvious difference, already discussed in Chapter 2, is that the reflectivity of x-rays drops catastrophically above the critical angle. Using Equation (2.28) and known values for δ, the refractive-index decrement, a plot of the critical angle as a function of energy is given for three commonly used elements for x-ray mirrors in Figure 5.16. A useful rule of thumb relating the maximum efficiently reflected incident grazing angle to the photon energy is

$$\alpha_{\max}[\text{degrees}](= \alpha_c) = \frac{\sqrt{Z}}{30}\lambda[\text{Å}]. \tag{5.21}$$

This approximation derives from Equations (2.24) and (2.28) and the linear fit shown in Figure 2.17. Hence, although ultraviolet photons at 50 eV (≈ 250 Å) are efficiently reflected up to a few tens of degrees incident angle, hard x-rays at around 1 Å can only be reflected at grazing angles of around 0.2° or smaller.

This highlights a major technical obstacle to producing high-quality reflective optics for hard x-rays. The length of the mirror that is illuminated by a grazing-incidence beam is h/α, where h is the beam height perpendicular to the mirror surface, and the incident angle α is given in radians. Hence, for a typical beam height of 1 mm and an incident angle of 0.2°, the mirror length must be at least 290 mm. The atomic roughness of the mirror over distances of the order of the longitudinal coherence length (approximately 1 μm) cannot exceed a few angstroms for the mirror to be usable.

Longer-range 'wobbles' in the flatness, given by the figure of merit known as the 'slope error', should be of the order of 1 μrad, or often substantially smaller. From Figure 5.17(a), one sees that Δ_{se}, the peak-to-valley height of the wobbles, is related to their typical angular slope error α_{se} and periodicity D_{se} by

$$\Delta_{se} = \frac{\alpha_{se}D_{se}}{2}.$$

The associated phase shift, or phase error, of those parts of the beam reflected off these features depends on the incident angle θ and is given by

$$\phi_{se} = \frac{2\pi}{\lambda}D_{se}\alpha_{se}\theta = \frac{4\pi}{\lambda}\Delta_{se}\theta. \tag{5.22}$$

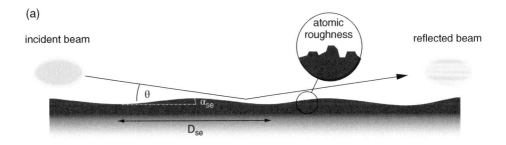

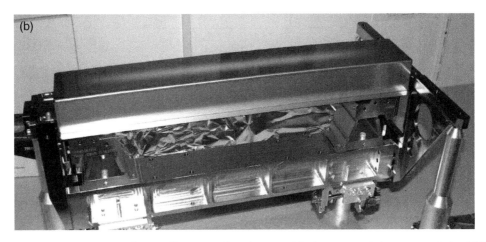

Figure 5.17   X-ray mirrors. (a) A reflecting flat x-ray mirror has a residual mesoscopic/macroscopic 'wobbliness' to it, referred to as the slope error, which causes the beam profile to become more irregular. In addition, the roughness on an atomic scale must not exceed a few angstroms. (b) A bendable silicon mirror at the Materials Science beamline. The usable length is 400 mm. The footprint of the grazing-incidence beam is shown as the narrow yellow ellipse. Courtesy Dominik Meister, Paul Scherrer Institute.

So, for example, in order to obtain phase errors smaller than π (stripes showing complete destructive interference), a mirror reflecting 1 Å radiation with an incident angle of 5 mrad must have $\Delta_{se} < 5 \times 10^{-9}$ m (5 nm). This corresponds to slope errors $\alpha_{se} < 10^{-6}$ for periodicities of $D_{se} \sim 1$ cm.

Small slope errors are thus essential in order to preserve the fidelity of the beam profile after reflection. This presents a formidable technological challenge, and mirrors can be the major contributor to an increase in the apparent source size and consequently lower effective brilliance of a beamline. The emergence of XFELs and DLSRs has driven a corresponding demand in improvements in mirror specifications to match these developments in source quality.

Mirrors can be flexed in order to provide dynamic focussing. Remember that meridional radii are typically measured in kilometres, hence the angle that needs to be subtended by the curved surface of a bent mirror is only of the order of 1/5000 radians, easily achieved using flexor systems. A more sophisticated setup is the 'bimorph' mirror, formed by several active layers of piezoelectric material (either sandwiched in between the substrate and reflecting surface, or, more recently, bonded to the sides of the substrate), allowing for fine local adjustments of the mirror profile on a smaller scale than the full mirror extent [3].

As described in Section 5.3.1, the theoretically most perfect reflective focussing element is a two-dimensional paraboloidal mirror, which, ignoring diffraction effects, can focus a parallel beam to a point. In reality, such surfaces are very hard to engineer, especially if one considers how large the surface needs to be. A simpler engineering

(a)

(b)

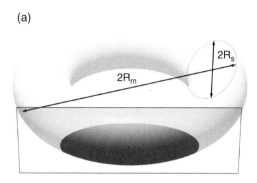
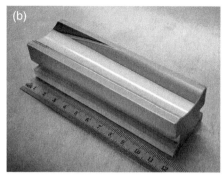

Figure 5.18 Toroidal mirrors. (a) A toroidal mirror is fabricated by grinding and polishing a surface to a radius of R_m in one direction, and to a radius R_s at 90° to this. (b) The toroidal mirror used at the ADRESS beamline of the SLS, for which $R_m = 79$ m and $R_s = 24.1$ mm. Courtesy Uwe Flechsig, Paul Scherrer Institute.

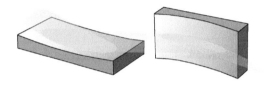

Figure 5.19 By using two cylindrical mirrors in series, vertical and horizontal focussing can be achieved independently in the so-called 'Kirkpatrick–Baez' mirror configuration.

task is to make the mirror toroidal, that is, its shape is like a section of the inner surface of a bicycle tube [see Figure 5.18(a)]. This approximates a paraboloidal surface to a good degree, and the resulting 'spherical aberrations', that lead to a smearing out of the focus, can be kept acceptably small.

A still simpler solution is to divide the jobs of horizontal and vertical focussing between separate optical elements – by using two cylindrical mirrors, horizontal and vertical focussing can be achieved independently (see Figure 5.19). This arrangement is known as a Kirkpatrick–Baez mirror, after its inventors, or, more simply, a K-B mirror [4]. It has the advantages that horizontal and vertical focussing are decoupled from each other, and that, like all focussing via reflection, the focal lengths are independent of the photon energy. A disadvantage is that the total length of the mirror system is effectively doubled and can easily exceed a metre.

5.4.2 Monochromators

Experiments performed at synchrotrons normally require a well-defined photon energy (Laue diffraction, described in Chapter 6, being one notable exception), and may also demand that the photon energy is changed one or more times within a given experiment (as in multiwavelength anomalous dispersion, MAD), or that it is scanned (as, for example, in x-ray absorption spectroscopy, XAS).

In many instances, radiation from insertion devices, even from a low-K undulator, is not sufficiently monochromatic to be used as a source for experiments without further energy dispersion, or monochromatization[5]. This

---

[5] With the advent of DLSRs and short-period undulators, techniques which do not require the best monochromacity, such as tomography or ptychography, can accept a full harmonic bandwidth, which may have a bandwidth approximately two orders of magnitude larger than for Si(111) and thereby increase the incident

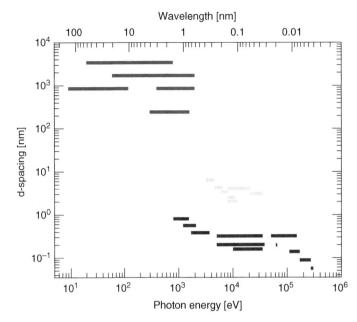

Figure 5.20 Different monochromator types. A selection of grating (red), multilayer (yellow), and crystal (blue) monochromator element periodic spacings *d* and the typical energy range that they serve.

is normally achieved by diffraction. For longer wavelength radiation, manufactured periodic gratings are used, in which the separation between grating lines can be several microns down to approximately 100 nm, depending on the wavelength range of interest. For wavelengths shorter than a few nanometres, one uses instead crystal monochromators, which exploit the 'natural' gratings of crystal planes, with periodicities measured in angstroms, or, alternatively, multilayer monochromators, which let through a larger bandwidth of wavelengths than do crystal monochromators, with a consequent increase in the transmitted number of photons per second[6]. Multilayer monochromators bridge the spacing gap between gratings and crystals, covering the range of a few nanometres (see Figure 5.20).

5.4.2.1 *Grating Monochromators*

Diffraction gratings used for synchrotron radiation are normally reflection gratings, at least for photon energies above the visible region of the electromagnetic spectrum, although a small fraction do utilize the transmitted beam. Gratings consist of a surface covered with a periodic array of many parallel grooves. There are several types of grating profiles and geometries. In particular, the most common profiles are lamellar, sinusoidal, and blaze, shown schematically in Figure 5.21, while the overall geometrical surface of the grating might be planar or concave (spherical, elliptical, or toroidal). Concave gratings are employed in order to combine focussing with dispersion, thereby dispensing with focussing mirrors. It lies beyond the scope of this book to detail all possible configurations, for which the reader is referred to reviews in the literature [5–7]. Here, we will consider one of the most efficient grating types, the so-called 'blaze' grating, in order to discuss the general properties of grating monochromators.

---

photon rate on the sample. Nonetheless, even here, some sort of energy selection is required in order to accept only one of the harmonics and suppress integer multiples of this. Multilayer monochromators are well suited for this purpose.

[6] Here, I have been careful not to state that using multilayer monochromators increases the *flux*, as, although the number of photons per second does indeed increase, the flux is unchanged, as the *bandwidth* is increased by the same factor.

Figure 5.21   Three common grating profiles used to disperse ultraviolet light and soft x-rays.

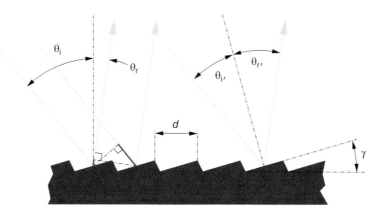

Figure 5.22   Schematic of a blaze grating, showing the groove periodicity d, incident angle θ_i, reflected angle θ_r, and 'blaze' angle γ. The optical path difference between rays reflected by adjacent facets is the difference of the lines shown in red and is given by Equation (5.23).

A schematic of a blaze grating and the relevant angles and parameters is shown in Figure 5.22. The optical path difference between reflections from adjacent facets is

$$d(\sin \theta_i - \sin \theta_r),\tag{5.23}$$

whereby d is the grating period (or 'pitch'), and θ_i and θ_r are the incident and reflected angles, respectively, relative to the *average* grating surface[7]. For constructive interference, Equation (5.23) must equal an integer number of wavelengths, $m\lambda$. The intensity of the monochromatized light will be highest when both Equation (5.23) is satisfied, and its direction is equal to that of light specularly reflected from the individual facets, i.e. when $\theta_{i'} = \theta_{r'}$. For a given incident angle θ_i and order m, this double condition is satisfied for only one wavelength, known as the 'blaze' wavelength of the grating.

The bandwidth of wavelengths that satisfies the diffraction condition depends on how many facets are illuminated by the incoming beam. Consider the Argand (complex-plane) diagram of Figure 5.23 (see also Appendix D). It should be immediately clear that the more facets are involved, the sharper is the drop-off from constructive interference – for any given phase difference between adjacent vectors, the change in the (squared) amplitude for the four facets in Figure 5.23 is more gradual than that involving eight facets. In mathematical terms, it can be shown that the resolving power, defined as the ratio of the wavelength λ to the bandwidth of the interference maxima $\Delta\lambda$, is given by

$$\frac{\lambda}{\Delta \lambda} = mN_f(\sin \theta_i + \sin \theta_r),\tag{5.24}$$

---

[7] Equation (5.23) is sometimes expressed as $d(\sin \theta_i + \sin \theta_r)$, whereby the sign of the angles is positive for a clockwise rotation relative to the grating normal, and negative for an anticlockwise rotation. In the above, the absolute values are assumed.

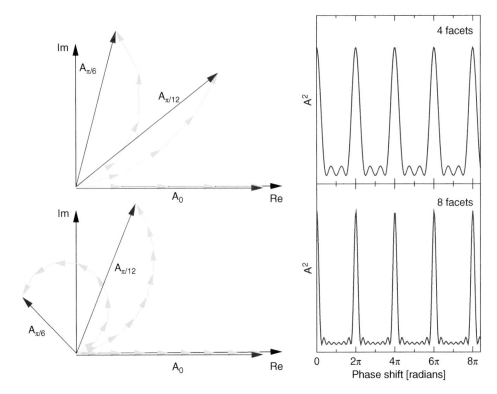

Figure 5.23 Interference between grating facets. Left: Argand diagram showing how the amplitude changes with phase shift between adjacent facets. The vector sum of reflections off 4 facets (top) and 8 facets (bottom) are shown for phase shifts of 0, $\pi/12$, and $\pi/6$. Right: the change in intensity ($= A^2$) as a function of phase shift, ignoring the drop off in intensity away from specular reflection off the facets.

whereby N_f is the number of (equally illuminated) facets. N_f can be of the order of several thousand, hence a resolving power of the order of 10^4 is readily achievable. Gratings with lower numbers of lines per mm (the units in which this property is normally expressed) have a larger bandwidth and therefore provide more flux, while gratings with higher line densities provide better resolution, though at the cost of the number of photons available for experimentation.

5.4.2.2   Crystal Monochromators

As we have already intimated, the size of the separation of scatterers in diffraction phenomena needs to be comparable to the wavelengths that are being diffracted. Hence, crystalline solids can provide the regular array of scattering centres for hard x-rays available at synchrotron facilities, with wavelengths typically between 0.05 and 10 Å.

A crystal monochromator uses Bragg's law

$$\lambda = 2d_{hkl} \sin \theta \tag{5.25}$$

to filter out a narrow band of radiation. d_{hkl} is the interplanar spacing between the (*hkl*) crystallographic planes, and θ is the angle of incidence on those planes. We will discuss the physics behind Bragg's law in more detail in Chapter 6. For the meantime, it is enough to know that Bragg's law predicts that if one were to have a perfect, non-absorbing, and very weakly scattering crystal of infinite depth, and a perfectly collimated incoming polychromatic

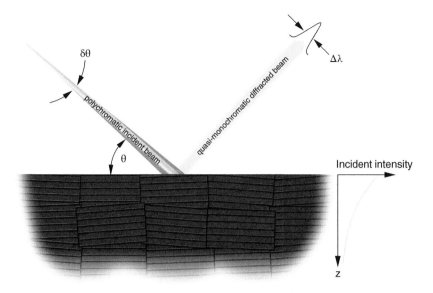

Figure 5.24 Factors effecting the bandwidth of radiation diffracted by a crystal include the degree of collimation of the incident beam ($\delta\theta$), the crystal perfection ('mosaicity'), and the depth to which x-rays penetrate the crystal (the extinction and absorption depths).

beam, the bandwidth of the selected radiation would be infinitely narrow. However, residual beam divergence, slightly misaligned crystal domains making up the (imperfect) single crystal (referred to as the 'mosaicity'), and finite absorption and extinction depths[8] all contribute to a given crystal always having a bandwidth $\Delta\lambda$ greater than zero (see Figure 5.24).

The choice of crystal is dictated by the available crystal quality, the ability of the crystal to dissipate the thermal load associated with monochromatization, its intrinsic minimum bandwidth [see Equation (6.21)], and its resistance to radiation damage (see Figure 5.25). By far the most commonly used material is silicon, and to a lesser extent also germanium. Silicon and germanium have reasonable thermal conductivities (though germanium's is only approximately 40% that of silicon), and can be efficiently cooled by liquid nitrogen, in order to minimize the mechanical strain induced by a local thermal 'bump' where the incoming, polychromatic, beam impinges on the crystal surface. Importantly, they can be produced as near-perfect single crystals with linear dimensions up to tens of centimetres, thanks to their use in the semiconductor industry, where large and exceedingly perfect wafers are produced *en masse*. Despite its high scattering efficiency (or structure factor, see Chapter 6), germanium is only used at very hard x-ray beamlines which offer radiation above approximately 16 keV, due to germanium's K-absorption edge at 11.1 keV (see Figure 5.26).

In the last decade, it has become possible to synthesize plates of diamond of high crystalline quality. Its very high thermal conductivity makes diamond particularly attractive for the use in high-flux beamlines, where heat loads can be substantial. The cost of synthetic diamond single crystals of sufficient size means that diamond-based monochromators are still a rarity [8, 9], while its Bragg reflections are over an order of magnitude weaker than those for silicon (see Figure 5.26). The extremely high transient power densities associated with XFELs seem to guarantee, however, the implementation of diamond as the only feasible material for many FEL-optics components.

---

[8] The attenuation length is the depth to which the incident radiation is attenuated to $1/e$ due to photoabsorption, and is equal to $1/\mu$ in Equation (2.40). The extinction depth, on the other hand, is associated with attenuation due to *elastic* scattering of the incident beam by the electrons in the crystal, and becomes markedly smaller when the Bragg condition is met [Equations (5.25) and (6.20)].

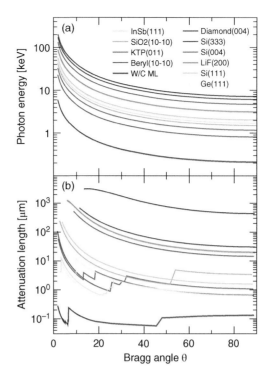

Figure 5.25   A selection of crystal types used in crystal monochromators. (a) The range of photon energies accessible for different crystals, which depends solely on the crystallographic interplanar spacing d_{hkl}. (b) The attenuation lengths perpendicular to the crystal planes of the same selection of crystals. Note the occasional jumps in attenuation length for those crystals that contain heavy elements (in particular InSb, KTP, and beryl), due to the associated photon energy becoming insufficient to access an absorption edge associated with that material. When possible, therefore, it is advisable to operate these crystals at Bragg angles above these features. The interplanar spacings are as follows: W/C multilayer 30 Å; Beryl(10$\bar{1}$0) (chemical composition $Be_3Al_2Si_6O_{18}$) 7.903 Å; KTP(011) (chemical composition $KTiOPO_4$) 5.484 Å; SiO_2(10$\bar{1}$0) (quartz) 4.255 Å; InSb(111) 3.741 Å; Ge(111) 3.267 Å; Si(111) 3.136 Å; LiF(200) 2.015 Å; Si(004) 1.358 Å; Si(333) 1.0452 Å; diamond(004) 0.8918 Å.

The two most common designs for crystal monochromators are the double-crystal monochromator (DCM) and the channel-cut monochromator (CCM).

Figure 5.27 shows a typical geometry for a double-crystal monochromator. The polychromatic light entering the monochromator is diffracted using a highly perfect single crystal. This first crystal absorbs that radiation which does not satisfy the Bragg condition, which normally corresponds to well over 99.9% of the flux. It is therefore actively cooled (using water or liquid nitrogen) and can additionally be dynamically flexed to compensate for the thermal bump caused by the high heat load.

The residual divergence of the radiation entering the DCM, crystalline imperfections in the monochromator crystals, and their natural extinction depth all contribute to there being a band of energies (or wavelengths) that are transmitted by the DCM. This beam is then allowed to impinge on a second crystal, which, because of the much reduced transmitted flux, does not need to be cooled[9]. Horizontal focussing is also possible using the second

---

[9] Note, however, that if there is a temperature difference between the first (cooled) and second (uncooled) crystal, this must be taken into account in calculating the Bragg angle, as the crystals' lattice constants *d* depend on the temperature. It can be shown that, for a monochromator using two identical crystal types and

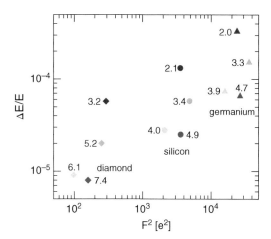

Figure 5.26   The bandwidths $\Delta E/E$ and Bragg intensities F^2 for diamond (diamonds), silicon (circles), and germanium (triangles) for the four most commonly used crystal orientations (111) (blue), (022) (green), (311) (yellow), and (004) (red). The number next to each symbol is the minimum accessible photon energies in keV for that material and crystal orientation, assuming a maximum practical Bragg angle of 70° ($2\theta = 140°$).

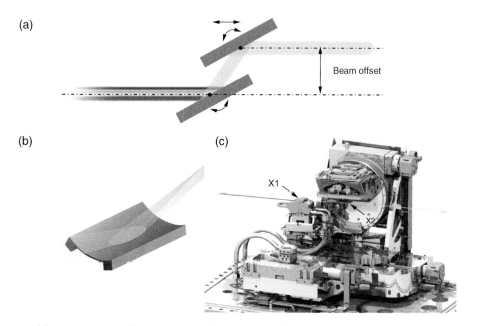

Figure 5.27   Double-crystal monochromators. (a) The geometry of a DCM. The first crystal monochromatizes the incoming polychromatic light, while the second crystal redirects the monochromatized beam parallel to the incoming beam. In order to keep the offset between the incoming and exit beam height constant for all photon energies (and monochromator crystal angles), the horizontal separation between the two crystals must be variable. (b) The second crystal can be dynamically flexed to 'sagittally' focus the beam in the horizontal plane. The bending radius of the crystal depends on the angle of incidence (in other words, the Bragg angle θ) and the desired focal position, and can be calculated using Equation (5.11). (c) A technical rendition of the DCM at the Materials Science beamline, Swiss Light Source. The path is shown of the incident polychromatic beam (shown in red) on the first crystal (X1) and the monochromated beam (in yellow) incident and diffracted off the second crystal (X2).

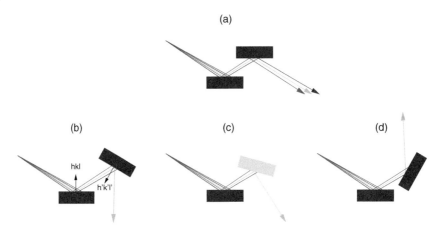

Figure 5.28 Nondispersive (a) and dispersive (b)–(d) modes in DCMs. In case (a), the incident angle for any particular component of the bandwidth is the same for the first and second crystals. If another crystal orientation (b), another crystal type (c), or the same sense of reflection (d) is used for the second crystal, the Bragg condition becomes more stringent – the bandwidth will become narrower and the photon intensity will decrease.

crystal, by bending it sagittally, i.e. with the plane of curvature perpendicular to the plane of the beam propagation and containing the surface normal.

According to the relative geometries, crystallographic orientation, and/or crystal types, the radiation emerging from the monochromator can be more or less monochromatic. Consider Figure 5.28. We assume that the crystal is perfect and that absorption and extinction are negligible. We allow, however, the incoming polychromatic beam to have a divergence of $\delta\theta$ in the plane containing both it and the crystal normal. According to Equation (5.25), the crystal will select longer wavelengths from that part of the beam that impinges more steeply (larger θ) on it than that part of the beam that strikes the crystal at a shallower angle.

In the normal 'nondispersive' configuration, shown in Figure 5.28(a), the monochromacity of the beam is not improved after the second crystal, as long as that second crystal is made from the same crystalline material and has the same orientation of its crystal planes as the first. In this case, the divergence of the beam defined by diffraction at the first crystal is then the same as that after diffraction at the second. There is therefore a gradient of energies across the beam (see Figure 5.29).

However, if there is a desire to further improve the monochromacity, this can be achieved in one of three manners [see Figures 5.28(b)–(d)] – by the second crystal either having a different crystal orientation than the first; by using a crystal of a different material (assuming it has a different lattice constant, which is invariably the case); or by the second crystal deflecting the beam in the same sense as the first crystal (both clockwise or both anticlockwise). Needless to say, all of these dispersive arrangements achieve this improved monochromacity at the expense of the number of photons transmitted per second. This is succinctly explained using so-called 'Du Mond' diagrams [10], outside the scope of this text.

In channel-cut monochromators, the nondispersive geometry of Figure 5.28(a) is obtained by cutting the crystal surfaces out of a single, monolithic crystal (see Figure 5.30). This simple design has the advantage that both crystal surfaces are maintained at the same temperature, and that the diffracting planes from the two surfaces are almost perfectly parallel, resulting in their 'automatic' alignment with each other. Only at high photon energies above around 20 keV can residual strains in the crystal and possible distortions due to the mounting mechanism begin

---

crystal orientations, the difference in their Bragg angles (in radians) is $\delta\theta = (\Delta d/d)\tan\theta$, where $\Delta d/d$ is the fractional difference in the lattice constants of the two crystals, due to thermal expansion.

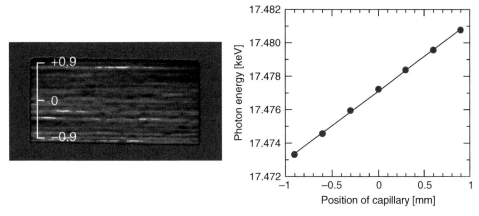

Figure 5.29   Dispersion across a beam emerging from a DCM. Left: image of the unfocussed but monochromatized photon beam at a nominal energy of 17.480 keV at the Materials Science beamline of the SLS. The vertical extent of the beam is approximately 2 mm at the detector. A 100-μm diameter LaB_6 capillary powder sample was placed in this beam. LaB_6 is a calibration-standard material in powder diffraction, with an exceedingly precisely known crystal structure. The capillary was shifted in vertical steps of 300 μm from $y = -0.9$ to $+0.9$ mm and the powder pattern recorded for each step. From this, the photon energy at each position could be determined by fitting the known structure to the powder pattern. Right: plot of the photon-energy fits as a function of the vertical capillary position. The best-fit linear curve is $h\nu = 17477.1 + 4.146y$, where $h\nu$ is in eV and y in mm. Courtesy Nicola Casati, Paul Scherrer Institute.

Figure 5.30   Channel-cut monochromators. (a) A channel-cut monochromator uses a monolithic crystal (normally Si) in which a channel is milled along a crystallographic plane (e.g. the [111]-plane), creating two opposing and parallel faces separated by a distance D. (b) A so-called four-bounce crystal monochromator with a constant beam exit height consists of two CCMs in series. (c) A fixed-exit single channel-cut monochromator, in which the channel profile has been tailored in such a manner that the exit height remains constant. Two angular positions are shown: $\theta = 22.5°$ and $\theta = 45°$. The parallel lines denote the crystal planes.

to cause the Bragg condition for the second crystal surface to drift outside the energy width of the beam produced by the first crystal face.

The 'beam offset' is the lateral shift in the x-ray beam between upstream of, and downstream from, the monochromator [see Figure 5.27(a)]. A beam offset of several millimetres to a few tens of millimetres is required, due to the problem of Bremsstrahlung radiation: there is a cone of Bremsstrahlung radiation emanating from the storage ring at the height of the incoming beam, caused by collisions of the electrons with residual gas particles in the ring. This Bremsstrahlung contains gamma-ray photons with exceedingly high energies, indeed up to that of the electrons in the storage ring (measured in GeV!). If not blocked, this presents a serious health hazard, and will also produce a significant amount of background signal. In order to overcome this, a block of tungsten several centimetres thick is placed in the path of these gamma rays, while an aperture in this block at the height of the outgoing beam allows through the energy-filtered synchrotron radiation.

In the case of CCMs constructed by milling a parallel channel in the monolithic single crystal, the beam offset varies with photon energy such that, for a Bragg angle θ and a channel height D, the beam offset is

$$\Delta y = 2D \cos \theta, \tag{5.26}$$

or, in other words, the exit-beam height is not constant for all selected photon energies. In addition, the beam impinging on the second surface of the channel drifts more downstream with increasing energy (decreasing θ) as $D/\tan \theta$. Hence, a large beam offset is associated with a necessarily large channel-cut crystal, especially at higher photon energies.

The problem of a variable exit beam height can be resolved in one of two manners. The first approach is to use two identical CCMs in series, as shown in Figure 5.30(b). Such a configuration is referred to as a 'four-bounce crystal-monochromator' (4BCM). Note, however, that a 4BCM is a dispersive setup (as the x-rays travel from the second to third bounce), which results in lower flux.

The second possibility is to profile the channel surfaces, as shown in Figure 5.30(c). The offset Δy is held constant by varying the channel height continuously to compensate the $\cos \theta$ term in Equation (5.26) [11]. Such profiles are technically challenging to manufacture, but have important advantages in rapid x-ray spectroscopy experiments in which a significant energy range (for example, an EXAFS spectrum, see Section 7.5) should be scanned in the second- to subsecond-range.

5.4.2.3   Multilayer Monochromators

Lastly, so-called multilayers can be used for monochromatizing hard x-rays. Multilayers are produced by growing periodic layers of two sorts of thin-film material, usually by a deposition process called magnetron sputtering. Each period normally consists of a thinner sublayer (thickness t_A) of high-Z material (e.g. tungsten) and a thicker sublayer (thickness t_B) of a low-Z (i.e. high transmission) material, such as carbon. Scattering of the incoming x-rays at the interfaces between the two sublayers results in diffraction maxima occurring in exactly the same manner as in crystals, whereby the 'lattice' spacing is not that between atomic planes in a crystal, but the multilayer period $\Lambda = t_A + t_B$. Bragg's law for multilayers can be expressed as

$$\lambda = 2(\Lambda/m) \sin \theta, \tag{5.27}$$

whereby the integer $m = 1, 2, \ldots$ is the order of the reflection maximum. Multilayers can therefore be thought of, in this context, as 'artificial crystals'.

An example of the reflectivity of a Ru/B_4C multilayer used at the BM5 bending-magnet beamline at the ESRF, Grenoble, is shown in Figure 5.31 [12]. Seventy multilayer periods with $\Lambda = 4.0$ nm were grown on a silicon substrate, and an extra 5.4 nm of B_4C was used to cap the multilayer structure. The thicknesses of the Ru and B_4C sublayers were equal. The simulated reflectivity curve in Figure 5.31(a) assumes an interfacial roughness of 0.3 nm.

The reflectivity curve has several interesting features. First, because there are relatively few scattering planes (at least compared to the typical number of scattering atomic planes in a crystal), multilayer monochromators are ideal for beamlines with less stringent requirements regarding the degree of monochromacity, and where the arrival rate of photons is at a premium, such as in high-speed x-ray tomography (see, for example, Section 8.2.5). The full-width at half-maximum (FWHM) of a multilayer peak (in radians) is

$$\Delta \theta = \frac{\lambda}{m\Lambda}, \tag{5.28}$$

whereby λ is the x-ray wavelength and m is the number of periods in the multilayer. In the example of Figure 5.31, this provides a bandpass of nearly 300 eV for the first maximum at 8010 eV, or $\Delta E/E \approx 3\%$, over 200 times larger than for a Si(111) crystal.

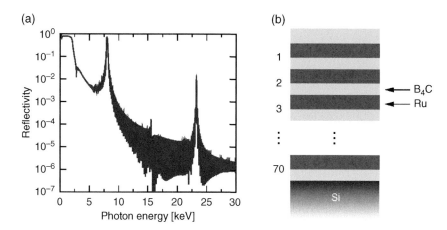

Figure 5.31 Multilayers used in monochromators. (a) The simulated reflectivity curve at an incident angle of 1.15° as a function of photon energy for the Ru/B$_4$C multilayer fabricated for the BM5 beamline at the ESRF. (b) A schematic of the structure [12].

Secondly, because the Ru- and B$_4$C sublayers are equally thick ($t_A = t_B$) interference effects cause the even harmonics to be forbidden. Hence the peak which should otherwise be seen at approximately 16 keV is almost completely suppressed. In more general terms, if the thickness ratio of the 'reflection' (high-density) sublayer t_A and the 'spacer' (low-density) sublayer t_B can be expressed as a simple integer ratio a/b, reflections which are integer multiples of $(a + b)$ are suppressed. This is simply the phenomenon of 'systematic absences', discussed in more detail in Section 6.4.3, for the case here of the one-dimensional 'artificial crystals' that are multilayers.

Lastly, the progression of the positions in energy of the reflectivity maxima deviate from being exactly integer multiples of the first-order peak (here, at 8010 eV) – for example, the third maximum lies at 23 270 eV, 760 eV lower than $3 \times 8010 = 24\,030$ eV. The reasons for this are twofold. First, due to refraction, the wavelength of the x-rays increases marginally inside the multilayer compared to in vacuum (see also Section 2.6.2); secondly, refraction also bends the x-rays in the multilayer compared to the incident angle, and this effect is most pronounced for the shallowest angles. Equation (5.27) is thus an approximation that ignores refraction. Once refraction is considered, Equation (5.27) is modified to

$$\lambda = 2(\Lambda/m)\sin\theta(1 - \kappa\Lambda^2/m^2), \tag{5.29}$$

where κ is a material-dependent constant, but always assumes values close to 2×10^{-3} nm^{-2}. Because Λ is an order of magnitude larger than typical lattice constants of most inorganic crystals, refraction effects thus become more pronounced than in crystal diffraction; and because the correction term $\kappa\Lambda^2/m^2$ is inversely proportional to m^2, it is the first maximum ($m = 1$) which is most affected.

The above two phenomena of systematic absences and shifts in the diffraction maxima in multilayers can both be exploited to suppress higher harmonics, discussed in more detail now.

5.4.3 Higher Harmonics

Referring once more to Equation (5.25) describing the Bragg condition for crystal monochromators, we see that for a fixed $2d\sin\theta$, there is a set of distinct wavelengths λ_1, $\lambda_2 = \lambda_1/2$, $\lambda_3 = \lambda_1/3$ etc., that satisfy the Bragg condition, whereby $m\lambda_m$ is a constant. λ_1 is known as the fundamental, and the others are referred to as higher harmonics.

In general, however, we want only one wavelength (or narrow band of wavelengths), which is normally (though not always) the fundamental. How can we suppress the higher harmonics?

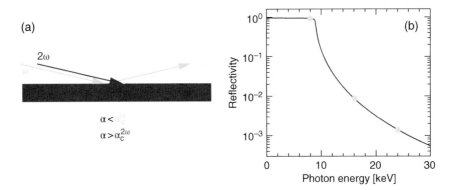

Figure 5.32 Higher-harmonic suppression. (a) Higher harmonic radiation can be suppressed by adjusting α, the angle of incidence of the x-rays on the mirror, so that only the fundamental frequency has a critical angle that is larger than α. (b) The reflectivity of silicon as a function of photon energy at a fixed incident angle of 0.2°. For a fundamental energy of 8 keV, the critical angle is $\alpha_c = 0.227°$, hence 8-keV photons are reflected with close to 100% efficiency. In contrast, the reflectivity of the second- and third harmonics at 16 keV and 24 keV, respectively, is 8.64×10^{-3} and 1.37×10^{-3}.

5.4.3.1   Suppression through Reflectivity

As we saw in Chapter 2, the critical angle for total external reflection increases as the square-root of the electron density of the material from which the x-rays are reflected and increases linearly with λ [Equations (2.24) and (2.28)]. One can therefore use an incident angle on an x-ray mirror which lies below the critical angle for the fundamental wavelength, but which is above the critical angles for all the harmonics (Figure 5.32). As one tunes the monochromator to higher fundamental photon energies, one can lower the incident angle on the mirror so that it remains below the critical angle for that photon energy and reflecting-surface material.

Above a certain photon energy, however, the incident angle will become so shallow that the footprint of the beam significantly spills over the front and back edges of the mirror, resulting in loss of flux. This of course depends on the size of the profile of the beam and on the mirror length.

In order to extend the range of energies that can be used for a given mirror size, one can coat it lengthwise with one or more stripes of different coatings, such as Au, Pt, Rh, or Cr – for a given photon energy, heavy materials with high electron densities have steeper critical angles [Equation (2.28)] – and translate the mirror laterally from one stripe material to another, depending on which photon energy range is desired. So, for example, by changing from an uncoated region of a silicon mirror to one that is coated with Pt, the critical angle can be increased by a factor of $\sqrt{\rho_{Pt}/\rho_{Si}} = 2.7$ (see Figure 2.17).

5.4.3.2   Suppression through Systematic Absences

Higher harmonics are further suppressed by the fact that, in general, the flux at λ/m is usually lower than that at the fundamental λ, and also because, for certain crystals, reflections of some of the higher harmonics are forbidden, due to destructive interference between the scattering atoms within the unit cell of the crystal. Such forbidden Bragg reflections are referred to as 'systematic absences', described in more detail in Section 6.4.3. So, for example, although the Si(111) and Si(333) reflections are allowed, the Si(222) Bragg peak is largely suppressed[10].

---

[10] The (222) reflection of silicon is not *entirely* suppressed, as, in generating the (222) structure factor from the eight individual atomic form factors per unit cell of silicon, it is assumed that the latter are independent of their respective atoms' orientation, that is, the electron density around each atom is perfectly spherically symmetric. Simply by considering the tetrahedral configuration of the bonding electrons in silicon (involving four of the fourteen electrons in total per atom), it should be apparent that the electron-density distribution deviates from spherical symmetry. Consequently, the Si(222) reflection is very weak, but not completely suppressed.

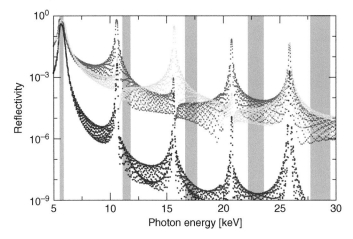

Figure 5.33 Simulated reflectivity curve at a fixed incident angle of 1°, as a function of photon energy for a DMM consisting of two 40-period multilayers, each having a periodicity of 6.9 nm, the first having a ratio of Si to Mo of 2:1 (red curve, with systematic absences for $m = 3$), the second a ratio of 1:1 (yellow curve, with systematic absences for $m = 2$ and 4). The total reflectivity is given by multiplying the two individual reflectivities, and is shown in blue. The green bands are integer multiples of the FWHM of the fundamental (5.55 to 5.90 keV). Note that, due to refraction effects, these do not overlap with the multilayer harmonic for $m \geq 2$, meaning that all higher harmonics of an undulator are suppressed to better than 10^{-6}.

Through a careful choice of the fabrication parameters of multilayers, suppression of undulator harmonics via systematic absences can be especially effectively exploited. As already stated in Section 5.4.2, if the thickness ratio of the 'reflection' (high-density) sublayer t_A and the 'spacer' (low-density) sublayer t_B is chosen so that it can be expressed as a simple integer ratio a/b, reflections which are integer multiples of $(a + b)$ will be systematic absences, that is, they will be suppressed. So, by using two multilayers with the same periodicity Λ in a double multilayer monochromator (DMM), but for which one multilayer has $t_A/t_B = 1$, and the other where $t_A/t_B = 1/2$, the second, third, and fourth harmonics can be effectively suppressed. The first harmonic above $m = 1$ that can in principle be transmitted is thus the fifth harmonic, which in most cases will be anyway weak, and which can also be very effectively suppressed through mirror reflection, or by use of the last suppression method described here, namely detuning.

Lastly, the impact of refraction in multilayers described in Section 5.4.2 and shown in Figure 5.33 means that the optimal angle for the first harmonic is significantly different than those for the higher harmonics.

5.4.3.3   *Suppression through Detuning*

A slight detuning of the Bragg angle of the second monochromator crystal will suppress higher harmonics significantly more efficiently than the fundamental wavelength. This is because the width of the diffraction peak (mh, mk, ml) of the mth harmonic is narrower than that of the fundamental reflection (h, k, l)[11], hence a slight detuning by rotating the monochromator crystal away from the Bragg condition by an amount $\delta\theta$ causes the flux of the higher harmonic to fall away more severely than for the fundamental. For example, the Darwin widths of Si(111) and its third harmonic Si(333) at 10 keV are 26 and 6.6 µrad, respectively.

---

[11] The width of a diffraction peak of a perfect crystal is given by the so-called 'Darwin width', which accounts for dynamical (multiple scattering) effects, the expression for which is given in Equation (6.21), while values for common crystal types and orientations are plotted in Figure 6.16.

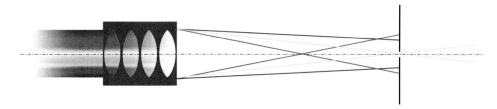

Figure 5.34   In-line monochromatization using CRLs and a pinhole aperture. The aperture should have approximately the same diameter as the Airy disc and be thinner than the depth of field of the focus.

5.4.3.4   Suppression through Refraction

Undulator spectra at DLSRs are considerably cleaner than at third-generation facilities (see Section 3.9.4 and Figure 3.42). Those experiments that are photon hungry and do not demand the narrowest spectral bandwidth could use entire undulator harmonics, which are, in DLSRs, devoid of the lobes on the low-energy flanks of the maxima produced by off-axis constructive interference [Equation (3.73)]. Refraction effects depend on the photon energy, which can be exploited to spatially separate the undulator harmonics at DLSRs. Two possible approaches are sketched here.

The first involves the use of compound refractive lenses (CRLs), described in detail in Section 5.5.1. The quadratic dependence of the focal length of CRLs on photon energy, often cited as a drawback, can in fact be exploited by using them in conjunction with small pinhole apertures as inexpensive and compact in-line monochromators (see Figure 5.34). The harmonic components of the incident 'white' beam are focussed at distances proportional to the square of their energy. A pinhole aperture placed at the harmonic of interest thus effectively cuts out the other components. For the tightest monochromatization, the aperture should have a diameter approximately equal to the Airy disk, or $2w_0$, the beam waist, and a thickness smaller than z_R, the Rayleigh length. Under these conditions, it emerges that the relative photon bandwidth is given by

$$\frac{\Delta E}{E} = \frac{z_R}{f}. \tag{5.30}$$

Such a system is most suited to gentle focussing of soft x-rays, as otherwise the beam waist and Rayleigh length become so small that a suitable aperture is impossible to manufacture. All other harmonics are spatially filtered except for small fractions of divergent radiation that pass through the pinhole, which can be further suppressed by a second, parallelizing CRL stack mirrored across the plane of the pinhole.

The second approach is to pass the white beam through a prism of low-Z material such as beryllium or diamond. Each harmonic is refracted through an angle that depends on its energy.

Undulator harmonic selection through refraction works most effectively for the lower harmonics, where the differences in focal length (CRLs), or refraction angle (prisms [13]) between neighbouring harmonics is largest.

5.4.4   Double-crystal Deflectors

Certain experiments require that the incident x-ray beam impinges on the surface of a liquid, and therefore has to be tilted downwards. In experiments such as x-ray reflectometry (XRR, discussed in Section 6.14), the incident angle α must be varied while keeping the liquid sample stationary. This is achieved using a double-crystal deflector (DCD) [14], which is essentially a DCM using dissimilar crystals, such as the geometries shown in Figures 5.28(b) and (c). Consider Figure 5.35(a). Two crystals X1 and X2 have lattice spacings d_1 and d_2 perpendicular to their optical surfaces. The doubly diffracted beam therefore crosses the axis of the initial beam axis at an angle $\chi = 2(\theta_2 - \theta_1)$, where θ_1 and θ_2 are the Bragg angles for X1 and X2, respectively, at the energy of interest. The entire DCD (i.e. the construction on which the two crystal holders/goniometers are mounted) is allowed to rotate

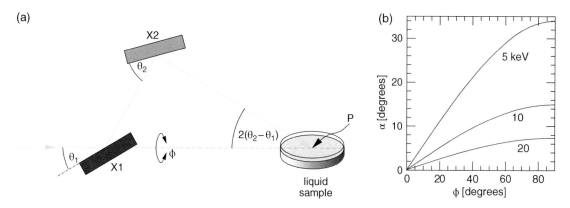

Figure 5.35 Double-crystal deflectors. (a) the Bragg angles θ_1 and θ_2 of two dissimilar crystals X1 and X2 are adjusted to select a certain photon energy. Because $\theta_2 > \theta_1$, the doubly diffracted beam crosses over the incident beam axis at P, where a liquid sample is placed. The incident angle α of the diffracted beam on the liquid surface can be adjusted by rotating the entire DCD around the original incident axis between $\phi = 0$ and 90°. The sample remains entirely stationary. (b) The change in incident angle α for a DCD using Si(111) and Si(220) crystals as a function of ϕ for three photon energies.

around the incident-beam axis, between $\phi = 0°$ and 90°, whereby the lower limit of ϕ is defined as being when the diffracted beams lie in the same plane as the liquid-sample surface, and the upper limit is when they lie in the same plane as the surface normal. In this manner, the diffracted beam will impinge on the sample sitting at the intercept P of the initial axis and the diffracted beam at an angle

$$\alpha = \arcsin(\sin \chi \sin \phi). \qquad (5.31)$$

The change in α as a function of ϕ for different photon energies and using Si(111) and Si(220) crystals is shown in Figure 5.35(b). Note also that γ, the orientation of the footprint of the beam on the sample surface relative to the initial beam direction, is given by

$$\gamma = \arctan(\tan \chi \cos \phi). \qquad (5.32)$$

5.5 Microfocus and Nanofocus Optics

X-ray beams with focal spots below a micron and extending to the few-nanometre regime are increasingly important experimental and characterization tools at synchrotrons and XFELs, and are used in scientific disciplines as diverse as macromolecular crystallography of submicron-sized crystals; scanning spectroscopic methods (in particular absorption and fluorescence, see Chapter 7) in the environmental sciences and studies of historical artefacts; and studies of the electronic properties of domains and domain boundaries using, for example, RIXS. The advent of DLSRs, with their much improved emittance, promises similar gains in focussing capabilities.

Demagnification by a judicious choice of the optics layout at the beamline can only provide focal spots down to the order of 1 µm in the vertical direction. Further demagnification to submicron dimensions normally requires secondary optics due to (a) the difficulty in obtaining a sufficiently large demagnification (i.e. short focal length) from the primary optics and (b) vibrations and lateral drifts in the position of the primary beam. The general approach to achieve submicron spot sizes consists of focussing down the beam upstream of the sample at a pinhole which has a diameter significantly smaller than that of the primary focal spot, and using this as a virtual source for secondary optics (see Figure 5.36). For the tightest focussing, it is important to be able to decouple the position

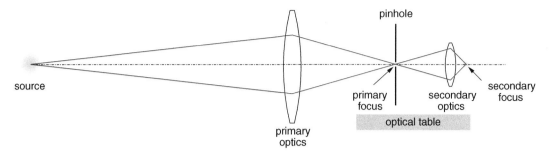

Figure 5.36 Micro- and nanofocus beams are formed by selecting a fraction of the primary focus using a pinhole, and demagnifying this 'virtual source' using secondary optics.

Figure 5.37 Compound refractive lenses. X-ray focussing elements can be constructed from high-transmission material drilled with arrays of holes, which act in much the same way as (very weak) cylindrical convex lenses do for visible light.

of secondary source from fluctuations in the position of the primary focus, caused by drifts and vibrations in the primary optics. This is achieved by mounting the pinhole, secondary optics, and sample (at the secondary focus) on a quasi-vibrationless optical table.

In the following, three of the most common lens types (excepting bendable mirrors and the Kirkpatrick–Baez system, detailed above) are described. They each use different optical phenomena, namely refraction, reflection, and diffraction, to focus x-rays [15].

5.5.1   Compound Refractive Lenses

The fact that the refractive index for x-rays in solid material is generally lower than in vacuum or air can be exploited in the manufacture of x-ray optical elements, either by reflection or refraction. So-called compound refractive lenses (CRLs) can be most easily manufactured by drilling a sequential array of holes into high-transmission material (Be is a popular choice) [16]. Cylindrical focussing is achieved if all the hole axes are coplanar, while quasi-spherical focussing requires that alternate holes are drilled at 90° to one another (see Figure 5.37).

The focal length of CRLs is given by

$$f = R/2n_h\delta, \tag{5.33}$$

where R is the hole radius, n_h is the number of holes in the array $= 2n_s$, the number of curved surfaces, and δ is the refractive index decrement (equal to 2.27×10^{-6} for Be at 12 keV, for example). A Be lens consisting of an array of thirty 200 μm-diameter holes therefore has a focal length of approximately 73 cm at 12 keV.

(a)

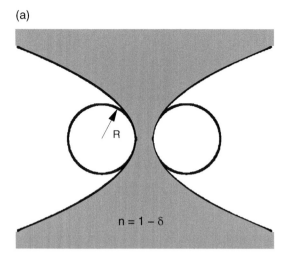

(b)

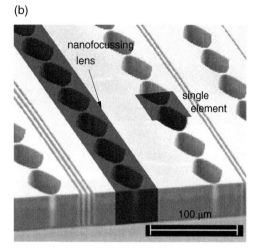

Figure 5.38 Nanofocussing parabolic CRLs. (a) Schematic of a single element in a paraboloid CRL-stack. The paraboloid has a profile $z = r^2/2R$, where r is the radius at a distance z along the symmetry axis from the paraboloid's apex. At the apex, the tightest radius of curvature is R. The divergence of the circle from the parabolic profile with distance from the symmetry axis provides a qualitative indication of the spherical aberrations associated with the former. (b) A scanning electron microscope image of a nanofocussing lens array fabricated by lithographic and ion-etching techniques. Reprinted from [19] with permission of the American Institute of Physics.

CRLs have the advantage that they are easy to align in the x-ray beam, they do not divert the beam axis, and elements can be stacked after one another to change the focal length, such as in 'transfocators' [17]. Their main disadvantages are that (a) the focal length is proportional to the square of the x-ray photon energy [see Equations (2.24) and (5.33)], which means that they must be repositioned, restacked, or replaced if the photon energy is changed; and (b) the transmission is in general fairly poor. For example, a 100-hole CRL made of Be, with 100 μm-thick walls separating adjacent holes, has a transmission of 0.5 at 12 keV.

The weak refraction of hard x-rays in condensed matter requires the radius of curvature R of the individual holes to be small. The spherical approximation to an ideal paraboloid lens holds only for lenses with acceptance apertures for the x-rays (diameter $= 2R_0$) that are much smaller than R. However, in most CRL designs, $2R_0$ is comparable to or larger than R, and the spherical approximation breaks down. For the highest-quality imaging applications, these spherical aberrations become unacceptable, limiting the performance of CRLs that use cylindrical or spherical surfaces.

This problem was overcome in 1999 by the fabrication of a train of lenses with a paraboloidal profile [18], thus enabling high-quality focussing in two directions with one element. The focal length follows Equation (5.33), whereby R is now the radius of curvature of the apex of the parabola, which itself has the form $z = r^2/2R$ [see Figure 5.38(a)].

Arrays of parabolic lenses in one plane have more recently been fabricated using two techniques, electron-beam microlithography (EBM) and reactive-ion etching (RIE) at the Technical University in Dresden [see Figure 5.38(b)]. Using these lenses, formed in a silicon wafer, it was possible to generate a focal spot of approximately 50 nm for 21 keV photons [19].

A further refinement of CRLs is illustrated in Figure 5.39. Here, one of the disadvantages of CRLs is effectively eliminated by decreasing the attenuation of the refracting medium via removal of sections of height Δ, chosen such that the phase difference ϕ between adjacent steps is 2π [20]. This is satisfied when

$$\Delta = m\lambda/\delta, \tag{5.34}$$

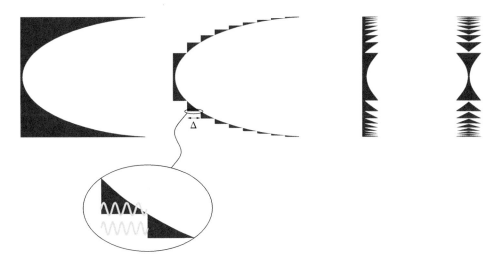

Figure 5.39 Kinoform lenses. A parabolic CRL can be modified by removing sections of depth Δ, thereby reducing the volume of attenuating material that does not actually refract the x-ray beam. The CRL can then be made to be more compact by collapsing all the elements to the same plane. The focussing power can be doubled by mirror imaging the CRL back-to-back. The inset shows that carefully choosing the cutaway depth Δ, the phase difference between rays passing through the refractive material and those passing through the adjacent vacuum is a multiple of 2π.

where m is an integer and δ is the refractive index decrement. For 1 Å radiation, $\delta \approx 3 \times 10^{-6}$ and hence Δ must be an integral multiple of 30 μm. Because δ is itself proportional to the square of λ [Equation (2.24)], the disadvantage of these so-called kinoform lenses is that Δ is inversely proportional to λ and so they function only for discrete wavelengths. Kinoform lenses have been more recently also manufactured out of diamond [21]. Only beryllium has a smaller absorption coefficient than diamond of the commonly used elemental materials (by a factor of approximately two to three across a large photon-energy range), while diamond has the largest refractive-index decrement, δ (see Figure 5.40), meaning the most compact CRLs can be fabricated from it. Importantly, diamond is relatively easy to etch using modern electron-beam lithographic techniques; it is very mechanically robust; has excellent thermal conductivity; and lacks the severe toxicity issues of beryllium.

5.5.2 Tapered Glass Capillaries

The capillary lens, first reported by scientists at Cornell University in 1994 [22], was the first lens type to produce submicron spot sizes; indeed a spatial resolution of 50 nm was achieved. Capillary lenses operate by 'funnelling' the light through a shallow cone by means of mulitiple reflections.

The simplest taper is conical in shape. Consider the conical tube with a half-opening angle α shown in Figure 5.41(a). Parallel rays entering the capillary will impinge on the surface at an angle α and be deflected by an angle 2α. This means that for a linearly tapered cone, the nth reflection impinges on the surface at an angle $(2n - 1)\alpha$, which will continue until this value exceeds α_c, the critical angle for total external reflection, after which, the reflectivity drops precipitously. Beams entering the capillary further from the capillary axis will undergo more reflections and, depending on the geometry, may be lost. Hence, the effective acceptance aperture of the capillary depends on α, α_c, and its length, and can be increased by coating the inner walls of the cone with a high-Z material with a large critical angle [Equation (2.24)].

Tapered capillaries with profiles exhibiting a gradually decreasing angle of the inner surface relative to the central axis with the x-rays propagation distance [Figure 5.41(b)] offer superior focussing properties and increased light-gathering power, on account of the increasing shallowness of the walls as the x-rays travel downstream.

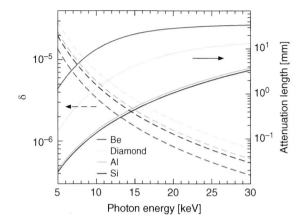

Figure 5.40   The refractive-index decrement (dashed curves) and attenuation lengths (solid curves) of the four most commonly used elements for CRLs.

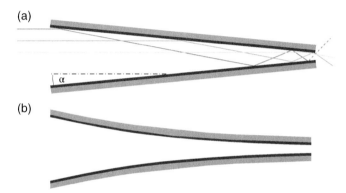

Figure 5.41   Schematics of (a) a conically tapered glass capillary lens and (b) a curved capillary lens. The inner surface of the capillary can be coated with a high-Z material to increase the critical angle for total external reflection. Note in (a) that the outermost (darkest) traced ray impinges on the conical surface at the fourth reflection at an angle that exceeds the critical angle for the inner coating, and is therefore lost.

More recently, single-bounce capillary lenses with a parabolic or elliptical profile, such as shown in Figure 5.5(c), have been developed, which can provide focal spots in the range of a few tens of nanometres with a high numerical aperture (that is, light-gathering power) and an efficiency of over 50% [23].

An advantage of capillary lenses is that, because they operate on the principle of reflection, the focal length is independent of the photon energy, although the light-gathering power decreases with increasing energy, because of the concomitant lowering of the critical angle for total external reflection [Equations (2.24) and (2.28)]. A drawback of these optical elements is that the sample has to be placed very near the exit tip, generally within 20 to 100 times its inside diameter.

5.5.3   Fresnel Zone Plates

Fresnel zone plates (FZPs) focus x-rays using diffraction. A FZP consists of a set of radially symmetric rings, called Fresnel zones, which alternate between being opaque and transparent (see Figure 5.42). Constructive interference

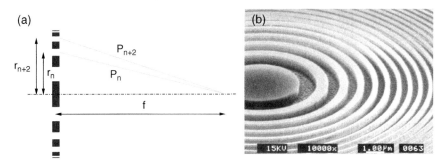

Figure 5.42   Fresnel zone plates. (a) The path difference between adjacent transparent rings, $P_{n+2} - P_n$, in a zone plate should be equal to the wavelength of the x-rays being focussed. (b) An electron microscopy image of a zone plate manufactured using electron-beam microlithography. Courtesy Christian David, Paul Scherrer Institute.

and thereby focussing occurs when the optical path difference between x-rays scattered through adjacent transparent rings is equal to an integer multiple of the wavelength. Using Pythagoras' theorem,

$$\sqrt{f^2 + r_{n+2}^2} - \sqrt{f^2 + r_n^2} = m\lambda. \tag{5.35}$$

Note here that n labels the ring and should not be confused with the refractive index. Let us consider the first-order focal length for which $m = 1$. If we assume that the radius of the largest Fresnel zone, r_N is much smaller than the focal length f, Equation (5.35) is accurately approximated by

$$\frac{r_{n+2}^2 - r_n^2}{2f} = \lambda. \tag{5.36}$$

This condition is met if

$$r_n = \sqrt{n\lambda f}, \tag{5.37}$$

from which it can be deduced that the areas of all the rings are constant and equal to

$$\pi(r_{n+1}^2 - r_n^2) = \pi f \lambda. \tag{5.38}$$

This is an important condition, needed in order to obtain complete constructive interference at the focus, as the integrated amplitude of the diffracted x-rays originating from each zone must remain the same for an evenly illuminated zone plate.

For plates containing many zones (i.e. large N), the focal length can be calculated from Equation (5.38) by remembering that $r_{n+1}^2 - r_n^2 = (r_{n+1} + r_n)(r_{n+1} - r_n)$. For an outermost zone of radius r_N with width Δr_N, the focus is given by

$$f = \frac{2r_N \Delta r_N}{\lambda}. \tag{5.39}$$

From Equations (5.37) and (5.39), it follows that

$$\Delta r_N = \sqrt{\frac{\lambda f}{4N}}. \tag{5.40}$$

Hence, for a given x-ray wavelength and desired focal length, the product of the outermost zone radius and its width is a constant. A consequence of this is that the fabrication of FZPs for hard x-rays has in the past been impractical – for a focal length of, say, 5 cm, a wavelength of 1.5 Å, and $N = 100$, the outermost zone width

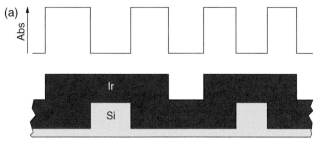 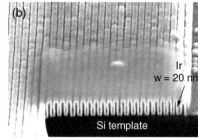

Figure 5.43 Doubling the ultimate resolution of FZPs. (a) FZPs with ultranarrow zone-ring widths can be manufactured by depositing high-Z material (e.g. iridium) very homogeneously on top of a low-Z structured substrate (e.g. Si). (b) The rings of another zone plate were imaged using this novel design in a scanning transmission x-ray microscopy experiment. Features as small as 15 nm could be resolved. Reprinted from [24] with permission of the American Physical Society.

would be smaller than 140 nm. One might therefore naïvely argue that the technological challenge to fabricate such narrow rings could be relaxed by choosing a smaller value for N. However, the ultimate resolution Δl of a zone plate depends on the zone widths according to

$$\frac{\Delta l}{\Delta r_N} = 1.22, \qquad (5.41)$$

thus the smallest sized object that can be imaged using a zone plate is directly proportional to (and very close to, in absolute value) Δr_N.

Advances over the last decade in the fabrication process called electron-beam microlithography, however, have allowed FZPs to be produced with outer rings well under 100 nm, so that hard x-rays in the angstrom range can now be focussed down to just a few tens of nm.

A further improvement, essentially resulting in a doubling of the possible resolution, has been achieved using a simple post-processing of FZPs [24]. Consider Figure 5.43. A zone plate is etched out of silicon by electron-beam lithography such that the opaque (thicker) regions are narrower than they should be, according to Equation (5.37), and their height is small, making them only marginally opaque. This FZP is then homogeneously coated, atomic layer for atomic layer, with a very dense (and therefore x-ray opaque) material, such as iridium to a depth of a few tens of nm. The resulting transmission profile exhibits an effective increase in the number of rings by a factor of two, thereby improving the resolution accordingly. Using such a device, it was possible to resolve features smaller than 10 nm.

The theoretical efficiency of FZPs is $1/\pi^2$, or approximately 10%. 50% is lost in the opaque rings, while the remaining 40% is diffracted into the higher orders. This is one of the most severe limitations of FZPs.

Instead of making FZPs by alternating opaque and transparent rings, one can simply make both transparent, but adjust the thickness of one set such that they retard or advance the phase by π radians relative to the other. Remember that the two sets in a 'standard' FZP anyway have a phase relation of π radians to one another, hence adding (or subtracting) another half-wavelength by refractive effects [the phase velocity of x-rays increases by $1/(1 - \delta)$ in the medium] will bring all ring contributions in phase. These so-called 'phase zone plates' (PZPs) increase the diffraction efficiency, which maximizes for weakly absorbing material when $t = \lambda/2\delta$. An increasingly commonly used material for this is diamond, which has a transmission of over 99% at 8 keV and a thickness of $t = 6$ microns.

The fraction F_{PZP} of the incoming beam intercepted by such a PZP diffracted to the first order is given by

$$F_{\mathrm{PZP}} = [1 + \exp(-2\eta\phi) - \cos\phi \exp(-\eta\phi)]/\pi^2, \qquad (5.42)$$

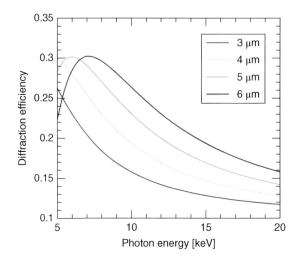

Figure 5.44   The theoretical diffraction efficiencies of the first order ($m = 1$) of diamond PZPs constructed with thick-
-ring heights of 3 to 6 μm, assuming the thin rings have zero thickness. Real PZP diffraction efficiencies must be
multiplied by the transmission spectrum of the (diamond) membrane support, typically of approximately 10 μm thickness. This has only an insignificant impact on the curves in the shown energy range. In practice, imperfections and
aberrations are more important, and real PZPs typically achieve 70 to 80% of the theoretically achievable values.

where $\eta = \beta/\delta$ is the ratio of the absorption index to the refractive-index decrement, $\phi = 2\pi\delta t/\lambda$ is the phase
shift, and t is the height of the thick-ring set responsible for retarding the phases relative to the thin-ring
set. An example of the theoretical efficiency of a diamond FZP constructed in this manner is shown in
Figure 5.44 [25].

Like CRLs, FZPs have the advantage that they are easy to align. In contrast, their focal length is only linearly
proportional to the photon energy [see Equation (5.39)]. In general, they can potentially provide the tightest focus,
although their best transmission (for PZPs) is typically only approximately 30 percent. Another technological
challenge is the very high aspect ratios that must be achieved for the outer rings of PZPs that focus down to a
few nanometres – in the hard x-ray regime, t needs to be a few microns in order to induce the π phase shift. For
a 10 nm focus, Δr_N must be no larger than 8 nm [Equation (5.41)], implying an aspect ratio of 750. Not only
is this very difficult to achieve using lithographic techniques, but also the high aspect ratio precludes a simple
'thin-mask' description of the scattering at the zones. This problem has been essentially overcome with the most
recent focussing elements to have been developed, namely multilayer Laue lenses.

5.5.4   Multilayer Laue Lenses

Multilayer Laue lenses (MLLs) can be thought of as the marriage of Fresnel zone plates and the multilayers used
in multilayer monochromators – wedged reflecting layers are stacked upon one another with the separations given
by the zone-plate conditions [Equations (5.37) and (5.38)] at angles such that rays parallel to the zone axis incident
on the MLL are reflected to the same focal spot at f. This happens when the reflecting-layer surfaces all converge
on a line at $2f$ (see Figure 5.45).

Because MLLs are fabricated by slicing a section from a wedged multilayer grown using the most sophisticated
deposition techniques, such as molecular-beam epitaxy and sputter deposition (which have controlabilities better
than a single atomic layer), exceedingly thin outer-zone layers with very large aspect ratios can be created. A
recent example was reported to be able to focus down to below 10 nm with a high numerical aperture (that is,
light-gathering power) [26].

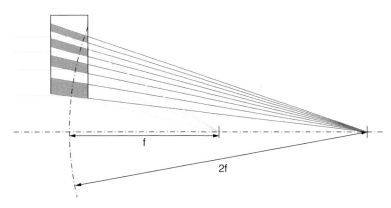

Figure 5.45   The geometrical conditions for MLLs. A second mirror image of the shown multilayer can be used to create a symmetrical lens. MLLs focus in one plane only, hence two crossed MLLs are needed to focus to a spot.

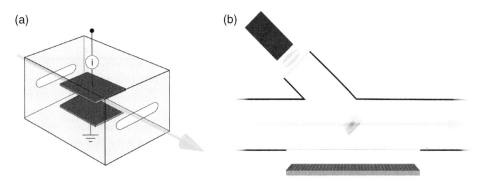

Figure 5.46   Beam-intensity monitors (a) The intensity of the incoming beam in an ionization chamber is proportional to the current induced by photoionization of the gas in the chamber. (b) The photon-counting Mythen microstrip detector provides a strong and very reliable signal over several orders of magnitude by recording scattered x-ray signal off a thin, nitrogen-doped, diamond foil placed in a vacuum tube. The detector is placed below an x-ray-transparent window. Simultaneously, the position and any drift of the x-ray beam can be monitored by observing the visible luminescence induced in the doped diamond foil, using a microscope objective and CCD camera.

5.6   Beam-intensity Monitors

It is important to monitor the incoming x-ray beam, in order to correct the recorded data for any fluctuations in incident beam intensity. The two most critical aspects of all devices for monitoring the incoming beam are that their response is linear with photon flux and that only an insignificant part of the beam is removed.

Although a high-priority goal of the Control Room at a synchrotron facility is to deliver a highly stable beam regarding both intensity and position, these can vary at the end-station of a beamline for a variety of reasons, including gradual electron-current loss in the storage ring and drifts in optics due to thermal or mechanical effects. The need for a reliable beam-intensity monitor is therefore obvious.

The most common form of beam-intensity monitor is the ionization chamber [Figure 5.46(a)]. This operates by filling a vessel containing two electrodes with a gas (typically N_2 or Ar). The vessel is sealed at opposite ends with a transparent material such as kapton foil, through which the direct x-ray beam can pass. One electrode is connected to a high-voltage source, of the order of a few hundred volts. An x-ray photon entering the chamber will be absorbed by the gas with a certain (low) probability. For example, about 0.5% of 16 keV photons are absorbed as they travel through 5 cm of N_2 at one atmosphere pressure. The gas molecule is thereby ionized and the photoelectron and ion accelerate to the anode and cathode, respectively. This process is measured as a current that is proportional to the x-ray flux[12]. The voltage should be adjusted to a value whereby avalanche effects are avoided, which would otherwise lead to nonlinear behaviour. The response of an ionization chamber depends on the photon energy, the gas type in the chamber, the high voltage, and the gain of the electronics [27]. It is therefore important to establish a suitable set of these parameters for which the response is linear over a significant intensity range, typically spanning one or two orders of magnitude.

In some experiments, the response time of standard ionization chambers is insufficient. This is because the drift velocity of the ions in the electric field of the ionization chamber, of the order of a few metres per second, means that the rise time to a step change in intensity is a few hundred microseconds for typical plate separations. This limits their bandwidth to approximately a kilohertz. Recently, a novel adaptation to ionization chambers has been implemented, in which a metal grid is placed between the ionizing radiation and the current-measuring anode [28]. In these so-called 'gridded ionization chambers' (GICs), the grid effectively screens the anode from the positive charge carriers in the ionization region. Only the three-orders-of-magnitude faster electrons can contribute to the signal. In this manner, these GICs exhibit bandwidths of the order of a MHz. GICs are now being implemented in QEXAFS experiments (see Section 7.5.3).

A particularly reliable detector based on a one-dimensional microstrip module, described in more detail in Section 5.7.8, has recently been developed at the Swiss Light Source. The module, containing 1280 elements, is either placed in air under the x-ray beam, or under a transparent window in a vacuum tube in which a thin scattering, nitrogen-doped, diamond foil has been placed [Figure 5.46(b)]. The photon-counting module measures the scattered signal, which typically provides detected photon rates of 10^8 s^{-1}, thus delivering a relative accuracy of one part in ten thousand. The doping in the diamond causes the diamond to luminesce upon irradiation by x-rays. This can be monitored using a microscope/CCD camera, thereby allowing one to simultaneously track drifts of the beam on a scale of a few microns.

There are several other methods of monitoring the direct beam intensity, for example, from the fluorescence or elastically scattered intensity from a nearly transparent foil, or from the photocurrent produced in metallic meshes or ultrathin foils.

5.7 Detectors

In this section, we review common types of detectors used to detect x-ray photons and photoelectrons. We begin, however, with a brief discussion of different sources of signal noise.

5.7.1 Sources of Noise in Detectors

5.7.1.1 *Shot Noise*

Any stochastic or random process composed of a series of discrete events, such as the arrival of x-ray photons on to a detector element, is subject to statistical, or 'Poisson', noise. In Poisson statistics, it is assumed that the occurrence of one event has no influence on the probability of other events occurring and that two events cannot happen at exactly the same instant. If the *average* number of events that occur within a given interval (e.g. the

---

[12] We can make an estimate of the current produced under these conditions as follows: Let us assume a photon rate of 10^{13} s^{-1}. Each absorbed photon produces a 'current' of $2e$, one for the electron, one for the ion. Hence the current is $2e \times 5 \times 10^{-3} \times 10^{13} = 16$ nA.

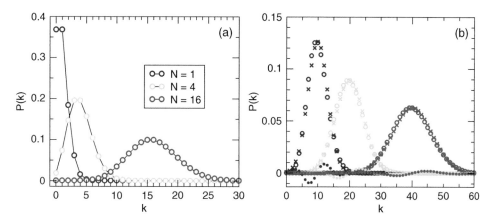

Figure 5.47 Poisson and normal distributions. (a) The Poisson probability distribution $P(k)$ as a function of detected number of events, k, for $N = 1$, 4, and 16 average number of events in an interval. (b) The Poisson distribution (open circles) for large N can be increasingly accurately approximated by the normal distribution (crosses), as shown here for $N = 10$ (blue), 20 (yellow), and 40 (red). The corresponding differences between the two distributions are shown by the solid circles.

number of photons detected by a pixel in a photon-counting device during a fixed length of exposure) is N, then the probability $P(k)$ that k events are actually measured in that time interval is

$$P(k) = \frac{e^{-N} N^k}{k!}. \tag{5.43}$$

The Poisson distributions for some values of N are plotted in Figure 5.47(a). Intuitively, it should be clear that the integer value k must always be greater or equal to zero[13] – one cannot observe fewer than no events!

The Poisson distribution can be accurately approximated, for large values of N, by the normal distribution, that is,

$$P(k) \approx \frac{e^{-(k-N)^2/2N}}{(2\pi N)^{1/2}}. \tag{5.44}$$

The scatter[14] in the data is $N^{1/2}$ [see Figure 5.47(b)]. So, for example, if the average intensity of a Bragg-peak maximum after many identical measurements is 10 000 counts, approximately 68% of the results will lie within $\pm\sigma = \pm100$ counts of this value.

For a static experiment, the accumulating signal strength increases linearly with time during acquisition. A consequence of Poisson statistics is that in order to achieve a factor X improvement in the signal-to-noise ratio $\Delta N/N$, one must record X^2 as long (Figure 5.48). The careful experimenter must therefore decide how best to invest his or her precious beamtime – does improving the signal-to-noise ratio of a given signal by, say, a factor of two, justify the necessary fourfold increase in acquisition times, or would this time be better invested in recording other data points?

5.7.1.2 Dark Noise

Dark noise is a stochastic signal (and as such, is a form of shot noise) generated by a photosensitive diode such as a charge-coupled device (CCD) pixel element, even in the absence of any photons. It is caused by

---

[13] Factorials (k!) of large numbers are most easily determined using the Stirling approximation, given in Appendix B.

[14] More precisely, the standard deviation σ, whereby the full-width at half-maximum scatter equals $(8 \ln 2)^{1/2} = 2.36\sigma$.

Figure 5.48   A shot-noise simulation of a 2D image. From left to right, the average number of photons per pixel over the entire image is (top row) 4, 25, 100, (bottom row) 400, 10 000, and 160 000. Each image has been scaled linearly from black to white. The insets are blowups of the most prominent crater in the image, close to the edge of the shadow. Note that the image quality improves significantly between the first and fourth image (corresponding to an increase in exposure time by a factor of 100), while subsequent increases in exposure time by first another factor of 25, and then a factor of 16 (that is, a factor of 400 with respect to the fourth image) reap only dubious benefits. Original image by the author.

crystallographic defects in the depletion region of the diode device. The level of dark noise can thus vary from pixel to pixel, depending on the defect concentration in each pixel. A dark-field profile can be generated by recording in the dark sufficiently long to produce a statistically reliable response array. It is noted that some modern area detectors have effectively zero dark noise, on account of the fact that they are not based on diode technologies.

5.7.1.3   Readout Noise

Readout noise is a source of spurious signal in CCD detectors produced in the on-chip amplifier used to convert charge to a voltage (via a capacitor) in the readout process after the exposure. This becomes particularly important when reading out very weak signals.

5.7.2   Photographic Plates

Although x-rays were first detected using photographic plates, which are still used in modern medical radiography, their use in the physical sciences has long waned, due to many drawbacks, including difficulty in obtaining quantitative relative intensities due primarily to a nonlinear response for the change in transparency of the photographic plate with signal intensity and a poor dynamic range; poor spatial resolution; and the very long readout times.

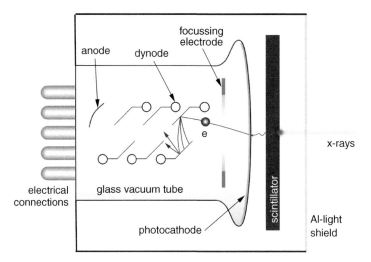

Figure 5.49   Scintillator counters. Photons produced by x-rays absorbed in the scintillator material strike photocathode material (shown in green) deposited on the inside of the entry window of an evacuated glass tube. Electrons produced by the photoelectric effect are directed by a focussing electrode towards an array of electrodes, called 'dynodes'. Each successive dynode is held at a more positive voltage than the previous one. The electrons are therefore repeatedly accelerated towards each dynode and arrive with high energy. On striking a dynode, each electron produces several secondary electrons, which, in turn, are accelerated toward the next dynode, resulting in a cascade production of electrons. Finally, the anode is reached, where the accumulation of charge results in a sharp current pulse corresponding to the arrival of the original x-ray photon at the scintillator.

5.7.3   Scintillator Detectors

Scintillation counters operate by the partial conversion of absorbed x-rays into visible or near-visible light, which can then be amplified using a photomultiplier tube (PMT). A schematic diagram of a typical detector is shown in Figure 5.49.

Typical inorganic scintillator materials are salts or metal oxides doped with high-Z materials. The host material is electronically excited by the absorption of an x-ray photon, and this excited state rapidly transfers its energy to energetically nearby states of the dopant ion. These relax efficiently by nonradiative processes (i.e. *not* involving the emission of a photon) to a much lower excited state only a few eVs higher in energy than the final relaxed state, which is accessed by the emission of a photon in the visible or soft ultraviolet range (see Figure 5.50).

There are many types of scintillator material, both organic and inorganic [29]. Here we mention just two inorganic scintillators, namely Tl-doped NaI [NaI(Tl)], and Ce-doped yttrium aluminum perovskite [YAP(Ce)]. The most commonly used material for point scintillator detectors is NaI(Tl). It has a good x-ray stopping power, and has a scintillation emission maximum at 415 nm. Its two main disadvantages are (a) its softness and sensitivity to moisture (it is highly hygroscopic), and (b) its relatively long 'dead time' of approximately 250 ns[15], which therefore limits the maximum reliable counting rate to about 1 MHz.

YAP(Ce) is a modern scintillator material with its peak scintillation peak centered at 350 nm. Although its efficiency is only about 40% of that of NaI(Tl), it is both mechanically robust and nonhygroscopic, and has a dead time of only 25 ns, making possible counting rates in the range of 10 MHz.

---

[15] The dead time of a system is the time after an event during which the system is not able to record another event if it happens, and can be thought of also as a 'recovery time'.

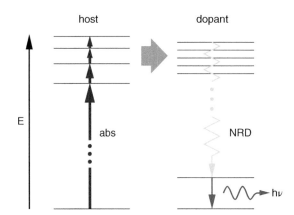

Figure 5.50   A simplified energy-level scheme for scintillation in inorganic crystals. An x-ray photon is absorbed by the host crystal. The excited energy is transferred to energy states of the dopant material, which rapidly relax via nonradiative decay (NRD) to one or more low-lying levels, which in turn decay via luminescence and the ejection of a visible or near-visible photon, back to the ground state or a low vibrational state.

Scintillator plates can be used to record x-ray images. For example, a scintillator plate can be placed in front of a two-dimensional array of photosensitive elements, such as a CCD array (see below), in techniques such as x-ray tomography (see Section 8.2). The thicker the scintillator material, the more efficient is the x-ray stopping power of the plate and hence the stronger the signal. Why not then make the scintillator layer as thick as possible? The reason why, in imaging applications, is that the spatial resolution is effected by the so-called 'point-spread function' of the device. We explain this briefly now.

5.7.4   The Point-spread Function

The point-spread function (PSF) describes the response of an imaging system to a point source or point object. Imagine, for example, an x-ray beam focussed down to a Gaussian spot of 1 μm full-width at half-maximum, imaged by an x-ray area detector consisting of an array of pixels, each of size 100×100 μm^2. The best possible imaging of the focus is, of course, a single illuminated pixel. In this limit, the PSF is a square the size of a pixel. In certain detector types, however, 'blooming' can cause the signal to leak across into the neighbouring area of the detector and increase the extent of the PSF still further, in which case, the PSF is intensity dependent. The degree of blooming and its causes vary from system to system. For example, blooming can occur in scintillator plates by multiple scattering of x-rays across the depth of the active layer, and also because the converted visible photons are emitted in all directions and are also multiply scattered (see Figure 5.51). The thicker is the active layer, the larger is this effect. Blooming in charge-coupled devices is discussed below.

5.7.5   Crystal Analysers

Scintillator detectors, in conjunction with slits, are known as point detectors. Consider such a system with the sensitive area defined by two pairs of slits (Figure 5.52).

The angular resolution in each orthogonal direction to the sample–detector axis $\Delta\theta_{x,y}$ is limited by three factors. The first originates from the incoming beam, which will have a nonzero divergence (typically of the order of 0.1 to 1 mrad). The divergence of the incident beam and the irradiated volume of the sample can be minimized by using sets of slits before the sample, but only at the cost of signal intensity. Let us assume, for argument's sake, that the beam *is* completely monochromatic and parallel. The signal width is then determined by the projected size of the

Figure 5.51 A sharp signal (left) can be smeared out due to various processes (right). This less sharp image is the convolution between the original object and the point-spread function (middle).

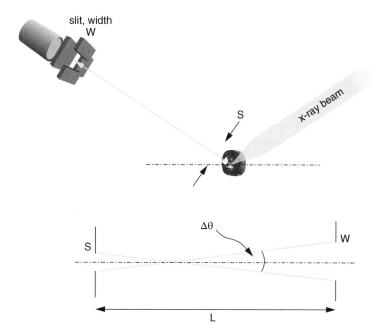

Figure 5.52 The angular resolution of a point detector is defined by the angles subtended by the irradiated sample volume and by the slit width at the detector.

irradiated volume as viewed by the detector and by the width of the slit aperture in front of the detector, also at the cost of signal strength. The angular resolution is given by

$$\Delta\theta = \frac{S + W}{L}, \tag{5.45}$$

where S is the irradiated sample size, W is the slit width, and L is the sample–detector distance. So, for example, a typical powder diffraction setup, consisting of a fully irradiated capillary sample of diameter $S = 200$ μm, a sample-to-detector slit distance of 1 m, and a detector slit opening of 2 mm, results in an angular resolution of 2.5 mrad, or 0.14°.

An additional problem associated with simple scintillator–slit point detectors is the fact that, to a first approximation, the scintillator does not distinguish between x-ray photons of different energies. Both this and the problem of loss of signal intensity with decreasing slit width can be resolved using an adaptation called the crystal-analyser detector.

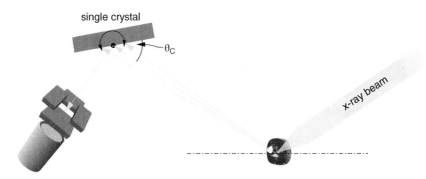

Figure 5.53   The introduction of a single crystal in the detector system to select the monochromatic diffracted signal can improve the resolution by precisely defining the angle between the incoming beam and the diffracted signal. In addition, it suppresses lower-energy fluorescence signal.

Consider Figure 5.53. A high-quality crystal such as Si(111) is mounted in the detector arm. The crystal planes parallel to the crystal surface have an interplanar distance d_C. The crystal is rotatable about an axis lying in its front surface. It is rotated to an angle θ_C subtended by the crystal surface and the line connecting the crystal rotation axis and the sample such that, from Bragg's law [Equation (5.25)],

$$m\lambda = 2d_C \sin\theta_C$$

$$\Rightarrow \sin\theta_C = \frac{mhc}{2Ed_C}, \tag{5.46}$$

where E is the photon energy. Note that, in diffraction techniques, we are dealing with elastic processes, and the photon energy is set by the incoming monochromatic beam – any fluorescent signal from the sample (which has lower photon energies) will therefore be filtered out by the crystal analyser. Only those elastically scattered rays impinging on the crystal-analyser surface with exactly the angle θ_C to the surface plane will be diffracted and there is hence no need to use mechanical slits – the angles are accurately determined by the strict diffraction condition set by the crystal. Quantitatively, the 'effective' slit width, i.e. the width of a slit that would have to be introduced into the detector system if the analyser crystal were absent (and ignoring the sample size), is given by

$$W = L\, w_D^C, \tag{5.47}$$

where L is the distance between the sample and analyser crystal and w_D^C is the analyser-crystal rocking-curve width (the 'Darwin width', see Section 6.4.4 and Figure 6.16), typically of the order of 10 µrad. Hence, the effective slit width of an analyser-crystal detector is of the order of 5 µm. The signal intensity, however, is much larger than that achieved with a 5 µm mechanical slit because, in the absence of a slit aperture, the full irradiated volume of sample can be exploited.

Systems for data acquisition using a point detector (such as a scintillator, crystal analyser, and/or slit system) are limited by their lack of 'peripheral vision' – scans such as rocking curves probe one dimension only. Optimization of the signal by iteratively scanning different directions in the neighbourhood of a signal of an unknown intensity profile can easily lead to the recording of a local maximum, while the real signal of interest may be quite different. By using an area detector, especially one with low background noise, data-acquisition rates can be increased significantly.

5.7.6   Image Plates

Image plates became popular in the mid 1980s for x-ray crystallography. The heart of image plates is a storage phosphor screen. The most common materials are the barium fluorohalides. Wherever x-ray photons are absorbed on the phosphor screen, they produce secondary electrons which are trapped in so-called 'F-centres'.[16] After the end of the x-ray exposure, these metastable centres can then be read by a finely focussed laser beam that scans across the image plate surface. The laser light causes the F-centres to release visible photons, which are detected by a PMT.

5.7.7   Charge-coupled Devices

Nowadays, image plates are much less popular, as their specifications, in particular regarding efficiency, read-out times, point-spread function, and dynamic range, lag behind those of charge-coupled devices (CCDs) [30]. The use of CCDs for synchrotron radiation was first described by Strauss *et al.* in 1988 [31]. The smallest pixel size at the time was approximately 20 μm, while the maximum imaging area was approximately 5×5 cm^2. Since then, the quality, resolution, and efficiency of CCDs have improved with developments in semiconductor technology, so that nowadays, pixel sizes of well under 10 μm are readily available, with active areas in excess of 500 cm^2.

A CCD records images by converting photons of light into electrons. The electrons are temporarily accumulated in individual picture elements (pixels) on a photosensitive semiconductor chip. At the end of an exposure, the accumulated charges are read off the chip and converted into an array of digital numbers, each number being proportional to the accumulated charge at the corresponding pixel. A digital image has been created. Because the signal is only read out after the exposure, CCDs are therefore termed *integrating* devices.

To record x-ray images, CCDs are used in conjunction with an overlayer of scintillator material typically a few tens of microns thick. The point-spread function is of the order of 10 to 50 μm.

The images stored in modern CCDs can be read out in as little as a millisecond or even faster for specialized devices, and also depends on the size of the CCD array. This is some five orders of magnitude faster than that possible using image plates.

During long exposures needed to record faint objects, any bright features also in the field can exceed the maximum capacity (i.e. the charge capacity) of the pixels on which they are being recorded. When this occurs, the excess charge spills over into adjacent pixels. This spillover causes blooming. This can often manifest itself as streaks, due to the readout structure of the CCD allowing charge to overflow more easily in one direction than the other (see Figure 5.54). Although a metallic so-called 'antiblooming grid' can be used to drain off excess charge from each pixel, this reduces the light-gathering capacity of the CCD and also makes the response nonlinear. Such nonlinearity is unacceptable for accurate scientific measurements, such as spectroscopy, crystallography, and photometry, but may be satisfactorily used for qualitative imaging purposes.

Another drawback of recording long-exposure images with CCDs is that of 'dark noise' – current and collected charge that is randomly generated within the diodes of a CCD device, even in the complete absence of photons. Averaged over a sufficiently long time, the production rate of dark noise can be calibrated pixel for pixel. The resulting dark-noise pattern, unique for each CCD, can be used to subtract dark noise, although the stochastic nature of this phenomenon (following Poisson statistics), means that some shot noise will still remain. Lastly, readout noise will further add to the noise. Modern scientific CCDs are designed to minimize these two noise sources, and, due to the small pixel sizes they can offer, CCDs are still highly competitive in synchrotron science.

---

[16] The 'F' in F-centre comes from the German word for colour, 'Farbe'. An F-centre is a crystallographic defect in which an anionic (e.g. oxygen) vacancy in a crystal is filled by one or more electrons, depending on the charge of the missing ion. Electrons in such a vacancy tend to absorb light in the visible range of the electromagnetic spectrum, such that a material that is otherwise transparent becomes coloured.

Figure 5.54   An example of streaking/blooming. Photograph by the author.

5.7.8   Pixel and Microstrip Detectors

A revolutionary new x-ray area detector type has been developed in the first decade of the new century. The two-dimensional hybrid pixel-array detector (HPAD) [32, 33] operates in *single-photon-counting mode*, i.e. counting registers for each pixel increment by one every time an x-ray photon above a set energy is detected [see Figure 5.55(a)]. The mode of operation is thus fundamentally different from integrating devices like CCDs. Moreover, HPADs directly detect x-ray photons without the intermediate step of converting them to visible light with a scintillator.

The photon-counting electronics comprise of a preamplifier, a comparator, and a counter. The preamplifier amplifies the charge generated in the sensor by the incoming x-ray; the comparator produces a digital signal if the charge exceeds a predefined threshold. Thus, together with the counter, one obtains a complete digital storage and readout of the number of detected x-rays per pixel without any readout noise. Because the pixel-detector architecture contains no diodes, HPADs are also free of dark noise. In addition, the use of a predefined but adjustable threshold means that these detectors are able to suppress signal deriving from photons with energies lower than a user-defined value. This is very important when fluorescent signal should be eliminated in diffraction experiments, where only the elastically scattered photons should be recorded [see Figure 5.55(b)].

The standard Si-sensor in the Pilatus and Eiger HPADs is 450 µm thick. The absorption length of silicon only exceeds this value above 15 keV (where it therefore has an efficiency of $1 - 1/e = 63\%$). The use of CdTe-sensors extends the quantum efficiency of HPADs to approximately 70% at 60 keV, and allows them to be used up to and beyond 100 keV.

HPADs feature several advantages compared to current state-of-the-art CCD and imaging-plate detectors (Figure 5.56). The main features include: no readout noise; superior signal-to-noise ratio; fast readout times, which can be as short as 20 µs (allowing a frame rate of up to 50 kHz, over an order of magnitude faster than that of most CCDs); dynamic ranges of up to 32 bits (i.e. the counter of any one pixel only saturates once $2^{32} \approx 4 \times 10^9$ events have been recorded); a high detector quantum efficiency; and the possibility to suppress fluorescence by an energy threshold that is set individually for each pixel. The short readout and fast framing times allow one to record diffraction data in a continuous mode without opening and closing a shutter for each frame. For the first time in 2017, more than 50% of structures deposited in the Protein Data Bank were recorded using HPADs; this trend looks to continue in the foreseeable future.

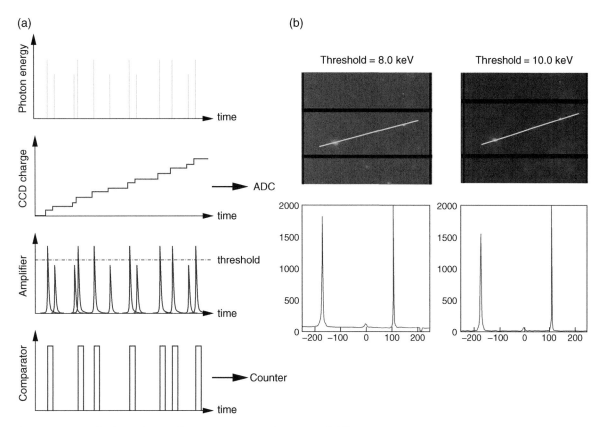

Figure 5.55 Hybrid pixel-array detectors. (a) Consider x-rays of different energies impinging on a detector. In the case of integrating detectors, such as CCDs, charge is accumulated and then converted using an analog-to-digital converter (ADC). In the case of single-photon counting detectors, the charge generated by individual photons is amplified and converted into a voltage pulse with a height that is proportional to the photon energy. This is then compared to a voltage corresponding to a given energy threshold for each pixel; only those pulses which are higher than the threshold result in an increment of one in the counter. (b) By setting a threshold of 10.0 keV, fluorescence emission at 8.1 keV from Cu-atoms in an Al-Cu-Fe quasicrystal can be suppressed, resulting in a much lower background signal and the emergence of fine detail in the elastically scattered signal, as demonstrated by the two line plots (below) through the yellow paths in the images (above). Courtesy Beat Henrich, Paul Scherrer Institute.

For each pixel, there is a recovery or dead time τ after detection of a photon in which no subsequent photon will be registered. In the case of the Pilatus detector, this is approximately 100 ns. Thus, any incoming signal with an arrival rate of absorbed photons at a given pixel of the order of τ^{-1} or larger will be partially suppressed. The loss in counting efficiency at high arrival rates can be corrected in online software using the equation

$$N_{\mathrm{obs}} = N_0 \exp(-N_0 \tau), \tag{5.48}$$

whereby N_{obs} is the detected rate and N_0 the true incident rate [33, 35] (see Figure 5.57). Nonetheless, it is generally advisable to limit the detected count rate to be substantially smaller than τ^{-1} per pixel in order to avoid counting errors.

Because of their specifications, HPADs offer important advantages over state-of-the-art CCD and imaging-plate detectors for various x-ray detection experiments. Major improvements have been reported in time-resolved

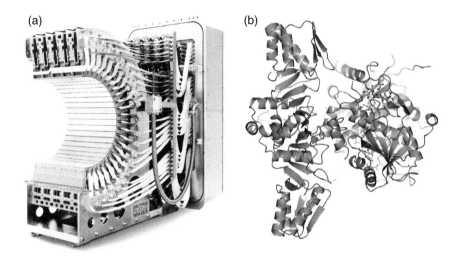

Figure 5.56 Example of a photon-counting HPAD. (a) The custom-made vacuum-compatible Pilatus 12M detector for the I23 low-energy macromolecular crystallography beamline at the Diamond Light Source, UK. The detector is composed of 5×24 modules, each module containing approximately 100 000 pixels. The detector covers an angular range of $2\theta = \pm100°$. Courtesy Christian Brönnimann, Dectris AG. (b) ThcOx, the first structure to be solved using the Pilatus 12M detector at I23, recorded at 4 keV. The resolution was 3.15 Å [34]. Rendered from the pdb file 5lq4.

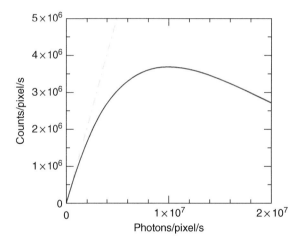

Figure 5.57 Measured count rate N_{obs} as a function of true incident photon rate for a detector with a 100 ns dead time. The yellow dot-dashed line shows the response of a detector with zero dead time.

experiments, in the study of weak diffraction phenomena (e.g. diffuse scattering), and in macromolecular crystallography.

An analogous one-dimensional detector, or strip detector, has also been developed at the Paul Scherrer Institute [36]. The Mythen detector consists of a line array of modules, each module containing 1280 silicon strips of width 0.05×8 mm (see Figure 5.58). Like HPADs, it also operates in a photon-counting mode. Its ultimate intrinsic resolution is of the order of 0.004°, given by the size of the individual strips and their distance from the sample. In reality, the effective resolution may be poorer, depending on the volume of the sample that is being

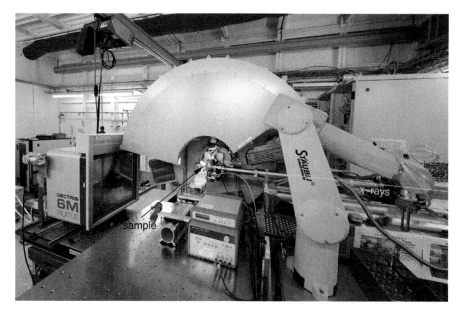

Figure 5.58   The 120° Mythen-II detector at the Powder Diffraction station of the Swiss Light Source, containing 30 720 readout elements over an angular range of 120°. A Pilatus 6M area detector can also be seen downstream of the sample.

illuminated with x-rays (see Figure 5.52). A Mythen detector is being used at the Powder Diffraction station of the Materials Science beamline at the Swiss Light Source, which spans an angle of 120° (24 modules), and is revolutionizing time-resolved diffraction studies – the minimum readout time for the entire detector is of the order of a millisecond, which should be compared to several hours scan time when using more conventional point detectors, especially when the latter are used in conjunction with a crystal analyser (see Chapter 6). This sensational increase in data-acquisition rates on the one hand opens new vistas of time-resolved powder-diffraction studies, while on the other, it allows one to record entire data sets within a fraction of a second, thereby avoiding, or at least minimizing, the deleterious effects of radiation damage on the sample.

An example of the benefits of such detectors is illustrated in Figure 5.59. The pharmaceutical compound bupivacaine hydrochloride can exhibit several polymorphs. Despite their chemical similarity, the biological impact of one polymorph over another can be fundamentally different, hence the need to identify and characterize each polymorph cannot be overstated.

Crystals of pharmaceutical products are particularly prone to radiation damage by x-rays, due to their relatively low bond strengths. A crystalline powder sample of the so-called D-form of S-bupivacaine was therefore investigated using a multicrystal analyser detector in a 'fast' scan over 60° lasting 15 minutes (4°/minute, step size of 0.004°). A small section of the scan is shown in plot AC1 of Figure 5.59. A second scan of the same sample using identical conditions was recorded immediately afterwards. This scan was significantly weaker (note the 5× scaling of plot AC2 in Figure 5.59) and showed different peak positions and intensities compared to the first run, indicative of radiation damage. A fresh sample was then investigated using the Mythen detector (lowest plot, Figure 5.59) using a 1 s exposure. The difference in the signal shape compared to *both* plots AC1 and AC2 is clearly identifiable in the highlighted regions, demonstrating that the sample suffered significant radiation damage even after recording the first 13° of the first scan, i.e. after approximately three minutes exposure! Therefore, obtaining reliable data using conventional detectors for subsequent structural refinement is all but impossible for such sensitive organic crystals; the use of one-dimensional detectors for parallel acquisition is indispensable.

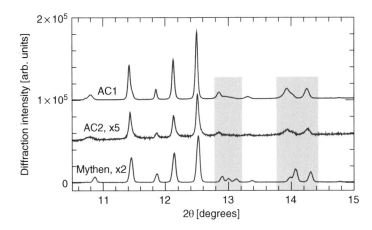

Figure 5.59 Parallel data acquisition using a microstrip detector can be essential in some compounds in order to avoid radiation damage from the synchrotron beam. Two successive 15-minute analyser-crystal scans (AC1 and AC2) of a powder sample of a polymorph of the pharmaceutical compound bupivacaine, plus a 1 s scan of a fresh sample using the Mythen microstrip detector. The regions around 13 and 14.1° highlight those parts of the scans which are most markedly different from one other. Courtesy Fabia Gozzo, Paul Scherrer Institute.

5.7.9   To Integrate or to Count?

The arrival rate of photons at a beamline precisely follows the pulsed time structure of the electron bunches within the storage ring. The peak rate after monochromatization is of the order of 10^{15} photons per second. If we assume the beam has not been focussed and is incident on 50 000 pixels of a HPAD, the incident rate per pixel is 2×10^{10} s^{-1}. Hence even a photon-counting detector with a dead time as low as 1 ns (two orders of magnitude shorter than that of the Pilatus) cannot possibly directly measure how many photons are being produced by any given electron bunch[17].

Photon-counting detectors can therefore not be used for XFEL experiments (see Chapter 4). A typical XFEL experiment records individual data for each XFEL pulse. The duration of these pulses is several orders of magnitude smaller than τ in all photon-counting devices, which can thus only generate outputs for any given pixel of zero (if no photon was detected) or one (if one *or more* photons were detected). In such experiments, therefore, integrating devices are mandatory.

As mentioned above, CCDs are integrating devices. In contrast to photon-counting detectors, CCDs only read out the accumulated charge after an exposure, and are not influenced by the photon arrival rate.

A novel charge-integrating detector called the Jungfrau has been developed at the Paul Scherrer Institute. It has a large dynamic range of over 10^4, thanks to an automatic adjustment of the gain to the amount of charge deposited on each pixel. This might seem at first blush an impossible feat – how can electronics adjust the gain during the few-femtosecond-long irradiation of an XFEL pulse? Consider Figure 5.60(a). The incident XFEL pulse generates free electrons in the surface region of a buffer layer on top of the pixel array. The electrons drift across the buffer to the pixels, which takes approximately 20 ns, which is long enough for electronics to respond. A schematic of the gain-switching preamplifier electronics in each pixel element is shown in Figure 5.60(b). The relationship between a charge Q collected in a capacitor C to the voltage V across that capacitor is $V = Q/C$.

---

[17] A word of warning here. If, after an exposure, one measures on the HPAD a signal of N_{obs} photons per second, and this appears to be significantly smaller than τ^{-1}, remember that the *peak* arrival rate is larger than N_{obs} by a factor $\Delta T/\Delta \tau$, the ratio of the temporal bunch separation to the bunch duration, of the order of 100.

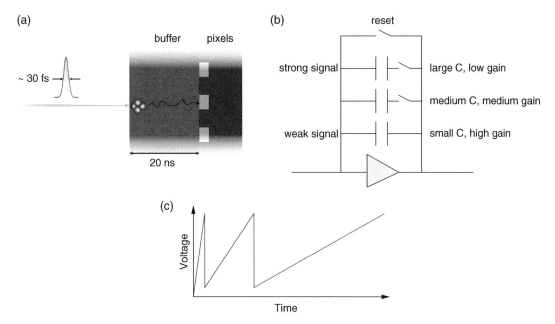

Figure 5.60 The operational principle of the Jungfrau detector. (a) The electrons produced by the XFEL radiation in the surface region of the buffer layer drift over approximately 20 ns to the pixel array. (b) Each pixel has an automatic gain-switching preamplifier, shown schematically here. (c) By successively closing the medium and large capacitors, the gain (i.e. the voltage response to the accumulated charge) can be reduced.

Also, the total capacitance C_{tot} of a set of capacitors connected in parallel is $C_{tot} = C_1 + C_2 + \ldots$. Hence, as the charge increases as the electrons arrive from across the buffer at the pixel, the voltage increases proportionally. At a threshold value of that voltage, the circuit to a second, larger, capacitor placed in parallel is automatically closed, the total capacitance therefore increases, and the voltage drops. A third, even larger, capacitor provides the last automatic gain change. The pixels are reset after readout by short-circuiting the capacitors [37].

The European XFEL and LCLS-II generate pulses with frequencies in the MHz range (see Table 4.2). The Adaptive-Gain Integrating Pixel Detector (AGIPD) has been developed to allow single-pulse imaging at a frame rate of 4.5 MHz with a dynamic range allowing single-photon-counting detection. Crucially, the architecture includes an analogue memory for each pixel, capable of storing data for up to 352 images. An external veto signal can be provided to maximize the number of useful images by overwriting the veto-selected images. In the case of the European XFEL, the data can be read out in the idle period of approximately 100 ms between the pulse trains [38].

5.7.10 Energy-dispersive Detectors

Spectroscopy experiments in which the energy spectra of photons or electrons originating from the sample are investigated (such as x-ray fluorescence or photoelectron spectroscopy) require energy-dispersive detectors. Here we discuss photon detectors in the range of 100 eV to several tens of keV, and electron detectors from approximately one meV to a few keV.

5.7.10.1 Photon Detectors

Spectral analysis of an x-ray spectrum, such as in x-ray fluorescence, is achieved in one of two ways. In wavelength-dispersive spectrometers (WDX or WDS), the photon energy is selected by diffraction on a single crystal or a grating in much the same way as in crystal or grating monochromators, described above. In their most rudimentary form, WDX spectrometers are both slow (because of the need to scan the Bragg angle) and produce weak signals, due to the fact that only a very small fraction of the fluorescence is emitted in a direction which meets the Bragg condition on the spectrometer crystal or grating. In most cases, however, the signal-to-noise ratio can be improved by using spherically or cylindrically curved diffraction crystals or gratings, while in the so-called von Hamos geometry, the diffraction signal is spatially dispersed, allowing parallel spectral acquisitions.

Here, we describe in more detail three WDX variations, namely the Johann [39], Johansson [40], and von Hamos [41] geometries. Before we do this, however, the concept of the Rowland circle is explained.

Consider Figure 5.61(a). A monochromatizing single crystal (or grating) is bent to a radius R_b. If a source of x-rays is placed on a circle with half this radius, that is, $R_R = R_b/2$ and if this smaller circle touches the dispersing element tangentially, then all those x-rays with the energy that satisfies the Bragg condition when impinging at the touching point of the two circles will also be diffracted at other positions on the surface. Moreover, the diffracted rays will converge on the mirror-image point of the source on the smaller circle. The circle with radius R_R is called the Rowland circle.

The advantages of using such a geometry are immediately apparent – the solid angle of x-rays captured by the bent element is far larger than that which can be collected using planar elements.

The above description of the Rowland circle is only approximate insofar that the diffracted beams do not converge exactly at the same point, but suffer a small amount of aberration. This only slightly imperfect setup is known as the Johann geometry [39]. However, bending the crystal to a radius R_b *and* also grinding the surface to a radius R_R allows precise focussing. This so-called Johansson geometry [40] is considerably more expensive to realize than the Johann spectrometer and is thus less common. With the advent of routine submicron spot sizes at DLSR beamlines, the use of Johannson-type spectrometers for high-resolution emission spectroscopy is likely to increase [42].

Thus, for any one position of the source on the Rowland circle, a particular photon energy will be selected. Energy scans in the Johann and Johansson geometries are performed by executing the movements shown in Figure 5.61(b). These geometries are thus superior in light-gathering power to the simple planar-element WDX setup, but require concerted mechanical motions.

In contrast, the von Hamos geometry, shown schematically in Figure 5.62, combines some of the light-gathering features of the Johann and Johansson setups, and, importantly, spatially disperses the spectrum, allowing parallel spectral data acquisition. This geometry thus lends itself to high-resolution, time-resolved x-ray emission studies [43].

In energy-dispersive spectrometers (EDX or EDS), a solid-state detector allows the determination of the energy of the photon when it is detected. Such semiconductor detectors are usually fabricated from silicon (sometimes doped with lithium) or germanium. The device is polarized with a high voltage. When an x-ray photon hits it, electron–hole pairs are created that drift in the applied high electric field. The charge is collected in a manner similar to the charging of a capacitor. Crucially, the voltage increment due to the collected charge is proportional to the photon energy, hence it is possible to determine the energy spectrum. The voltage is reset regularly to avoid saturation. Such detectors also capture large solid angles of signal and therefore have high sensitivity. They are more suited to experiments where signal is weak, such as when trace elements are being investigated.

EDX spectrometers are smaller, cheaper, and measurements are faster. However, the resolution, typically between 100 and 200 eV, is far lower than those of WDX spectrometers [see Figure 5.63(b)].

A subclass of energy-dispersive photon detectors (so-called low-temperature detectors, LTDs) operate at cryogenic temperatures and offer an order of magnitude better resolution than conventional solid-state detectors, though with lower detector efficiency. This is summarized in Figure 5.63(a).

The speed of data acquisition using EDX spectrometers permits the user to rapidly obtain chemically resolved two-dimensional maps with a spatial resolution down to the micron range and sensitivities in the sub-mg kg^{-1} range (see the fluorescence map of a fossilized fish in Figure 5.64).

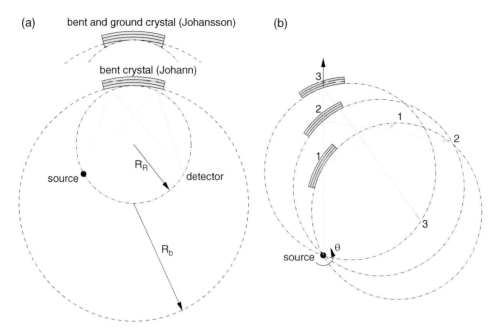

Figure 5.61 The Rowland circle and bent-crystal-geometry spectrometers. (a) A particular photon energy from a source of radiation sitting on the Rowland circle of radius R_R is selected by a crystal bent to a radius $2R_R = R_b$ sitting tangentially on the Rowland circle, and is focussed to the mirror-image point of the source. Slight aberrations in this 'Johann' geometry can be removed by also grinding the crystal surface to a radius R_R, in the 'Johansson' geometry. (b) The spectrum of the source can be probed by a concerted movement of the crystal position and angle, and the detector.

Which detector type one uses depends, needless to say, on the application. Which is more important – sensitivity, speed, or spectral resolution? In resonant inelastic x-ray scattering (RIXS, see Section 7.6.3), for example, line separations are typically of the order of a few eV or smaller, marginally below the present capabilities of modern LTDs. Future improvements in LTDs might therefore potentially open up new vistas of possible experiments in techniques such as RIXS, where signal intensity is often at a premium and thus the deciding factor in determining the feasibility of a given experiment.

5.7.10.2 Electron-energy Analysers

Electron-energy analysers are necessary for photoemission-spectroscopy experiments. How does one measure the kinetic energies of electrons? There are two common detector types, the cylindrical mirror analyser (CMA) and the concentric hemispherical analyser (CHA). Even though the resolving power of CMAs can be improved using a double-pass configuration, most electron-spectroscopy instruments nowadays use CHAs, and it is these that are now described in some more detail.

A schematic of a CHA is shown in Figure 5.65. Two metallic hemispheres of radii R_1 and R_2 ($> R_1$) are positioned concentrically. The inner and outer hemispheres are floated to the negative potentials $-V_1$ and $-V_2$, respectively.

An electron entering the CHA approximately parallel to the equipotential surfaces with a kinetic energy E will be focussed at the exit to the detector if

$$e\Delta V = E \left(\frac{R_2}{R_1} - \frac{R_1}{R_2} \right), \tag{5.49}$$

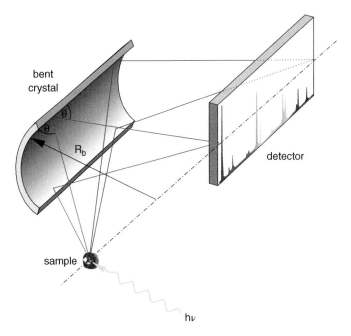

Figure 5.62   Schematic of a von Hamos spectrometer. The emission source and a strip or area detector lie on the cylinder axis of a crystal bent to a radius R_b. The spectrum of the source is spatially dispersed on to the detector by the different incident Bragg angles θ on the crystal surface. Bending the crystal improves the light-gathering power.

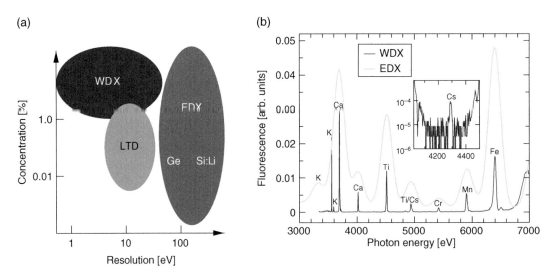

Figure 5.63   Energy-dispersive photon detectors. (a) A broad classification of different detector types, according to their sensitivity and spectral resolution. (b) The fluorescence spectra of a clay mineral recorded with EDX and WDX, containing approximately 100 parts per million Cs. The Cs Lα line is just detectable using WDX (see inset), but is completely lost using EDX. Courtesy Daniel Grolimund, Paul Scherrer Institute.

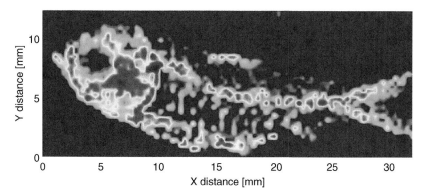

Figure 5.64 An energy-dispersive x-ray fluorescence map taken at the Th Lα emission line (12 969 eV) of a fossil fish found in the Green River Formation of Fossil Lake, Wyoming. Warm regions depict high concentrations of thorium. The incident x-ray beam was tuned to 17.2 keV. The high levels of thorium are caused by its substitution for calcium in the phosphate mineral that composes the bone, fluorapatite. By studying such paleontological specimens, it is hoped to better understand the chemistry of lanthanides and actinides in the natural environment. Adapted from [44] and courtesy of Troy Rasbury, Stony Brook University, with permission of Elsevier.

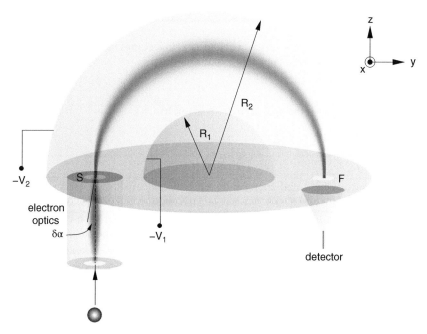

Figure 5.65 Schematic of a concentric hemispherical analyser (CHA). Electrons enter the CHA through the aperture *S* and are focussed at the exit *F* to the detector.

where $\Delta V = V_2 - V_1$. Equation (5.49) can be re-expressed as $E = \kappa e \Delta V$, where κ is known as the spectrometer constant. Electron-energy spectra can therefore be recorded by sweeping ΔV.

Note that the electrons can enter the spectrometer at any angle in the *x-z*-plane (see Figure 5.65) and still arrive exactly at the exit to the detector. This is not so for the perpendicular *y-z*-plane. Suppose that the electrons are injected at an angle $\delta\alpha$ to the entrance-aperture normal in the *y-z*-plane. For entrance and exit apertures *S* and *F*

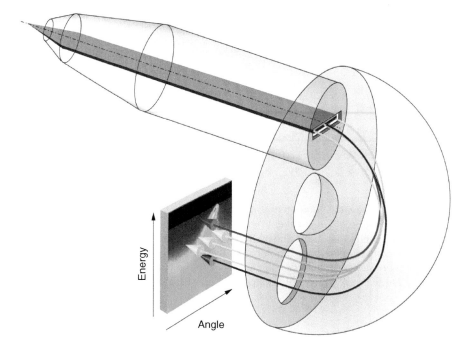

Figure 5.66   Data spanning different electron kinetic energies and trajectories can be simultaneously recorded using a multichannel plate and CCD camera as the detector system in a CHA.

of the same widths W, the resolution of the spectrometer is given by

$$\frac{\Delta E}{E} = \frac{2W}{R_1 + R_2} + (\delta\alpha)^2. \tag{5.50}$$

Therefore the resolution improves for small aperture dimensions and small electron divergences, though at the expense of signal intensity.

Spectra of electron energies can be recorded with either a constant $\Delta E/E$ or a constant ΔE. Spectra taken with a constant $\Delta E/E$ therefore show signals with widths that are directly proportional to their electron energies. More commonly, it is desired that the instrumental peak widths remain constant in any given spectrum, and the constant ΔE mode is chosen. This is achieved by accelerating or decelerating the electrons within the electron optics immediately before entering the CHA to a fixed energy value, known as the pass energy, which typically lies at about 50 to 100 eV. Hence, the energy selection is not performed by varying ΔV in the CHA. This is kept constant according to Equation (5.49) for $E = E_{\text{pass}}$. Instead, the voltage used to change the electron energy is scanned. For example, if one chooses a pass energy of 70 eV, and is measuring electrons with kinetic energies of 1000 eV, a negative voltage of −930 V must be applied with respect to the electrons as they enter the electron-optics unit.

The efficiency, resolution, and sophistication of CHAs has recently increased by the replacement of the point detector at F with a multichannel plate (MCP) and CCD camera. The entrance aperture S is a slit which allows up to a 30° acceptance angle in the x-z-plane. This is projected and imaged along one axis of the MCP (three yellow paths in Figure 5.66), while electrons of different energy (the blue and red paths in Figure 5.66) impinge on the MCP at different positions in the perpendicular direction. At the time of writing, such detectors have an energy resolution as small as 1 meV.

5.8 Time-resolved Experiments

In recent years, more and more synchrotron studies involve time-resolved experiments. This might be on the subsecond to millisecond scale, using the HPADs and microstrip detectors described above, or go down to the range of less than 1 ns using fast electronic devices such as avalanche-photodiode-detector (APD) arrays. Increasingly. experiments are designed to exploit the time structure of the storage-ring current, with a timescale measured in a few tens of picoseconds. Indeed, x-ray free-electron lasers provide unsurpassed temporal resolution, potentially down to a few femtoseconds (see also Chapter 4).

Measurements requiring time resolutions below a few picoseconds require the use of streak cameras, which have been demonstrated to be able to probe timescales down to under 250 fs [45]. Here, we briefly describe the operating principles of streak cameras.

5.8.1 Streak Cameras

Streak cameras are devices which convert a short signal over time into one across space. The signal is recorded on an area detector, whereby one coordinate corresponds to time and the orthogonal axis to space. Consider Figure 5.67. A part of an incident transient x-ray signal is selected using a narrow slit aperture. This impinges on a negatively biased photocathode, resulting in the ejection of a packet of photoelectrons, which is then accelerated to a partially transmitting metal (gold) grounded grid. The electrons now have a well-defined energy, given by the potential difference between the photocathode and the anode, typically of the order of 10 kV. As they begin to drift between a pair of sweep electrodes, a ramp voltage is applied across these. The start of the ramp is triggered by the incident signal such that the leading edge of the electron package sees a different electrical field than the trailing edge and is therefore deflected by a different amount. This effectively skews the package perpendicular to both the slit and the streak camera axis. The skewed signal is then magnified using an electrostatic or magnetic lens and imaged on an area detector. The axis parallel to the slit aperture corresponds to the spatial coordinate, while the orthogonal axis represents time. Indeed, it may be possible to scan the source of the transient signal (in Figure 5.67, this would be in the vertical direction), so that

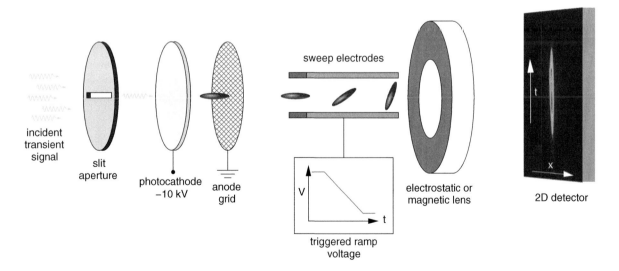

Figure 5.67 The operating principle of a streak camera.

different parts of it are selected by the slit aperture, thereby obtaining information about the second spatial coordinate.

Streak cameras can be used for several applications, from the temporal form of x-ray pulses to transient x-ray emission phenomena from samples. Time resolutions of less than 250 fs using a streak camera with a particularly large magnetic solenoid lens have been demonstrated [45].

5.8.2   X-ray Streaking at XFELs

XFELs produce extremely intense and very short pulses, down to a few femtoseconds. A key application is in ultrafast, time-resolved studies of the dynamics of matter after excitation by a 'pump' pulse. No detector has the required time resolution to track this process and, hence, such experiments typically involve repeating multiple pump-probe cycles at different delay times in a stroboscopic manner. The success of such an approach is predicated on the assumptions that either the sample returns to an identical or very similar state after each pump-probe cycle, or a sufficient number of identical samples exist. In most cases, therefore, the investigation of ultrafast irreversible processes is precluded. In addition, the best temporal resolution is normally significantly longer than the pulse duration, due to the inherent jitter of XFEL radiation.

A novel solution to this problem has been recently reported, in which a time trace of the dynamical response from a single pump pulse is achieved with fs resolution and is entirely jitter-free [46].

The approach is one of 'split and delay'. Consider Figure 5.68. Fractions from a single XFEL pulse are split away using a set of gratings (S_n) into a fan of beams. Each grating in S_n has a different periodicity p_n, resulting in different splitting angles 2θ. These beams are then intercepted half way between S_n and the sample under

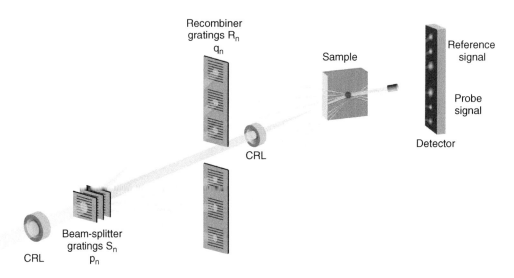

Figure 5.68 The XFEL streaking principle. A fan of weak, diffracted beams is produced from an XFEL pulse using beam-splitter gratings S_n. These are intercepted by two sets of recombiner gratings R_n at a distance $L/2$ from S_n, whereby L is the distance between S_n and the sample. The upper set is made to converge on the sample within the illumination region of the strong, undiffracted, direct 'pump' beam. The delay of these beams relative to the pump, induced by their increasing optical path length with diffraction angle, provides the set of probe signals on a detector. In contrast, the lower, 'reference', set of diffracted beams is made to converge by the lower set of R_n *outside* the pump region on the sample surface, thus providing the reference signal for comparison with the probe beams. Although only three probe and three reference beams are shown here for the sake of simplicity, in reality, there are several more, typically of the order of 15. Adapted from [46] with permission from Macmillan Publishers Ltd.

investigation by a second set of gratings, R_n, which diffracts them back to a convergent spot on the sample, along with the direct, undiffracted, beam. To achieve this, these 'recombining' gratings must have a periodicity q_n half that of the corresponding grating in S_n. The increasing optical path length of the split beams with diffraction angle 2θ provides the delay with respect to the direct beam. It is the intense direct beam (which arrives at the sample first) which pumps the sample. It is overfocussed to provide a large enough illumination on the sample to guarantee overlap with the top set of convergent diffracted beams. These beams probe the dynamics of the sample at delay times relative to the direct-beam pump of

$$\Delta t_n = \frac{L\lambda^2}{2c\, p_n^2} = \frac{L\lambda^2}{8c\, q_n^2}, \tag{5.51}$$

whereby L is the distance between the splitter gratings S_n and the sample, λ is the x-ray wavelength, and c is the speed of light.

A careful inspection of Figure 5.68 reveals that, in contrast to the upper 'probe' set, the lower set of diffracted beams converges on the sample not at the same position as the pump beam, but is, through slight readjustment of the lower set of recombining gratings, incident a little below the pumped region. These reference beams can be compared against the signals from the probe beams.

Depending on the grating periodicity, delays as small as 20 fs and as long as 1.3 ps have been achieved for the innermost and outermost diffracted beams, respectively.

5.9 Concluding Remarks

Beamlines can appear to be very bewildering and intimidating to the uninitiated user, and at one level they are complex, insofar that there are usually many components to consider. Each component, however, can be understood by scientists from all disciplines. It is therefore hoped that the above breakdown of the commonly found equipment in beamlines given in this chapter will help ameliorate the trepidation felt by most first-time users.

Needless to say, this chapter has not been exhaustive in the listing of beamline equipment, or in describing all the possible variations that can arise, both of individual components and also how they are combined, which are innumerable – there are seldom two beamlines that are clones of one another. Indeed it is this possibility of 'mixing and matching' to the user's needs which makes synchrotron experiments so diverse and exciting.

Problems

1. In the one-to-one focussing configuration shown in Figure 5.7, demonstrate that the meridional bending radius R_m of a cylindrical mirror approximating the central section of an ellipse of major and minor semi-axes a and b, respectively, is given by

$$R_m \approx \frac{a^2}{b}.$$

   Make an estimate of the order-of-magnitude size of this radius. (3P)
2. Beginning with Equation (5.18), prove Equation (5.19). Hint: calculate the differential dw_z/dz and assume $z \gg z_R$. (3P)
3. An x-ray beam is to be focussed on to a sample 40.5 m downstream from the x-ray source using a Fresnel zone plate with a focal length of 725 mm. Calculate how far upstream from the sample the FZP should be positioned. (2P)

4. A parabolic compound-refractive lens (CRL) is free from spherical aberrations for on-axis focussing of a near-perfect parallel beam [Figure 5.5(d)]. It is claimed in the text that a circle with radius R has approximately the same curvature as that of a paraboloid of the form $y = r^2/2R$ close to its apex [Figure 5.38(a)]. Demonstrate this to be true. Hint: differentiate both forms and consider the limit for $r/R \ll 1$.   (2P)

5. Derive from first principles the relationship

$$f = \frac{a}{\delta}$$

between the focal length f, the characteristic size a, and refractive-index decrement δ, for a single paraboloid refractive lens described by a profile $z = r^2/2a$.   (4P)

6. The two most commonly used crystals in double-crystal monochromators (DCMs) are silicon and germanium, and the most common crystal orientation is (111). With the advent of high-power insertion-device beamlines at third-generation synchrotron facilities, the x-ray optics are subjected to unprecedentedly high absorbed-power densities, an issue which has become even more problematic in the design of x-ray optics for XFELs. As a result, diamond is becoming more commonly used as the crystal material in DCMs, due to its unparalleled thermal conductivity and consequent ability to handle high thermal loads.

The Darwin width w_D of a single crystal with a diamond-like unit-cell structure reflecting off the (111) planes is given by

$$w_D = \frac{2r_0 N |F(Q)| \lambda^2}{\pi \sin 2\theta},$$ (5.52)

whereby $r_0 = 2.82 \times 10^{-5}$ Å is the classical Thomson scattering length, N is the number density of unit cells per unit volume, and $|F(Q)| = 4\sqrt{2} f^0(Q)$ is the structure factor for reflections involving exclusively odd Miller indices for diamond-like crystals containing atoms with atomic form factors $f(Q)$.

Prove that, for the (111) reflection, Equation (5.52) simplifies to

$$w_D = \frac{16\sqrt{2}\, r_0 |f^0(Q)|}{3\pi a} \tan\theta.$$

The atomic form factors $f^0(Q)$ of diamond, silicon, and germanium are plotted in Figure 5.69. Determine the Darwin widths and their ratios for these three materials for the (111) reflection at $\lambda = 1$ Å (12.3984 keV). The lattice constants of diamond, Si, and Ge are $a = 3.5672$ Å, 5.4307 Å, and 5.6569 Å, respectively.

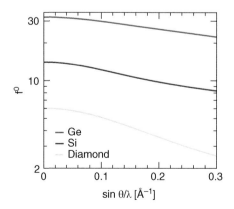

Figure 5.69   The atomic form factors of diamond, silicon, and germanium as a function of $\sin\theta/\lambda$.

Determine the expression describing the bandwidth of transmitted photon energies $\Delta E/E$ after reflection by the (111) planes and demonstrate that this is independent of the photon energy E. Hint: express E using Bragg's law and differentiate this with respect to θ.

Based on the above, comment on drawbacks in using diamond instead of Si or Ge. (7P)

7. In the previous problem, it was shown that, by differentiating Bragg's law (when expressed as the dependence of the photon energy on angle θ), we arrived at the expression

$$\left|\frac{dE}{E}\right| = \frac{d\theta}{\tan\theta}.$$

Given that the capillary sample at the powder station of the Materials Science beamline at the Swiss Light Source is 32.797 m downstream from the undulator source, demonstrate that the spatial distribution of energies across the beam shown in Figure 5.29 is consistent with this. The monochromator uses Si(111) crystals, for which $d = 3.1356$ Å. (3P)

8. An x-ray mirror is fabricated by depositing a thick coating of rhodium ($Z = 45, A = 102.91$) on an atomically flat silica block. The mirror is installed in the optics hutch at a fixed incident angle for the x-rays of 0.2°. What is the maximum photon energy that the mirror can efficiently reflect? The mass density of Rh is 12.4 g cm^{-3}, and Avagadro's number is 6.022×10^{23}/mole. (2P)

9. A bendable x-ray mirror is situated 25 m downstream from the x-ray source and is used to vertically focus 1 Å radiation onto a sample positioned 40 m from the source. The x-rays impinge on the mirror at an angle of of 0.16°. What is the bending radius R_m required to achieve this? (2P)

10. In the same setup and at the same photon energy, horizontal focussing is achieved by sagittally bending a Si(111) crystal in the monochromator, 23 m from the x-ray source. What is the bending radius R_s? The separation between adjacent Si(111) lattice planes is 3.1356 Å. (2P)

11. A capillary tube is tapered to have a cone half-angle of 0.025° and is coated on its inside with platinum ($Z = 78, A = 195.08$). How many times will 5 keV x-rays that are perfectly parallel and propagating parallel to the capillary axis be reflected by the capillary? The mass density of Pt is 21.45 g cm^{-3}. (2P)

12. Bragg's law for crystal diffraction ignores refraction effects due mainly to the steepness of the Bragg angles θ employed as a result of the normally comparable size of the interplanar spacings relative to the x-ray wavelength. In the case of multilayer diffraction, however, the wavelengths are two orders of magnitude smaller than typical multilayer periodicities Λ, and hence refraction effects cannot be neglected. Using Snell's law, theory shows that the corrected Bragg equation can be written as

$$m\lambda = 2\Lambda \sin\theta \left[1 - \frac{2\bar{\delta} - \bar{\delta}^2}{\sin^2\theta}\right]^{1/2}, \tag{5.53}$$

whereby $\bar{\delta}$ is the mean value of the refractive index decrement given by

$$\bar{\delta} = (t_A\delta_A + t_B\delta_B)/\Lambda,$$

t_A and t_B are the two sublayer thicknesses, and $t_A + t_B = \Lambda$. Beginning with Equation (5.53), derive Equation (5.29), and determine an order-of-magnitude value for κ. (5P)

13. A double multilayer monochromator has one multilayer with $\Lambda = 60$ nm and $t_A = 30$ nm, $t_B = 30$ nm, and a second multilayer with the same periodicity Λ, but with $t_A = 20$ nm, $t_B = 40$ nm. What motivations lie behind this design? (2P)

14. Prove Equation (5.30). Given a CRL focal length of 1 m and an incident parallel-beam diameter of 0.2 mm, determine the Rayleigh length z_R, the beam-waist diameter $2w_0$, and the relative energy bandwidth $\Delta E/E$ for $\lambda = 1$ nm.

The CRL used to focus the 1 nm at 1 m uses $n_{ps} = 6$ paraboloidal surfaces and is fabricated from beryllium, for which $\delta = 2.272 \times 10^{-4}$ at 1 nm. Calculate a, the minimum radius of curvature of the paraboloid, given by $z = r^2/2a$. (6P)

15. A filter set is installed at an undulator beamline to attenuate the beam to differing degrees. In order to record the direct beam without damaging the x-ray detector, one of the filters was chosen to be a dielectric with a 10^{-7} transmission. The filter is insulating, with poor thermal conductivity, and a relative emissivity of $\varepsilon_r = 0.7$. A 10 keV beam incident on the filter has a cross-section of 1.25×10^{-2} mm^2 and a flux of 2×10^{13} photons per second. Assuming the absorbed energy can only be dissipated via blackbody radiation, calculate the temperature of the part of the filter surface in contact with the x-rays. Based on this result, make suggestions of materials which would be more suited for this application. (4P)

16. The River Thames floods on average three times every century. Calculate the probability that no floods occur this century. (2P)

17. An experiment requires that a weak signal is to be recorded on a relatively strong but constant background using an area detector. The peak intensity of the true signal is 20 times smaller than the background, and spatially separated features within the signal should be distinguishable to 10% or better of the signal maximum. If the background signal on average accumulates at 100 counts per second per pixel, calculate the minimum integrated exposure in order for the essential signal features to be discerned. (2P)

18. A beamline receives from its undulator an *average* photon arrival rate of 10^{13} photons per second. The storage ring is powered by an RF Klystron running at $f = 200$ MHz and produces pulses with a temporal width of $\Delta t = 50$ ps. The direct beam is attenuated using a filter and allowed to impinge on a photon-counting detector with a maximum tolerated true count rate of $N_0 = 10$ MHz per pixel over an area of 16 pixels. If one makes the simplification that both the temporal and spatial forms of the pulses are top-hat profiles, determine the maximum filter transmission that can be employed, assuming 100% counting efficiency. (3P)

19. A Bragg peak of a crystalline sample is recorded using a photon-counting detector and a filter with a transmission of 2×10^{-5}, yielding an integrated count rate of N_{obs} counts per second. The experimenter considered the signal to be too weak, so she removed the filter. The integrated Bragg-peak intensity only went up by a factor of 1×10^4. The dead time of the detector was known to be $\tau = 170$ ns. Determine the true incident count rate N_0 and the observed count rate, N_{obs}, without the filter. (2P)

20. The sensitive elements of an area x-ray detector consist of silicon devices which are 300 μm thick. The attenuation length of silicon at 8 keV is 69.6 μm. Calculate the quantum efficiency of such a detector at 38 keV and how this improves for detectors using 1 mm thick Si sensors. (3P)

21. Prove Equation (5.51). (3P)

References

[1] C. Schulze-Briese, B. Ketterer, C. Pradervand, et al., "A CVD-diamond based beam profile monitor for undulator radiation," *Nucl. Instrum. Methods A*, vol. 467, pp. 230–234, 2001.

[2] E. Abbe, "Beiträge zur Theorie des Mikroskops und der mikroskopischen Wahrnehmung," *Arch. Mikrosk. Anat.*, vol. 9, pp. 413–418, 1873.

[3] S. G. Alcock, I. Nistea, J. P. Sutter, et al., "Characterization of a next-generation piezo bimorph x-ray mirror for synchrotron beamlines," *J. Synchrotron Rad.*, vol. 22, pp. 10–15, 2015.

[4] P. Kirkpatrick and A. V. Baez, "Formation of optical images by X-rays," *J. Opt. Soc. Am.*, vol. 38, pp. 766–774, 1948.

[5] R. L. Johnson, "Grazing-incidence monochromators for synchrotron radiation – a review," *Nucl. Instrum. Methods A*, vol. 246, pp. 303–309, 1986.

[6] H. Petersen, C. Jung, C. Hellwig, W. B. Peatman, and W. Gudat, "Review of plane grating focusing for soft x-ray monochromators," *Rev. Sci. Instrum.*, vol. 66, pp. 1–14, 1995.

[7]   P. Melpignano, S. Di Fonzo, A. Bianco, and W. Jark, "Optimization of spherical grating monochromators operating with variable included angle for different applications," *Rev. Sci. Instrum.*, vol. 66, pp. 2125–2128, 1995.

[8]   L. E. Berman, J. B. Hastings, D. P. Siddons, et al., "Diamond crystal x-ray optics for high-power-density synchrotron radiation beams," *Nucl. Instrum. Methods A*, vol. 329, pp. 555–563, 1993.

[9]   Y. V. Shvyd'ko, S. Stoupin, A. Cunsolo, A. H. Said, and X. R. Huang, "High-reflectivity, high-resolution x-ray crystal optics with diamonds," *Nature Phys.*, vol. 6, pp. 196–199, 2010.

[10] J. W. M. DuMond, "Theory of the use of more than two successive x-ray crystal reflections to obtain increased resolving power," *Phys. Rev.*, vol. 52, pp. 872–883, 1937.

[11] S. Oestreich, B. Kaulich, and J. Susini, "Single-movement fixed-exit channelcut x-ray monochromator based on profiled surfaces," *Rev. Sci. Instrum.*, vol. 70, pp. 1921–1925, 1999.

[12] T. Bigault, E. Ziegler, C. Morawe, et al., "Double multilayer monochromator to tailor bending magnet radiation spectrum," *Proc. SPIE*, vol. 5195, pp. 12–20, 2003.

[13] I. Inoue, T. Osaka, K. Tamasaku, et al., "An X-ray harmonic separator for next-generation synchrotron X-ray sources and X-ray free-electron lasers," *J. Synchrotron Rad.*, vol. 25, pp. 346–353, 2018.

[14] V. Honkimaki, H. Reichert, J. S. Okasinski, and H. Dosch, "X-ray optics for liquid surface/interface spectrometers," *J. Synchrotron Rad.*, vol. 13, pp. 426–431, 2006.

[15] G. E. Ice, J. D. Budai, and J. W. L. Pang, "The race to x-ray microbeam and nanobeam science," *Science*, vol. 334, pp. 1234–1239, 2011.

[16] A. Snigirev, V. Kohn, I. Snigireva, and B. Lengeler, "A compound refractive lens for focusing high-energy x-rays," *Nature*, vol. 384, pp. 49–51, 1996.

[17] G. B. M. Vaughan, J. P. Wright, A. Bytchkov, et al., "X-ray transfocators: focusing devices based on compound refractive lenses," *J. Synchrotron Rad.*, vol. 18, pp. 125–133, 2011.

[18] B. Lengeler, C. Schroer, J. Tümmler, et al., "Imaging by parabolic refractive lenses in the hard x-ray range," *J. Synchrotron Rad.*, vol. 6, pp. 1153–1167, 1999.

[19] C. G. Schroer, O. Kurapova, J. Patommel, et al., "Hard x-ray nanoprobe based on refractive x-ray lenses," *Appl. Phys. Lett.*, vol. 87, p. 124103, 2005.

[20] K. E. Evans-Lutterodt, J. M. Ablett, A. Stein, et al., "Single-element elliptical hard x-ray micro-optics," *Optics Express*, vol. 11, pp. 919–926, 2003.

[21] A. F. Isakovic, A. Stein, J. B. Warren, et al., "Diamond kinoform hard x-ray refractive lenses: design, nanofabrication and testing," *J. Synchrotron Rad.*, vol. 16, pp. 8–13, 2009.

[22] D. H. Bilderback, S. A. Hoffman, and D. J. Thiel, "Nanometer spatial resolution achieved in hard x-ray imaging and Laue diffraction experiments," *Science*, vol. 263, pp. 201–203, 1994.

[23] X. Zeng, F. Duewer, M. Feser, et al., "Ellipsoidal and parabolic glass capillaries as condensers for x-ray microscopes," *Appl. Optics*, vol. 47, pp. 2376–2381, 2008.

[24] K. Jefimovs, J. Vila-Comamala, T. Pilvi, et al., "Zone-doubling technique to produce ultrahigh-resolution x-ray optics," *Phys. Rev. Lett.*, vol. 99, p. 264801, 2007.

[25] J. Kirz, "Phase zone plates for x-rays and extreme UV," *J. Opt. Soc. Am.*, vol. 64, pp. 301–309, 1974.

[26] A. J. Morgan, M. Prasciolu, A. Andrejczuk, et al., "High numerical aperture multilayer Laue lenses," *Sci. Reports*, vol. 5, p. 09892, 2015.

[27] R. F. Pettifer, M. Borowski, and P. W. Loeffen, "The physics of ionization chambers – or how to improve your signal-to-noise ratio for transmission EXAFS measurements," *J. Synchrotron Rad.*, vol. 6, pp. 217–219, 1999.

[28] O. Müller, J. Stötzel, D. Lützenkirchen-Hecht, and R. Frahm, "Gridded ionization chambers for time-resolved x-ray absorption spectroscopy," *J. Phys. Conf. Ser.*, vol. 425, p. 092010, 2013.

[29] M. Nikl, "Scintillation detectors for x-rays," *Meas. Sci. Technol.*, vol. 17, pp. R37–R54, 2006.

[30] G. E. Smith, "Nobel Lecture: The invention and early history of the CCD," *Rev. Mod. Phys.*, vol. 82, pp. 2307–2312, 2010.

[31]   M. G. Strauss, I. Naday, I. S. Sherman, et al., "CCD sensors in synchrotron x-ray detectors," *Nucl. Instrum. Methods A*, vol. 266, pp. 563–577, 1988.

[32]   C. M. Schlepütz, R. Herger, P. R. Willmott, et al., "Improved data acquisition in grazing-incidence x-ray scattering experiments using a pixel detector," *Acta Crystallogr. A*, vol. 61, pp. 418–425, 2005.

[33]   P. Kraft, A. Bergamaschi, C. Brönnimann, et al., "Performance of single-photon-counting PILATUS detector modules," *J. Synchrotron Rad.*, vol. 16, pp. 368–375, 2009.

[34]   A. F. Bent, G. Mann, W. E. Houssen, et al., "Structure of the cyanobactin oxidase ThcOx from Cyanothece sp. PCC 7425, the first structure to be solved at Diamond Light Source beamline I23 by means of S-SAD," *Acta Crystallogr. D*, vol. 72, pp. 1174–1180, 2016.

[35]   J. E. Bateman, "The effect of beam time structure on counting detectors in SRS experiments," *J. Synchrotron Rad.*, vol. 7, pp. 307–312, 2000.

[36]   B. Schmitt, C. Brönnimann, E. F. Eikenberry, et al., "Mythen detector system," *Nucl. Instrum. Methods A*, vol. 501, pp. 267–272, 2003.

[37]   A. Mozzanica, A. Bergamaschi, S. Cartier, et al., "Prototype characterization of the JUNGFRAU pixel detector for SwissFEL," *J. Instrum.*, vol. 9, p. C05010, 2014.

[38]   B. Henrich, J. Becker, R. Dinapoli, et al., "The adaptive gain integrating pixel detector AGIPD a detector for the European XFEL," *Nucl. Instrum. Methods A*, vol. 633, pp. S11–S14, 2011.

[39]   H. H. Johann, "Die Erzeugung lichtstarker Röntgenspektren mit Hilfe von Konkavkristallen," *Z. Phys.*, vol. 69, pp. 185–206, 1931.

[40]   T. Johansson, "Über ein neuartiges, genau fokussierendes Röntgenspektrometer," *Z. Phys.*, vol. 82, pp. 507–528, 1933.

[41]   L. von Hamos, "Röntgenspektroskopie und Abbildung mittels gekrümmter Kristallreflektoren," *Ann. Phys.*, vol. 17, pp. 716–724, 1933.

[42]   K. Klementiev, I. Preda, S. Carlson, K. Sigfridsson, and K. Norén, "High performance emission spectrometer at Balder/MAX-IV beamline," *J. Phys. Conf. Ser.*, vol. 712, p. 012018, 2016.

[43]   J. Szlachetko, M. Nachtegaal, E. de Boni, et al., "A von Hamos x-ray spectrometer based on a segmented-type diffraction crystal for single-shot x-ray emission spectroscopy and time-resolved resonant inelastic x-ray scattering studies," *Rev. Sci. Instrum.*, vol. 83, p. 103105, 2012.

[44]   J. M. Cole, J. Nienstedt, G. Spataro, et al., "Phosphor imaging as a tool for in situ mapping of ppm levels of uranium and thorium in rocks and minerals," *Chem. Geology*, vol. 193, pp. 127–136, 2003.

[45]   J. Feng, H. J. Shin, J. R. Nasiatka, et al., "An x-ray streak camera with high spatio-temporal resolution," *Appl. Phys. Lett.*, vol. 91, p. 134102, 2007.

[46]   C. David, P. Karvinen, M. Sikorski, et al., "Following the dynamics of matter with femtosecond precision using the X-ray streaking method," *Sci. Reports*, vol. 5, p. 7644, 2015.

6

Scattering Techniques

6.1 Introduction

What conditions allow one to 'see' individual atoms within a molecule or condensed-matter structure? We have already argued in Section 5.3.3 that, in order to obtain a certain desired resolution, one requires a probe (be it an electron beam or electromagnetic radiation) with a wavelength of similar magnitude. As interatomic distances are only a few angstroms, the study of structure with atomic resolution thus requires the use of either hard x-rays or electron beams (the latter needing to have an energy of 100 eV or greater, see Section 7.3). The penetration depth of electrons is very limited, and although cryogenic electron microscopy ('cryoEM', see Appendix A) is an exceedingly exciting and burgeoning field in the determination of macromolecular structures, x-rays are most often the first probe of choice, in particular for studies that require penetration depths of several nanometres or larger.

This begs the question, can we fabricate a full-field x-ray microscope analogous to conventional optical microscopes? This would require a magnification M such that angstrom separations in the object would be resolvable in the image using a detector (see Figure 6.1). As the pixel sizes of modern detectors are of the order of a micron, and often substantially larger, this would require $M \geq 10^4$. No hard x-ray lens exists that even comes close to such a magnifying power, on account of the near-unity refractive index of hard x-rays (for refractive optics), or the unrealistic required dimensions of any diffractive optics (e.g. Fresnel zone plates). We must therefore conclude that full-field x-ray microscopy on the angstrom scale is impossible.

Fortunately, the information provided by x-rays scattered by an object before they are refocussed by a lens can in principle reveal the atomic-scale structure. Such scattering experiments, be they diffraction by a crystal, or diffuse scattering by an individual (nonperiodic) structure, are the subjects of this chapter.

Diffraction is one of the most established and venerable synchrotron methods – virtually no modern synchrotron facility is without one or more macromolecular crystallography[1] beamlines, for example. Insights from diffraction data taken at synchrotrons continue to surprise and fascinate the scientific community, as illustrated by the recent spate of Nobel prizes in Chemistry in 2003, 2006, 2009, and 2012 for research into fundamental biological processes using macromolecular crystallography. In the multidisciplinary investigations that are now so common in synchrotron research, many users who would gain invaluable information from diffraction studies, are not hard-core crystallographers, and indeed may have little or no background in solid-state physics or materials science. Such scientists do not necessarily feel the need or drive to plough through any of the many excellent crystallography monographs that are available [1–4] in order to gain the perceived necessary level of understanding to conduct a successful diffraction experiment; they can thus sometimes be deterred from pursuing this line of research. In

---

[1] The term 'crystallography' was first coined in 1723 by the Swiss physician, Maurice Capeller.

An Introduction to Synchrotron Radiation: Techniques and Applications, Second Edition. Philip Willmott.
© 2019 John Wiley & Sons Ltd. Published 2019 by John Wiley & Sons Ltd.

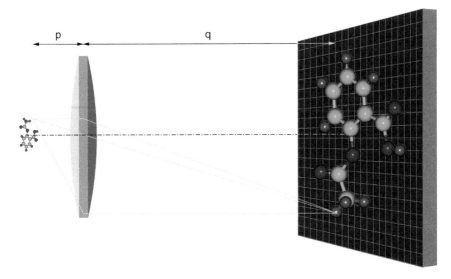

Figure 6.1   Magnification of an object by x-ray full-field microscopy so that atomic-scale features are resolved requires a magnification $M = q/p$ equal to or larger than the detector spatial resolution (pixel size) divided by the typical interatomic distances, corresponding to $M \geq 10\ 000$. No such x-ray lenses can be fabricated.

this chapter, I will attempt to convey the most important aspects of diffraction, provide useful insights that assist in obtaining a physical understanding from diffraction data, and explain experimental and technical aspects of synchrotron-based diffraction methods, in a manner such that the reader does not require a prior deep knowledge of crystallography.

X-ray diffraction (XRD) is a technique that enables one to determine the atomic structure of crystals with unsurpassed precision. The word 'diffraction' has its etymological roots in the past participle of the Latin verb 'diffringere', which means 'to break up into pieces'. The 'pieces' that are observed in diffraction are those parts of an incoming beam that constructively interfere after being scattered by the individual particles (in x-ray diffraction, atoms, or more precisely, electron clouds around the atoms) within the scattering body. XRD has many important applications in science and technology, for example, in determining the structure and functionality of proteins; in identification of changes in crystal structure due to external influences or crystallographic defects, which can induce property changes and are of great importance in electronics and sensor technology; in tracking the dynamics of chemical transitions; and in innumerable other technological fields.

When x-rays impinge on a crystal, they are scattered by the electrons within that crystal. The electron density has the same periodicity as that of the crystal lattice (see below), and hence, it acts like a diffraction grating in just the same manner as does an optical grating for visible light. In certain directions, the scattered x-rays add up constructively to produce diffraction peaks. The principal goal in XRD is to record this diffraction pattern and then convert it back into the periodic electron-density distribution that produced it.

This goal is hampered by the fact that x-ray detectors do not measure the electric-field amplitudes and phases of the x-radiation,[2] but rather the *intensity*, i.e. the average absolute square of the amplitude, in which the phase information is lost. For relatively simple crystalline systems, containing fewer than a few hundred atoms per unit cell, the brute calculating power brought to bear by modern computers normally allows one to overcome this so-called 'phase problem' and so reconstruct the atomic structure using reasonable physical and chemical constraints. This

---

[2] This should come as no surprise, as typical frequencies of electromagnetic radiation in the x-ray region lie around 10^{18}–10^{19} Hz. The fastest electronic equipment has a bandwidth of approximately 10^{12} Hz, corresponding to the far-infrared region.

becomes quite impractical, however, when attempting to determine more complicated structures, such as proteins, which may contain several tens of thousand atoms per unit cell. In this case, additional experimental information is required. We discuss the phase problem in a semi-quantitative manner in Section 6.6.

In this chapter, we will study the diffraction of x-rays by regular (three-dimensional and two-dimensional) periodic structures, and also scattering from noncrystalline objects. We will only concern ourselves with elastic (Thomson) scattering of x-rays by electrons.

As is usual in all diffraction phenomena, the wavelength of the x-rays used for diffraction studies is of the same order of magnitude as the distance between the scattering centres used to produce the pattern, and therefore lies in the broad region between approximately 0.1 and 4 Å, corresponding to typical photon energies between 100 keV and 3 keV, respectively. The exception to this is resonant soft x-ray scattering, which uses photon energies somewhat below 1 keV.

6.2 Diffraction at Synchrotron Sources

The tuneability and high brilliance of x-ray radiation at diffraction facilities at modern synchrotron sources offer the user greater possibilities compared to laboratory-based sources (see Figure 6.2). What problems might motivate a potential user to consider using synchrotron radiation, and what factors determine the choice of the wavelength to be used? The answers for any one experiment may be manifold and include subtle aspects too specialized to be generalizable. However, one can usually say that one or more of the following play a rôle:

- Many experiments require samples to be probed deeply – for example, stress measurements and studies of granular structure, strain, and microcracks in welds in mechanical structures such as automotive or aeronautical

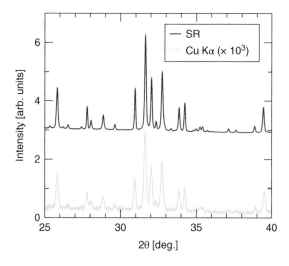

Figure 6.2 Comparison of a powder-diffraction pattern of calcium phosphate [$Ca_2(PO_4)_3$], recorded using a laboratory Cu-Kα source, with that recorded at a synchrotron facility. The former required 1500 steps (0.01° per step and 6.5 s per step) of a point detector, thus totalling a recording time of 9800 s; the latter was recorded using a Mythen strip detector (with an intrinsic resolution of 3.6×10^{-3} degrees) with a total exposure time of 5 s. Note also that the pattern recorded in the lab is multiplied in this figure by 1000 in order to adjust it to a scale comparable to that of the SR-recorded pattern; while the peak widths of the SR pattern, at 0.08° are half that of the lab-source pattern. Much of the small (but important) signal is lost in the noise of the lab-source pattern. Courtesy M. Fisch, University of Bern and N. Casati, Paul Scherrer Institute.

components. In such cases, very high photon energies are used, where the absorption coefficient is low, and the properties of the sample even some centimetres below the surface can be probed.

- The very low beam-divergence required, especially for structure determination, may only be achievable at laboratory-based diffractometers after compromising the flux so much that the experiment becomes impractical.
- Far from any absorption edges of the contributing atoms, the ratios of the intensities of the diffraction spots do not significantly differ with modest changes in photon energy. This is not true, however, as one crosses an absorption edge, for which there can be large changes between the relative scattering intensities These provide additional information in the battle to solve the phase problem. This phenomenon of anomalous scattering can only be exploited at a synchrotron facility, where the photon energy can be tuned to lie anywhere around an absorption edge. These techniques are indispensable and are now standard tools in the determination of the structures of complex protein molecules in macromolecular crystallography, covered later in this chapter.
- The intense flux provided by synchrotron sources enables one to measure small volumes of material. This may occur, for example, if (a) individual micron-sized crystallites are to be investigated; (b) when an ultra-thin film must be measured; (c) in the investigation of nanomaterials; or (d) when the surface region of a sample, or interface region between two or more heterogeneous crystalline materials, differs subtly from the bulk structure.
- Some experiments require special and bulky sample environments, such as *in situ*, nonambient setups including high-pressure anvil cells, high-temperature furnaces, film-deposition chambers, or cryogenic coolers; or large 1D or 2D detectors. Most of these cannot be mounted on a conventional laboratory-based diffractometer and require the larger experimental 'elbow room' available at most synchrotron-based diffraction stations.
- Time-resolved studies with characteristic time-scales in the millisecond regime or below, where there is a premium on high flux, require the high beam intensities offered by synchrotron radiation. Laue and powder diffraction are especially suited to such studies, as both are capable of recording large volumes of diffraction data in parallel.

Here, we will present, as much as is possible, concepts of x-ray diffraction in qualitative and pictorial terms, restricting the use of mathematical descriptions to a minimum. For the more mathematically inclined reader, however, we describe XRD using the powerful concept of the convolution theorem at the end of the theoretical section. We then move on to practical applications. No attempt is made to cover all the menagerie of experimental diffraction techniques available, but instead we concentrate on those that exploit the advantages of synchrotron radiation. Before all this, however, we begin with a brief recap of periodic crystal structures.

6.3　Description of Crystals

An ideal crystal consists of an infinite array of identical units separated evenly from each other in three directions in space. Clearly, there exists no ideal crystal in reality, as all crystals are finite in size. Moreover, all real crystals contain crystallographic defects of some sort. More often than not, these define the limits of periodicity. Let us, however, begin with the ideal crystal model as a mathematical construct for understanding diffraction effects.

6.3.1　Lattices and Bases

An ideal crystal can be described in terms of an infinite regular lattice of points in space, the so-called 'Bravais lattice', which maps out the crystal's periodicity, and a group of atoms called the 'basis' that is anchored to each and every Bravais-lattice point (Figure 6.3). Together, the lattice and basis form the crystal structure. As we will see, Bragg's law depends only on the Bravais lattice, but cannot predict the intensity of the diffraction peaks. For that, the basis must also be known.

One can move from one Bravais-lattice point to another by a translation vector $\mathbf{T}$ such that

$$\mathbf{T} = u_1\mathbf{a} + u_2\mathbf{b} + u_3\mathbf{c}, \tag{6.1}$$

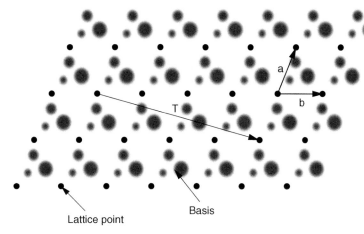

Figure 6.3   A portion of a two-dimensional crystal structure. All lattice points are separated from one another by a crystal translation vector **T**, which is itself composed of an integral number of the primitive translation vectors **a** and **b** (in this example, **T** = −**a** + 4**b**). Within the two-dimensional 'volume' |**a** × **b**| resides the basis, consisting of one or more atoms.

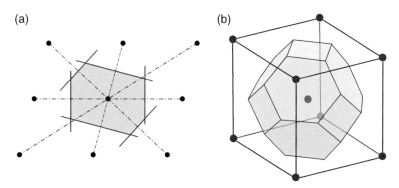

Figure 6.4   Wigner–Seitz cells. (a) Construction of a two-dimensional primitive Wigner–Seitz cell. Planes which bisect lines joining neighbouring atoms define the boundaries of the Wigner–Seitz cell. (b) The Wigner–Seitz cell (shown in yellow) of a body-centred cubic (bcc) conventional unit cell (black lines) is a truncated octahedron. The square facets derive from the planes bisecting lines between the body-centre lattice points, while the hexagons are from the planes bisecting corner lattice points to the body-centre lattice point. The volume of the bcc Wigner–Seitz cell is half that of the conventional bcc cell.

where u_1, u_2, and u_3 are integers and **a**, **b**, and **c** are the three primitive translation vectors which give the directions and minimum distances required to translate the crystal, so that it is indistinguishable from its original position. These vectors form three edges of a parallelepiped of volume

$$V_c = |\mathbf{a} \cdot \mathbf{b} \times \mathbf{c}|. \tag{6.2}$$

They describe a primitive cell for which there are lattice points only at the cell's corners. All crystals can be mapped out using such primitive cells. The associated primitive basis contains the minimum number of atoms that can be used to describe the crystal structure and one lattice point each. An important type of primitive cell, called the Wigner–Seitz cell, is constructed by the volume enclosed by planes that perpendicularly bisect lines between pairs of lattice points, shown in Figure 6.4.

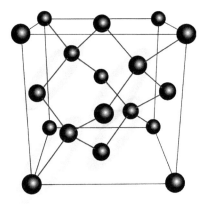

Figure 6.5   The crystal structure of diamond. Each atom is bonded tetrahedrally to four nearest neighbours. The face–centred unit cell contains four lattice points, each associated with a pair of carbon atoms, highlighted by the yellow ellipses.

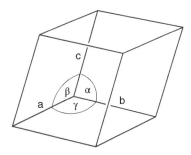

Figure 6.6   A unit cell of a crystal can be defined by the parameters a, b, c, α, β, and γ.

Often, however, nonprimitive cells and bases are chosen, for reasons of convenience and high symmetry. So, while the simple cubic cell is primitive, the 'conventional' body-centred cubic (bcc) cell is nonprimitive and has twice the volume of the associated primitive cell for a bcc crystal [see Figure 6.4(b)].

The basis associated with the crystal structure consists of N atoms within the volume defined by **a**, **b**, and **c**, such that

$$\text{basis} = \{x_j\mathbf{a} + y_j\mathbf{b} + z_j\mathbf{c}, \quad j = 1 \cdots N\}, \tag{6.3}$$

where $0 \leq x_j, y_j, z_j \leq 1$. The diamond lattice, shown in Figure 6.5, is face-centred cubic with a basis of two identical atoms at $(0,0,0)$ and $(1/4, 1/4, 1/4)$.

Note that the lattice parameters **a**, **b**, and **c** are vectors, and therefore their relative orientations are implicitly given. Another way of representing a unit cell, be it primitive or nonprimitive, is by using the parameters $a = |\mathbf{a}|$, $b = |\mathbf{b}|$, and $c = |\mathbf{c}|$ and α, β, and γ, as shown in Figure 6.6. In this formalism, the unit-cell volume is given by the general equation for a parallelepiped, that is,

$$V_c = abc(1 + 2\cos\alpha\cos\beta\cos\gamma - \cos^2\alpha - \cos^2\beta - \cos^2\gamma)^{1/2}. \tag{6.4}$$

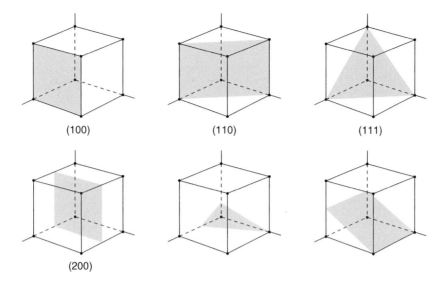

(100) (110) (111)

(200)

Figure 6.7 Miller indices of crystal planes in a cubic lattice. The indices of the last two examples are left to the reader to determine.

6.3.2 Crystal Planes

A plane can be defined by any three points in a volume, provided they are not collinear. The orientation of a crystal plane defined by three lattice points is specified by the so-called 'Miller indices' (*hkl*), which are themselves determined in the following manner.

- Find the intercepts of the plane on the crystal axes in units of their respective lattice constants *a*, *b*, and *c*.
- Take the reciprocals of these numbers and then reduce these to the smallest three integers that have the same ratio. The result (*hkl*) is called the index of the plane. Some examples are given for a cubic crystal in Figure 6.7.

An important quantity to determine is the spacing d_{hkl} between (*hkl*) planes, as shown schematically for a two-dimensional crystal in Figure 6.8. This is in general given by

$$d_{hkl} = \frac{X}{Y}, \tag{6.5}$$

whereby

$$X = [1 - \cos^2\alpha - \cos^2\beta - \cos^2\gamma + 2\cos\alpha\,\cos\beta\,\cos\gamma]^{1/2} \tag{6.6}$$

and

$$\begin{aligned}
Y = \Bigg[&\left(\frac{h}{a}\right)^2 \sin^2\alpha + \left(\frac{k}{b}\right)^2 \sin^2\beta + \left(\frac{l}{c}\right)^2 \sin^2\gamma \\
&- \frac{2kl}{bc}(\cos\alpha - \cos\beta\,\cos\gamma) - \frac{2lh}{ca}(\cos\beta - \cos\gamma\,\cos\alpha) \\
&- \frac{2hk}{ab}(\cos\gamma - \cos\alpha\,\cos\beta)\Bigg]^{1/2}.
\end{aligned} \tag{6.7}$$

In general, the interplanar spacing decreases as the Miller indices increase, and the density of lattice points in a plane – i.e. the number of lattice points per unit area of the plane – decreases as the Miller indices increase. Note that Equations (6.6) and (6.7) become significantly simpler for high-symmetry systems such as orthorhombic, tetragonal, and cubic unit cells, for which $\alpha = \beta = \gamma = 90°$.

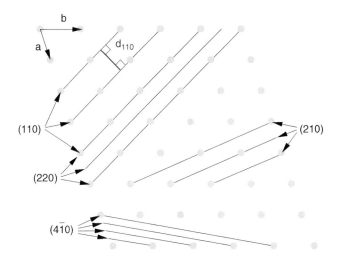

Figure 6.8   For any given unit cell, the spacing between planes defined by the Miller indices (*hkl*) can be determined using Equations (6.5), (6.6), and (6.7).

6.3.3   Labelling Crystallographic Planes and Axes

A specific crystal axis, normal to the crystal plane, is labelled [*uvw*]. Sets of equivalent axes, due to lattice symmetry, are labelled ⟨*uvw*⟩. Hence, in a cubic crystal, the set of axes ⟨100⟩ include [100], [010], [001], [$\bar{1}$00], [0$\bar{1}$0], and [00$\bar{1}$], whereas in an orthorhombic cell ⟨100⟩ only includes [100] and [$\bar{1}$00].

As described above, specific crystal planes are described by the Miller indices (*hkl*). Equivalent planes are denoted by {*hkl*}.

6.4   Basic Tenets of X-ray Diffraction

6.4.1   Introduction

Perhaps the earliest example of diffraction that a student encounters is Young's double-slit interference experiment. This involves two identical sources and results in the familiar double-slit fringe pattern. What is the difference between interference and diffraction? Essentially, there is none, though one tends to talk of interference effects when a small number of scatterers are involved and diffraction when there is a large array of scatterers[3].

A diffraction pattern is in fact nothing more than the square of the Fourier transform (FT) of the scattering electron-density distribution, $\rho(\mathbf{r})$. Formally,

$$\{F_{hkl}\} = \mathscr{F}\left[\rho(\mathbf{r})\right], \tag{6.8}$$

$$I_{hkl} = |F_{hkl}|^2, \tag{6.9}$$

where $\{F_{hkl}\}$ is the set of scattering amplitudes explained in Section 6.4.3, and I_{hkl} is the observed intensity of the diffraction peak at (*hkl*). To the uninitiated, this may appear to be an unhelpful statement (it isn't!) and immediately begs the question, what *is* a Fourier transform? An introduction to Fourier analysis is given in Appendix C.

A highly ordered periodic scatterer such as a crystal will scatter x-rays in very specific directions to produce a scattering pattern with sharp maxima, the diffraction peaks. The diffraction peaks together make up the (absolute

---

[3] The verbs are more distinct: objects diffract; waves interfere.

square of the) Fourier transform of the electron-density distribution within the crystal's unit cell. Each peak, a so-called Fourier component, represents a sinusoidal wave of electron density with a certain frequency, amplitude, and direction determined by the peak's position [its (*hkl*)-values] within the pattern. The phase of the wave is unknown, as only the intensity, proportional to the absolute square of the amplitude, is measured. Note that, as each wave component of this pattern must have a spatial frequency equal to a multiple integer of the structure's periodicity, the diffraction maxima (variously also referred to as the 'Fourier components' or 'structure factors') are distributed in a regular, equally spaced array in so-called 'Fourier space'[4]. Once the phase relationship between all these Fourier components is known through the process of 'phase retrieval' detailed in Section 6.6, the sinusoidal waves they represent, when superimposed upon one another, reconstruct the electron density within the unit cell.

As discussed in Appendix C, nonperiodic structures have Fourier transforms which are no longer a series of evenly separated and precisely defined components, but contain instead a continuous distribution of spatial frequencies. They are, however, no less deterministic in nature. We will discuss the investigation of noncrystalline structures by scattering at the end of this chapter in Section 6.15.

Once Fourier transforms are understood, they help considerably in interpreting the meaning of the profiles of diffraction peaks, the overall envelope of intensity change in a diffraction pattern, and the general shapes of the crystallites contributing to a given diffraction pattern.

To begin, however, we will avoid the formal use of Fourier transforms – we will be using them, but for the innocent bystander this may not be obvious. Instead, we consider the conditions which lead to constructive interference. We will discover that although the periodic spacings in the crystal can be derived from the positions of the peaks in the diffraction pattern using simple geometrical considerations (the Bragg law), the peak intensities are determined by the positions and types of the scattering centres (i.e. the atoms) within the unit cell of the crystal, and that it is this information that enables us (sometimes only after considerable effort) to regenerate the atomic basis.

In Figure 6.9, basic examples of scattering from arrays of scattering centres plus their resulting diffraction patterns are given. What general features do diffraction patterns have in common? A few of the most important points are listed here.

- The scattering vector **Q**, i.e. the vector joining up the incoming beam $\mathbf{k}_{in}$ and the diffracted beam $\mathbf{k}_{out}$, always lies perpendicular to the scattering planes (see below).
- The sharpness of the diffraction signal is proportional to the number of scattering planes that are involved.
- The separation between the peaks in the diffraction pattern is inversely proportional to the separation of the diffracting planes in real space.
- The maximum number of accessible reflections N in a diffraction pattern is directly proportional to the unit cell volume V_c and the cube of the photon energy. Precisely,

$$N = 33.5 \frac{V_c}{\lambda^3}.$$
(6.10)

The maximum potential volume of data for protein crystals, which have unit cell dimensions of the order of 100 Å, can therefore easily exceed ten million structure factors, though geometrical considerations always means that this number is never achieved.

6.4.2 The Bragg Law and Reciprocal Lattice

The peak positions of an x-ray diffraction pattern were explained by W. L. Bragg and his father W. H. Bragg in 1913 by their famous eponymous equation

$$m\lambda = 2d \sin \theta.$$
(6.11)

---

[4] Fourier space is also called 'reciprocal space' (as the axes have dimensions of inverse length), 'k-space', or 'momentum space'.

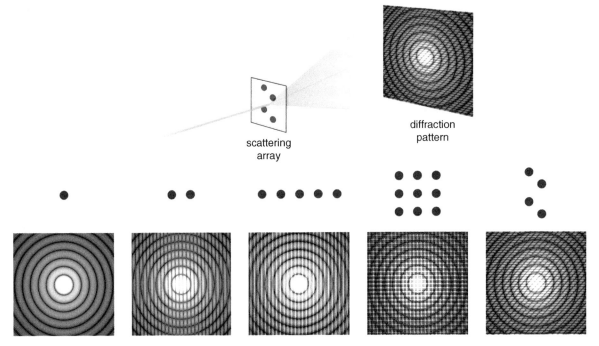

Figure 6.9 Diffraction from regular one- and two-dimensional arrays of scatterers. The diffraction maxima become sharper with increasing number of scatterers. Two other important aspects are highlighted in the array on the far right: an increase in distance between the scatterers causes the diffraction spots to move closer in an inversely proportional manner, while an acute angle θ in the monoclinic real-space arrangement results in a diffraction pattern with an obtuse angle of $\pi - \theta$.

Here, λ is the wavelength of the x-ray light, d is the interplanar spacing of the (hkl) planes, θ is the angle of incidence above the plane surface, and m is an integer (Figure 6.10). The law describes the difference in the optical path length between reflections from adjacent crystal planes, which must be an integer multiple of wavelengths for constructive interference to occur. This therefore also implies that the phase difference between scattering from adjacent planes is a multiple of 2π. We will come back to this shortly in Section 6.4.3, in which we discuss the influence of the basis.

We can re-express Bragg's law in terms of the photon energy by substituting λ with hc/E. This yields

$$\sin\theta = \frac{6.1992}{d_{hkl}[\text{Å}] \, E[\text{keV}]}, \tag{6.12}$$

where d_{hkl} is the spacing of the (hkl)-scattering plane.

One of the most important conclusions of Bragg's law is that when Equation (6.11) is satisfied, *the scattering vector* $\mathbf{Q}$ *always lies perpendicular to the scattering planes, or in other words, the angle subtended by* $\mathbf{k}_{in} = 2\pi/\lambda$ *(or* $\mathbf{k}_{out}$*) and the scattering planes is* θ.

Moreover, when the condition for detecting a diffraction maximum is met, the scattering vector $\mathbf{Q}$ always connects the (000) diffraction spot (in other words, the spot produced by the direct beam) to another maximum in the diffraction pattern.

How are the diffraction maxima positioned relative to one another? We have already argued above that they should be regularly spaced in a three-dimensional periodic array with the (000) direct spot at the centre. The three

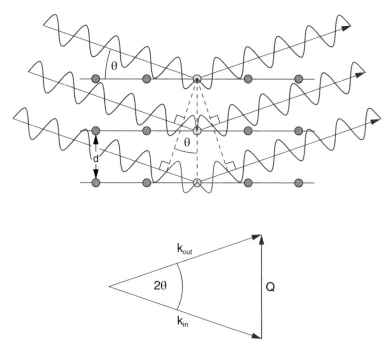

Figure 6.10 The Bragg law describes interference between rays elastically scattered off successive atomic planes, separated from one another by a distance *d*. When the optical path difference (shown here in red) between adjacent rays is an integer multiple *m* of the x-ray wavelength, interference is constructive, and a diffraction peak will be seen at that angle. The scattering vector **Q** is the vectorial difference between the incoming wavevector $\mathbf{k}_{in}$ and the outgoing wavevector $\mathbf{k}_{out}$. The magnitudes of $\mathbf{k}_{in}$ and $\mathbf{k}_{out}$ are equal to $2\pi/\lambda$.

periodicities describing this array, known as the 'reciprocal lattice', are called the 'reciprocal-lattice vectors' and are related to the lattice vectors in real space by

$$\mathbf{a}^* = 2\pi \frac{\mathbf{b} \times \mathbf{c}}{\mathbf{a} \cdot (\mathbf{b} \times \mathbf{c})};$$

$$\mathbf{b}^* = 2\pi \frac{\mathbf{c} \times \mathbf{a}}{\mathbf{b} \cdot (\mathbf{c} \times \mathbf{a})};$$

$$\mathbf{c}^* = 2\pi \frac{\mathbf{a} \times \mathbf{b}}{\mathbf{c} \cdot (\mathbf{a} \times \mathbf{b})}. \tag{6.13}$$

The denominators in the above three equations are all equal to the unit-cell volume V_c, a scalar quantity without a direction. The numerators, on the other hand, are the cross-products of two of the lattice vectors, and therefore have their direction perpendicular to the planes defined by those two vectors. So, for example, the reciprocal lattice vector **a*** is perpendicular to the plane containing the two vectors **b** and **c**.

The above rules can be very simply and intuitively represented by the so-called 'Ewald construction' in reciprocal space, shown in Figure 6.11: the incident wavevector $\mathbf{k}_{in}$ must end, and the scattering vector $\Delta k = \mathbf{Q}$, must begin at the (000) diffraction spot of the direct beam, while, for constructive interference to occur, **Q** and $\mathbf{k}_{out}$ must end at another diffraction maximum (a 'reciprocal-lattice point'). As x-ray diffraction is an elastic process, this means that these two points must lie on the surface of a sphere (the 'Ewald sphere') of radius $|\mathbf{k}|$ and whose centre lies at the base of the $\mathbf{k}_{in}$ and $\mathbf{k}_{out}$ vectors. This therefore geometrically defines the value of θ (and 2θ) by

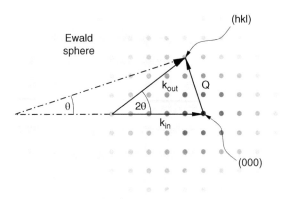

Figure 6.11   The Ewald construction. The incident beam $\mathbf{k}_{in}$, ending at (000), and the diffracted beam $\mathbf{k}_{out}$, ending at (*hkl*), define the Ewald sphere.

the magnitude of $\mathbf{Q}$. If one rotates the crystal in real space, the array of diffraction maxima is rotated around the (000) point by the same amount in reciprocal space.

Diffraction patterns are all too often plotted out as a function of the scattering angle 2θ. In order to determine the scattering plane separations, however, one must also know the photon energy (or wavelength). It is more convenient, therefore, to plot the pattern out as a function of $\mathbf{Q}$. It is clear from Figure 6.10 that

$$\mathbf{Q} = 2|k| \sin\theta = \frac{4\pi}{\lambda} \sin\theta. \tag{6.14}$$

Note that for angles close to $\theta = 90°$, the maximum accessible scattering vector expressed in reciprocal angstroms is approximately equal to the photon energy given in keV, that is,

$$\mathbf{Q}[\text{Å}^{-1}] = 1.0135 \, E \, [\text{keV}] \sin\theta. \tag{6.15}$$

Substituting for $\sin\theta/\lambda$ using Equation (6.11), we obtain

$$|Q| = \frac{2\pi}{d_{hkl}}, \tag{6.16}$$

which, importantly, is independent of the photon energy. We are able to drop the order m from the equation, as this is implicitly given by the Miller indices (*hkl*).

As we have already discussed, each Bragg peak represents a Fourier component of the electron density within the unit cell, that is, a wave with periodicity d_{hkl}. Thus, the best resolution obtainable from any given diffraction pattern is determined by the largest recorded value of Q and is equal to $2\pi/Q$.

The Bragg law reflects the periodicity of the lattice, but tells us nothing of the basis within the unit cell. It is the composition of the basis, however, which determines the relative intensities of the diffraction peaks. We turn now to this problem.

6.4.3   The Influence of the Basis

The earliest structures to be investigated by x-ray diffraction by the Braggs, including rock salt and diamond, could be solved using symmetry arguments and the Bragg law, without having to resort to knowing in detail the

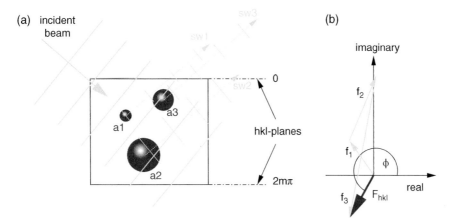

Figure 6.12 (a) The intensity of a Bragg peak depends on the positions and atomic scattering factors (or 'form factors') of the atoms making up the basis. In this schematic, there are three atoms. The scattered waves sw1, sw2, and sw3 are, of course, spherical, but here we draw them as plane waves in the direction of 2θ for reasons of clarity. (b) The structure factor F_{hkl}, of the wave scattered at the Bragg angle is the vector sum of the contributions of each atom (i.e. f_1, f_2, and f_3), taking into account their phases relative to the incident beam.

scattering strength of each individual atom in the unit cell. Unravelling the structure of more complex crystals would not be so trivial, and it was the son, W. L. Bragg, who developed the necessary methods to tackle these more complex systems, first shown for the mineral diopside $CaMg(SiO_3)_2$ with Warren in 1928 [5]. A beautiful review of this seminal work and the early history of x-ray analysis is given by one of Bragg's most luminary students, Max Perutz [6].

Consider Figure 6.12. An incoming x-ray beam is weakly scattered by the electron clouds around the three atoms a1, a2, and a3 of a unit cell. We want to find out how these scattered waves, sw1, sw2, and sw3, add up in the scattering direction 2θ which satisfies Bragg's law. The amplitude of each scattered wave is proportional to the vector sum of the scattering strengths of all the electrons attached to the scattering atom, that is, to its atomic scattering factor $f(Q)$, while the phases between the scattered waves depend on the relative positions of the atoms in the unit cell. This is schematically illustrated in Figure 6.12(b) in an Argand diagram. The total scattering amplitude, or so-called 'structure factor' $\mathbf{F_{hkl}}$ is the vector sum of the individual atomic scattering factors $\mathbf{f_1} + \mathbf{f_2} + \mathbf{f_3}$, whereby the phases are determined by their relative orientations: precisely, the phase of each scattered wave is given by the associated atom's position in the direction perpendicular to the scattering planes, varying linearly from 0 at one plane to $2m\pi$ at the next. This is described mathematically as

$$F_{hkl} = \sum_j f_j \exp[-i2\pi(hx_j + ky_j + kz_j)],\tag{6.17}$$

where f_j is the atomic form factor of the jth atom, (x_j, y_j, z_j) is its position in the unit cell expressed in fractions of the unit-cell lattice vectors, and the summation is over the j atoms within the unit cell. The scattered intensity I_{hkl} is the absolute square of $\mathbf{F_{hkl}}$.

There may arise a situation whereby the magnitude F_{hkl} is zero, i.e. the scattering vectors from the individual atoms within the unit cell cancel one another out, in which case no Bragg peak is observed. This is referred to as a '*systematic absence*'. It is a straightforward exercise to demonstrate that some crystal types have systematic absences for certain values of (hkl), due to their internal symmetry. So, for example, all reflections are allowed for a simple cubic structure; for a body-centred cubic (bcc) cell, reflections for which $h + k + l$ is odd are missing; for face-centred cubic (fcc), reflections in which there are both even and odd values of h, k, and l (i.e. of mixed

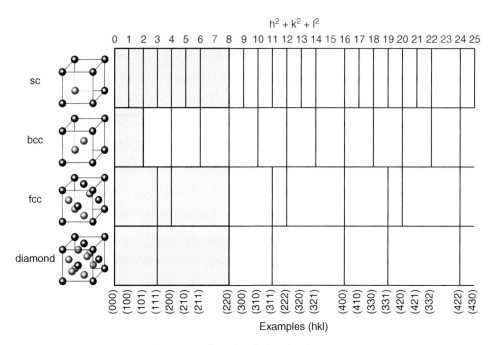

Figure 6.13   Allowed reflections as a function $h^2 + k^2 + l^2$ for the four unit cell types simple cubic (sc), body–centred cubic (bcc), face-centred cubic (fcc), and diamond. Notice that no reflections exist even in principle if $h^2 + k^2 + l^2 = 8m - 1$, where m is an integer greater or equal to unity (shown here in red font). The repeat pattern unit for each cell type is highlighted in yellow.

parity) are missing; and for diamond-like structures, only those reflections for which h, k, and l are all odd, or in which they are all even *and $h + k + l$ is* a multiple of 4, are present. A summary of which reflections are present or not in these lattice types is shown in Figure 6.13.

The phenomenon of systematic absences highlights the shortcomings of the simple Bragg law. The reason systematic absences occur, and why some diffraction peaks are strong while others are weak, is because we must consider the positions and scattering strengths of the atoms between the crystal planes, and not just the planes' separation.

In general, in order to determine the relative intensities of Bragg peaks, we must know the atomic form factors f_j of the j atoms making up the basis as a function of scattering angle, as first mentioned in Chapter 2. It is beyond the scope of this text to give a detailed description of their dependence here. As a rule of thumb, however, the atomic form factor increases approximately linearly with the atomic number Z (i.e. the number of electrons) and decreases quasi-monotonically with the scattering vector **Q** (see also Figure 2.10). Values for f can be found in the *International tables for x-ray crystallography*, Volume 3 [7].

Finally, we discuss a property of the structure factors of crystals exhibiting centrosymmetry, as this is relevant to anomalous-diffraction phenomena, covered in Sections 6.6.3 and 6.6.4. Centrosymmetric crystals are those which are invariant after the inversion of each atomic position from **r** to −**r**. Consider Figure 6.14. The contributions to the structure factor from pairs of atoms at **r** and −**r** mirror each other across the real axis, and hence their imaginary components cancel each other. If there is an atom at the inversion centre, conventionally given as being at the centre of the unit cell, its phase is 0, and thus only has a real component. The conclusion to be drawn from this is that crystals exhibiting centrosymmetry have structure factors which have no imaginary component and are therefore always real for all (*hkl*) reflections. This argument is predicated on the assumption that f_2, the imaginary component of the structure factor of any given atom, is negligible. Close to an absorption edge, however, the

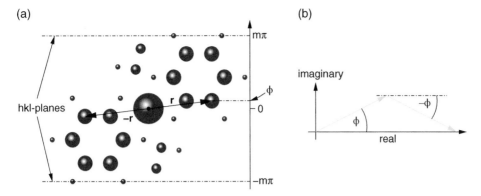

Figure 6.14 Structure factors of centrosymmetric crystals. (a) A centrosymmetric crystal is one which remains invariant when all atoms are translated from **r** to −**r**. For convenience, the zero phase is set to be that of the centre of inversion. In this case, the phase at the (*hkl*)-planes are ±*m*π, where *m* is integral. (b) The vector addition of any pair of atoms connected by ±**r** lies on the real axis, as their atomic form factors are equal in magnitude but opposite in phase, and hence the imaginary components cancel.

associated atoms will have both real and imaginary components; nonetheless, the vector sum of the two atomic form factors associated with absorbing centrosymmetric pairs will then always have f_1 still on the real axis, and f_2 exactly on the imaginary axis, independent of where they reside within the unit cell.

6.4.4 Dynamical Diffraction

Until now, it has been tacitly assumed that the interaction between the x-rays and the crystal planes has been weak and, hence, that the intensity of the incoming beam at each successive plane is the same. No consideration was made for a drop in intensity of the incident beam due to elastic scattering or photoabsorption, or multiple-scattering effects. This simplifies analysis considerably and is known as the *kinematical approximation*. When multiple scattering becomes important in macroscopic crystals, the change in intensity of the incident beam as it penetrates the material can no longer be ignored, in which case, one talks of *dynamical diffraction*. It lies outside the scope of this text to discuss these effects in detail. The concepts of the Darwin width and extinction depth are, however, important, not least for monochromator design, and are thus briefly presented now.

One of the consequences of dynamical diffraction is that the number of crystal planes in a near-perfect crystal required to scatter the incoming beam so it has an intensity $1/e$ of its initial value (known as the extinction length, Λ_{ext}) can be significantly smaller than the photoabsorption attenuation length when the crystal has been oriented so it is close to the Bragg condition. Precisely at the Bragg condition,

$$\Lambda_{\text{ext}} = \frac{V_c}{2r_0\lambda|F|}, \tag{6.18}$$

where $|F|$ is the magnitude of the unit-cell structure factor at the Bragg condition and depends on both Q and the photon energy. Λ_{ext} is of the order of a few microns for the (111) reflection of diamond. In contrast, the photoabsorption length in the hard x-ray regime is measured in millimetres (see Figure 6.15).

Note that Λ_{ext} is the length travelled by the x-rays at an angle equal to the Bragg angle relative to the scattering planes. The extinction *depth* Λ'_{ext} is, in the case of the diffracting planes being parallel to the crystal surface, the perpendicular depth penetrated in the crystal by the x-rays after travelling Λ_{ext} and is thus given by

$$\Lambda'_{\text{ext}} = \Lambda_{\text{ext}} \sin\theta = \frac{V_c \sin\theta}{2r_0\lambda|F|}. \tag{6.19}$$

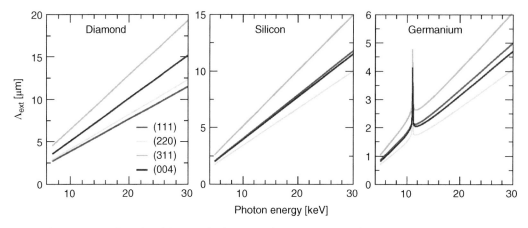

Figure 6.15   The extinction lengths of diamond, silicon, and germanium at selected Bragg peaks. These are in general two or more orders of magnitude smaller than the photabsorption attenuation length $1/\mu$. Even at the K-edge of germanium, $\Lambda_{\text{ext}} = 4.7$ µm for the (311) reflection and $1/\mu = 9.3$ µm.

But $2\sin\theta/\lambda = 1/d_{hkl}$ and so

$$\Lambda'_{\text{ext}} = \frac{V_c}{4d_{hkl}r_0|F|}. \tag{6.20}$$

Note that Λ_{ext} is proportional to the photon energy while Λ'_{ext} is, to a first approximation, energy independent. This is not entirely correct, however, as $|F|$ is energy dependent, especially when close to absorption edges.

As the number of scattering planes is finite, there is therefore also a nonvanishing spread of angles over which interference between the scattered signals from all the involved planes is still constructive, resulting in a top-hat profile for the diffraction maximum. This flat region has a width w_D called the 'Darwin width', given by

$$w_D = \frac{2r_0\lambda|F_{hkl}|d_{hkl}}{\pi V_c \cos\theta}. \tag{6.21}$$

The Darwin widths of diamond, silicon, and germanium are plotted for different Bragg reflections in Figure 6.16.

Lastly, condensed matter in general does not have the same refractive index as air [albeit only by a small amount, see Equation (2.20)] and thus the wavelength of the x-rays in the medium is marginally larger than in air. This results in a very slight, but often measurable, increase in the Bragg angle compared to the value one would expect if refraction effects were ignored.

6.5   Diffraction and the Convolution Theorem

Before we move on to practical examples of diffraction experiments performed at synchrotron sources, we briefly introduce the more mathematically inclined student to the interpretation of diffraction patterns in terms of the convolution theorem, which can provide useful insights later in the chapter.

6.5.1   The Convolution Theorem

What is a convolution? Convolutions arise in many fields of science and engineering, such as in image and signal processing, but are also very important in understanding diffraction theory. Firstly, a convolution involves at least two functions to produce a third function. Essentially, it describes the action of measuring the weighted mean of a

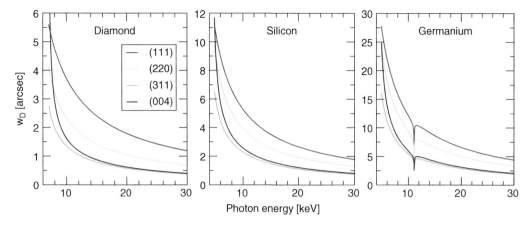

Figure 6.16   The Darwin widths of three diamond-like crystals (diamond, silicon, and germanium) for the first four allowed reflections (see also Figure 6.13). The conversion from arcseconds to microradians is $1'' = 4.85 \, \mu\text{rad}$. Note the anomalous dip at approximately 11 keV for germanium, due to its K-absorption edge and associated dip in the complex atomic form factor $f(Q, \hbar\omega)$.

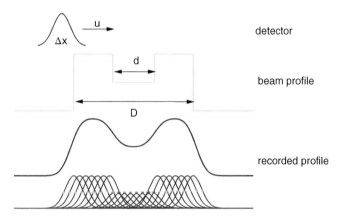

Figure 6.17   Sketch of a signal resulting from the convolution of a top hat x-ray beam profile of width D illuminating a partially transparent homogeneous block of width d and the spatial response of a photodiode of resolution Δx that has been scanned across the beam along the coordinate u.

physical quantity over a certain range. The measuring device is described by the first function, while the measured quantity is the second function. Still confused? Consider the following example. A photodiode having a detection full-width-half-maximum Δx is scanned (the coordinate u) across an x-ray beam with a perfect 'top hat' profile of width D in which a partially opaque block is inserted (see Figure 6.17).

The recorded profile from the output of the photodiode is the convolution of the beam profile and the spatial response function of the photodiode. It should be clear to the reader that if $\Delta x \ll D$ and d, then the recorded profile is an accurate approximation of the true profile. In general, the closer the range of the response function of a detecting instrument is to what it is trying to detect, the more convolved will be the output.

Mathematically, a convolution is defined as being the integral of one function at x multiplied by another function at $u - x$, whereby the integration is taken over the variable x (which may be a 1D, 2D, or 3D variable), typically between plus and minus infinity for all the involved dimensions. Hence, the convolution $C(u)$ between two functions $f(x)$ and $g(x)$ is given by

$$C(u) = f(x) \otimes g(x) = \int f(x)g(u-x)dx, \tag{6.22}$$

where $a \otimes b$ means 'convolved with', and the variable u represents the 'scanning' parameter.

Importantly, the convolution theorem states that if one of the original functions is known, Fourier analysis of the convoluted (measured) signal can be applied to obtain the form of the other function. Precisely, the convolution theorem states that *the Fourier transform of a convolution of two functions is equal to the product of their individual Fourier transforms*. This can also be read backwards as *the Fourier transform of the product of two functions is the convolution of their individual Fourier transforms*. Mathematically, this is written as

$$f(x) \otimes g(x) \leftrightarrow \mathscr{F}(k)\mathscr{G}(k), \tag{6.23}$$

where $\mathscr{F}(k)$ and $\mathscr{G}(k)$ are the FTs of $f(x)$ and $g(x)$, respectively. Interchanging f and g has no effect on the convolution (i.e. they commute).

6.5.2   Understanding the Structure Factor

How can we apply the convolution theorem in diffraction theory? The first thing to keep in mind is that a perfect crystal consists of an infinite array of lattice points (the Bravais lattice). Being lattice *points*, each one is of zero extent, i.e. a delta function. Each point of the Bravais lattice 'anchors' the electron density within the volume of the unit cell that represents the basis of the crystal, as we have already described at the beginning of this chapter. The effect of convolving any function g with a delta function δ is merely to shift g by an amount equal to the distance of δ from the origin (if you are having problems seeing this, consider again Figure 6.17 and think of the photodiode as having point-like resolution, i.e. $\Delta x \ll D$). One can therefore think of the electronic distribution of the crystal as the convolution of the Bravais lattice with the electronic distribution of the basis, as shown schematically in Figure 6.18(a).

Let us now look at these two aspects, the lattice array and basis, one at a time. The lattice array is a three-dimensional comb-array function (that is, a set of equally spaced, infinitely narrow spikes), which we argue in Appendix C to also be another comb array with separations inversely proportional to the separations in the real-space comb array. Using the convolution theorem, the diffraction pattern of an infinite crystal (assuming the kinematical limit) is the product of the comb-function FT of the Bravais lattice and the FT of the electronic distribution within a unit cell. This latter FT is the 'structure factor' and therefore determines the intensities of the peaks in the diffraction pattern [Figure 6.18(b)].

Lastly, it is briefly mentioned that we will return to the convolution theorem in Section 6.12 in helping us understand features of surface diffraction. It is therefore recommended that all but the most extreme mathematics-phobe should take five (or ten) minutes to understand some of this subsection.

6.6   The Phase Problem and Anomalous Diffraction

6.6.1   Introduction

Given an electron density in real space within a unit cell of a crystal, the calculation of the resulting diffraction pattern is routine. However, the bread-and-butter work of a crystallographer consists of the reverse process – determining the electron density in real space from the recorded diffraction pattern.

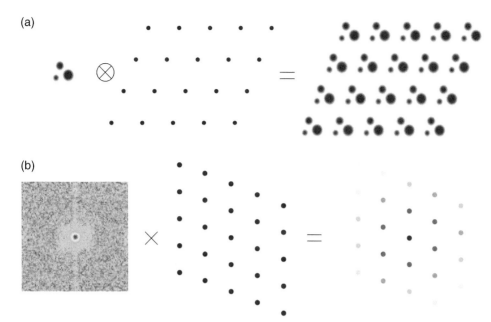

Figure 6.18 Diffraction-pattern intensities explained using the convolution theorem. (a) An infinite crystal can be thought of as the convolution of the electronic distribution within the unit cell with the Bravais lattice. (b) Using the convolution theorem, it becomes apparent that the diffraction pattern is the product of the FT of the electronic distribution in the unit cell (called the 'structure factor') and the FT of the Bravais lattice, which, like the Bravais lattice itself, is another infinite 3D comb function (see text).

It has already been mentioned in the introduction to this chapter that, because we measure the *intensity* of x-ray diffraction signals ($|F_{hkl}|^2$, a scalar quantity), and not the amplitude (i.e. the *complex* structure factor, which has both a magnitude *and* a phase), we lose half the information about the system we are measuring, namely the relative phases of the elastically scattered waves from the different atoms within the unit cell.

Referring back to Figure 6.12, we know that the structure factor that results in the measured diffracted intensity $I_{hkl} = F_{hkl}F_{hkl}^*$ must lie in the Argand diagram on a circle of radius $\sqrt{I_{hkl}}$, as shown in Figure 6.19.

The information we have at our disposal includes the integrated intensities of the Bragg peaks of our diffraction pattern, I_{hkl}, the atomic scattering factors of the constituent atoms as a function of $\sin\theta/\lambda$, and the angles of the crystallographic planes relative to the incoming beam and Bragg spots. For a given set of atomic positions within a unit cell, the phase relationship between the atomic scattering factors (given by the angles between the red, green, and yellow arrows in Figure 6.19) changes in a predetermined way from Bragg peak to Bragg peak. Our task is to find the atomic positions that satisfy the Bragg intensities for all recorded diffraction spots. The more Bragg peaks are recorded, the tighter the set of conditions becomes which must be satisfied to achieve a self-consistent model (see also the concepts of redundancy and oversampling, Section 8.4.4).

For unit cells containing only a few atoms, a trial-and-error procedure using the brute force of modern computer power can rapidly deliver the unit-cell structure. (This is often in conjunction with some reasonable physical constraints, such as, for example, that atom centres cannot be closer to one another than approximately 1 Å – one of the smallest atomic separations in condensed matter is approximately 1.25 Å, between adjacent ions in LiF. The triple bond of nitrogen gas is only 1.1 Å.) This approach becomes more impractical as the complexity of the unit cell increases. In the case of macromolecular unit cells, such as protein crystals, other supplementary information must be provided. This is discussed in more detail in Section 6.11.

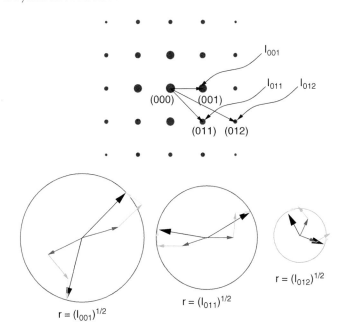

Figure 6.19 The phase problem. Different phase relationships between the constituent atoms of a unit cell can lead to the same total scattering amplitude $r = |F_{hkl}| = \sqrt{I_{hkl}}$, although the direction may be different.

6.6.2 The Patterson Map

However, all is not lost. We can obtain valuable information about the unit cell even without knowing the phases. It turns out that the inverse Fourier transform (IFT) of the diffraction-pattern intensities yields the *relative* distances and angles between the atoms in a unit cell, referred to as the Patterson map (PM) or Patterson function [8], as shown schematically in Figure 6.20[5]. One can easily demonstrate that the Patterson map is in fact the convolution of the unit cell with its inverse.

In addition to providing the relative distances and angles of atoms within a unit cell, the PM exhibits two other important features. Firstly, the number of maxima in the PM of a unit cell containing N atoms, including the central maximum, is equal to $1 + N(N - 1)$, which for large N is approximately equal to N^2 – a PM can soon become a crowded affair. Note also that because the vector connecting a given atom A to another atom B is the inverse of that connecting atom B to atom A, a Patterson map is centrosymmetric.

Importantly for the techniques of multiple isomorphous replacement (MIR), multiwavelength anomalous dispersion, and single-wavelength anomalous diffraction, all of which are used extensively in protein and macromolecular crystallography to help overcome the phase problem and are discussed in more detail in Section 6.11, the intensity of the maximum in a PM associated with two atoms A and B is proportional to $Z_A Z_B$, the atomic numbers of those atoms. Thus, peaks corresponding to the distances and angles between heavy atoms dominate PMs[6].

---

[5] This is not a treatise on crystallography, and the reader is referred to the many excellent textbooks available in the literature to understand the derivation of the Patterson function.

[6] MIR and PMs were extensively used by Dorothy Crowfoot-Hodgkin in her pioneering work on crystallography of organic and biological compounds, most famously in her determination of the structures of insulin (in which the native zinc atom in the core of the insulin molecule was substituted with lead, uranium, and mercury), and penicillin, for which sodium, potassium, and rubidium derivatives were synthesized. This latter piece of work, carried out during World War II, provided the key information for the large-scale production of penicillin at a time when it was most sorely needed. It was the discovery that penicillin contains the

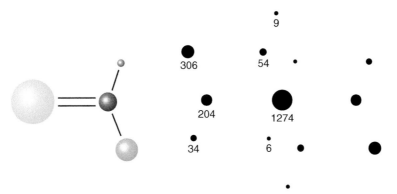

Figure 6.20   A Patterson map of a fictitious crystal containing within its unit cell one planar molecule of fluorose-lenoformaldehyde, SeCHF. Se: yellow, $Z = 34$; C: black, $Z = 6$; F: green, $Z = 9$; H: blue, $Z = 1$. The intensities of the PM are given as the product of the atomic numbers Z of the contributing pair of atoms, except for the central spot, which is the sum over all four atom types of Z^2. The position of each spot relative to the central spot is given by the vector connecting the two contributing atoms.

6.6.3   Friedel's Law and Bijvoet Mates

George Friedel, a french crystallographer in the nineteenth and twentieth centuries, discovered that the Fourier transform of a *real* object (i.e. one without imaginary components) has Fourier components $F(k)$ and phases $\phi(k)$ with the following properties:

$$F(k) = F^*(-k),$$

$$|F(k)|^2 = |F(-k)|^2,$$

$$\phi(k) = -\phi(-k), \tag{6.24}$$

whereby F^* is the complex conjugate of F. This is Friedel's law. Far from absorption edges, the electron cloud within the unit cell of a crystal is indeed a real quantity, hence all crystals, even those lacking a centre of symmetry, will have diffraction patterns exhibiting centrosymmetry, that is

$$I(hkl) = I(\overline{hkl}). \tag{6.25}$$

$I(hkl)$ and $I(\overline{hkl})$ are referred to as Friedel pairs, or Friedel mates. This can also be understood by considering Figure 6.21.

The relative phases of the scattered x-rays from the component atoms of the unit cell (ignoring for the time being absorption phenomena) are directly proportional to the differences in their distances away from the (hkl)-planes [Equation (6.17)]. Hence, the phase diagram for a Bragg peak from an incident beam from below the unit cell in Figure 6.21 will mirror that of a Bragg reflection from a beam approaching from above. The scattering vectors Q_{hkl} and $Q_{\overline{hkl}}$ are equal in magnitude but opposite in direction.

For unit cells with internal symmetry, there will be additional reflections with the same intensity, such as the (hkl) and $(hk\overline{l})$ reflections in cubic, tetragonal, orthorhombic, and hexagonal systems. These equivalent reflections are referred to as Bijvoet mates[7].

---

four-membered, so-called 'β-lactam', ring, C_3N, assumed until this study to be too chemically unstable to exist, that proved to be crucial – indeed, the instability of the C_3N-ring allowed it to open and deactivate enzymes responsible for cell-wall growth in bacteria, causing them to rapidly expire. Note the use of the past tense – bacteria soon developed resistance to this line of attack.

[7] 'Bijvoet' is pronounced 'bay-foot'

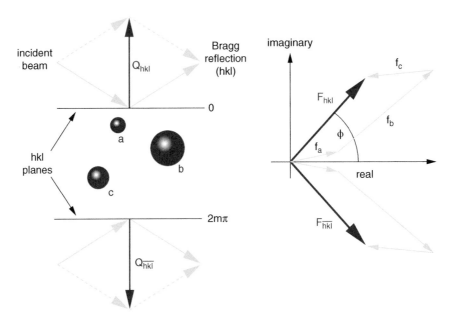

Figure 6.21   Friedel's law. A Bragg reflection (*hkl*) is shown for a set of three atoms a, b, and c, between (*hkl*)-planes. The same atoms will produce the Friedel mate ($\overline{hkl}$) for the same diffraction condition, but this time for a beam incident from below. The phases of the atomic form factors f_a, f_b, and f_c change signs between the Friedel pairs.

As one approaches the energy of an absorption edge, however, Friedel's law breaks down. We discuss this now in more detail.

6.6.4   Anomalous Diffraction

Here, we will look more closely at the change in the structure factors of crystalline materials in the region of an absorption edge. We begin by studying the illustrative example of the compound semiconductor, GaAs. Next, we show how the breakdown of Friedel's law in this region can provide valuable extra information about the atomic structure. This will prepare us for methods used for structure determination in macromolecular crystallography, described in Section 6.11.

6.6.4.1   *Tuning the Structure Factors*

We have already argued that the intensity of a Bragg peak depends both on the relative positions of the atoms in the unit cell, and on the complex form factor, $f_1 + if_2$, of each atom (see Figure 6.12). Because the atomic form factors change with photon energy, especially in the neighbourhood of an absorption edge, the intensities of the Bragg peaks will change.

The elemental semiconductor germanium and the compound semiconductor GaAs are in many ways very similar. They both have the same number of electrons per unit cell, and also have almost identical lattice constants (5.6461 Å and 5.6533 Å, respectively) and, thus, mass densities. Importantly, however, from the perspective of x-rays, their crystal structures are distinct in an important aspect – germanium has a diamond structure [face-centred cubic, with Ge atoms at $(0, 0, 0)$ and $(\frac{1}{4}, \frac{1}{4}, \frac{1}{4})$ relative to each of the four lattice points], while that of GaAs is zincblende, also fcc, but in which the Ga atoms sit at $(0, 0, 0)$ and the As atoms at $(\frac{1}{4}, \frac{1}{4}, \frac{1}{4})$ [see Figure 6.22(a)].

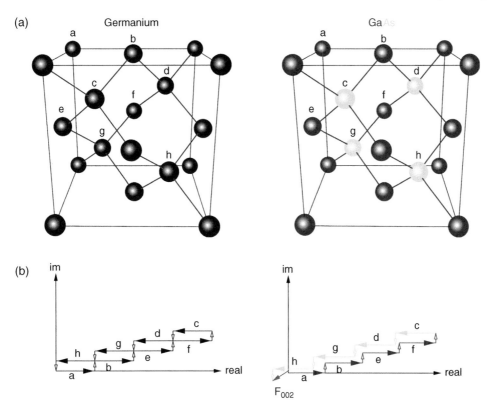

Figure 6.22   Structure factors of germanium and GaAs. (a) The face-centred cubic unit cells of germanium and GaAs. (b) Argand diagram of the structure factors of the Ge and GaAs (002) and (311) Bragg reflections (red arrows), including both the real components f_1 (shown as arrows with solid arrow heads) and imaginary components f_2 (open headed arrows) of the atomic form factors.

It can be easily demonstrated that the structure factor for the (004) reflection of germanium is equal to $8f_{Ge}$, and $4(f_{Ga} + f_{As})$ for GaAs. The (002) reflection for germanium is zero, i.e. it has a systematic absence [see Equation (6.17) and Figure 6.13]. However, because the atomic form factors of Ga and As are unequal, the (002) reflection of GaAs is small but nonzero and given by

$$F_{002}^{GaAs} = 4(f_{Ga} - f_{As}).$$ (6.26)

This is shown in the Argand diagram of Figure 6.22(b).

Close to an absorption edge, both f_1 and f_2 vary significantly, as can be seen from the experimentally determined values for Ga and As shown in Figure 6.23. The difference between f_{Ga} and f_{As} are maximal at the absorption edges at 10.367 and 11.867 keV and we should thus expect the (002) reflection here to be strongest. In contrast, at approximately 11.4 and 12.6 keV, the curves for $f_{1,Ga}$ and $f_{1,As}$ cross one another, and thus we can expect a particularly weak GaAs (002)-reflection [Equation (6.26)]. Indeed, at the higher-energy crossing point at about 12.6 keV, the imaginary components, $f_{2,Ga}$ and $f_{2,As}$, are also almost identical, differing by only approximately 0.4 of an electron. From Figure 6.22(b), therefore, we can expect the intensity here to be the weakest.

A plot of both the theoretically expected intensities of the (002) and (004) Bragg peaks (based on the curves shown in Figure 6.23) and the experimentally determined values, is shown in Figure 6.24 as a function of photon energy. Note that at approximately 12.6 keV, the (002) reflection is three orders of magnitude weaker than at the Ga K-absorption edge at 10.4 keV. A quick back-of-the-envelope calculation can tell us why this should be so – at

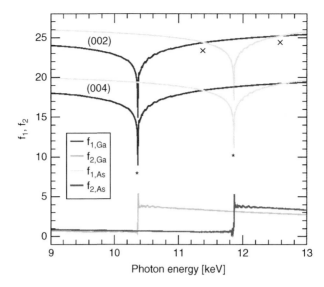

Figure 6.23 Variation of the atomic scattering factors f_1 and f_2 for Ga and As near their K-absorption edges. The curves for $f_{1,Ga}$ and $f_{1,As}$ differ for the (002) and (004) Bragg reflections because they have different values of $f^0(Q)$. Four energies of particular interest are highlighted – two are at the absorption edges of Ga and As (asterisks at 10.367 and 11.867 keV, respectively), while the other two are where f_1 for Ga and As are the same, at 11.391 and 12.584 keV (crosses). The data was obtained by the author by performing a Kramers–Kronig analysis on the absorption spectrum of a GaAs powder sample.

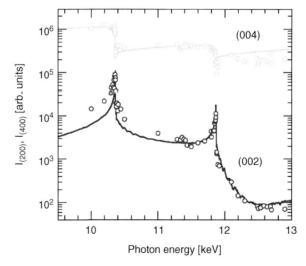

Figure 6.24 The calculated and experimentally recorded Bragg intensities of the (002) and (004) Bragg peaks of GaAs as a function of photon energy. Experimental data recorded by the author at the Surface Diffraction station of the Materials Science beamline, Swiss Light Source.

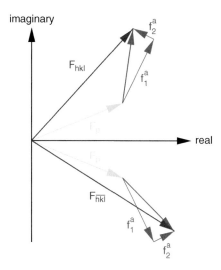

Figure 6.25 The effect of anomalous scattering. A unit cell includes a set of atoms with absorption edges far from the used photon energy; the vector addition of their atomic form factors is represented by F_p (in yellow). In addition, there is a single anomalously scattering atom with components f_1^a and f_2^a at right angles to each other, shown in red. The total structure factor F_{hkl} is the vector addition of F_p, f_1^a, and f_2^a (blue). The vectors resulting in the Friedel mate to this structure factor, $F_{\overline{hkl}}$, are mirrors about the real axis of those of F_{hkl}, with the important exception of f_2^a, whose phase with respect to f_1^a is preserved at $+90°$. Therefore, if we were to mirror the vectors associated with $F_{\overline{hkl}}$ across the horizontal (real) axis, the two structure factors F_{hkl} and $F_{\overline{hkl}}$ would differ by a vector of length $2f_2^a$.

10.4 keV, the values of the differences between Ga and As of f_1 and f_2 are approximately 11 and 4.5, respectively, and hence their vector sum is approximately 12 [see Figure 6.22(b)]. In contrast, at 12.6 keV, Δf_1 and Δf_2 are 0 and 0.4, respectively, and hence the vector sum (0.4) is 30 times smaller. But the intensities are proportional to the square of the structure factors, and hence the ratio of these is $30^2 = 900$.

This illustrative study highlights the large relative effects that can be induced due to anomalous changes in the atomic form factors, and how these are in general most pronounced for weak structure factors for which the contributing atomic form factors largely cancel one another out.

6.6.4.2   *The Complex Electron Density – Breaking Friedel's Law*

In reasoning that the intensities of Friedel pairs are equal, we were careful to include the caveat that the x-ray photon energy should lie far from any absorption edge of the atoms within the unit cell. The reason why the individual atomic form factors f_a, f_b, and f_c in the Argand diagram of Figure 6.21 have different angles and imaginary components is entirely due to the positions of the respective atoms between the Bragg planes, and the consequent phase differences between the scattered waves. It is not due to absorption. So what happens when we *do* introduce absorption?

Consider the atom selenium exactly at its K-edge at 12.658 keV. As we have already seen, the real part of the atomic form factor f_1 dips at the absorption edge, while the imaginary component f_2 increases sharply. In the case of Se, f_1 decreases from approximately 34 (the number of electrons in a Se atom) to less than 22, while f_2 increases from approximately 0.3 to 3.9. In other words, far above the absorption edge, $f_2/f_1 \approx 0.01$, while at the K-edge, it is 0.18. Crucially, the absorption term of the atomic form factor f_2 always lies at $+90°$ to f_1, and, in contrast to the phase differences caused by the atoms' positions within the unit cell, the sign of this phase remains invariant for both Friedel mates. The result is that the magnitudes of the total structure factors F_{hkl} and $F_{\overline{hkl}}$ are no longer equal and Friedel's law breaks down. This is shown schematically in Figure 6.25.

There is an important exception to this breakdown of Friedel's law, and that is for crystals exhibiting centrosymmetry. We argued in Section 6.4.3 that, far from absorption edges, centrosymmetric crystals have structure factors that lie along the real axis, that is, they have no imaginary component (see Figure 6.14). Therefore, any introduction of an anomalous contribution f_2 must point exactly perpendicular to the 'normal' structure factor, that is, parallel to the imaginary axis. The intensities of the Friedel pairs, although they change in magnitude with photon energy as one moves across the absorption edge, remain equal to each other and the diffraction pattern thus maintains its centrosymmetry.

The derivation of Friedel's law is based on the assumption that the electron-density distribution is real. Its breakdown therefore implies the introduction of an imaginary component to the electron density, which is merely a manifestation of absorption [remember from Equations (2.19) and (2.23) that the refractive index has both a real part and an imaginary part]. This breakdown increases the number of inequivalent Bragg reflections and hence also the amount of information available to solve the crystal structure.

Note that only the tuneability of synchrotron radiation allows one to maximize anomalous effects by selecting photon energies close to the absorption edge of the anomalous scatterer(s). Indeed, those experiments such as multiwavelength anomalous dispersion (MAD) are only possible at synchrotrons, as they require one to record data sets at three or more photon energies. Exactly how one extracts this added information from data sets recorded at different photon energies across an absorption edge is discussed in more detail in Section 6.11.4.

6.6.5   Direct Methods

In the last part of this section, we briefly discuss so-called 'direct methods' for retrieving the phase information. There are many variants to direct methods, but they all share the common property that they are iterative algorithms that search for self-consistency between *a priori* available information (so-called 'constraints') in real and in reciprocal space. It is briefly noted that at the present level of sophistication, direct methods in crystallography are limited to unit cells containing at most approximately a thousand atoms, and, despite some successes with small proteins, are not usually capable of solving larger macromolecular structures.

Traditionally, structure determination from diffraction data follows model-fitting methods, most notably Rietveld refinement in powder diffraction [9, 10]. Model fitting suffers from several limitations, which become increasingly severe with the complexity of the unit cell under investigation. Perhaps the most serious problem lies in the fact that these methods normally apply a 'goodness-of-fit minimization' algorithm, which means that they can become trapped in local minima within the fitting-parameter space being used. There are a few 'clever' search techniques which allow one to escape local minima and find the global solution, such as so-called 'simulated annealing', or Monte-Carlo sampling. This comes, however, at very considerable cost in computing time.

Another problem with model fitting is that the starting model is, to an extent, a subjective quantity. Based perhaps on previous chemical or physical knowledge of similar systems or information gleaned from complementary techniques such as nuclear magnetic resonance, infrared spectroscopy, or Raman spectroscopy, an eminently reasonable starting guess may be chosen that in fact lies far from the true solution. The likelihood of this happening increases with the complexity of the system. It would therefore be advantageous to use structural-solution methods which are model-independent.

Iterative phase-retrieval methods (a subclass of direct methods) provide such a possibility [11]. They attempt to retrieve the lost phase information in diffraction data in an iterative and model-independent manner. Although they vary in detail, they all rely on a common basic strategy of alternating between real and reciprocal space, whereby 'constraints' are imposed on both the diffraction data and the real-space electron density, until self-consistency is achieved [12–14]. Starting with *any* initial set composed of the known (that is, experimentally determined) amplitudes of the structure factors $\{|F_{hkl}|\} = \{\sqrt{I_{hkl}}\}$, to which a random set of phases $\{\phi\}$ is assigned, the basic concept of these algorithms follows four simple steps (see also Figure 6.26):

1. Inverse Fourier transform the data (including the phases) to yield a real-space model;
2. Apply real-space constraints to this model. Examples of such constraints might include replacing any negative electron densities (which are self-evidently unphysical) that the IFT has generated with positive values, or

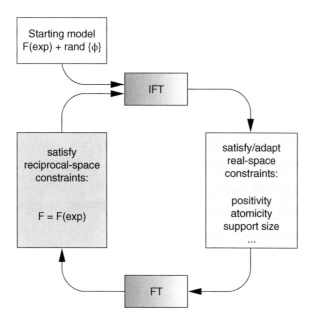

Figure 6.26   Iteration scheme for iterative direct-methods algorithms. Note that in more sophisticated algorithms, the output from the inverse Fourier transform (IFT) must not strictly comply with the real-space constraints, but can be adapted in an appropriate manner based on the shortcomings of the previous iteration.

that any electron density outside the known unit-cell volume is removed. In more sophisticated algorithms, such as the so-called 'hybrid input-output' algorithm first proposed by Fienup [13], the input to the real-space domain need not comply strictly with the constraints (e.g. electron-density positivity or the maximum allowed size and shape of the real-space object), but can instead be adapted by using information about the known shortcomings of the input used in the previous iteration. In this manner, it is possible to avoid being trapped in local solution-minima.

3. Fourier transform this modified real-space model;

4. Keep the resulting phases, but replace the calculated moduli of the FT by the square-root of the measured diffraction intensities (this is the constraint in reciprocal space).

These steps constitute a single iteration of the algorithm, and are repeated until a state of self-consistency is reached between the real-space and reciprocal-space domains.

An example of a popular direct-methods algorithm called 'charge flipping' is shown in Figure 6.27 [15]. The structure was determined from powder-diffraction data of a zeolite material called ZSM-5, used in the petroleum industry as a catalyst for hydrocarbon isomerization. The data was recorded using a Mythen microstrip detector up to 6.5 Å^{-1}, corresponding to a resolution of marginally better than 1 Å. The unit cell of ZSM-5 has a volume of approximately 5400 Å^3, and contains 96 silicon or aluminium atoms and 192 oxygen atoms.

The essence of charge flipping is in the corrections made in the real-space part of each iterative cycle. First, a threshold level for the electron density (δ in Figure 6.27) is selected. This should be somewhat lower than the lowest electron density expected to emerge from the structure (in the case of ZSM-5, this would be the electron density of an oxygen atom). The real-space output after performing an IFT using the set of phases from the last iteration and the experimentally determined structure-factor magnitudes (see Section 6.6.5) is then modified so that any electron density that is below δ has its sign reversed (or 'flipped'). Note that small positive *and* negative electron densities alike undergo this change, not only the originally negative values. The iterative algorithm is then repeated until self-consistency is achieved between reciprocal and real space.

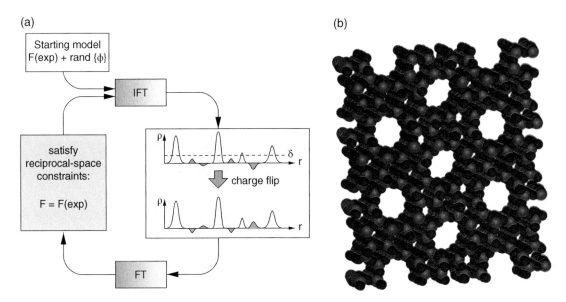

Figure 6.27   Structural determination of ZSM-5. (a) The charge-flipping phase-retrieval algorithm used to analyse the powder-diffraction data. (b) The unit cell of ZSM-5 was successfully generated using the charge-flipping algorithm on powder-diffraction data.

Using charge flipping, an electron density emerged, which corresponded precisely to the known unit-cell structure of ZSM-5, shown in Figure 6.27(b). It was thereby demonstrated that the atomic structure of a unit cell containing nearly three hundred atoms could be solved using no *a priori* knowledge (and, incidentally, using a single, one-second exposure of a Mythen microstrip detector). A decade ago, this achievement was unthinkable.

6.7   Types of Crystalline Samples

Before discussing different types of synchrotron-based diffraction experiments, it is worth to first pause and consider the myriad types of samples that can be investigated using XRD, as an understanding of these helps in deciding the type of experiment to undertake (see Figure 6.28).

We can divide sample types into three broad classifications, namely (and in increasing degree of disorder) single-crystal, textured, and powder samples. Within each class, however, there are strikingly different manifestations. Take, for example, single crystals. Even the highest-quality silicon single crystal still contains crystallographic defects (including vacancies, interstitials, dislocations, and mosaic domains), hence in reality there is no such thing as a perfect single crystal. At what level of imperfection does one begin to regard a system as being no longer a single crystal? Unsurprisingly, there is no well-defined boundary, and depending on experimental methods, preconceptions, and 'what you're used to', the degree of perfection may be very different. So high-quality large crystals such as diamonds, micron-sized protein crystals (which, incidentally, size-for-size, can be far more valuable than diamond!), or heteroepitaxial thin films grown on single-crystal substrates and fabricated for microelectronic applications are all, in their own fields, regarded as being single crystals. This fuzziness in boundaries extends all the way to nanocrystalline powder samples and semi-amorphous structures.

A well-known phenomenon in single crystals is that of twinning, in which two crystals share crystal lattice points at a boundary in a symmetric manner. The resulting diffraction pattern of the twins (there can be more than two twins, depending on the crystallographic system) becomes more complicated. For example, each diffraction spot of an untwinned system may be split into two or more closely lying features. One can therefore consider twinning as the first step towards disorder and eventually to powder samples.

single crystal | twinned crystal | crystal with mosaic spread | textured sample | powder sample | nanocrystalline powder

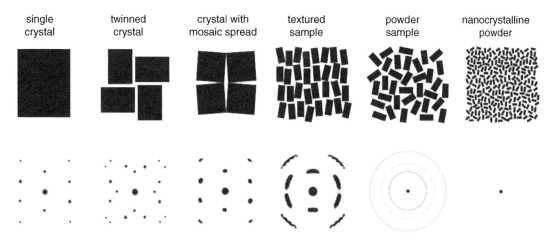

Figure 6.28   Schematic of different crystalline sample types and their diffraction patterns.

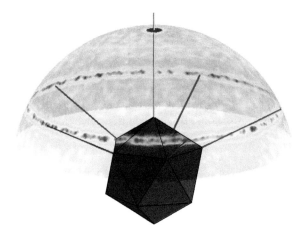

Figure 6.29   Pole figure of a quasicrystal film mapped on a spherical surface. An icosahedron is shown at the centre as an aid to understanding the pattern: the fivefold symmetry axis pointing perpendicularly out of the film is shown in yellow, while the five other fivefold axes at 63.435° relative to each other and the out-of-plane axis, and shown in blue, produce the ring feature, as the icosahedron is rotated around the yellow axis. Courtesy Christian Schlepütz, Paul Scherrer Institute.

Textured samples exhibit an intermediate degree of order, with preferred directions for certain crystallographic planes. This may be due, for example, to a 'brushing' effect in the fabrication process of the sample (such as in polymer fibre pulling), to the synthetic process, such as in natural fibres such as flax, or because of energetic reasons. Figure 6.29 shows texturing in a so-called 'quasicrystalline' thin film of a Ti-Ni-Zr alloy exhibiting icosahedral symmetry. The angular distribution of the Bragg peak associated with the fivefold crystalline-axis was probed in a so-called 'pole figure'.

A true powder sample consists of crystalline grains that have no preferred orientation in space, and are significantly smaller than the illuminated volume. For some samples, this latter condition may be only poorly met, in which case the sample can be spun, in order to record a spatially averaged signal. Note that for undulator radiation, the focussed beam cross-section might be only of the order of a square micron, and that the sampled volume may be fewer than 100 μm^3.

When the characteristic size of the individual grains is reduced to only a few unit cells, such as in nanocrystals, the widths of the diffraction peaks broaden accordingly, producing diffuse ring patterns. This lower limit in crystallinity often best represents the grim reality of 'real' systems, and as such is a burgeoning area of research in twenty-first century crystallography.

6.8   Single Crystal Diffraction

6.8.1   Laue Diffraction

Although the majority of x-ray diffraction methods use (quasi-)monochromatic x-rays and a sample which one has to orient in space in order to satisfy the diffraction condition, it is also possible to record diffraction patterns of stationary single crystals using a broad spectrum of x-rays, in a technique known as the Laue method. Indeed, the Laue method is historically the oldest and was used in the original discovery of x-ray diffraction and is named after its founder, Max von Laue (see Figure 1.10). After the development of monochromatic sources using characteristic radiation from x-ray tubes, Laue diffraction fell somewhat into abeyance, except to orient single crystals and to determine their crystal quality. However, with the advent of synchrotron radiation, and enormous improvements in computing power and detector technology, Laue diffraction enjoyed a renaissance as a technique which provided a fast and efficient means to record diffraction data, and has been used to great effect in dynamical studies of transient crystalline states, not least in macromolecular crystallography [16]. With the emergence of XFELs, however, Laue diffraction is again on the wane.

Consider a static single crystal illuminated by a broad and continuous spectrum of x-rays between two energies E_{min} and E_{max} (Figure 6.30). Each set of crystal planes (hkl) is at a well-defined angle θ_{hkl} relative to the incident beam. From Bragg's law, Equation (6.11), it is evident that, given a sufficiently broad spectrum of x-rays, there will be a specific wavelength λ_{hkl} that satisfies the Bragg condition, such that

$$\lambda_{hkl} = 2d_{hkl} \sin \theta_{hkl}. \tag{6.27}$$

Hence, each diffraction peak has a different 'colour'. Importantly for time-resolved studies, all possible Bragg reflections can in principle be simultaneously recorded.

In modern synchrotron-based Laue-diffraction studies, the mode of operation is almost always that of transmission (in contrast to the 'back-reflection' mode, for which $2\theta > 90°$). The region of reciprocal space that can be

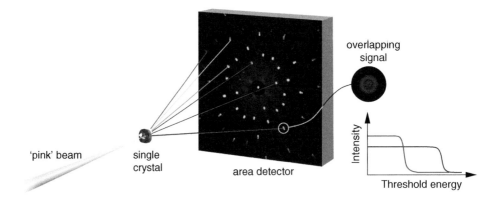

Figure 6.30 A polychromatic 'pink' beam is focussed on to a stationary single crystal sample. Because there is a continuum of wavelengths in the beam, there will always exist certain wavelengths which satisfy the Bragg condition for any given set of crystal planes and orientation. Higher-order reflections can overlap, whereby the photon energies are integer multiples of that of the first-order diffraction spot. These overlapping signals can be separated by scanning the lower threshold energy in modern pixel detectors (see Section 5.7.8).

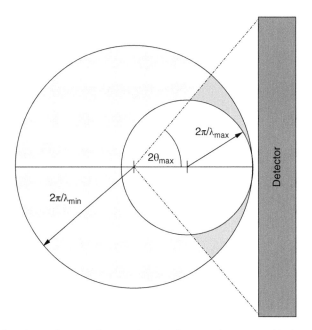

Figure 6.31 The volume of reciprocal space that can be simultaneously accessed in Laue diffraction, shown here in bright yellow, depends on the range of photon energies of the polychromatic beam and the maximum angle that can be subtended by the area detector.

accessed for a given setup is defined by the wavelength range of the polychromatic beam (i.e. $\lambda_{\min} = hc/E_{\max}$ and $\lambda_{\max} = hc/E_{\min}$) and the half-angle $2\theta_{\max}$ subtended by the area detector. This is shown succinctly in the Ewald reconstruction of Figure 6.31.

Laue diffraction therefore provides a lot of structural information in a very short time. However, it is not as well-suited as monochromatic scattering for determining the full atomic structure of a crystal, due on the one hand to the often complex and unknown intensity distribution of the 'pink' incident x-ray beam[8].

In addition, families of lattice planes that are parallel to one another, for example the (111), (222), (333) … planes, have Laue diffraction maxima overlapping at the same position, resulting in a loss of information. This is called the 'energy overlapping problem'.

Recently, however, this positional degeneracy has been lifted by the availability of energy-dispersive pixel detectors, as described in Section 5.7.8 – by tuning the threshold energy, reflections below a chosen value are suppressed. A new recording mode thus presents itself. The threshold energy of the area detector is scanned and successive images are recorded. The difference in intensity between patterns with different thresholds therefore yields the contributions to the pattern from within the energy range bracketed by the two thresholds (Figure 6.30).

6.8.2 Single Crystal Diffraction with Monochromatic X-rays

6.8.2.1 The Rotation Method

In the large majority of single-crystal diffraction experiments, monochromatic radiation is used and the so-called 'rotation' or 'oscillation' method is applied. This is shown schematically in Figure 6.32.

---

[8] The polychromatic beam is referred to as being 'pink' (rather than 'white') because it has been reflected at low angles from mirrors and/or focussing optics, which filter out the more energetic photons with their associated lower critical angles for total external reflection.

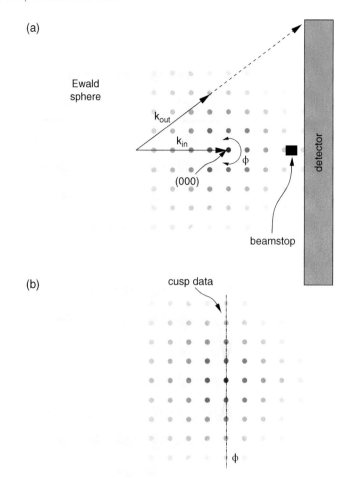

Figure 6.32   Schematic figure showing data collection in the rotation method. (a) By rotating the crystal around an axis perpendicular to the incident beam (ϕ), diffraction maxima pass through the surface of the Ewald sphere and are registered on a 2D x-ray detector. Refer also to Figure 6.11. The highly intense (000) reflection is prevented from damaging the area detector by the use of a beam stop. (b) When viewed from above the plane containing the ϕ-axis, one sees that for a given crystal orientation relative to the axis, some data cannot be accessed (known as 'cusp' data and shaded blue here) as it never passes through the surface of the Ewald sphere. However, by reorienting the crystal axis (typically by 90°), this data can also be recorded.

Data collation runs as follows. Firstly the crystal symmetry, the unit-cell parameters, the crystal orientation, and the resolution limit must be ascertained. Armed with this information, a data-collection strategy is derived which will maximize both the resolution and completeness of the data set.

The method used is to rotate the crystal, while simultaneously recording the x-ray diffraction pattern. Until the development of detectors with fast readout times in the first decade of the twenty-first century, data acquisition was 'stop-and-go' – it consisted of ramping up the rotation velocity of ϕ, opening an x-ray shutter, gating the detector over $\Delta\phi$ (of the order of 1° in macromolecular crystallography), closing the shutter again, then ramping down the ϕ motor while reading out the image (hence the term 'oscillation method'). The angular position of the motor had then be reset so that the next 'slice' of the ϕ range could be recorded.

Advances in detector technology have made possible a novel operational mode, referred to as 'fine phi-slicing' [17, 18], which is now standard in macromolecular crystallography. In this mode, the ϕ motor is allowed to operate continuously and no x-ray shutter is used. Images are continuously recorded over small $\Delta\phi$ ranges. This is made possible by the very short readout times, typically measured in a few milliseconds, compared to exposure times of the order of 0.1 to 1 s in the past.

What constitutes a fine phi slice? $\Delta\phi$ can be considered to be fine if it is smaller than the angular width of the diffraction peaks in the direction being sampled by the Ewald sphere. In nearly all instances of protein crystals, the diffraction-peak widths are primarily determined by the crystal mosaicity, or mosaic spread, introduced in Section 5.42, which is often large, due to the necessity to flash-cryocool. This inevitably produces nano- or micro-crystals of ice, increasing the crystals' degree of imperfection. Typical values for Bragg-peak widths lie between 0.1 and 1°, although repeated temperature cycling between cryo and ambient temperatures can reduce the mosaic spread by approximately a factor of two [19, 20] (see also Section 6.11).

Importantly, in the fine phi-slicing mode, not only are the integrated Bragg-peak intensities accurately recorded, but also so are their profiles. This not only precisely determines the position of the diffraction peak, but also yields information on properties such as strain and domain size.

In many cases, it is necessary to record a complete data set in order to determine the crystal structure. However, from Figure 6.32, it is apparent that those Bragg peaks that lie very close to the ϕ rotation axis never cross the Ewald sphere and therefore are not recorded. To acquire these so-called 'cusp', or 'blind' data, two methods can be applied. For crystals with low symmetry, for example triclinic systems in which the only symmetry partners (i.e. equivalent reflections with the same intensities) are the Friedel pairs (hkl) and ($\overline{hkl}$), if one Bragg peak lies outside the cusp volume, so must its partner. In this case, a second set of data must be recorded for which the crystal is rotated around a second axis, normally orthogonal to the first. For systems exhibiting higher symmetry, one can offset the ϕ rotation axis from a high-symmetry axis, so that although there will be reflections that still lie permanently in the cusp volume, there will be one or more symmetry (Bijvoet) partners that will pass through the Ewald sphere.

6.8.2.2 *The Selected Bragg-peak Method*

In relatively rare cases, one is interested in recording only a subset of the diffraction data. An example might be orbital ordering of electronic states which increases the crystal periodicity and thereby introduces additional (and usually very weak) Bragg peaks at nominally nonintegral positions in reciprocal space. In such cases, the signal of interest might be swamped by signal from the more intense 'main' Bragg peaks, and one therefore records across individual Bragg peaks by the use of a crystal goniometer (such as a so-called 'kappa goniometer stage') and a point detector (or equivalently, a small-area detector), as shown schematically in Figure 6.33.

6.9 Textured Samples

As we have already discussed, crystalline samples may on the one hand be polycrystalline, with small crystallites with random orientations, or, on the other, be single crystals with a well-defined set of crystal axes in the laboratory frame of reference. Textured samples are a half-way house between these two extremes – the crystallites do not all point exactly in the same direction, although there is a preferred direction (to a greater or lesser degree). There are many possible reasons why a preferred direction arises, for example due to strain, 'brushing' of crystallites by an external force, or energy minimization of exposed surfaces, to name only three.

In pole figures, the detector is fixed to a certain 2θ-value of interest. The sample is then rotated azimuthally by as much as 360°, and the polar axis is tilted by up to 90°. In this manner the orientations of the crystal planes associated with the 2θ-value are probed over as much as a hemisphere (2π-steradians). Pole figures (see, for example, Figure 6.29) are employed to measure texture. Traditionally, these are done using a point detector, in order to highlight a particular material and its orientation. With the advent of fast area detectors, it is now possible

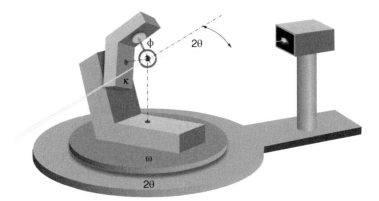

Figure 6.33   A kappa goniometer stage. The detector angle 2θ moves only in one plane (here, the horizontal plane), hence the scattering vector associated with a given Bragg peak (*hkl*) must also lie in this plane. To achieve this, the diffracting planes must be vertical (by rotating the sample around κ and ϕ) and subtend an angle θ with the incident beam, by adjusting ω.

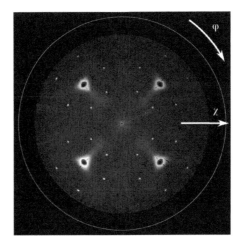

Figure 6.34   Pole figure acquired over a range of 2.25 to 2.35 Å^{-1} and at a photon energy of 16.49 keV. The pattern encompasses three families of reflections associated with a heteroepitaxial thin-film system [21]. Reprinted with permission from the American Chemical Society.

to record a range of diffraction angles (and hence Q-values) simultaneously [22, 23]. In this manner, more than one crystal type or orientation can be investigated in parallel, significantly accelerating the data acquisition times. An example over a range of 2.25 to 2.35 Å^{-1} is shown in Figure 6.34.

6.10   Powder Diffraction

6.10.1   Introduction

For some materials, it may prove to be very difficult, or indeed impossible, to grow macroscopic crystals of sufficient crystallographic quality to be investigated using single-crystal diffraction. In such instances, powder diffraction (also called the 'Debye–Scherrer' method) is an invaluable technique, from which it is often possible to obtain an unambiguous unit-cell structure.

When a 'powder sample' (that is, a collection of many small crystallites of random orientation) is irradiated with x-rays, the resulting diffraction pattern is not one of discrete Bragg spots, but instead, consists of a sets of rings. Powder diffraction suffers from the fact that because the microcrystallites making up the powder sample are randomly oriented, the angularly separated information one obtains in single-crystal diffraction is projected on to a single coordinate, and distinct reflections can overlap[9].

It should not be thought, however, that powder diffraction is a 'last-resort' technique only to be used when large, high-quality, single crystals are unavailable. Powder diffraction not only provides a rapid and nondestructive means to identify the composite parts in multicomponent mixtures or complex system in fields as disparate as metallurgy, archaeology, pharmaceutical sciences, mineralogy, and condensed-matter physics, but is also indispensable in extreme environmental studies, where phase changes are studied as a function of temperature and/or pressure. In such cases, information can often only be extracted from powder samples, which, due to the microcrystallites' large surface area-to-volume ratio, can accommodate phase changes under these severe conditions by relaxation and propagation of crystallographic faults far better than can larger single-crystal samples, which most often degrade unacceptably, and in an anisotropic manner. In addition, powder diffraction lends itself to time-resolved studies, whereby the projection of the crystallographic information into one dimension can be considered in this case to be positively beneficial.

The time resolution of such studies has been extended into the sub-millisecond regime since the emergence of novel detector types such as the Mythen microstrip detector (see Section 5.7.8) [24].

Lastly, it should be mentioned that the most widespread use of powder diffraction is in the identification of crystalline samples, whose patterns can be used as 'fingerprints' and compared with the enormous reserve of data of known materials to be found in the 'International Centre for Diffraction Data' (ICDD) database [25]. The use of synchrotron light for such routine experimental checks is, however, frowned upon.

6.10.2   Basics of Powder Diffraction

Consider Figure 6.35. A sample is irradiated with quasi-monochromatic x-rays of energy $E = \hbar c k$. A detector is placed at an angle 2θ to the incoming x-ray beam. It will detect a diffracted signal only if the diffraction condition is met, i.e. there must be crystal planes with Miller indices (hkl) at an angle $\theta = \arcsin(\lambda/2d_{hkl})$ to the incoming beam.

Imagine now that the sample consists of a large number of randomly oriented crystallites with an interplanar spacing d_{hkl}. The detector is positioned at an angle 2θ that satisfies the Bragg law for the x-ray wavelength λ and interplanar spacing d_{hkl}. The subset of the crystallites with their (hkl) planes at an angle $\theta \pm \delta\theta$ to the incoming beam and the normal to these planes in the plane defined by the incident beam and the detector will satisfy the diffraction condition (see Figure 6.35). The magnitude of $\delta\theta$ depends on the one hand on the beam parallelism and monochromacity, and on the other on the size and crystallographic perfection of the crystallites.

Due to the cylindrical symmetry of the setup about the incident beam axis, cones of diffracted signal, called Debye–Scherrer cones, are produced (Figure 6.36). A scan is recorded in one of three ways – by sweeping a point detector in a plane that bisects these Debye–Scherrer cones about an axis situated at the sample; by using a one-dimensional strip detector, as described in Section 5.7.8; or by using an area detector.

In the latter case, the cone signal is integrated azimuthally to maximize the count rates. This can improve the signal-to-noise ratio significantly. For example, the Mythen detector installed at the Materials Science beamline of the Swiss Light Source has element widths of 7 mm. The circumference of a Debye–Scherrer cone at a 2D detector, however, is $2\pi R \tan(2\theta)$, where R is the distance between the sample and the centre of the detector. This can be

---

[9] As a trivial example, the signals for the {430} and {500} reflections in a simple cubic system overlap, as they both have interplanar spacings of $a/5$. See Equations (6.5) to (6.7).

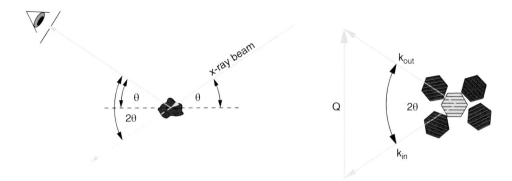

Figure 6.35   Conditions for diffraction in a powder sample. A detector will only see a diffracted signal if the d_{hkl} spacing, the orientation of the crystallite, and the angle of the detector 2θ to the incident x-ray beam lead to the diffraction condition being satisfied. This is fulfilled by the yellow-highlighted crystallite.

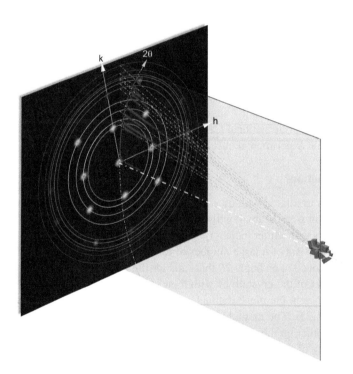

Figure 6.36   A schematic of a powder diffraction experiment. Those crystallites with crystal planes (*hkl*) at an angle θ given by the Bragg law to the incoming beam will diffract. The cylindrical symmetry of the experimental setup about the incident beam axis means cones of diffracted signal are produced. A diffraction pattern is obtained by scanning radially out from the beam axis with a detector in a plane that contains that axis.

of the order of a metre or more for $2\theta \approx 20°$, and $R \sim 500$ mm, well over two orders of magnitude larger than the Mythen microstrip detector width. Azimuthal integration of data recorded by an area detector thus allows one to employ a more modest dynamic range per pixel, which in turn enables higher frame rates. This is an important feature for dynamical studies (see Figure 6.37).

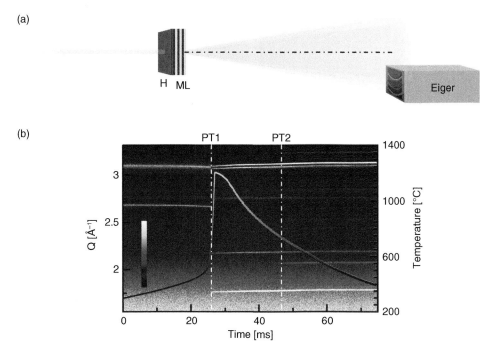

Figure 6.37 Following reaction dynamics using powder diffraction. (a) a 2-μm-thick metallic multilayer (ML) was heated using a membrane-based heater (H) transparent to x-rays (12.6 keV) at a rate of 4000 K s^{-1} up to 500 °C, in order to form an alloy. The powder pattern was recorded using an Eiger 500k detector running at 20 kHz (4-bit operation, dead time of 4 μs). The signal was integrated azimuthally (that is, along the powder rings). (b) Temporal evolution of the powder pattern, along with the measured temperature profile. A runaway reaction heats the sample rapidly from 500 °C to 1200 °C and drives the alloy formation. Two distinct phase transitions PT1 and PT2 are observable. The colour-scale-bar spans 0 to 0.24 average counts per pixel in the azimuthally integrated direction. Courtesy Group of Karsten Woll, Karlsruhe Institute of Technology, especially Tobias Neuhauser.

Hence, in powder diffraction, in contrast to single-crystal diffraction, the information is all contained in one degree of freedom. Lastly, it should be clear from Figure 6.35 that no valuable information is obtained by performing rocking curves on powder samples, whereby the detector angle 2θ is held constant, and the sample is rotated about its axis (out of the paper in Figure 6.35), apart from establishing that the crystallites are indeed isotropically oriented in space. On the other hand, it is highly advantageous to spin powder samples that have relatively few crystallites, in order to spatially average out any resulting 'graininess' in the signal (see Figure 6.38). The shorter the acquisition time, the faster should be the spinning rate.

6.10.3 The Pair-distribution Function

Important structural information can be directly extracted from powder-diffraction data under certain conditions. As we have argued above, a one-dimensional powder pattern is essentially the orientational average of the 3D pattern of a single crystal. In Section 6.6.2, it was argued that the direct Fourier transform of the diffraction intensity distribution of a single crystal delivers a 3D map of the relative vectorial separations of the atoms within the unit cell in a Patterson map. Similarly, through a specialized Fourier transform that takes into account the orientation averaging in powder samples, we can directly Fourier transform a powder pattern to obtain a density profile of the interatomic distances – this is the so-called 'pair-distribution function', or PDF [26]. In order to

(a)   (b)

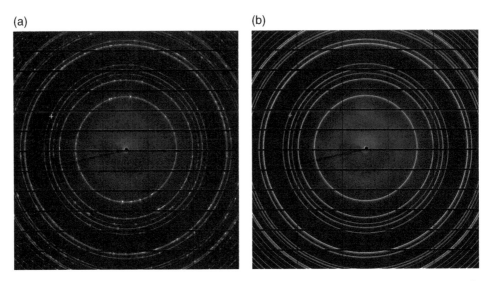

Figure 6.38   The effect of spinning or shaking a powder sample during data acquisition. (a) The powder pattern of a stationary rutile powder sample with crystallite sizes varying between 20 and 45 μm after a 2 s exposure on a Pilatus 6M detector. (b) Pattern of the same sample after the same exposure time, but shaken at 70 Hz using a piezo drive. Courtesy Nicola Casati, Paul Scherrer Institute.

obtain useful information from a PDF, the resolution in real space must be good enough to clearly distinguish different atomic distances as separate maxima. This requires accurate measurement of the powder pattern up to as high a scattering-vector value $Q = 4\pi \sin\theta/\lambda$ as possible and is therefore facilitated by using small wavelengths (0.05 to 0.5 Å) and sufficiently high counting rates at large Q-values, where the x-ray scattering efficiency is strongly suppressed. Generating a high-quality PDF is thus very photon hungry, making it almost exclusively a synchrotron-based technique.

PDF is complementary to EXAFS, discussed in Section 7.5. The former uses the scattering vector, Q, the latter the photoelectron energy, $\mathcal{E}_e$, to probe local structure [27].

An example of a PDF of silicon powder is shown in Figure 6.39. The maximum 2θ-value was $120°$ and the pattern was recorded with 25.3 keV photons ($\lambda = 0.49$ Å). The maximum scattering vector was therefore 22.2 Å^{-1}, which provides a spatial resolution of 0.283 Å.

6.11   Macromolecular Crystallography

6.11.1   Introduction

A young and ambitious student contemplating which line of research to pursue for a PhD program could do worse than choose macromolecular crystallography (MX). Of the 22 Nobel Prizes awarded to studies or inventions involving x-rays as a tool, up to and including 2018, ten have been in the field of biomolecular structure, including all six prizes which have used synchrotron radiation (see Table 1.1). No other discipline comes close to the recognition MX has so deservedly earned by the Nobel Committee for its impact on society; it can also be argued that MX is the *raison d'être* synchrotrons are built at all …

Macromolecular crystallography is the study of macromolecules such as proteins and nucleic acids (DNA and RNA) and their complexes, and is often also called protein crystallography (PX), as proteins make up the majority of the types of studied structures. The transition from 'small molecules' to macromolecules is somewhat vague,

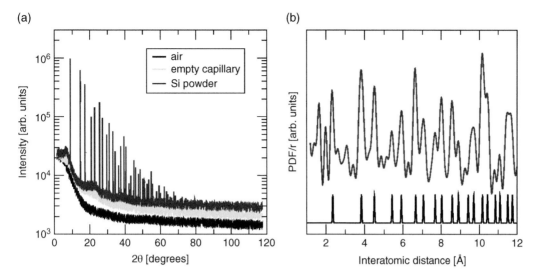

Figure 6.39 The pair-distribution function of silicon powder. (a) A powder pattern of silicon, plus the patterns formed by air and an empty capillary of the sort used to hold the Si powder, recorded at 25.3 keV. After subtraction of the two background signals, the PDF was calculated and is shown in (b), weighted by $1/r$, whereby r is the interatomic distance. This can be compared to the theoretical interatomic distances for the known structure of silicon. Note the excellent positional agreement. Courtesy Antonio Cervellino, Paul Scherrer Institute.

but lies at around a few thousand Daltons. Some solved macromolecular structures lie orders of magnitude above this threshold; for example, the 80S yeast ribosome is composed of 45 peptide chains and many RNA units and has a molecular weight of over 1.8 MDa [28]. Proteins are polymers of amino acids, of which there are 20 naturally occurring types. If the number of amino acids in a polymer chain is below about 20, the chain tends not to fold, and one normally refers to this as being a polypeptide rather than a protein.

It is the folding and thus the three-dimensional structure of proteins that lend them their catalytic activity that drives metabolic reactions, genetic-code replication, and the body's response to internal or external stimuli, among other functions. The study of protein structure is, therefore, of paramount importance in understanding the foundations of molecular biology and thus efficient drug design. Applications of MX extend to other fields, such as the design through mutation of organisms that produce efficient macromolecules for environmental applications [29] (Figure 6.40).

MX is a form of single-crystal diffraction which uses the rotation method. What, therefore, is the justification for discussing this separately from Section 6.8?

As mentioned above, the determination on an atomic scale of the three-dimensional structures of proteins and related biological molecules is fundamental to understanding biological processes at the most basic level; how macromolecules interact in a 'lock-and-key' manner, and how enzymes catalyse reactions. The economic drive to synthesize new and increasingly effective drugs based on this detailed knowledge means that MX beamlines are the most ubiquitous in modern synchrotron facilities. Despite this and the burgeoning number of facilities being built, there remains a high overbooking factor for MX experiments. Although complementary methods such as Fourier-transform infrared spectroscopy and nuclear magnetic resonance can yield indispensable information, particularly regarding the presence and local environment of functional groups of low-Z atoms, MX provides a comprehensive overall picture [30]. A burgeoning new method to study the structures of large biomolecules is cryogenic electron microscopy, or cryoEM, in particular for the larger structures. A brief review of cryoEM is provided in Appendix A.

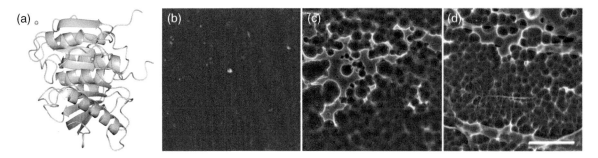

Figure 6.40   PETase, an enzyme evolved to break down poly(ethylene terephthalate) (PET). (a) The structure of PETase determined to a resolution of 0.92 Å, rendered from the pdb file 6eqe. (b)–(d): Degradation of PET by PETase. (b) Buffer-only control sample on a PET coupon. (c) PET degradation after exposure to wild-type PETase. (d) Degradation is enhanced when exposed to doubly mutated PETase. Scale bar = 10 μm. Adapted from [29] with permission from the Proceedings of the National Academy of Sciences.

An immediately obvious qualitative difference between MX and XRD of crystals with smaller unit cells is exactly that – the complexity of the basis within the unit cell. When the basis contains several thousand atoms, conventional approaches to solving the structure through approaches such as direct methods become entirely impractical, and other more sophisticated techniques must be brought to bear. This obstacle is further compounded by the fact that, because diffraction-peak intensities scale with N^2, that is, the square of the number of coherently illuminated unit cells, and because the unit cells in MX can have linear dimensions up to two orders of magnitude larger than those of small inorganic structure, the signal intensity in MX is concomitantly smaller for a given crystal size. The large unit-cell dimensions also mean that the diffraction spots in MX are separated by much smaller angles and that many more reciprocal-lattice points will lie on the Ewald sphere (and thus satisfy the Bragg condition) for any one crystal orientation. For example, a protein crystal with a unit-cell size of 200 Å, irradiated with 12.7 keV x-rays, will have angular separations of approximately 0.3°. As many protein crystals exhibit a mosaicity of similar magnitudes, this can lead to overlapping signals. We discuss these and other practical issues in Section 6.11.2.

Macromolecular crystallography is the leading method of choice when studying bio-macromolecular structures – indeed over 85% of all the presently known protein structures have been solved using MX. The growth of depositions of protein structures into the Protein Data Bank (PDB, [31]) has followed an increasingly steep path[10] since its inception in 1971 [32] and shows no sign of slackening (see Figure 6.41).

The birth of macromolecular crystallography was signalled in 1934 when Desmond Bernal and Dorothy Crowfoot reported on the first high-quality diffraction patterns of pepsin crystals [33]. Their breakthrough was to recognize that the crystals should not be allowed to dry during their exposure to x-rays, but remain in their 'mother liquor'. Eleven years later, Crowfoot and her colleague C. H. Carlisle would publish the first three-dimensional, atomically resolved biomolecular structure determination of the steroid cholesterol iodide, $C_{27}H_{45}I$ [34]. Dorothy Crowfoot (later Crowfoot-Hodgkin, or simply Hodgkin) also pioneered the application of fast electronic computers in x-ray calculations, using the first IBM analog computers during World War II, to establish the structure of penicillin.

The viability of using synchrotron radiation for MX was first quantitatively investigated by Phillips *et al.* in 1976 [35]. In this seminal paper, not only was it demonstrated that the speed of data recording was improved by two orders of magnitude compared to that recorded using a lab source, but also that the decay rate per dosage was substantially smaller. In addition, the application of the tuneability of synchrotron light for anomalous scattering was demonstrated in principle.

---

[10] I have avoided using the term 'exponential' here, as the growth in deposited structures fits poorly to an exponential fit.

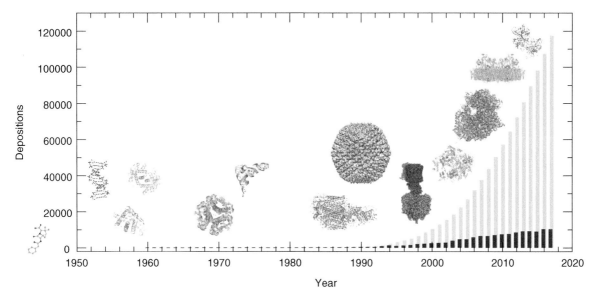

Figure 6.41   Plot of deposited macromolecular structures derived by x-ray diffraction in the Protein Data Bank. Blue bars are the number deposited each year; yellow, the accumulated number. Also included are images of some of the most important breakthroughs in biological structures determined using x-rays. Their positions correspond roughly to their times of discovery. From left to right: penicillin, DNA, myoglobin, haemoglobin, insulin, tRNA, photosynthetic reaction centre, adenovirus, ATP-synthase, RNA-synthase, 80S ribosome subunit, photosystem II, cathepsin B.

Most proteins do not spontaneously form crystals in nature. In order to determine the arrangement of atoms within a protein, the crystallization of proteins and protein-like substances is thus a necessary evil; only the repetitive nature of a crystal allows one to extract the necessary information to determine the angstrom-scale structure. Protein-crystal growth has been a science (some call it an art!) for over a century, and is generally recognized as being the bottleneck in structure determination in MX. Van Deen (1864) believed (perhaps somewhat optimistically) that all naturally occurring organic substances can be crystallized when manipulated effectively. Schimper reported on the first active crystallization attempt by Maschke in 1859, who evaporated the solution from a preparation of Brazil nuts in order to obtain crystals. Although some proteins such as lysozyme (the 'drosophila' of MX) easily form single crystals of mm-sized dimensions, others are far less accommodating – one of the most ubiquitous types of protein, the so-called 'membrane proteins', which are the most common targets for modern pharmaceutical drugs, are notoriously difficult to produce as three-dimensional crystals of sufficient size and quality. Interestingly, the first membrane-protein structure to be solved was the photosynthetic reaction centre, which resulted in the Nobel Prize for Chemistry in 1988 (see Figure 6.42).

The pace of discovery in molecular biology has been exceptionally rapid in the last few decades and has greatly enriched our understanding of the molecular basis of life. The sequencing of the human genome; the construction, substitution, and cloning of new combinations of genes; the elucidation of DNA transcription; and the unravelling of the structure and functionality of the ribosome are just some of the highlights of recent research.

The progress in structural biology has been made possible by the explosion in computer technology, improvements in x-ray equipment, and an increasing expertise in obtaining pure protein crystals[11] in macroscopic quantities using recombinant DNA techniques[12]. As a result, the structures of many biologically significant

---

[11] In the following, I use the term 'protein crystal' to mean crystals of all macromolecular structures, not just proteins.

[12] A *recombinant* system is one which has been artificially engineered by combining DNA or RNA sequences that do not naturally occur.

Figure 6.42   The structure of the photosynthetic reaction centre from *Rhodopseudomonas viridis*. The red and blue planes represent the outer and inner planes of the membrane across which the protein spans. Rendered from the pdb file 1prc.

proteins have been determined. The main stumbling block, as ever, is the phase problem, covered earlier in this chapter, although in MX, this problem is inordinately larger than for small-unit-cell structures, and additional methods must be brought to bear.

Another problem is that preparation of protein crystals is a complex procedure. Samples of sufficient crystalline quality may be very small, containing only micrograms of material or often far less. Most crystals contain between 10^{10} and 10^{15} molecules, each unit cell consisting of one or more molecules, with lattice constants typically between 50 and 500 Å. Linear crystal dimensions span less than a micron to hundreds of microns – we are dealing with exceedingly small amounts of material!

Organic crystals tend to be considerably less robust than inorganic substances: the single C–C bond and the C–OH bond have dissociation energies of approximately 300 and 380 kJ mol^{-1}, respectively, which should be compared, for example, with the Ti–O bond in crystalline anatase of over 670 kJ mol^{-1}. It is therefore inevitable that the protein crystal under investigation will suffer more radiation damage as a result of exposure to the x-ray beam – radicals are produced by photoabsorption and rapid dissociation, which can then migrate in the crystal lattice and wreak havoc to the local structure.

There are several ways to minimize this problem. Firstly, the crystal is normally cooled to about 100 K using a flow of nitrogen from a so-called 'cryojet' (see Figure 6.43), thereby kinetically hindering the diffusion of the photo-induced radicals [36, 37]. Secondly, it is possible to scan the x-ray beam across the crystal so that fresh parts are constantly being exposed. The rate of scanning depends on the sensitivity of the crystal to radiation damage – early systematic quantitative studies indicated that most cryocooled protein crystals become unacceptably damaged after a dosage of approximately 4×10^7 Gy, known as the Henderson limit, defined as the absorbed

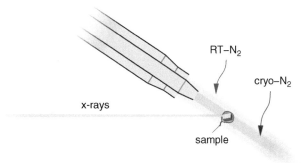

Figure 6.43   Cryojets in MX. A central jet of cryocooling N_2 is generated from a reservoir of liquid nitrogen and directed onto a protein crystal. This is surrounded by an annular sheath of dry, room-temperature N_2 in order to avoid ice formation on the sample.

dose that results in the diffraction intensities falling by 50% [38, 39][13]. More recently, a more conservative dosage of 3×10^7 Gy, called the Garman limit, suggests a maximum intensity reduction by a factor of $\ln 2 = 0.693$ [40, 41].[14] The maximum dose also depends weakly on the dose rate, with a 10% decrease observed for a tenfold dose-rate increase. Lastly, specific damage to particularly sensitive parts, like disulfide bonds and carboxyl groups of acidic residues, happens significantly earlier than the global average decay [42]. Note that the integrated dose limit for noncooled crystals is approximately 100 times lower.

Scanning the beam across the protein crystal requires the focus to be considerably smaller than the sample size. This sets a premium not only on growing as large as possible high-quality crystals, but also on focussing the x-ray beam to micron-sized dimensions (with the attendant increase in flux and rate of radiation damage), and recording diffraction data as quickly and efficiently as possible. Microfocussing improves the signal-to-noise ratio for microcrystals, and allows one to select the best diffracting part of larger crystals. It is best achieved at undulator beamlines, exploiting their naturally low beam divergence. If we assume that the attenuation length is at least two or three times larger than the crystal dimensions, we can make an estimate of the exposure time t_{Gm} before the irradiated volume reaches the Garman limit. This turns out to be

$$t_{\mathrm{Gm}} = 3 \times 10^7 \frac{A\rho}{nh\nu\mu}, \tag{6.28}$$

whereby A is the cross-section of the x-ray beam incident on the crystal, ρ is the mass density of the crystal (of the order of 1.35 g cm^{-3}), $nh\nu$ is the incident photonic power on the area A, and μ is the attenuation coefficient (the reciprocal of the attenuation length). A plot of the attenuation length and Garman times for different focal spots for an 'average' protein is shown in Figure 6.44. These can be of the order of 10 ms for spot sizes of a square micron, and be as little as 100 µs for noncooled samples, important facts to consider when performing room-temperature serial synchrotron crystallography (RT-SSX) at DLSRs, as described in Section 6.11.3.

---

[13] The SI unit of radiation dosage, the Gray (Gy) is equal to one Joule of absorbed radiation per kilogram of material. A dose of 5 Gy (about 350 to 400 Joule of hard x-rays for a grown adult) is usually lethal for humans. Note that this same total amount of energy is required to heat a cup of tea by approximately 0.5 K, or is the amount of sunlight absorbed by a sunbather on a sunny day after approximately half a second. From this, it is apparent that investing energy into ionization through ejection of core electrons from atoms is immensely more deleterious to health than thermal heating.

[14] A useful rule of thumb is that a crystal diffracting to X Å can accept a dose of approximately $10X$ MGy.

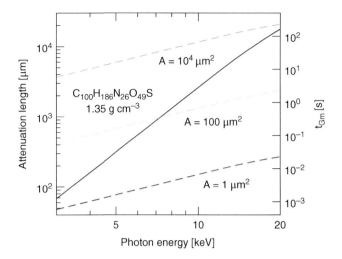

Figure 6.44   The attenuation length (blue curve) and Garman times t_{Gm} (dashed curves) as a function of photon energy for different focal spots A on a 'typical' protein crystal. The Garman times assume the thickness of the crystals is at least two or three times smaller than the attenuation length, which, for the lowest photon energies, is only true for thicknesses below approximately 10 µm; and that $n = 10^{13}$ photons per second across the entire photon-energy range.

6.11.2   Geometries and Photon Energies used in MX

As discussed below in Section 6.11.4, tuneability of the photon energy in MX is indispensable in order to exploit anomalous effects in solving the phase problem. In principle there is no upper limit to the photon energy in MX. However, because any one hard x-ray beamline can offer only a limited range of energies covering less than one order of magnitude, the upper photon energy is normally around 20 keV. In addition, the drop-off of the structure factor with the scattering factor Q and the fact that proteins are made from low-Z atoms both conspire to limit the practically usable upper photon energy.

The minimum photon energy useful for MX depends on the sample size, absorption strength, and the desired spatial resolution. Below approximately 5 keV, the attenuation length of many protein crystals becomes comparable to the crystal dimensions[15]. This results not only in a loss in signal intensity, but also increased radiation damage. Moreover, the maximum accessible scattering vector Q for a given experimental geometry is directly proportional to the photon energy – for 5 keV photons and a maximum 2θ-value of $60°$ (a large value), this corresponds to an electron-density map with a 2.5 Å resolution. Nonetheless, as we will see in Section 6.11.4, recent efforts to perform MX at energies as low as 3 keV are showing great promise.

The highest resolution a_{min} with which the electron-density distribution of a crystal unit cell can be mapped is determined by the largest scattering vector Q_{max} that can be recorded. Remember from Equation (6.14)

$$Q = \frac{4\pi}{\lambda} \sin \theta.$$

Consider the geometry shown in Figure 6.45. Here, therefore

---

[15] Protein crystals vary in their mass densities [43, 44] but are typically about 1.35 g cm^{-3} for a 'void' volume fraction of water of approximately 35% (note also that the attenuation-length curves of water and pure protein are anyway very similar, being almost identical in the tender x-ray regime, and differing only by 10% at 25 keV). If one uses the chemical formula $C_{100}H_{186}N_{26}O_{49}S$ for the 'average' chemical formula for protein structures (see Section 6.11.4), then the attenuation length of a protein crystal at 5 keV is approximately 200 µm.

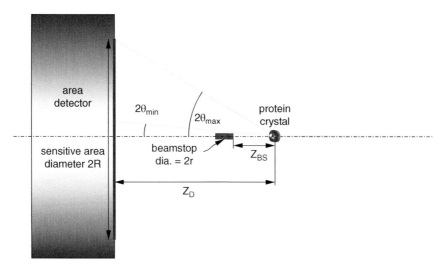

Figure 6.45 The maximum subtended angle on a circular area detector is $2\theta_{max}$, while a beamstop used to protect the detector from the direct beam occludes diffraction maxima at angles smaller than $2\theta_{min}$.

$$Q_{max} = \frac{2\pi}{a_{min}} = \frac{4\pi}{\lambda} \sin\theta_{max}, \tag{6.29}$$

whereby

$$\theta_{max} = \frac{\arctan(R/Z_D)}{2}. \tag{6.30}$$

The smallest features which can be resolved therefore have a periodicity, or characteristic length, of

$$a_{min} = \frac{\lambda}{2\sin\theta_{max}}. \tag{6.31}$$

So, for example, data from an experimental setup in which a detector with a radius of 200 mm is at a distance of 300 mm from a protein crystal illuminated with 12.66 keV x-rays, can provide electron-density maps with a resolution of 1.69 Å [see Figures 6.46(a) and (b)]. Features of the electron density with characteristic lengths smaller than this cannot be resolved. Another factor to consider is that, as the scattering angle and Q_{max} increase, the amplitudes of the form factors drop off. The most common three elements in proteins are carbon, nitrogen, and oxygen, all of which have similar atomic numbers. The signal intensity is proportional to the square of the atomic form factors, plotted for carbon in Figure 6.46(a).

Equation (6.10) describing N, the theoretically maximum number of accessible reflections (including equivalent reflections and systematic absences), assumes that the detector can access reflections as far as $2\theta = 180°$ from the incident beam (that is, those which are perfectly back-scattered). In any given MX experiment, the geometry is fixed with $2\theta_{max} < 180°$. In this case

$$N = 33.5\frac{V}{\lambda^3}\sin^3\theta_{max}. \tag{6.32}$$

In general in MX, resolutions poorer than 4 Å preclude assigning any individual coordinates, while the types of folding and the rotamers on side chains can remain ambiguous for resolutions between 2 and 4 Å. Below this, very few residues are assigned incorrect rotamers and identification of folding becomes reliable. The best resolutions to date lie between 0.5 to 1 Å, although such measurements normally require high photon energies of the order of 20 keV [Figures 6.46(c) and A.3(c)].

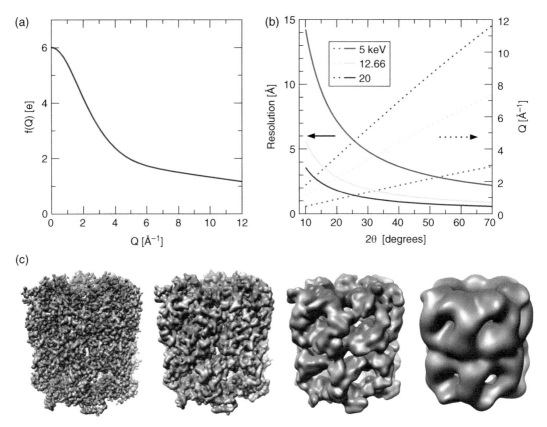

Figure 6.46   Resolutions of structural solutions obtained by the rotation method. (a) The atomic form factor of carbon as a function of the scattering vector Q. Note that at $Q = 12$ Å^{-1} (equating to a resolution of approximately 0.5 Å), $f(Q)$ is almost six times smaller than at small scattering vectors. (b) Resolution of structures obtained (solid lines) and maximum scattering vectors (dotted lines) for different maximum scattering angles (2θ) and photon energies. (c) The structure of the folding molecule GroEL at different resolutions (from the original pdb file 1j4z). From left to right: 4 Å, 8 Å, 16 Å, and 32 Å. The source data were from the Protein Data Bank, reprinted from Wikipedia, Copyright 2008 Free Software Foundation, Inc.

Similarly, the *largest* features which can be resolved a_{max} are given by the low-resolution limit, determined by the smallest scattering vector Q_{min}, which is limited by the shadow of the beamstop in the detector plane (Figure 6.45). Using the same arguments as above for the high-resolution limit, we can calculate the largest features that can be resolved for a beamstop of diameter 200 µm, placed 100 mm downstream from the sample, will be approximately 1000 Å, larger than all but the biggest unit-cell linear dimensions of protein crystals.

Because the lattice constants in protein crystals can be up to several hundred angstroms, the angular separation between diffraction spots can be very small, of the order of a few mrad (a few tenths of a degree). This sets stringent upper limits on the beam divergence at MX beamlines in order to avoid signal overlap and is another reason why undulator radiation, with its intrinsic high brilliance and low divergence, is indispensable.

6.11.3   Opportunities for MX at DLSRs

Despite the fact that, at the time of writing, well over 120 000 atomic-resolution structures of biological molecules and their complexes have been deposited in the Protein Data Bank, one important class of protein – membrane

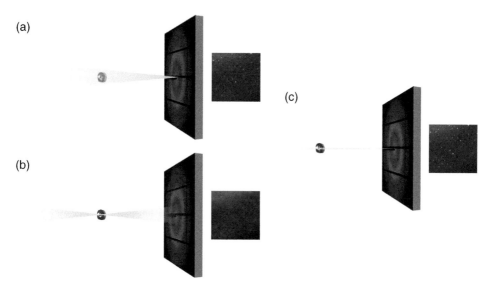

Figure 6.47 Focussing scenarios for small crystals at MX beamlines. At third-generation facilities, focussing at the detector (a) will provide the sharpest diffraction data, but will mean that, at the crystal itself, the beam cross-section may be unacceptably large. By focussing on the crystal instead (b), the diffraction pattern may become blurred and might even result in overlap of signal for crystals with large unit-cell dimensions. (c) The low emittance of DLSRs allows one to both focus on the detector and make efficient use of the beam cross-section at the crystal.

proteins – which accounts for one third of all proteins and two thirds of medicinal drug targets, is extremely under-represented (1 to 2%) [45]. This is, in large part, due to the hydrophobic nature of membrane proteins, which makes crystallization difficult and often limits the crystal size to the micron-scale. This problem is mitigated to an extent by using lipidic cubic phases (LCPs) as the crystallization medium. LCPs consist of lipid, water, and protein in appropriate proportions; this mixture forms a three-dimensional lipidic array permeated with an network of aqueous channels. LCPs provide seeding centres for crystallization and promote growth by diffusion of protein molecules in the medium [46].

Such small crystals require micron or submicron beam focussing, which, until the advent of DLSRs, meant divergences could be 0.2°, or even larger, in the horizontal plane [Equation (3.32)], weakening the diffraction signal at high resolution. The angular separation between Bragg peaks is equal to $2\lambda/d$, which is only 0.6° for patterns recorded at 1 Å for crystals with typical unit-cell sizes of $d = 200$ Å, uncomfortably close to the beam divergence (see Figure 6.47).

Only a limited amount of the diffraction signal can be obtained from any given crystal before radiation damage sets in. Individually, microcrystals are insufficient for a complete data set, hence diffraction data from multiple crystals must be merged for structure determination [47]. Prompted by successes in serial femtosecond crystallography (SFX) at XFELs, there has been a concerted effort in the last five years towards similar approaches using synchrotron radiation in a technique coined serial millisecond crystallography (SMX), or synchrotron serial crystallography (SSX, the acronym I will adopt from hereon in) [48–53]. SSX can be carried out at both room temperature and cryogenic conditions, requiring novel techniques in sample preparation, delivery, data collection, and processing.

Room-temperature crystallography is experiencing a renaissance through SSX [53]. Despite the one to two orders of magnitude reduction in the highest tolerable dose compared to cryo-MX, RT-SSX offers several advantages, including sampling conformational landscapes [54], dispensing with cryoprotectants, and the possibility of investigating dynamic processes down to the microsecond time scales. Moreover, and very importantly with regards to the tight but quasi-parallel beam focussing attainable with DLSR-radiation, it has been demonstrated that, in contrast to cryo-MX [41], there is a positive correlation between dose rate and maximum tolerable integrated dose in room-temperature MX, which enables the measurement of more useful diffraction data by approximately a factor of six when increasing the dose rates from 0.5 to 5 MGy s^{-1} [55, 56].

In parallel, the well-established technique of cryo-MX at third-generation synchrotron beamlines has been extended to SSX. In addition to a two-orders-of-magnitude larger tolerable x-ray dose ($\sim$ 20 MGy), cryogenic methods offer advantages in sample preparation, storage, and transportation, allowing the preservation of crystals in their optimal state prior to beamtime. The advantages of cryo-SSX have already been demonstrated using radiation from third-generation synchrotrons, with the structure determination of both soluble and membrane proteins from ensembles of crystals as small as a few microns [51, 57]. It is confidently predicted that SSX will function far more efficiently still with the micron- and submicron-sized beams promised by DLSRs.

The use of small micron-sized crystals furnishes one further benefit: it appears that radiation damage per unit volume is further reduced compared to larger crystals, as the photoelectrons produced by the initial absorption process and the subsequent secondary electrons can escape with a significant probability [58, 59]. The threshold size below which this begins to bestow an experimentally significant advantage appears to be a few microns [60].

In SSX, the microcrystals are delivered to the beam via one of several methods: by a fast-moving liquid jet; a slow-moving LCP 'toothpaste' extrusion; mounted on a membrane; or as a collection of crystals in a crystallization well, specifically designed to produce a low x-ray background for *in situ* measurements. Fresh material can then be delivered at controlled speeds adjusted to administer the maximum tolerable dose for each crystal. For example, even in the first (far from optimized) proof-of-principle measurements using LCP streams, 300 µg of lysozyme were required, far less than needed in serial femtosecond crystallography at XFELs [48].

Since many membrane proteins are expected to be novel, experimental *'de novo'* phasing is required to reveal their structures (that is, molecular replacement will not suffice, see Section 6.11.4). Recent progress in native-SAD phasing has led to great advances in *de novo* phase determination [61]. The development of DLSRs provides a timely opportunity to optimize MX beamlines for native-SAD experiments, which require x-rays down to the tender regime at 3 keV, and sample environments with minimum background scattering and absorption [62, 63] (see also Figure 5.56).

Even higher photon intensities are possible by increasing the relative bandwidth of the incident radiation. Presently, this is typically $\Delta\nu/\nu = 1.4 \times 10^{-4}$ for Si(111) monochromatized radiation. The relationship between this and the angular spread $\Delta\theta$ (assuming perfectly parallel radiation and zero mosaicity) is

$$\frac{\Delta\nu}{\nu} = \frac{\Delta\theta}{\tan\theta}. \tag{6.33}$$

So, for example, at a value of $2\theta - 60°$, bandwidth smearing of the Bragg peak from the Darwin width of Si(111) would amount to approximately $2\Delta\theta = 0.01°$. Except for the very largest-unit-cell samples, an increase in the energy bandwidth and hence also photon intensities by a factor of 10 could therefore be easily tolerated. Combined with detectors capable of recording at hundreds of Hz, SSX at DLSR-MX beamlines will not only enable biologists to study structure and function of largely unexplored protein families, but also pave the way for high-throughput, structure-based drug discovery of membrane proteins.

6.11.4   Solving the Phase Problem in MX

The gamut of available tools to solve the phase problem in MX (also simply called 'phasing') has progressed in parallel with transformative innovations in x-ray sources, an exponential increase in computer power, and, in a self-reinforcing manner, an ever-expanding database [31, 64]. The evolution of the preferred phasing methods reflects, in particular, developments at synchrotrons (see Figure 6.48). By 1997, several dedicated MX beamlines were operating at third-generation sources, promoting the use of multiwavelength anomalous dispersion (MAD), a technique only possible at tuneable x-ray sources. Single-wavelength anomalous diffraction (SAD)

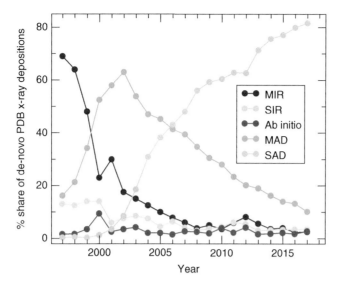

Figure 6.48 Recent trends in phasing methods. MIR includes MIRAS, SIR includes SIRAS, *ab initio* includes direct methods, and SAD includes native SAD. Adapted from and expanded on Figure 15 in [65], with permission from Cambridge University Press.

requires exceedingly exact measurements of Bragg-peak intensities, and as such, has profited from developments in detector technology, especially since the first decade of the twenty-first century.

The first step in obtaining the unit-cell structure in all crystallography is indexing the diffraction pattern. In indexing, the dimensions and symmetry of the unit cell are identified. There are 243 possible combinations of unit-cell type and internal symmetry. Fortunately, because proteins are polymers of amino acids, and all but one of the twenty naturally occurring amino acids are chiral, essentially all protein crystals are chiral[16] which excludes all but 65 symmetry types. It also means that protein crystals are not centrosymmetric and thus they *do* exhibit differences between Friedel pairs due to resonance (absorption) effects, an essential prerequisite for SAD experiments, as discussed below.

In order to visualize the protein structure we need to solve the phase problem. Traditional approaches used for small molecules, such as direct methods, are unsuitable for these very large structures, often containing in excess of 100 000 atoms (not including hydrogen). Instead, we can obtain the structure using one or more of several different approaches.

6.11.4.1 Molecular Replacement

Although Nature seems to provide infinitely varied solutions to the biochemistry of life, certain structural motifs crop up again and again, which may or may not be used for the same biological activity. Indeed, there has been a ten-year gap between the most recently discovered fold[17] in 2017 (pdb file 5vtg) and the previous one to be deposited in the Protein Data Bank in 2007 – the number of distinct folds has thus levelled off at around 1400. It seems therefore possible that Nature's library of how the secondary structures of proteins arrange themselves in

---

[16] A chiral structure is one which cannot be superposed on its mirror image. Examples include hands (indeed, the Greek for 'hand' is the etymological root for the term chirality), and the molecule C H Cl Br I (chloro-bromo-iodomethane).

[17] A protein fold describes how secondary structures such as alpha helices and beta sheets arrange themselves relative to each other in space. Note that any given fold may be produced by more than one amino acid sequence – the constraints imposed by Nature on the allowed three-dimensional structure are thus much tighter than those enforced on the amino acid sequence.

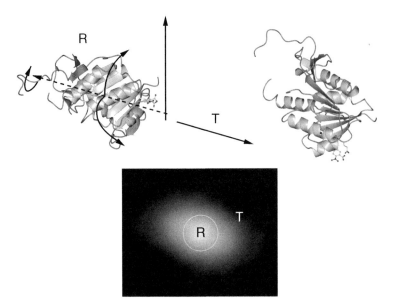

Figure 6.49   The molecular replacement method. Above: The angular orientation of a molecular fragment is varied via a three-coordinate rotation function R (composed of a rotation of the fragment around its own axis, and polar and azimuthal rotations of that axis around an arbitrary fixed axis), and is guided by a comparison of the Patterson map of the experimental data compared to that of a known structure containing the scrutinized fragment (below). Once the rotational orientation has been established, the motif is then translated (T) until the best agreement between the diffraction patterns is achieved.

space may have been fully determined, although new surprises may emerge as the, until now, relatively unexplored family of membrane proteins becomes more amenable to investigation through SSX and SFX (Section 6.11.3).

Molecular replacement (MR) exploits this repeated use of similar or identical molecular 'chunks' to solve the phase problem. The degree of resemblance between two macromolecular structures is strongly correlated with their sequence identities. This can be used to assess in advance whether molecular replacement is likely to succeed or not.

Consider Figure 6.49. A part 'A' of a solved structure is suspected to be also present in some unknown position and orientation in a protein structure for which a set of diffraction patterns has been recorded. In principle, one could then directly compare this pattern to that of the known structure for all possible orientations and positions of that structure within the unit cell. A good best match between the two would provide invaluable information about the unknown structure. Such a procedure would require varying three rotational and three translational parameters with sufficient resolution (given, respectively, by the rotation angles between diffraction images in the data set, and by the resolution of the electron-density map of the known structure). This six-dimensional search is, for anything other than the smallest structures, computationally too demanding.

Instead, a 'divide-and-conquer' strategy is brought to bear on the problem with the help of the Patterson function. As we have seen in Section 6.6.2, the Patterson map provides information on the *relative* distances and directions between atom pairs. Consider two structures containing a common molecular chunk, but this has different rotational orientations and absolute positions within the unit cells. We define an *intra*molecular signal in the Patterson map as being that between atoms within the molecular motif of interest; and an *inter*molecular signal being produced between atoms within the fold (or motif) and atoms elsewhere in the unit cell.

The intramolecular signal depends only on the orientation of the motif under investigation, and not on its position in the unit cell. This will mostly affect the centre of the Patterson map out to a radius given by the characteristic size of the motif, on account of the fact that only the local vectors between atoms within the motif are involved. It is therefore this part of the Patterson map (the region R within the yellow circle in Figure 6.49) that is used to determine the angular orientation, which involves varying the three rotational degrees of freedom until the Patterson map of the unsolved structure agrees best with that of a known structure containing the tested motif. In contrast, *inter*molecular vectors between atoms within and outside the protein fold depend both on its orientation *and* on its position relative to the rest of the structure. Thus, in the second step, the intermolecular vectors can be exploited in the outer parts of the Patterson map (region T in Figure 6.49) to determine the translation vector of the entire substructure. This procedure can be repeated iteratively until no further improvements in agreement are ascertained.

By exploiting this property of Patterson maps, the six-parameter search can thus be reduced to two independent, three-dimensional searches of first a rotational operation [66], and then a translation within the bounds of the unit cell. It is important to appreciate just how beneficial this simplification is. Critically, the complexity of optimization search routines increases not linearly with the number of parameters, but exponentially. As an example, if a one-parameter search spanning one hundred values of that parameter requires one second of CPU time, adding a second parameter with the same range of values increases the CPU time to 100 seconds. A similar six-parameter search would require three hundred years! In contrast, two three-parameters searches would be complete after only six hours.

The power of modern computers and the ever-growing database of high-quality, atomic-resolution protein structures means that the majority of protein structures are solved at least partially using MR. Indeed, this predominance of MR is likely to increase with computing power, the introduction of novel software tools to automatically perform the rotational searches, and, in a self-fulfilling process, the expanding volume of reliable structural data [67].

Modern phase-solving techniques in macromolecular crystallography which cannot be solved by molecular replacement are often based on changing the form factors of specific atoms within the unit cell. According to how the structure factor is perturbed by this, deductions can be made about the values of the phases. Changes in the structure factor can be induced by (among others) so-called multiple isomorphous replacement (MIR), or by multiwavelength anomalous dispersion (MAD), both of which we discuss now.

6.11.4.2 Multiple Isomorphous Replacement

Multiple isomorphous replacement (MIR) was first used for the class of minerals known as alums as long ago as 1927 by J. M. Cork, a postdoctoral fellow of W. L. Bragg [68]. Its first successful use in solving the structure of large macromolecular structures was famously performed by Max Perutz in his pioneering work in determining the structure of haemoglobin [69], although Dorothy Hodgkin had also been using MIR very effectively for over two decades in her studies of smaller structures.

MIR exploits the relative importance to the scattering intensities of heavy atoms that have either been additionally inserted in a well-defined position in the unit cell, or that substitute for a native atom (such as selenium substituting sulfur, see Figure 6.50) [70]. In order for the technique to succeed, however, the addition of the heavy atom(s) or their substitution for lighter atoms should not significantly perturb the structural form (hence the term 'isomorphous', which means 'same shape')[18]. The success of this technique is rarely guaranteed *a priori* and is usually not known until more x-ray data have been collected.

Commonly used heavy atoms in MIR are Hg^{2+} ions, which bind to thiol (SH) groups; uranyl salts, which bind between carboxyl groups in aspartic acid and glutamic acid; Pb, which binds to cysteine; $PtCl_4^{2-}$, which binds to histidine; and Se, which replaces sulfur in methionine to produce selenomethionine.

How exactly does one retrieve the phases of the structure factors and thereby allow the structure to be solved in MIR? To understand this, we must familiarize ourselves with so-called 'Harker constructions'. In the following, we

---

[18] W. L. Bragg, Perutz and Kendrew's boss, succinctly, if perhaps overly confidently, summed up the power of MIR by stating that 'The molecule takes no more notice of such an insignificant attachment than a maharaja's elephant would of the gold star painted on its forehead'.

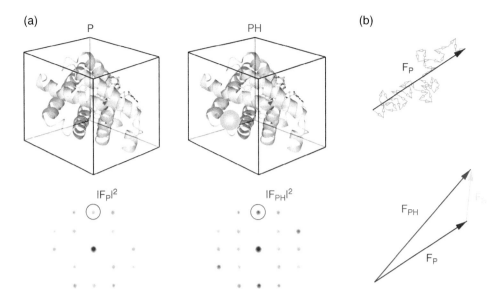

Figure 6.50 Multiple isomorphous replacement. (a) A crystal of a native protein P produces a certain diffraction pattern. The intensities of the diffraction pattern will change if an additional or substitute heavy atom is inserted in a well-defined position in the unit cell in a heavy-atom derivative, *PH*. (b) The structure factor F_P of a given (*hkl*) Bragg peak is composed of the many vector contributions of the individual atomic form factors of the native-protein atoms. The corresponding structure factor of the derivative protein, F_{PH}, has an additional (and large) contribution from the heavy atom, F_H.

assume we have measured diffraction data for the native protein, P, and two heavy-atom derivatives, *PH*1 and *PH*2. We know the magnitudes of the set of structure factors of the native protein, $\{F_P\}$, and the heavy-atom derivatives, $\{F_{PH1}\}$ and $\{F_{PH2}\}$ from this data. Additionally, we know the magnitudes of the heavy-atom contribution $\{F_{H1}\}$ and $\{F_{H2}\}$ from tabulated values as a function of Q. Importantly, we also assume we know the phases $\{\phi_{H1}\}$ and $\{\phi_{H2}\}$. This information is absolutely necessary in the following description; exactly how we determine these phases will be explained shortly.

Let us begin by looking at a certain structure factor (*hkl*). Consider Figure 6.51. Because we don't yet know the phase ϕ_P of the native-protein structure factor F_P (though this is our ultimate goal), we begin by drawing a circle on an Argand diagram with a radius equal to F_P to represent this present state of knowledge [Figure 6.51(a)].

Now, we have claimed we know both the magnitudes *and* phases of the heavy-atom contributions, so in Figure 6.51(b) we add the vector F_{H1} *to the base of* F_P, that is, the origin of the Argand diagram. We still do not know ϕ_P, but we do know the magnitude F_{PH1}. If we draw a circle of radius F_{PH1} with the base of F_H as its origin, we see that the circles intersect at two points. This yields two possible solutions for the vectors F_P and F_{PH1}. This ambiguity is lifted by performing the same operation using the data for *PH*2, as shown in Figure 6.51(c). This is the Harker construction [71].

As already stressed, the above is predicated on a knowledge of the phases of the sets $\{F_{H1}\}$ and $\{F_{H2}\}$. This information is obtained as follows. We define $|\Delta F_{\mathrm{iso}}|$ as being the difference in magnitudes between the structure factor of the native crystal and the heavy-atom derivative, that is

$$|\Delta F_{\mathrm{iso}}| = |F_{PH}| - |F_P|. \tag{6.34}$$

The set $\{|\Delta F_{\mathrm{iso}}|\}$ can be immediately determined from the measured data sets $\{F_P\}$ and $\{F_{PH}\}$. How does this help?

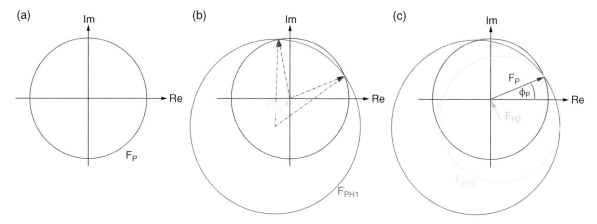

Figure 6.51   The Harker construction. (a) The structure factor of a native protein crystal, F_P, at a particular Bragg peak, (*hkl*), connects the origin of the Argand diagram to somewhere on a circle of radius F_P. (b) If we know both the magnitude *and* phase of the heavy-atom contribution F_{H1}, we can include this as shown. We then draw a circle of radius F_{PH1}, the experimentally determined structure-factor magnitude of the same (*hkl*) Bragg peak of the first heavy-atom-derivative data set. This reduces the possible vectors for F_P (i.e. including the phase) to the two values given by dashed arrows ending at the circles' two intersects. The colour coding in (a) and (b) is the same as used in Figure 6.50. (c) The ambiguity between the two solutions in (b) can be lifted by including the data from the second heavy-atom-derivative, namely F_{H2}, including its phase (given in green), and the magnitude of F_{PH2} (cyan circle).

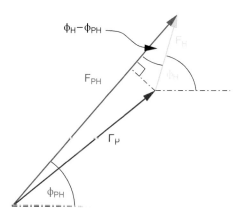

Figure 6.52   The difference in magnitudes between the structure factor of a native protein and a heavy-atom-derivative protein $|\Delta F_{\text{iso}}|$.

Consider Figure 6.52. From this, we can state that

$$|\Delta F_{\text{iso}}| \approx |F_H| \cos(\phi_H - \phi_{PH}). \tag{6.35}$$

This approximation is reasonable either if $\phi_H - \phi_{PH}$ is small, or F_H and F_{PH} are both substantially larger than F_P. We square Equation (6.35) to obtain

$$|\Delta F_{\text{iso}}|^2 = |F_H|^2 \cos^2(\phi_H - \phi_{PH}). \tag{6.36}$$

But $\cos^2 A = [\cos(2A) + 1]/2$, and hence

$$|\Delta F_{\mathrm{iso}}|^2 = \frac{|F_H|^2}{2} + \frac{|F_H|^2}{2}\cos[2(\phi_H - \phi_{PH})]. \tag{6.37}$$

Importantly, there is no correlation of $(\phi_H - \phi_{PH})$ from Bragg peak to Bragg peak, hence, on average, the second term on the right-hand side in Equation (6.37) produces only noise. In contrast, the first term is a usable Fourier component. The set $\{|\Delta F_{\mathrm{iso}}|^2\}$ can then be used to generate a 'difference Patterson function' of the heavy atoms only, thereby yielding the phase angles $\{\phi_H\}$, the information we require to determine $\{\phi_P\}$.

The stricter of the two assumptions required for the validity of Equation (6.35) is that of F_H being substantially larger than F_P, as this implicitly includes the weaker requirement that $\phi_H - \phi_{PH}$ is small. Before we continue with descriptions of other phasing techniques that exploit the anomalous behaviour of f_1 and f_2 close to absorption edges, we consider the magnitudes of the real and imaginary contributions of native proteins and added heavy atoms.

6.11.4.3   Magnitudes of Structure Factors in MX

What is the average composition of a protein? This can be calculated by taking the chemical compositions for the twenty naturally occurring amino acids, weighting them by their natural abundance in vertebrates, then normalizing the sum of these to one sulfur atom. This yields a chemical formula of $C_{100}H_{186}N_{26}O_{49}S$ for 'average' protein structures and has a molecular weight of 2566 Da[19]. Note also that only two of the twenty amino acids actually contain sulfur (cysteine and methionine) and that the natural abundance of these amino acids in proteins is such that, on average, the incidence of sulfur in proteins is only approximately one atom in 25 amino acid blocks.

For unit cells of protein crystals containing such large numbers of atoms, and with little or no internal symmetry on the local atomic scale, one can broadly assume that the atomic positions bear no systematic relationship to the x-ray waves passing over them, and thus that the phases are, statistically speaking, entirely stochastic[20]. The vector addition of N_j atomic form factors, $f_j = f_{1j} + if_{2j}$ for a given atom type j, will therefore on average follow a random path. In statistics, this is described by 'Einstein's drunkard's walk', for which the magnitude of the vector sum is on average equal to the length of a single form factor multiplied by the square-root of the number of contributing atoms [although no predictions can be made on the direction of the sum, see Figure 6.53(a)], that is,

$$F_j = \sum_{N_j} f_j = (N_j)^{1/2} f_j. \tag{6.38}$$

Note here that the summation is vectorial, not simply an addition.

Each atom type j thus produces a structure factor F_j. These also bear no phase relationship to one another, and must likewise be vectorially added in a random manner. This leads to a total structure factor for the protein, F_P, given by

$$F_P = \left(\sum_j F_j^2\right)^{1/2} = \left(\sum_j N_j f_j^2\right)^{1/2}. \tag{6.39}$$

For a protein of molecular weight MW, that has the average chemical composition given above, this becomes

$$F_P = \left(\frac{MW}{2566}\right)^{1/2} (100f_C^2 + 186f_{Hy}^2 + 26f_N^2 + 49f_O^2 + f_S^2)^{1/2}. \tag{6.40}$$

---

[19] The largest discovered protein, titin, has a molecular weight of 2 990 047 Da, or 1165.25 times the normalized 'average' protein molecular weight calculated here. Its chemical formula is $C_{132983}H_{211861}N_{36149}O_{40883}S_{693}$, which, when divided by 1165.25 yields, to the nearest whole atom, a formula of $C_{114}H_{182}N_{31}O_{35}S_1$, fairly close to the calculated chemical average.

[20] This is, of course, not at all true, as the positions are determined by the molecular structure and are not truly random. For our purposes, however, a stochastic model is sufficiently close to the truth.

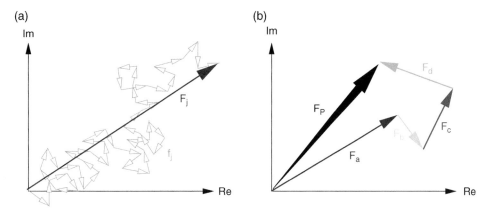

Figure 6.53 Addition of atomic form factors in proteins. (a) For a given atom type *j*, the total structure factor F_j (dark blue thick arrow) is, on average, equal to $N_j^{1/2}$ times the atomic form factor of that atom type f_j (small light blue arrows). (b) The structure factors for each atom type F_j (here, $j = a, b, c, d$) also add up randomly to produce the total protein structure factor F_P.

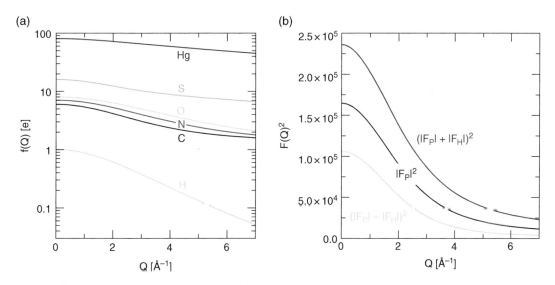

Figure 6.54 The relative contributions to structure factors in proteins as a function of the scattering vector *Q*. (a) The atomic form factors far from absorption edges of the five most commonly found elements in proteins, plus that of mercury, up to 7 Å^{-1}. (b) The structure factor F_P of a protein with molecular weight 50 kDa and the average chemical composition described in the text. By including a single mercury atom ($Z = 80$), the structure factor can be impacted considerably. The two limiting cases of the atomic form factor of mercury pointing exactly out of phase (gold curve) and exactly in phase (dark blue curve) with F_P are plotted.

Note that the atomic form factor for hydrogen has been labelled f_{Hy} in order to avoid confusion with contributions from the heavy atom in MIR.

The atomic form factors of the five commonest elements in proteins, plus that of mercury, are shown as a function of the scattering vector *Q* in Figure 6.54(a). How these combine in a protein of 50 kDa molecular weight are plotted in Figure 6.54(b). For example, F_P at 3 Å^{-1} is equal to 230, while that of mercury is 65. Thus, the approximation needed to obtain the difference Patterson function is only partially valid, but in practice normally sufficiently so over enough data points to obtain reliable values for ϕ_H, especially for Bragg peaks $|F_P|^2$ that are weak in the native-protein pattern.

The popularity of MIR has waned, due first to the advent of MAD in the mid-1990s, and, more recently, because of the increasing popularity of SAD, especially 'native' SAD (see Figure 6.48).

6.11.4.4   *Multiwavelength Anomalous Dispersion*

Multiwavelength anomalous dispersion (MAD) is an elegant and often very effective method that only works using tuneable synchrotron or XFEL radiation. It relies entirely on the measurement of the differences produced by one or more anomalously scattering atoms in the crystal as a function of photon energy [65, 72, 73].

The beauty of MAD over isomorphous replacement is that (a) only one heavy-atom structure is required, and (b) isomorphism is guaranteed, insofar that no physical atomic substitution or addition is needed to change the structure factors. Instead, this is achieved by changes in the atomic form factor of one or more heavy atoms as one passes across their absorption edge.

MAD experiments require one to record three or more diffraction data sets around the absorption edge of the heavy atom. F_H will change its magnitude and phase, causing F_{PH} to do likewise. In addition, for any one photon energy, the Friedel pairs will also show differences for the reasons outlined in Section 6.6.4 (see in particular Figure 6.25).

The pertinent features of selenium's atomic form factor are plotted in Figure 6.55. Selenium was used as an example, as it is the most commonly used substitutional element into protein structures for MAD and SAD, via selenomethionine, the selenium-substituted version of otherwise sulfur-containing methionine, one of the human essential amino acids (which, incidentally, has a natural abundance of 1.8%) [65].

MAD data is handled in much the same manner as for MIR, using Harker constructions. Note that there are *six* different data sets available for three photon energies, due to the unequal intensities of the Friedel pairs at any one energy. This provides increased redundancy and confidence in the phases.

6.11.4.5   *SAD and SIRAS*

The MAD method described above exploits the change in structure factors as a function of photon energy across an absorption edge. In contrast, single-wavelength anomalous diffraction (SAD, often also known under the misnomer 'single-wavelength anomalous *dispersion*', despite the fact that dispersion features are not exploited) operates at a single photon energy and takes advantage only of the fact that symmetry between Friedel pairs is broken (see Sections 6.6.3 and 6.6.4). Consider once more Figure 6.25. The difference in the structure factor of Friedel pairs is $2f_2^a$, where f_2^a is the imaginary component of the anomalous correction to the atomic form factor of the absorbing atom. SAD is therefore carried out at energies above the relevant absorption edge. f_2 is plotted as a function of energy for some pertinent atoms in Figure 6.56. The amplitude of $2f_2$ for the heavier of these elements, such as mercury, is of the order of 20 electrons.

SAD was developed almost serendipitously by Martha Teeter and Wayne Hendrickson in 1973. They had recorded diffraction data of the relatively small protein, crambin, down to 0.88 Å. In order to phase the data, they had attempted MIR, but crambin had resisted all attempts to introduce heavy atoms. In desperation, they wondered whether the anomalous contributions of sulfur, which has its K-edge at 2472 eV, would be sufficiently large to detect differences in the intensity of Friedel pairs at 8.04 keV, the photon energy of their Cu-Kα lab source (see Figure 6.57). At this energy $f_2 = 0.55$, hence the difference in the structure factors of Friedel pairs was only a scattering amplitude similar to that provided by a single helium atom – detecting this did indeed seem like a long shot. Nonetheless, a weak difference was detected, and successfully used to determine the positions of the six sulfur atoms in the asymmetric unit of the unit cell [74]. The difficulty in extracting this information is reflected by the fact that there was an eight-year lapse between the experiments and publication.

After this remarkable result, however, SAD went into abeyance for some eighteen years before serious efforts were again made [65]; because most protein structures have a smaller fraction of anomalous scatterers than

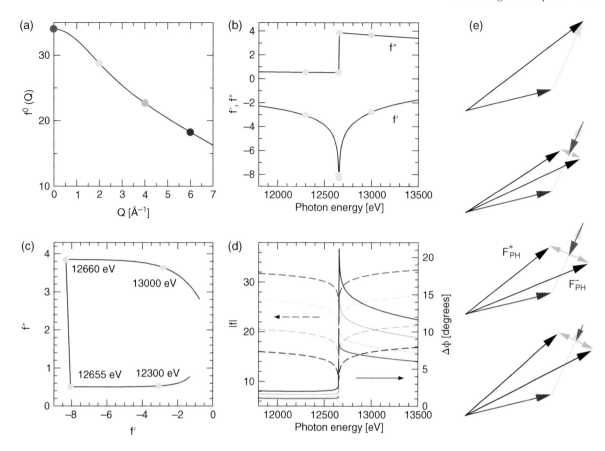

Figure 6.55 The atomic form factor of selenium. (a) The change of $f^0(Q)$ with Q. Four values are highlighted with the coloured circles. (b) The change of f' and f'' with photon energy across the K-edge of selenium. The four circles highlight typical photon energies used in MAD – below the edge, at the minimum of f', at the maximum of f'', and well above the edge. (c) The reason for these choices becomes more apparent when f'' is plotted as a function of f', as they represent the largest separations from each other, and hence produce the most marked changes in the diffraction intensities. (d) The change in the magnitude of $f = f_1 + if_2$ as a function of photon energy for different Q-values (dashed curves); and the phase change $\Delta\phi$ induced in $f = f_1 + if_2 = F_{Se}$ as a consequence of the anomalous contributions. $\Delta\phi = \arctan(f_2/f_1)$. (e) Summary of changes with increasing photon energy of F_{PH}^+ and F_{PH}^- (black arrows), the two Friedel pairs of the heavy-atom derivative crystal. The $(\overline{hkl})$ Friedel mate, F_{PH}^-, has been mirrored across the real axis for simpler comparison with the (hkl) Friedel mate, F_{PH}^+. Blue arrow: structure factor of protein, excepting the heavy atom(s); yellow arrow: $f^0(Q)$ of the heavy atom; red arrow: f' of the heavy atom; green arrow: f'' of the heavy atom.

does crambin, and because the weak anomalous signal demands exceedingly accurate determinations of the diffraction-peak intensities, SAD was deemed in most cases to be impractical up until the last year or so of the twentieth century, especially compared to the then-popular MAD approach. The breakthrough occurred primarily thanks to the direct-methods phasing technique of 'solvent flattening' (described in Section 6.11.4 [75]). Nowadays, with the high intensities available at third-generation facilities and the high reliability and accuracy of modern pixel-array detectors, SAD is becoming the 'first-choice' method, for native and for heavy-atom-substituted proteins alike. The former exploit the relatively weak anomalous scattering signal from naturally occurring atoms below $Z = 20$, in particular, phosphorus, sulfur, potassium, and calcium, in so-called 'native-SAD' [76] (also

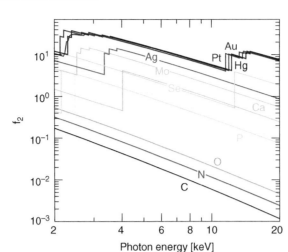

Figure 6.56   The imaginary component of the atomic scattering factors f_2 of various relevant elements in macromolecular crystallography, including the six heavy elements Se, Mo, Ag, Pt, Au, and Hg, often used in isomorphous replacement.

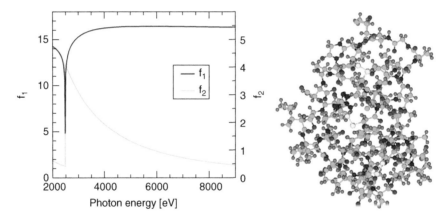

Figure 6.57   Sulfur SAD was first used by Teeter and Hendrickson in 1973 using 8.04 keV x-radiation [74]. Left: the change in $f_1(0)$ and f_2 as a function of photon energy. Note that at 8.04 keV, f_2 is little more than half an electron. Right: the crambin molecule, showing the six sulfur atoms, depicted in yellow.

called sulfur-SAD, or S-SAD). Native-SAD therefore comes 'for free', even if other phasing approaches are being employed.

This brings us to an important aspect when one records SAD data for a weak anomalous scatterer, such as sulfur in crambin. One reason why the first SAD experiment on crambin was successful was the fact that, at 4.1% in molecular weight, there is more than three times the average incidence of sulfur in crambin than in an 'average' protein molecule[21].

---

[21] The chemical formula for crambin is $C_{200}H_{313}N_{55}O_{65}S_6$ (approximately 4700 Daltons).

Table 6.1 Anomalous contributions from C, N, O, and S in an 'average' protein of 50 kDa molecular weight at 6 keV.

Atom type j	Number of atoms N_j	$\sqrt{N_j}$ (rounded)	f_{2j} [e]	$\sqrt{N_j} f_{2j}$ [e]
C	1949	44	0.0183	0.805
N	507	23	0.0346	0.796
O	955	31	0.0630	1.953
S	20	4	0.950	3.80

Consider the values of f_2 for carbon, nitrogen, oxygen, and sulfur provided in Figure 6.56. Although these change with photon energy, their ratios are fairly constant (as attested by their constant vertical separation in the logarithmic plot of Figure 6.56). Indeed, we can estimate their ratios from the fact that f_2 scales approximately with Z^4 (assuming no nearby absorption edges, see Section 2.5), to obtain $f_2^C:f_2^N:f_2^O:f_2^S = 6^4 : 7^4 : 8^4 : 16^4 = 1 : 1.853 : 3.160 : 50.57$. The actual ratio is very close to this and is in fact equal to $1 : 1.891 : 3.443 : 51.91$.

We have already argued that the contribution of a certain atom type j in a protein to the total structure factor is the product of the atom's form factor f_j with the square-root of the number of those atoms in the protein, N_j [Equation (6.38)]. But $f_j = f_{1j} + if_{2j}$, that is, it has both a real and an imaginary component. It follows that the direction of the summed imaginary contribution $N_j^{1/2} f_{2j}$ will remain at 90° relative to $N_j^{1/2} f_{1j}$.

Although the Friedel-pair-breaking scattering contributions f_2 from individual C, N, and O atoms are clearly much smaller than that from sulfur, there are many, many more of them. The question is, therefore, whether the Z^4-dependence on f_2 acts in favour of the sulfur-contribution dominating the SAD signal, or whether the approximate $N_j^{1/2}$-dependence on the vector sum of individual atoms is sufficient that the anomalous signals of the C, N, and O atoms swamp that of the sulfur atom(s).

Clearly, each case must be individually assessed – those molecules like crambin, which contain an unusually large fraction of sulfur atoms, obtain their SAD signal primarily from that atom. The average molecular weight of a protein structure lies at around 50 kDa. The chemical formula for an 'average' protein posited above has a molecular weight of 2566 Da, and thus an 'average' protein will contain $50\,000/2566 = 19.5$ such units, or in other words, $N_C = 1949$ C atoms, $N_H = 3624$ H atoms (which we ignore from here due to their exceedingly weak scattering strength), $N_N = 507$ N atoms, $N_O = 955$ O atoms, and $N_S = 20$ S atoms, whereby we have rounded up to integer numbers. We summarize the anomalous contributions for this 'average' protein at 6 keV in Table 6.1.

From this table, we can draw some important conclusions; first, the oxygen contribution to the SAD signal is on average over twice as strong as those from nitrogen and carbon. Secondly, the contribution from all atom types are of the same order of magnitude, and because there are so fewer S atoms than C, N, or O, their contributions to the anomalous signal can deviate much more strongly from the statistical average, that is, they may often dominate, or contribute almost nothing. Thirdly, and crucially, because the summed contributions of C, N, and O, are similar in magnitude to that of sulfur, the high accuracy with which structure factors can be recorded using modern detectors and recording techniques (e.g. fine phi-slicing) means that nowadays SAD signal is possible to detect *even in proteins lacking any elements heavier than oxygen*. The flipside of this is, of course, that ignoring the anomalous signal from C, N, and O, will lead to inaccuracies in the magnitude of f_2 for sulfur.

We now make an estimate of the largest difference in intensities in native-SAD between Friedel pairs for a macromolecule having our 'average' chemical composition (see above). This occurs when the structure factor F, including f_1^S, but excluding the contribution from f_2^S, is (anti)parallel to f_2^S [see Figure 6.58(a)]. The maximum of the so-called 'Bijvoet ratio' is

$$\frac{\Delta F^\pm}{|\overline{F}|} = \frac{2f_2^S}{F}. \tag{6.41}$$

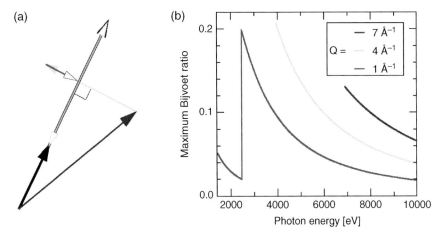

Figure 6.58 Maximum relative differences of Friedel pairs (the Bijvoet ratio) in a macromolecule of chemical composition $C_{100}H_{186}N_{26}O_{49}S$ and statistically random phases. (a) The condition for the maximal Bijvoet ratio. F_P, the total structure factor except for the sulfur form factor, is shown in blue. The yellow arrow is $f_S^0(Q)$ for sulfur (that is, its form factor without anomalous contributions). This is reduced by f_S' (red arrow) due to anomalous effects. $\pm f_2^S$ for sulfur for the Friedel pair is added in opposite directions (green arrows). The magnitude of the maximal relative difference in Friedel-pair intensities is $\Delta I/|\bar{I}| = 4Ff_2^S/[F^2 + (f_2^S)^2]$, whereby F is the vector sum of F_P, $f_S^0(Q)$, and f_S'. F^+ and F^- are shown as black arrows. (b) The maximal Bijvoet intensity ratio as a function of photon energy for three different scattering vectors Q. Note that the curves for $Q = 4$ and 7 Å^{-1} start at approximately 4000 and 7000 eV, respectively, as below these energies, these scattering vectors become inaccessible [Equation (6.15)].

The maximum relative difference in Friedel-pair intensities is given by

$$\frac{\Delta I}{|\bar{I}|} = \frac{2(I^+ - I^-)}{(I^+ + I^-)} = \frac{2[(F^+)^2 - (F^-)^2)]}{[(F^+)^2 + (F^-)^2)]}, \tag{6.42}$$

whereby F is the vector sum of $F_P, f_S^0(Q)$, and f_S' [the blue, yellow, and red arrows in Figure 6.58(a)]. Substituting in $F^\pm = F \pm f_2^S$, we obtain

$$\frac{\Delta I}{|\bar{I}|} = \frac{4Ff_2^S}{F^2 + (f_2^S)^2}. \tag{6.43}$$

The anomalous effects are most marked just above the K-edge of sulfur, and for large Q, for which the nonanomalous contributions are small (because they drop off with increasing Q) and therefore, the relative impact of the Q-independent anomalous components is largest. There is thus a trade-off between maximizing the Bijvoet ratio by tuning the photon energy as low as possible above the absorption edge, and obtaining sufficient resolution in the electron-density maps by using a photon energy high enough to access the desired Q-values. Measuring differences in Friedel pairs at high Q sets a premium on having a good signal-to-noise ratio, requiring that spurious signal from air scatter or from any optics components such as slits should be kept to a minimum[22].

Because SAD provides only two circles in the Harker construction (Figure 6.51), alone it cannot unambiguously determine the phases, although additional post-experimental techniques such as molecular replacement and

---

[22] Note that at 8.04 keV and the maximum accessed Q-value of $2\pi/0.88 = 7.14 \text{ Å}^{-1}$ in the study of crambin by Teeter and Hendrickson, the Bijvoet ratio can be as large as 10% for an 'average' sulfur content. As crambin has three times the average sulfur content of proteins, the Bijvoet ratio could have been as high as 17% ($\sqrt{3}$ times higher) [74].

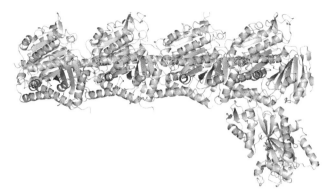

Figure 6.59 The multiprotein–ligand complex T$_2$R-TTL, solved using native-SAD [77]. Rendered from the pdb file 4wbn.

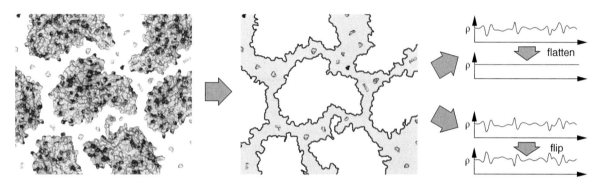

Figure 6.60 Solvent flattening. Once the general outline of the protein structure is discernible in the electron-density map, the void volume in between occupied by randomly positioned solvent molecules can be extracted. This generally noisy signal can either be smoothed, or flattened, to a constant electron density, or flipped, before the next iteration of phasing. This procedure accelerates the path to the final structural solution.

solvent flattening (see below) can be brought to bear to determine which of the two phase sets is most probable. Another approach is to also obtain experimental data from a single isomorphous derivative, a technique called single isomorphous replacement with anomalous scattering (SIRAS).

SAD, in particular native-SAD, is now the method of choice for *de novo* phasing where at all possible – it requires just one data set and no heavy-atom derivatives [61]. Recently, the structure of a multiprotein–ligand complex called T$_2$R-TTL, with a molecular weight of 266 kDa, was solved using native-SAD [77] (see Figure 6.59).

6.11.4.6 Solvent Flattening

Solvent flattening is an iterative procedure to facilitate the determination of the phases from MX experimental data. An initial electron-density map is produced from the experimental data and a 'guess' set of phases. Once the choice of phases is sufficiently good, the outline of the general structure of the protein becomes discernible. At this juncture, a trick can be employed – solvent flattening – to further improve the phases.

The concept of solvent flattening is based on the fact that, although the protein structure is repeated faithfully from unit cell to unit cell (i.e. it has translational symmetry and hence is crystalline), there is a large volume of solvent molecules in the voids between the protein molecules (see Figure 6.60). The solvent molecules assume random positions and orientations and, averaged over many unit cells, should therefore not produce a modulated electron density. Hence all the electron density outside the profile of the protein structure can be set to a small constant value representing the background homogeneous signal produced by an amorphous matrix of low-density solvent (the exact value of the average solvent electron density can actually be varied as an added variable). This modified structure with a 'flat solvent background' is then Fourier-transformed. The resulting scattering amplitudes are replaced by the experimental data, and the new phases used to perform the inverse Fourier transform back to a real-space structure. The solvent-flattening modification is then iteratively repeated until self consistency is achieved [75]. Solvent flattening is therefore very similar to charge flipping, described in Section 6.6.5, and indeed a variation of the solvent flattening procedure has been developed that does not simply flatten the solvent 'noise', but instead flips it around a certain threshold value [78].

6.11.5   MX Studies at XFELs

Macromolecular crystallography is, at the time of writing, in an unprecedented state of flux. Conventional cryogenic-MX held sway over structural biology for some three decades before, firstly, the advent of serial crystallography at XFELs [79] and shortly thereafter at third-generation synchrotrons (and now at DLSRs) and, secondly, game-changing advances in cryoEM (see Appendix A).

The amount of absorbed energy from the XFEL radiation in protein crystals means that they 'Coulomb explode' almost immediately. The 'almost' is critical, as it turns out that the electrons remain essentially in their ground-state configuration just long enough that the diffraction signal is generated – a 30 fs XFEL pulse has an extent of 9 μm, a similar order of magnitude to the sizes of protein crystals used at XFELs. This concept of 'outrunning radiation damage', or 'diffraction before destruction' using XFEL fs radiation was investigated theoretically as early as 2000 [80] and was demonstrated experimentally at the atomic scale in 2011 [79] for the fully hydrated model membrane protein 'photosystem I' (having a solvent fraction by volume of 78%). Crystals containing such a large fraction of water are often destroyed by cryocooling due to the formation of nanocrystals of ice, thus precluding them from investigations using conventional cryo-MX. Indeed, it was shown that the longest XFEL pulse length before the diffraction data began to suffer from the onset of a Coulomb explosion of the microcrystals was approximately 70 fs, well over a factor of two longer than predicted by theory. The technique of serial femtosecond crystallography (SFX) was born.

A schematic of a typical SFX setup is shown in Figure 6.61. Because of the enormous power density delivered by XFELs, a fresh crystal must be provided for each and every XFEL pulse. The delivery of microcrystals may be through one of several methods, including some sort of liquid jet, a thicker paste-like substance (the LCP also mentioned in Section 6.11.3), an aerosol, or on a fixed membrane system. Each method has its advantages and drawbacks. For example, liquid jets are prone to freezing as they expand into a vacuum, plus only a very small fraction of the crystallites are ever irradiated – the jet moves at a few metres per second, meaning that, at LCLS-I, which runs at 120 Hz, it has moved by approximately 5 or 6 cm between XFEL pulses, a stretch of material that can contain several thousand or even some tens of thousands of crystals. Increasing the repetition rate helps only so far, as the jet itself is ripped apart by the XFEL jet and requires some time to heal itself and flow continuously once more (see Figure 4.12). On the other hand, narrow liquid jets have a relatively low background caused by the water or other solvent the crystallites are travelling in.

Because the crystal orientation is random, each diffraction pattern must be indexed individually, and sophisticated programs have been developed to perform this automatically. In order to obtain a complete diffraction pattern with a resolution that allows identification of individual atoms, several hundred thousand or even more partial patterns must be recorded.

Accurate data interpretation is hampered by many issues, however. Firstly, there are large fluctuations in the energy of individual pulses, which can vary by several tens of percent, although this aspect is slowly improving. Secondly, the central energy of any single pulse may fluctuate by a few tens of eV, due to the random nature of the SASE process in XFELs. This means that the shell of the Ewald sphere increases in thickness with Q, varying

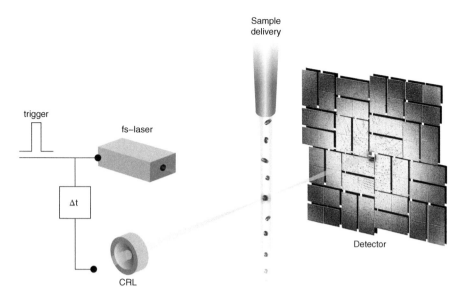

Figure 6.61   Schematic of a typical setup for SFX. The XFEL pulses are focussed using a CRL or PZP on to one of an ensemble of microcrystals (here, shown to be a stream of a liquid transport medium such as the mother liquor or LCP, although it could also be a target of crystals on a fixed support). Time-resolved experiments additionally involve a femtosecond laser, whereby the timing is such that there is a variable delay Δt between first laser illumination, which initiates a photoinduced process, and the XFEL pulse. The integrating detector has a central aperture to allow through the XFEL pulse, which may be refocussed and used for a second parallel experiment downstream.

approximately as $\sin^2\theta$ between 0 in the forward direction and $2\Delta k$ in the backward scattering direction, whereby Δk is the difference in the minimum and maximum wavevectors associated with the energy variations.

Because each diffraction pattern is recorded over a few femtoseconds, there is no possibility to rotate the crystal during exposure. This means that diffraction spots represent only a partial integration over the entire diffraction Bragg peak – this is most obvious in the forward-scattering direction, where the Ewald sphere is still very thin, and less so as Q increases (though these peaks become increasingly weak). Lastly, the size of the crystallite responsible for any one diffraction pattern is unknown – is the signal strong because the crystallite was big, the XFEL pulse-energy large, or due to a coincidental alignment of the Ewald sphere with the diffraction pattern in reciprocal space? Actually, this randomness in crystallite size can be considered advantageous – after many, many shots, numbering in hundreds of thousands, each diffraction peak will have been partially sampled many times at different positions relative to the Ewald sphere and originating from many different crystal sizes. The crystal can thus be modelled as having a kind of 'Gaussian cloud' form, representing an average of all the crystallite sizes and orientations.

The rôle of XFELs in MX is only now becoming clearer. Firstly, cryocooling can be entirely dispensed with, as the crystals are anyway destroyed within a few tens of femtoseconds, regardless of their temperature [80]. Any increased crystal mosaicity through the cryocooling process is therefore also avoided. However, it is expected that room-temperature SSX will also be able to solve static structures with similar efficiency as SFX. Where SFX is unrivalled is in time-resolved studies on time-scales of tens of femtoseconds to microseconds.

Time-resolved SFX can be divided into two classes – photon-triggered (typically by a femtosecond laser) [81, 82] or triggered via chemical mixing [83], the latter's temporal resolution being limited by mixing times, which, at the time of writing is a few hundred microseconds. Both methods allow reversible and irreversible reactions, in contrast to earlier Laue-diffraction-based methods, which were limited to reversible processes only [16, 84].

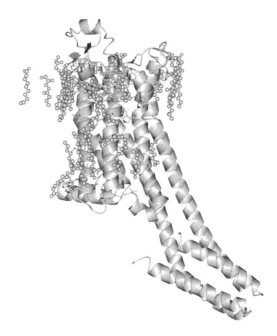

Figure 6.62 The structure of the human A_{2A} adenosine receptor to a resolution of 1.9 Å, solved using native-SAD SFX [85]. Image generated using the pdb file 5k2d.

Phasing XFEL data for *de novo* structures for which no homologous structures exist presents a particular challenge. Often, the differences in intensities between Friedel pairs in SAD and between the Bragg peaks at different energies in MAD are small. The inherent shot-to-shot intensity fluctuations of SASE pulses means that extracting these differences is highly challenging [86]. Nonetheless, native-SAD-SFX using 6 keV (to 3.2 Å resolution) and 9.8 keV (to 1.9 Å) photon energy, for which the average anomalous differences were smaller than 1.5%, has resulted in the *de novo* structure determination of the GPCR human A_{2A} adenosine receptor [85] (Figure 6.62). The structure contained 17 identifiable sulfur atoms.

6.12 Surface Diffraction

6.12.1 Introduction

Surface x-ray diffraction (SXRD) is concerned with the study of the structure of crystalline surfaces and interfaces [87, 88]. SXRD studies are important because the surface regions of crystals and crystalline interfaces between two heterogeneous materials rarely have the same structure as the bulk, and it is here that many physical and chemical processes take place. Indeed, we define the surface and interface regions as being those which differ from those of the bulk, which typically involve a depth of three or four monolayers (approximately 2 nm), the extent of (screened) Coulomb interactions in covalent or ionic solids. We would therefore like to characterize the structure of these surface regions to better understand and possibly exploit their physical and chemical properties.

In creating a surface, the near-surface atoms invariably reposition themselves to minimize the electrostatic energy of the electrons of the uppermost atoms that have failed to bond (because they have fewer neighbours than do bulk atoms). Concerted movements of whole atomic layers perpendicular to the plane alone are called relaxations, while larger periodic structures in the plane at or near the surface are called reconstructions (see Figure 6.63).

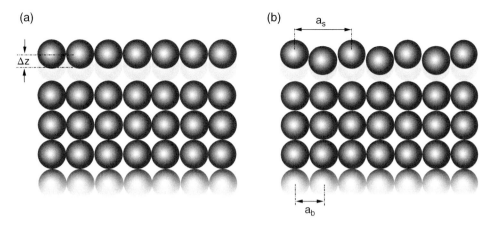

Figure 6.63   (a) Movements perpendicular to the surface of the uppermost atomic layer(s) Δz produce surface structures referred to as relaxations. (b) Rearrangements of atoms in the plane, unequal movements out-of-plane, or different chemical compositions of the surface atoms can lead to periodic structures in the plane a_s that are not equal in size to the bulk periodicity a_b, in which case one speaks of reconstructions. In this simple schematic, the surface 'net' periodicity is twice as large as that of the bulk.

Over and above this, any atoms chemically or physically absorbed on the surface will also affect the surface properties. It is therefore imperative in SXRD that the surface is well-defined, so that reliable conclusions can be drawn from the experimental data. This means that most surface-diffraction experiments are performed in ultra-high-vacuum chambers equipped with beryllium windows, which are reasonably transparent to x-rays above energies of about 4 keV. Notable exceptions are studies of interfaces between solids or a solid and a liquid[23]. Analysis of SXRD data is often facilitated by complementary information gleaned by other surface techniques such as scanning tunnelling microscopy, reflection high-energy electron diffraction, low-energy electron diffraction, and x-ray photoelectron spectroscopy.

6.12.2   Crystal Truncation Rods

In the simplest derivation of diffraction patterns, two assumptions are made – firstly, that one is operating in the kinematical limit (that is, single-scattering), and secondly, that the crystal is infinitely large. This results in the diffraction peaks being infinitely narrow (known as 'delta functions'). In reality, of course, all diffraction spectra are smeared out to a certain degree because there is partial absorption and extinction; moreover, dynamical multiple scattering cannot be ignored close to the Bragg peaks (see Section 6.4.4).

In addition, crystals are finite in extent and one therefore measures a finite sample volume. The diffraction pattern of a finite crystal can be generated by convolving the Fourier transform of an infinitely large crystal structure (i.e. its 'ideal' diffraction pattern) with the Fourier transform of the function describing the boundary of the real crystal (called the 'shape function'). In most cases (for example, in bulk single-crystal diffraction), the boundary function is irregular on an atomic scale and hence its Fourier transform, the shape function, is exceedingly narrow – broadening effects other than that due to convolution with the shape function will determine the linewidths of the diffraction peaks.

---

[23] On a technical note, however, it is normally desirable to immerse the sample in either a modest vacuum or in a He-gas atmosphere, in order to minimize air scatter immediately above where the x-rays impinge on the sample.

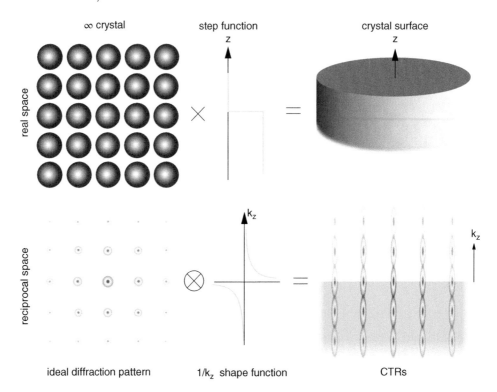

Figure 6.64   The generation of CTRs, pictorially explained using the convolution theorem. An atomically flat crystal surface can be mathematically described as a *semi*-infinite crystal, whereby an infinitely large crystal is multiplied by a step function, having a value of unity for $z \leq 0$ and zero for $z > 0$. In reciprocal space, the idealized diffraction pattern of the infinite crystal is convoluted with the Fourier transform of the step function, which has a $1/k_z$ dependence. The result of truncating the crystal in this manner is that the diffraction signal perpendicular to the surface is continuous, hence crystal truncation rods (CTRs).

However, a single crystal terminated with an atomically flat surface has a step function as the boundary function (see Figure 6.64). This has an FT showing a $1/k_z$ relationship that extends significantly in reciprocal space. Hence, the convolution of this with the 'ideal' diffraction pattern results in the latter being smeared out to produce a continuous signal in the direction perpendicular to the sample surface. These are crystal truncation rods (CTRs).

The signal intensity in between Bragg peaks is so weak because the vector sum of the scattering amplitudes from the different atomic layers is almost completely destructive. One can easily show that the intensity exactly in between Bragg points is of the same order of magnitude as that from a single atomic layer. The Argand diagram of Figure 6.65 shows this schematically. Any shifts in the atomic positions of the upper layers from their bulk positions, due to surface reconstructions and/or relaxations, will have a marked effect on the form and magnitude of the scattered amplitudes in portions of the CTRs away from the Bragg maxima. For large numbers of scattering planes (N in Figure 6.65), interference becomes very quickly destructive away from the Bragg maxima, and a great deal of information can be gained in almost all the region along CTRs in between Bragg peaks.

Recording CTRs therefore provides an exceptionally sensitive method for unravelling the structure of crystalline surfaces and interfaces. But here's the rub – because most of the useful information on the near-surface structure of a material is extracted from the weakest portions of the CTR in between the Bragg maxima, this information is particularly difficult to record – there might be intensity ratios between the Bragg maxima and the weakest portions

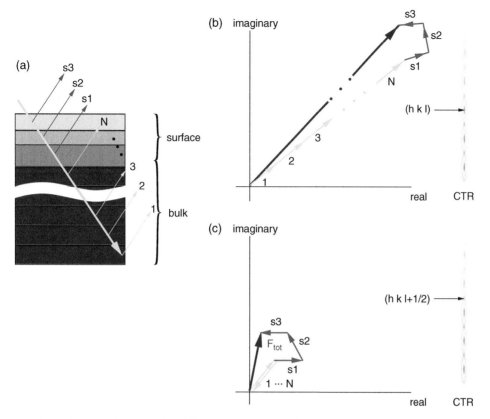

Figure 6.65 A phase diagram depicting the differing contributions from the surface region at Bragg maxima and in between Bragg maxima. (a) Consider scattering from a crystal with a thickness of N monolayers, where N is a large number. (b) In the kinematical approximation, each monolayer of the bulk crystal contributes the same amount (the yellow arrows) to the total scattering amplitude (shown as the blue vector), and at Bragg peaks, they all have the same phase, i.e. they add up in a straight line. Because N is large, the relative impact of scattering from the uppermost layers s1, s2, and s3 (shown in red, and which contribute differently from the bulk monolayers in both magnitude and phase, due to their different structure) on F_{tot}, is minimal. (c) At the anti-Bragg point, bulk contributions from successive monolayers are π out of phase with each other and cancel each other out. Therefore, deviations from this destructive interference, as in the vectors s1, s2, and s3, have a large *relative* effect on the magnitude of F_{tot}. Thus, where signal is weakest, the relative effect of surface changes in the structure is most prominent.

of the CTR of as much as eight orders of magnitude[24]. Hence, most surface diffraction experiments require the high flux available at synchrotrons and detectors with a high dynamic range and low background noise. Even then, great care must be taken to minimize diffuse scatter, typically by working in vacuum or a He atmosphere, and by a judicious use of beam-defining slits. On the other hand, because x-rays are only weakly scattered by a monolayer of crystalline material, this has the benefit that SXRD readily satisfies the kinematical approximation, simplifying analysis compared, for example, to surface-analytical techniques using electrons, such as low-energy electron diffraction (LEED), for which a dynamical approach and an associated much greater computational effort is required, in order to obtain quantitative data from the signal intensities.

---

[24] This value can be estimated using a back-of-an-envelope calculation and the fact that the intensity of the Bragg peak is approximately proportional to the square of the number of involved scattering planes. This is typically of the order of $N = 10^4$. Remembering that the weakest signal has an intensity of the order of that produced by a single scattering plane, we obtain a ratio of $(10^4)^2 = 10^8$.

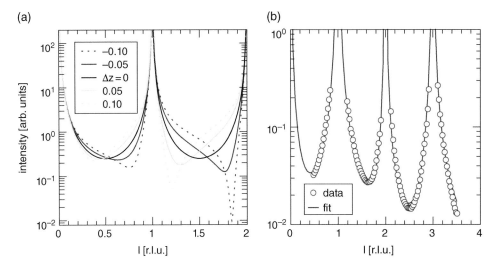

Figure 6.66 Surface relaxations detected by SXRD (a) The effect of the displacement of the uppermost layer (Δz, given as a fraction of the interlayer spacing) becomes more pronounced as Δz is increased. (b) Relaxation of the topmost atomic layer of Ru(0001) is best fit by a contraction of $2.2 \pm 0.1\%$ relative to the bulk value of 2.141 Å. Open circles: experimental data, solid line: fit. r.l.u. = reciprocal lattice unit. Adapted from [89] with permission of the Institute of Physics.

CTRs become interesting when the top few layers deviate from the structure of the bulk in the direction of the surface normal. This causes the CTR to become skewed. As the deviation of the uppermost layer z increases, so does the amount of skewing. This is shown for $\Delta z/a = 0, \pm 0.05$, and ± 0.1 in Figure 6.66(a).

CTRs of thin-film systems contain other important information about the film properties apart from the detailed atomic structure, including degree of crystallinity, film thickness, density, and interface and surface sharpness. An example of a 4-unit-cell-thick $YBa_2Cu_3O_{7-x}$ film grown on $SrTiO_3$ is shown in Figure 6.67.

6.12.3   Superstructure Rods

Superstructure rods (SSRs) are a signature of reconstructions. A surface structure which is, for example, $m \times n$ times larger in the plane of the surface than the bulk structure, will have SSRs in reciprocal space separated by $2\pi/ma$ and $2\pi/nb$, whereby m and n are integers and a and b are the in-plane lattice constants of the bulk material (see Figure 6.68).

The intensity and form of SSRs are in general very different from those of CTRs. Because they are manifestations of surface effects, the number of scatterers compared to those contributing towards CTRs is much smaller and SSRs are therefore weak, having intensities of the same order of magnitude as that of the weakest regions of the CTRs (see, for example, Figure 6.69). Moreover, the relative variation in intensity is much more modest than that for CTRs, as only a few layers contribute to the signal.

6.12.4   Data Acquisition

Typically, SXRD measurements are carried out using a grazing incident angle α, with respect to the sample surface. The choice of α depends on the experiment, sample quality, and scattering strength. For high-quality crystals consisting of heavy elements, such as metal oxide perovskites, one can normally choose α to be significantly above the critical angle. This has the advantage that the x-rays penetrate deeply into the sample and one can easily subtract the contribution from the bulk during subsequent analysis. For incident angles close to the critical

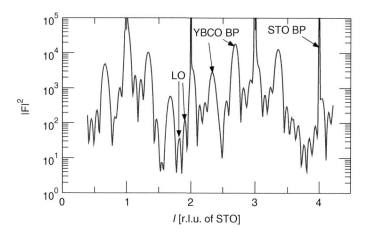

Figure 6.67   The 21*l*-CTR of a 4-unit-cell thick film of YBa$_2$Cu$_3$O$_{7-x}$ (YBCO) grown on SrTiO$_3$ (STO). The substrate Bragg peaks (STO BP) are very sharp and lie on integer values of the abscissa to within the accuracy of the experiment of 10^{-4} reciprocal lattice units (r.l.u.). Because the YBCO film is so thin (4 unit cells, approximately equivalent to 12 unit cells of STO), its Bragg peaks (YBCO BP) are correspondingly broad. In between the YBCO Bragg peaks are finite-size oscillations, known as Laue oscillations (LO). In contrast to Kiessig fringes, discussed in Section 6.14, these only occur in nonspecular CTRs if the film is crystalline, as they require in-plane momentum transfer, and therefore in-plane periodicity. Adapted from [90] with permission of the American Physical Society.

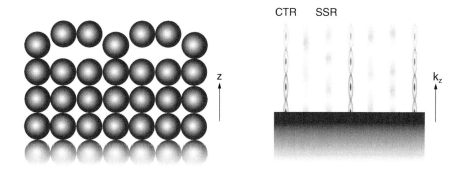

Figure 6.68   Left: a surface reconstruction with a periodicity three times that of the bulk. Right: in reciprocal space, the CTRs are interspersed with SSRs, whereby the separation between adjacent SSRs is three times smaller than that between CTRs.

angle α_c, the surface sensitivity increases rapidly. Exactly at α_c, the reflected wave is perfectly in phase with the incident wave, and the evanescent wave amplitude is approximately twice that of the incident wave. The evanescent *intensity* therefore approaches four times that of the incident beam. The penetration depth is low, and so the bulk contribution is suppressed. Hence, at α_c, the surface sensitivity is highest, which is advantageous for weakly scattering surfaces and also for systems with crystallographically poor bulk material (such as some metal crystals) which produce significant diffuse background. The main disadvantage of impinging on the sample at α_c is that even angular variations as small as a 200 μrad (0.01°) in the incident angle will result in significant systematic errors in the measured intensities.

SXRD can be divided into two activities – in-plane diffraction and out-of-plane measurements. The first is used to study the in-plane positions of the atoms in the uppermost layers, and reveals features such as surface

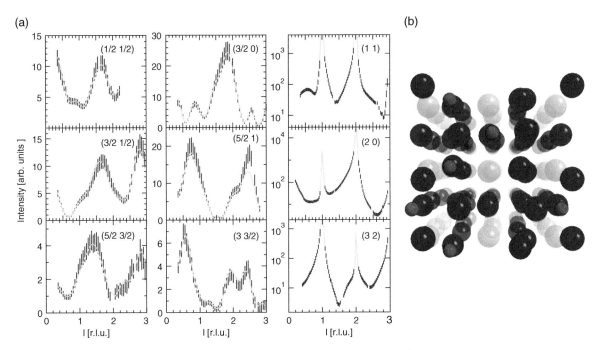

Figure 6.69   The surface of HF-etched and thermally annealed, TiO_2-terminated, strontium titanate. (a) Six representative SSRs and three CTRs (blue), plus fits (yellow solid lines), taken from a set of SXRD data recorded of the surface of $SrTiO_3(001)$. Note that the characteristic width of oscillations in the SSRs is approximately 0.5 r.l.u., indicating thereby typical deviations from the bulk structure down to two unit cells. (b) The structural model of the (2 × 2) reconstruction that generates the fits for the half-integral–half-integral SSRs. Ti: red, O: blue, Sr: yellow. This best-fit model contains a double surface TiO_2 atomic layer, in which the Ti atoms in the upper layer form an unusual zigzag motif. Adapted from [91] with permission of the American Physical Society.

reconstructions. By recording a large set of in-plane peaks, it is possible to obtain a two-dimensional Patterson map (see Section. 6.6.2), which often provides important information as to the probable configuration of the atoms in the surface plane. Nonintegral features yield information on the periodicity and orientation of any reconstructed surface net. In this context, qualitative information from LEED measurements or similar techniques can also provide invaluable prior information regarding where one should search for signal in reciprocal space.

The in-plane points can then be used as the starting points for out-of-plane measurements, in the form of CTRs, and also SSRs if the surface is reconstructed. The intensity distribution along a CTR or SSR depends on the structure of the near-surface region perpendicular to that surface. CTR and SSR scans are therefore complementary to in-plane diffraction studies (see Figure 6.69).

6.13   Resonant X-ray Scattering

As we have already seen in Section 6.11.4, elastic anomalous scattering has long been employed by crystallographers to help solve the phase problem. It can also be exploited in the technique of resonant x-ray scattering (RXS) to probe unoccupied states close to the vacuum continuum in a second-order process. Consider Figure 6.70. An x-ray photon of energy $h\nu = E_f - E_i$ is absorbed, causing an electron occupying the core level $|i\rangle$ to be excited to

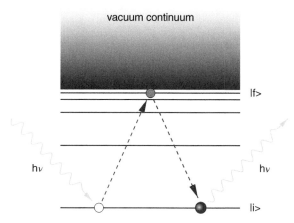

Figure 6.70 Schematic of the second-order resonant scattering process, which occurs via the intermediate state $|f\rangle$. The photon energy $h\nu = E_f - E_i$.

the unoccupied state $|f\rangle^{25}$ with a certain probability, which then promptly relaxes again to the original core level $|i\rangle$ via emission of an x-ray photon having the same energy as the incident photon. This process is therefore elastic. Crucially, both the absorption strength of the incoming photon and the emission probability of the final photon depend directly on the dipole transition between the lower state and upper state. Quantum selection rules between the two states and the polarization of the electric field of the x-ray beam determine the dipole strength. RXS is therefore a sensitive spectroscopic probe for the upper state and provides both spatial and site-selective (chemical, spectroscopic) information. This is not normally the case in conventional diffraction – the relative contribution of valence-state electrons in the first-order process of direct scattering is negligibly small, as the x-rays are scattered by all the electrons around the atom.

Although the probed state is unoccupied, it has related symmetry properties to occupied valence states – an example might be the so-called $d(3z^2 - r^2)$ and $d(x^2 - y^2)\,e_g$ states in manganite perovskites [92]. The importance of valence-state electrons can hardly be overstated, as they determine most, if not all, of the physical, structural, and electronic properties of condensed matter.

The material $La_{1-x}Sr_xMnO_3$ [LSMO, see Figure 6.71(a)] is a ferromagnetic compound which exhibits so-called 'colossal magnetoresistivity' (CMR) in the range $0.15 \leq x \leq 0.5$, whereby its resistivity drops by orders of magnitude, driven by the application of an external magnetic field. Importantly, La and Sr are always trivalent and divalent, respectively, but Mn can assume several different valencies. In the parent compound $LaMnO_3$ $(x = 0)$, all the Mn ions must be trivalent (Mn^{3+}) to maintain charge neutrality. For each substitution of La with Sr, however, one Mn atom becomes tetravalent (Mn^{4+}). Hence, LSMO:$x = 1/2$ contains one half Mn^{4+} and one half Mn^{3+}. The valence states and the orbitals of the Mn^{3+} ions for this compound are arranged in an ordered manner, as shown schematically in Figure 6.71(b).

The unit cell of LSMO is normally considered to be quasi-cubic, with one chemical unit of $La_{1-x}Sr_xMnO_3$ occupying a simple-cubic unit cell (although in reality, very slight distortions mean that the unit cell is, strictly speaking, orthorhombic with a four-times larger volume). Even if one assumes a pure cubic unit cell, however, charge ordering increases the unit-cell size by a factor of two, which takes into account the ordered arrangement of the different valence and chemical states of the Mn ions (though this is all too often neglected). More subtle is the unit cell which also takes into account the *orbital* ordering – here the Mn^{3+} ions are all chemically identical, and only the direction of the orbitals determines the cell size, which is now *four* times larger than the cubic cell.

---

[25] Pauli's exclusion principle dictates that the intermediate state $|f\rangle$ must be unoccupied.

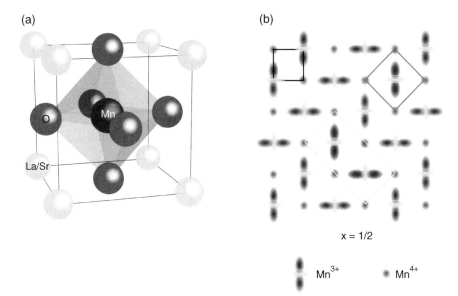

Figure 6.71 Charge and orbital ordering in the compound $La_{1-x}Sr_xMnO_3$. (a) The perovskite quasi-cubic structure of LSMO. (b) The charge and orbital order for LSMO ($x = 1/2$). Note that the size of the unit cell is larger than the 'chemical' unit cell (shown in black) when the charge ordering (blue) and orbital ordering (yellow) are taken into account.

Hence, charge ordering and orbital ordering cause the unit-cell size to increase compared to that of the 'chemical' unit cell, which results in Bragg reflections at noninteger scattering vectors (expressed in units of the quasi-cubic structure) that one would normally not see in nonresonant diffraction, due to the low scattering strength of the valence electrons. However, the RXS signal is detectable exactly because it specifically probes these upper valence states. Ordering of electron *spin* is yet another degree of freedom that can cause reconstructions.

Note that in this example it is the $3d$ states of the Mn atom that are responsible for charge, orbital, and spin ordering. The selection rules for electronic dipole transitions induced by the absorption of a single photon dictate that the orbital angular-momentum quantum number l changes by ± 1 (although it is briefly noted that in electric *quadrupole* transitions involving the absorption of two photons, $\Delta l = 0, \pm 2$ and the $3d$ orbital can be accessed from the $1s$ state). Hence, the $3d$ state can only be accessed by a $2p$ state, in other words, through an L absorption. Such experiments investigating the orbital properties of the $3d$ transition metals therefore use photon energies of around 500 to 1000 eV [93]. As a result, resonant *soft* x-ray scattering (RESOXS) is at the extreme limit of x-ray diffraction, where absorption by low-Z material is exceedingly high (see Figure 5.14): the entire beamline must be windowless (including the optics and diffractometer), and the vacuum must be better than 10^{-8} mbar. In addition, the Ewald sphere is very small, with a radius of the order of $2\pi/\lambda \sim 0.5$ Å^{-1} and may contain only a single Bragg reflection.

The nature and orientation of the probed state regarding the electron distribution (dipole, quadrupole, etc.) can be distinguished in ordered systems by its dependence on the polarization of the incident x-ray beam.

RXS experiments employ the selected Bragg-peak method (Section 6.8.2). Typical procedures involve moving the diffractometer to the Bragg condition for the suspected superlattice reflection [e.g. to the (3/4, 3/4, 0) position for orbital order in $La_{1/2}Sr_{3/2}MnO_4$] and then scanning the photon energy across the absorption edge of interest.

RXS is a relatively photon-hungry technique, due to the weakness of the signal. The first results on charge ordering and orbital ordering in the so-called 'half-doped' layered manganite $La_{1/2}Sr_{3/2}MnO_4$ were reported by

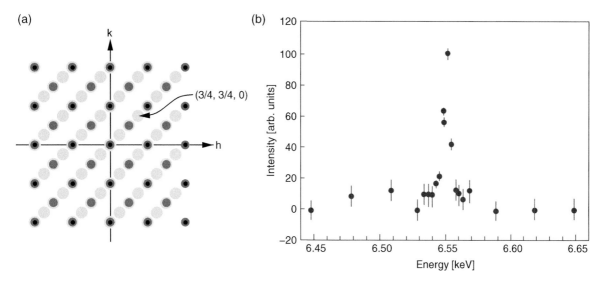

Figure 6.72 RXS data for the layered perovskite $La_{1/2}Sr_{3/2}MnO_4$. (a) The superlattice reflections due to charge order-ing (blue) and orbital ordering (yellow) are shown relative to the reciprocal lattice of the conventional 'chemical' unit cell (black). The real-space configuration of the charge- and orbital-ordering is the same as that shown in Figure 6.71(b). (b) The energy dependence of the orbital-ordering superlattice peak at (3/4, 3/4, 0) near the Mn K-edge in $La_{1/2}Sr_{3/2}MnO_4$. Adapted from [94] with permission of the American Physical Society.

Murakami *et al.* in 1998 [94] and showed charge ordering with a unit cell $\sqrt{2} \times \sqrt{2}$ times larger in the *a-b* plane (and rotated by 45°) and orbital ordering with a unit cell that was $2\sqrt{2} \times \sqrt{2}$ times larger in the *a-b* plane and also rotated by 45°. The energy spectrum of the latter at the (3/4, 3/4, 0) reflection is shown in Figure 6.72.

6.14 X-ray Reflectometry

6.14.1 Introduction

X-ray reflectometry (XRR) is a nondestructive technique used to precisely determine the thickness, roughness, and density profiles of surfaces, thin films, interfaces, and multilayers. XRR is performed simply (and with synchrotron radiation also rapidly) by measuring the specularly reflected x-ray intensity as a function of the grazing incidence angle, typically up to approximately 2°. This allows one to determine film thicknesses, multilayer periodicities, and electron-density profiles over a range spanning approximately 1 to 500 nm, whereby the lower limit is determined by the maximum recorded incident angle and the upper limit by the divergence of the incident x-ray beam in the plane perpendicular to the sample surface.

XRR works equally well for amorphous, polycrystalline, or indeed liquid materials as for crystalline materials, as it essentially probes contrast in the average electron density rather than scattering from individual atoms; it can be thought of as a special form of small-angle x-ray scattering, discussed in detail in Section 6.15.

Most XRR experiments start at angles slightly below α_c, the critical angle for total external reflection [Equation (2.28)], typically somewhere near 0.1°. The footprint on a sample of an x-ray beam with a cross-sectional height of 0.1 mm at this incident angle is therefore of the order of 5 mm. Typical linear dimensions of XRR samples are hence a few mm to a few cm. XRR scans can often span five orders of magnitude or more in the recorded reflectivity intensity, and are as such ideally suited to intense x-ray sources like synchrotrons.

6.14.2   Reflection of X-rays and the Fresnel Equations

Consider Figure 6.73. We describe the incident, reflected, and transmitted plane waves as

$$\psi_i = a_i e^{i\mathbf{k}_i \cdot \mathbf{r}},$$

$$\psi_r = a_r e^{i\mathbf{k}_r \cdot \mathbf{r}},$$

$$\psi_t = a_t e^{i\mathbf{k}_t \cdot \mathbf{r}}, \tag{6.44}$$

whereby i, r, and t refer to the incident, reflected, and transmitted beams, respectively. Note that $|\mathbf{k}_i| = |\mathbf{k}_r| = k$, the wavevector magnitude in vacuum, while $|\mathbf{k}_t| = nk$ is the magnitude of the wavevector in the medium of refractive index n.

We have already reasoned in Chapter 2 that below α_c, x-rays are totally reflected. From Equations (2.24) and (2.28), we see that

$$\alpha_c = \lambda \sqrt{\frac{\rho r_0}{\pi}} \tag{6.45}$$

and hence α_c scales with the square root of the electron density, and as such is an accurate measure of this quantity. Interestingly, almost all condensed matter has an average electron density somewhere between 0.5 and 5 electrons per cubic angstrom (see Figure 2.17), so a useful rule of thumb is that the critical angle expressed in degrees lies somewhere in between 0.12 and 0.4 times the wavelength given in angstrom. Because the electron density is also very approximately linearly related to Z (despite large excursions from this general trend, see Figure 2.17), a more accurate rule of thumb for α_c is provided by Equation (5.21).

At all angles, the reflectivity is determined by the boundary conditions that the normal component of the electric field of the electromagnetic wave and its derivative at the interface must be continuous functions. From Equations (6.44), we therefore obtain for $\mathbf{r} = 0$

$$a_i + a_r = a_t$$

and

$$a_i \mathbf{k}_i + a_r \mathbf{k}_r = a_t \mathbf{k}_t,$$

from which the Fresnel equations for reflectivity and transmission can be derived. Here we are primarily interested in the reflectivity amplitude, which above the critical angle is given by

$$r = \frac{a_r}{a_i} = \frac{\alpha - \alpha'}{\alpha + \alpha'}, \tag{6.46}$$

where α and α' are the incident and transmitted angles, respectively (see Figure 6.73).

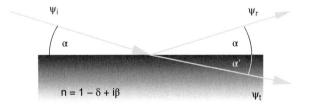

Figure 6.73 X-ray reflectivity from a surface. The incident beam ψ_i impinges on the surface with an angle α and is reflected specularly (ψ_r). The transmitted beam ψ_t is refracted to a lower angle α' because the refractive index n of the medium is less than unity.

From Snell's law, it can be derived that for shallow incident angles above the critical angle, the relationship between α and α' can be very accurately approximated by

$$\alpha^2 = \alpha'^2 + 2\delta. \tag{6.47}$$

The reflectivity amplitude r changes subtly, depending on the polarization of the x-ray electric field relative to the reflecting surface. The difference between parallel or perpendicular polarization is very small, and we limit ourselves here to the case whereby the electric field is quasi-perpendicular to the surface[26]. In this case, the reflection amplitude is

$$r = \frac{\sin\alpha - (n^2 - \cos^2\alpha)^{1/2}}{\sin\alpha + (n^2 - \cos^2\alpha)^{1/2}}. \tag{6.48}$$

We use the simplifications $n = 1 - \delta$, $\sin\alpha \approx \alpha$, and $\cos\alpha \approx 1 - \alpha^2/2$ for small angles and thereby obtain

$$r = \frac{1 - (1 - 2\delta/\alpha^2)^{1/2}}{1 + (1 - 2\delta/\alpha^2)^{1/2}}. \tag{6.49}$$

The measured reflectivity *intensity* is $R = |r|^2$. We see immediately that at the critical angle $\alpha_c = \sqrt{2\delta}$, both terms within the brackets of Equation (6.49) are equal to zero and the reflectivity is 100%, as expected. Above α_c, however, the reflectivity drops off very rapidly.

Once the incident angle becomes significantly larger than α_c, that is, when $2\delta/\alpha^2 \ll 1$, we can expand the terms in brackets in Equation (6.49) by recognizing that

$$\left(1 - \frac{2\delta}{\alpha^2}\right)^{1/2} \approx 1 - \frac{\delta}{\alpha^2}.$$

The reflectivity is then accurately approximated by

$$R = r^2 = \frac{\delta^2}{4\alpha^4} = \left(\frac{\alpha_c}{2\alpha}\right)^4 \tag{6.50}$$

and the reflectivity (a) falls off with the inverse fourth power of incident angle, and (b) is proportional to the square of the electron density ρ [see Equation (6.45) and Figure 6.74].

As soon as the surface has any roughness on a scale similar to or larger than the wavelength of x-radiation being used, the reflectivity drops still more sharply. For a Gaussian distribution of the roughness with a root-mean-square value σ, the reflectivity curve for a flat surface R_f is modified for $\alpha \gg \alpha_c$ such that

$$R_r = R_f \, \exp[-(2k\alpha\sigma)^2], \tag{6.51}$$

whereby R_r is the reflectivity for the rough surface, and is shown in Figure 6.74[27].

In the above, we have ignored two important phenomena, namely absorption effects (by assuming that $|\beta| \ll |\delta|$) and multiple scattering of the x-rays, which can be significant for very low incident angles. It lies outside the scope of this book to describe these effects quantitatively, and the reader is referred to the literature for rigorous discussions [95].

---

[26] This case is referred to as *s*-polarized light, whereby an electric field which is exactly perpendicular to the surface must by definition have its direction of propagation parallel to the surface, that is, $\alpha = 0°$. Hence. the need to qualify with the expression *quasi*-perpendicular for XRR, whereby the incident angle is always small, if not exactly zero.

[27] Again, one can apply the convolution theorem (Section 6.5.1) to understand this: the rough surface is the convolution of the flat surface with the roughness function (normally a Gaussian distribution), hence the reflectivity curve for the rough surface is simply that for the flat surface multiplied by the Fourier transform of the roughness function.

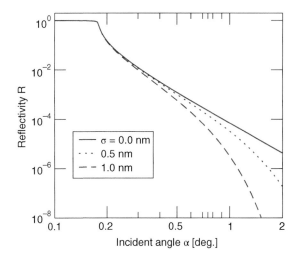

Figure 6.74 Reflectivity curves for a silicon surface with different roughnesses. Note that on this double-logarithmic plot, the reflectivity curve of the flat surface and for $\alpha \gg \alpha_c$ has a gradient of -4, as predicted by Equation (6.50).

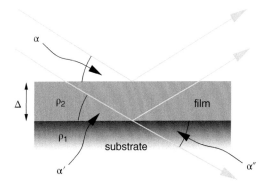

Figure 6.75 Kiessig fringes occur when that part of the incident beam reflected from the upper surface of a thin layer interferes with that part reflected from the layer–substrate interface.

6.14.3   Thin Films and Multilayers

Consider a system consisting of a thick, perfectly flat slab of density ρ_1 (the 'substrate') covered with an equally flat thin layer of thickness Δ and density ρ_2 (the 'film', see Figure 6.75). As in the case of a homogeneous slab on its own, x-rays will be partially reflected from the upper film surface. In addition, if $\rho_1 \neq \rho_2$, the interface between the film and substrate will provide contrast for any x-rays impinging on it, and they will be partially reflected here, too. Interference between the two reflected beams cause the reflectivity curves of such systems to be modulated with so-called 'Kiessig fringes' [96].

The reflectivity of a single layer on a substrate is given by

$$R = \left| \frac{r_1 + r_2 \, \exp(i4\pi\alpha'\Delta/\lambda)}{1 + r_1 r_2 \, \exp(i4\pi\alpha'\Delta/\lambda)} \right|, \tag{6.52}$$

whereby

$$r_1 = \frac{\alpha - \alpha'}{\alpha + \alpha'} \, \exp\left[-2\left(\frac{2\pi}{\lambda}\right)^2 \alpha\alpha'\sigma_1^2\right],$$

(6.53)

$$r_2 = \frac{\alpha' - \alpha''}{\alpha' + \alpha''} \, \exp\left[-2\left(\frac{2\pi}{\lambda}\right)^2 \alpha'\alpha''\sigma_2^2\right].$$

(6.54)

α'' is the angle of the x-ray beam transmitted into the substrate, while σ_1 and σ_2 are the roughnesses at the surface and substrate, respectively.

The optical path difference OPD is, to a high degree of accuracy, given by

$$\text{OPD} = 2\Delta \sin\alpha'.$$

(6.55)

This should equal an integral number of wavelengths for constructive interference. From Snell's law, we also know that

$$\cos\alpha = n\cos\alpha'$$

and, using the first two terms of the Taylor expansion for the cosine function, we obtain

$$1 - \alpha^2/2 = (1 - \delta)[1 - \alpha'^2/2].$$

If we multiply this out and ignore the term $\alpha^2\delta/2$, we obtain

$$\alpha' \approx (\alpha^2 - 2\delta)^{1/2},$$

(6.56)

from which it follows that constructive interference occurs when

$$\alpha'^2 = \frac{\lambda^2}{4\Delta^2} \, m^2 + 2\delta,$$

(6.57)

whereby m is an integer. Hence a plot of α'^2 against m^2 yields a straight line with a gradient $\lambda^2/4\Delta^2$ which crosses the ordinate at 2δ. This provides both the film thickness and density. One potential problem, particularly for thicker films, is obtaining a reliable value for m, as the Kiessig fringes only occur above α_c, for which m can be much greater than unity. However, if we consider those fringes at relatively high incident angles, that is, where $\alpha \gg \alpha_c$, Equation (6.57) reduces to

$$\alpha' = \frac{m\lambda}{2\Delta}$$

(6.58)

and the angular separation between adjacent fringes is simply $\lambda/2\Delta$.

Both the modulation depth and the phase of the Kiessig fringes are affected by the contrast (that is, the difference in electron densities) of the film and substrate. Consider the simulated reflectivity curves shown in Figure 6.76(a). The modulation depth of Si on Au is smaller than that of Au on Si, as scattering from the film's top surface is weaker in the former case. Note also that the Kiessig fringes are π out of phase relative to each other. This is explained by the fact that, when reflected off an interface to a medium with lower electron density (e.g. the Au-film/Si-substrate interface), the phase of the reflected beam is preserved, while reflections off surfaces to denser material (Si-film/Au-substrate) undergo a phase flip of π radians. When the difference in contrast between film and substrate is smaller, the Kiessig fringe amplitude drops accordingly, as also shown in Figure 6.76(a) for C on Si (with electron densities of 0.66 and 0.70 electrons per cubic angstrom, respectively.)

Figure 6.77 shows an example of the reflectivity curve of a 120 nm-thick quasicrystalline icosahedral Ti-Ni-Zr thin film grown on Al_2O_3, exposed to air for several days [97]. In addition to the high-frequency Kiessig fringes produced by the relatively thick film, there is a much lower-frequency modulation. This is caused by the top 4 nm of the film becoming oxidized.

(a)   (b)

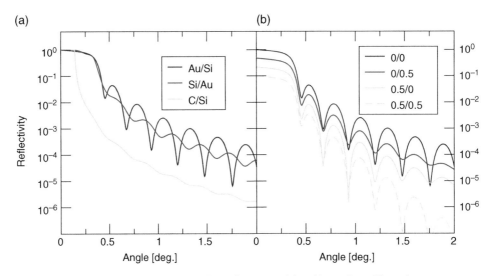

Figure 6.76   (a) The effect of differing contrast on the reflectivity of thin films. Three film/substrate systems are shown: Au on Si; Si on Au; and C on Si. All films are 10 nm thick and therefore have the same Kiessig-fringe periodicity. There is no substrate- or film roughness, and 1 Å radiation was used. (b) The effect of roughness on the reflectivity of a 10 nm thin film of gold on silicon. The roughnesses in the legend are ordered substrate/film and given in nm.

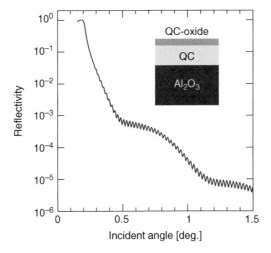

Figure 6.77   An example of an experimental reflectivity curve of a quasicrystal (QC) thin film grown on sapphire (Al_2O_3). In addition to the high-frequency Kiessig fringes produced by the bulk film, an additional low-frequency modulation can be observed, which is caused by the uppermost 4 nm of the QC-film having been oxidized by exposure to air. Adapted from [97] with permission of Elsevier.

The important parameters of periodic multilayer structures are shown in Figure 6.78(a). From a scattering perspective, these systems can be thought of as being two-dimensional 'artificial crystals', with lattice dimensions given now by Δ instead of the crystal interplanar spacing d, and the basis (which for crystals is given by the arrangement of the atoms within the unit cell) being determined by ρ_1, ρ_2, Δ_1, and Δ_2. Their reflectivity curves

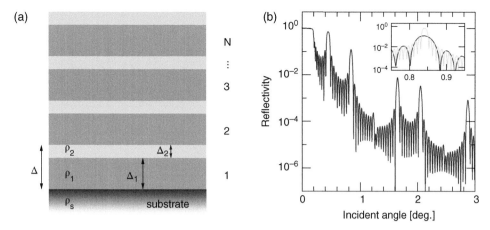

Figure 6.78   Multilayer x-ray reflectivity. (a) Schematic figure of a multilayer, showing the defining parameters. (b) Simulation of a multilayer reflectivity curve at 12.398 keV for a ten-layer Mo/Si multilayer. $\Delta = 7\,\text{nm}$, $\Delta_2/\Delta = 1/3$. Note that every third reflectivity peak is missing. Inset: a blowup of the second reflectivity peak near 0.85°. By increasing the number of layers from ten (blue curve) to 40 (yellow curve), the peak becomes both narrower and more intense.

therefore have similar characteristics to those of regular θ–2θ diffraction spectra of heteroepitaxial films [see Figure 6.78(b)]. The more multilayer periods there are, the sharper and more intense are the reflection maxima.

Multilayers can therefore be used as Bragg reflectors at angles far above the critical angle of either of the constituent materials. As already touched upon in Chapter 5, a particularly promising application of multilayers is the fabrication of monochromator elements and compact x-ray mirrors in x-ray optical systems. Conventional mirrors at synchrotron beamlines have critical angles of a small fraction of a degree, and need to be of the order of a metre long in order to capture the footprint of the entire x-ray beam. As can be seen in Figure 6.78(b), interference maxima with reflectivities close to unity can be found at incident angles of the order of one degree, meaning that multilayer mirrors can be an order of magnitude smaller. In general, such systems consist of layers composed of a thicker, low-Z material (to avoid excessive absorption) and a thinner high-Z material (to provide the necessary contrast for efficient reflection).

6.14.4   XRR Monitoring of Thin Film Growth

XRR can be used to study changes in the thickness or morphology of a thin film, particularly important for *in situ* studies during heteroepitaxial film growth.

As discussed in Section 6.12, the *relative* change in the specular-reflection intensity due to atomic-scale changes of the surface region is largest where the absolute signal is weakest, exactly in between Bragg peaks [for example, at the (0 0 0.5) position, see Figure 6.65(c)]. The reason for this is that most of the signal cancels out, as successive layers provide scattering vectors that are 180° out of phase with one another. Hence, any changes in one of those layers (in this case, the top, growing, layer, see Figure 6.79) will have a large *relative* effect on the signal. So-called roughness oscillations, with a periodicity equal to the time needed to grow a single monolayer, will therefore occur. This phenomenon is also regularly exploited in thin-film growth chambers using laboratory-based reflection high-energy electron-diffraction (RHEED) equipment.

In XRR, a second type of oscillation will also occur, due to the passage of Kiessig fringes across the detector. Let us assume we are at the (0 0 0.5) position of the specular rod. From Bragg's law, this corresponds to an angle

$$\sin \theta_{(0\ 0\ 0.5)} = \frac{\lambda}{4d_{(001)}}. \tag{6.59}$$

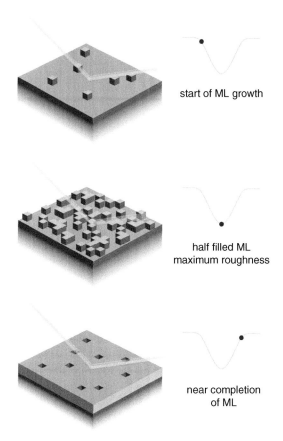

start of ML growth

half filled ML
maximum roughness

near completion
of ML

Figure 6.79   The intensity of the specularly reflected x-ray beam is affected by atomic-scale roughness. Monolayer–for-monolayer growth can therefore be monitored using the specular reflection, whereby each oscillation corresponds to the deposition of a single monolayer.

But Kiessig fringe maxima occur at this detector angle when

$$m\lambda = 2Nd_{(001)} \sin \theta_{(0\ 0\ 0.5)}, \qquad (6.60)$$

where m is an integer and Nd is the film thickness (N being the number of monolayers). Hence, constructive interference occurs when

$$N = 2m, \qquad (6.61)$$

i.e. when N is even, while destructive interference occurs for odd N. Note that these thickness-growth oscillations cannot be observed in RHEED, as the high-energy electrons cannot penetrate below the upper surface of the film to the interface. Hence, films which grow in a step-flow mode (i.e. one in which the impinging atoms diffuse rapidly to terrace edges compared to the time-scale for the growth of a complete monolayer and therefore exhibit no changes in surface roughness) will show thickness-growth oscillations but no roughness oscillations in XRR, and indeed no oscillations at all in RHEED.

An example of the presence of both types of oscillation is shown in Figure 6.80 for the heteroepitaxial growth of $La_xSr_{1-x}MnO_3$ on $SrTiO_3(001)$ using pulsed laser deposition [98].

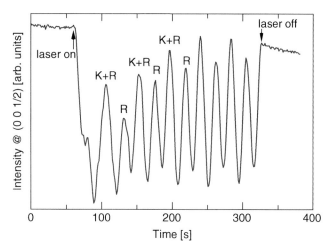

Figure 6.80 The specular-reflectivity intensity at the (0 0 0.5) position during pulsed laser deposition of $La_xSr_{1-x}MnO_3$ on $SrTiO_3(001)$. Note that the roughness oscillations (R) have a frequency twice that of the Kiessig (K) oscillations. Adapted from [98] with permission of the American Physical Society.

6.15   Small-angle X-ray Scattering

6.15.1   Introduction

Small-angle x-ray scattering (SAXS) is a technique used to determine the general structural features of systems that have typical dimensions from a few nanometres up to a micron. SAXS information can be extremely valuable in disciplines such as colloidal science, liquid-crystal technology, biochemistry, and cell biology (see Figure 6.81). In particular, it allows one to study systems which lack long-range order, and is therefore complementary to x-ray diffraction [99, 100]. Indeed, because proteins in crystals are forced into a regular repeat configuration, SAXS provides important information on the overall envelope of the structure in (close to) native conditions.

We have seen that in XRD, the scattering angles, given by the Bragg equation, are typically from a few tenths of a degree to several tens of degrees. For radiation wavelengths of the order of an angstrom, and objects of, say, a few to a few hundred nanometres, the scattering angles are typically smaller than $0.1°$. This is the regime of SAXS. Because the scattering angles involved are so small, the distances between the direct beam and the scattered signal at the detector are, to a high degree of accuracy, directly proportional to the scattering vector ($\tan 2\theta \approx \sin 2\theta \approx 2\theta = \lambda Q / 2\pi$).

The scattering characteristics of objects at these larger-size scales are qualitatively different from those produced in x-ray diffraction at atomic scales, the latter being characterized by high degrees of order. In contrast, on the nano to micron scale, the structures are generally less well organized and may be composed of complex and nonuniform building blocks. This is especially true when considering biological samples. The resulting scattering patterns in SAXS can therefore appear to be comparatively nondescript and diffuse (Figure 6.81). Nonetheless, important information can be extracted from SAXS data, with regards to characteristic size, surface-area-to-volume ratios, and cross-sections. As hard x-rays can penetrate water to significant depths, SAXS is particularly interesting for *in vivo* and biological samples in ambient liquid environments, or in embedded systems, which are excluded to most other nanoscale imaging techniques, such as transmission-electron microscopy and scanning-electron microscopy[28].

---

[28] Even in the tender x-ray regime at 4 keV, the attenuation length of water is 120 μm.

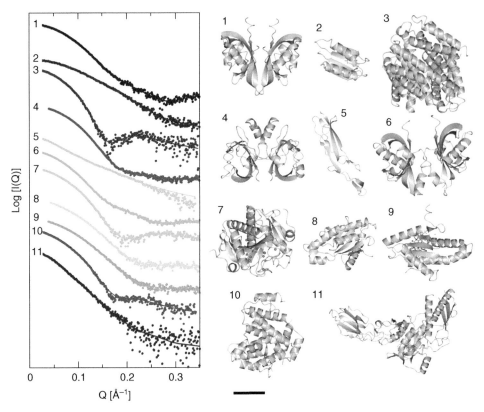

Figure 6.81   Examples of SAXS data, plus structural fits to the envelope shapes for biomolecules with known crystallographic structures. The scale bar is 2 nm. Adapted from [101] with permission of the American Physiological Society.

Other techniques are also capable of probing the nanometre to micron range of SAXS, most notably transmission electron microscopy and scanning electron microscopy. Both, however, sample such exceedingly small volumes, that it is very difficult to collect statistically significant information regarding size distributions. SAXS, on the other hand, samples many orders of magnitude larger volumes.

6.15.2   Theory

Unlike x-ray diffraction, SAXS is purely a contrast technique, whereby the scattering signal is generated by differences in the average electron density on scales significantly larger than that of atomic radii.

Let us consider the simplest example of scattering from a homogeneous object (Figure 6.82). If we consider scattering from all points in the object, it is self-evident that, for large enough angles 2θ, there will be no preferred phase difference between the scattered waves, and their superposition will lead to destructive interference. For small enough angles, however, the phase differences will become smaller, and eventually, below a certain angle, even the largest phase difference originating from scattering from the points furthest away from each other in the object will be less than 2π, and the component waves will begin to reinforce one another, until, for forward scattering ($2\theta = 0$), all waves are exactly in phase.

Hence, SAXS signal typically lies within an angular range of the order of $2\theta = \lambda/a$, where a is the characteristic linear dimension of the object. In general, small fluctuations of the electron density within the object over distances

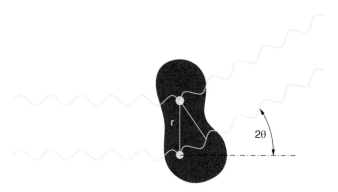

Figure 6.82   Scattering from a particle. At large scattering angles 2θ, the superposition of scattered waves from all points in the particle results in destructive interference, as all possible phases have equal probability. At low enough 2θ-values, however, the phase differences become smaller, resulting in increasingly constructive interference.

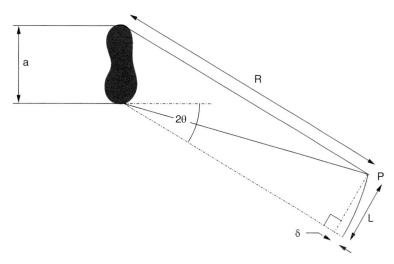

Figure 6.83   The Fraunhofer condition: the difference δ in the OPD between two converging rays on the detector at point P a distance R from the object of size a and that of two parallel rays should be significantly smaller than the wavelength of the x-rays being employed.

much smaller than a will not affect the scattering curve, and as a result, one can often assume the object to have a homogeneous density distribution.

6.15.2.1 *Far-field Diffraction and the Q-range*

How far must the detector be away from the object in order that the true (square of the) Fourier transform is recorded? Consider Figure 6.83. In the so-called 'Fraunhofer', or 'far-field', regime the difference δ in the optical path difference (OPD) between (a) two wavelets emanating from opposite sides of the object converging on the detector at a given point P and (b) that between two parallel waves propagating in the same direction, should be much less than the wavelength of light being used.

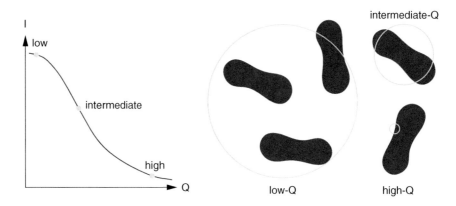

Figure 6.84   The three Q-domains of SAXS.

We first make the assumption that 2θ is small, and hence that $L \approx a$. We then use Pythagoras' theorem to obtain

$$(R - \delta)^2 + a^2 = R^2$$

$$\Rightarrow \left(1 - \frac{\delta}{R}\right)^2 = 1 - \frac{a^2}{R^2}.$$

But $\delta/R \ll 1$, and so

$$\left(1 - \frac{\delta}{R}\right)^2 \approx 1 - \frac{2\delta}{R}$$

and therefore

$$\delta \approx \frac{a^2}{2R}.$$

Because we want $\delta \ll \lambda$, we finally obtain the condition that

$$\frac{a^2}{2R\lambda} \ll 1. \tag{6.62}$$

If we insert typical values of $a = 1$ μm, $\lambda = 1$ Å into Equation (6.62), we find that the Fraunhofer condition is met at distances even as short as a centimetre. However, typically, object–detector distances of the order of several metres are more common. Why should this be so? The answer lies with detector technology – for a given pixel size, the further away the detector, the better resolved are elements of the diffraction pattern separated by δQ.

A measurement at a given Q value allows one to investigate electron-density fluctuations in the sample on a distance scale of $D = 2\pi/Q$. Consider Figure 6.84. At low Q, the observation window is very large, and only information about general sizes and interactions between the particles in the system can be investigated. At intermediate Q, the window is of the order of the particle size, while at high Q, in the so-called 'Porod' region (see below), contrast occurs primarily at the interface between the particles and their surroundings.

6.15.2.2   Scattering Curves

The scattering curve $I(Q)$ for an ensemble of N objects of shape $f(r)$, volume V, and electron-density contrast $\Delta\rho$ with the background medium is given by

$$I(Q) = NV^2(\Delta\rho)^2[\mathscr{F}(Q)S(Q)]^2 + B, \tag{6.63}$$

whereby $\mathscr{F}(Q)$ is the form factor, equal to the Fourier transform of $f(r)$, $S(Q)$ is the structure factor describing coherent scattering between individual particles, and B is a background signal, which can be subtracted experimentally.

For an ensemble of identical, anisotropically shaped but randomly oriented objects, one must calculate the average scattering signal for all possible orientations. In the following, we will assume that the objects are sufficiently separated from one another (i.e. they are 'dilute'), that coherent scattering by two or more objects is negligible, and that one can simply add up incoherently the intensities of scattering from individual objects. In this instance $S(Q) = 1$. In the following mathematical description of scattering-curve intensities, we ignore B and assume in the forward direction $NV^2(\Delta\rho)^2 = I_0$.

Often, the scattering curves for differently shaped objects appear at first blush to be somewhat nondescript and very similar. It is possible, however, to obtain important structural parameters directly from them. Many biological structures can be modelled by simple geometric shapes, in particular spheres, ellipsoids, rods, and flat discs[29]. These all have individual, distinguishable, features in their scattering curves.

We discuss these now in more detail, beginning with the scattering curve of a uniform sphere of radius r. This is given by

$$I(Q, \text{sphere}) = I_0 \left[\frac{3j_1(Qr)}{Qr} \right]^2, \qquad (6.64)$$

where $j_1(x) = (\sin x - x\cos x)/x^2$ is the *spherical* Bessel function of the first kind [and should not be confused with the more commonly encountered *cylindrical* Bessel function $J_1(x)$]. From this, we obtain

$$I(Q, \text{sphere}) = I_0 \left\{ \frac{3[\sin(Qr) - Qr\,\cos(Qr)]}{(Qr)^3} \right\}^2. \qquad (6.65)$$

Note that this function oscillates and has minima at $\tan(Qr - m\pi) = Qr$ (for integer $m \geq 1$). This is satisfied for $m = 1$ when $Qr = 4.493$ radians, which allows one to accurately determine the sphere radius. An example of such a system is given in Figure 6.85 for an ensemble of identical (so-called 'monodisperse') SiO_2 spheres, which were found to have a diameter of 82.3 nm from the position of the first minimum in the scattering curve[30]. These oscillations in curves for ensembles with a range of sizes (and/or shapes) become increasingly damped with increasing 'polydispersity' (see below).

The scattering curve of a hollow spherical shell of inner radius r_1 and outer radius r_2 is given by

$$I(Q, \text{shell}) = I_0 \left\{ \frac{3}{(r_2^3 - r_1^3)} \left[\left(\frac{\sin(Qr_2) - Qr_2\cos(Qr_2)}{Q^3} \right) \right.\right.$$
$$\left.\left. - \left(\frac{\sin(Qr_1) - Qr_1\cos(Qr_1)}{Q^3} \right) \right] \right\}^2. \qquad (6.66)$$

This is plotted in Figure 6.86.

Spheres are particularly simple to describe, as they have infinite three-dimensional symmetry, obviating the need to average orientationally. Most objects of interest have lower symmetry, however. If an ensemble of non-spherically symmetric particles is randomly oriented, calculating the scattering curve necessitates an averaging over all possible orientations.

Many objects can be approximated by an ellipsoid with semimajor axes a, b, and c. Such an ellipsoid oriented with these axes at angles α, β, and γ, respectively, to the Q-vector [see Figure 6.87(a)], has an effective radius r_e given by

---

[29] Cocci are bacteria with spherical or ovoid forms; bacilli are rod-shaped; coccobacilli, ellipsoidal. Red blood cells can be approximated by flat discs. Moreover, many protein structures and viruses are close to being spherical.

[30] A more biologically relevant example would be the third protein shown in Figure 6.81. From the scale bar, we can estimate that this most spherical of the 11 shown molecules has a diameter of approximately 53 Å. The minimum in the corresponding scatter curve is at $Q = 0.16$ Å^{-1}, from which a diameter of $2 \times 4.493/0.16 = 56$ Å is obtained, very close to our visual estimate.

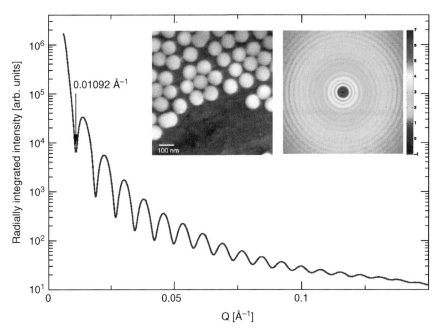

Figure 6.85 The SAXS signal from an ensemble of SiO_2 spheres, all with the same nominal diameter of 80 nm and shown in the inset scanning electron micrograph, shows clear oscillations. Analysis of the curve produced by radial integration of the image shows that the first minimum lies at 0.01092 $Å^{-1}$, corresponding to an average diameter of 82.3 nm. Courtesy of Dillip K. Satapathy, Confined Fluids Group, Paul Scherrer Institute.

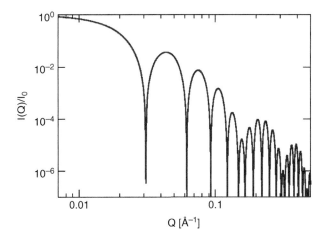

Figure 6.86 The normalized scatter curve of an ideal hollow spherical shell with an outer radius of 120 Å and an inner radius of 80 Å. Note the partial destructive interference, or 'beating' envelope to the curve, due to reflections from the inner and outer shell surfaces at integer multiples of $Q = 2\pi/\Delta r = 0.0157$ $Å^{-1}$.

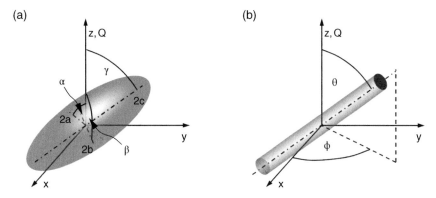

Figure 6.87 The geometries of (a) a uniform ellipsoid with semimajor axes a, b, and c, at angles α, β, and γ to the Q-vector, respectively; and (b) a uniform rod of length L and radius r, at a polar angle θ and azimuthal angle ϕ relative to the Q-vector.

$$r_e = (a^2\cos^2\alpha + b^2\cos^2\beta + c^2\cos^2\gamma)^{1/2}.$$

The scattering curves of either a solid or shell ellipsoid follow Equations (6.65) and (6.66), using r_e. The curve for an ensemble of randomly oriented ellipsoids is

$$I(Q, \text{ellipsoid, rand}) = \frac{1}{2}\int_{-1}^{1} d(\cos\gamma)I(Q, \cos\gamma). \tag{6.67}$$

Integration is performed numerically.

Scattering from cylinders with length L and radius r follows the general equation

$$I(Q, \theta) = I_0\left[\frac{\sin(QL/2 \ \cos\theta)}{QL/2 \ \cos\theta}\right]^2\left[\frac{2J_1(Qr\sin\theta)}{Qr\sin\theta}\right]^2, \tag{6.68}$$

where θ is the angle subtended by the rod axis and the Q-vector, and $J_1(x)$ is the *cylindrical* Bessel function of the first kind, given by

$$J_1(x) = \frac{x}{2}\sum_{k=0}^{\infty}\frac{(-x^2/4)^k}{(k+1)!(k+2)!}.$$

Again, if the rods are randomly oriented, averaging over θ by integration is performed manually, analogously to Equation (6.67). This is shown in Figure 6.88.

The two limiting cases for cylinders of $L \gg r$ (one-dimensional, very slender rod), and $r \gg L$ (two-dimensional, very thin disc) are worthy of note, as the expressions for their scattering curves, both oriented and random, become simpler, due to their reduced dimensionality. These are

$$I(Q, \text{disc}, \theta) = I_0\left[\frac{2J_1(Qr\sin\theta)}{Qr\sin\theta}\right]^2, \tag{6.69}$$

$$I(Q, \text{disc, rand}) = I_0\frac{2}{(Qr)^2}\left[1 - \frac{J_1(2Qr)}{Qr}\right], \tag{6.70}$$

$$I(Q, \text{rod}, \theta) = I_0\left[\frac{\sin(\frac{QL\cos\theta}{2})}{\frac{QL\cos\theta}{2}}\right]^2, \tag{6.71}$$

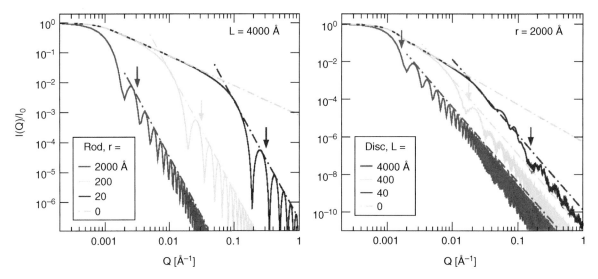

Figure 6.88   Scattering curves for ensembles of randomly oriented cylinders of different aspect ratios r/L, employing Equations (6.68) (red, yellow, and blue curves), (6.72) (cyan curve, left graph), and (6.70) (cyan curve, right graph). Note the average gradients (dashed lines) of -4 for the regions $Q \sim 2\pi/a$ and above, in this double-logarithmic plot (highlighted by the coloured arrows), where a is the smallest dimensional parameter. These regions follow Porod's law $I(Q) \propto Q^{-4}$. No such region exists for the infinitely slender rods and thin discs. The intensity of the one-dimensional rods exhibits a Q^{-1} dependence beyond the low-Q regime, while, for the two-dimensional discs, $I(Q)$ falls off as $1/Q^2$. Note that the red curves in both graphs are identical.

$$I(Q, \text{rod, rand}) = \frac{2\text{Si}(QL)}{QL} - \frac{\sin^2(QL/2)}{(QL/2)^2}. \tag{6.72}$$

Here, $\text{Si}(x)$ is the sine integral, defined as

$$\text{Si}(x) = \int_0^x \frac{\sin u}{u} du.$$

Finally, the scattering curve for a randomly configured[31] thin polymer chain of n monomers, each of length a, is given by

$$I(Q) = \frac{2I_0}{\phi^2}(e^{-\phi} + \phi - 1), \tag{6.73}$$

whereby $\phi = Q^2 a^2 n/6$.

6.15.2.3   *The Radius of Gyration and Guinier's Law*

The radius of gyration r_g is identical to the radius of inertia in mechanics about the centre of mass (not to be confused with the moment of inertia around a rotation *axis*). It is defined as the root-mean square of the distances of all the electrons in the object from their centre of gravity. On a more intuitive level, one can think of it being the radius of a hollow spherical shell of infinitesimal shell thickness, having the same mass and moment of inertia

---

[31] The orientation of each monomer is assumed to be entirely random, only that it is attached at the end of the previous monomer unit. This model is called a 'Gaussian chain'.

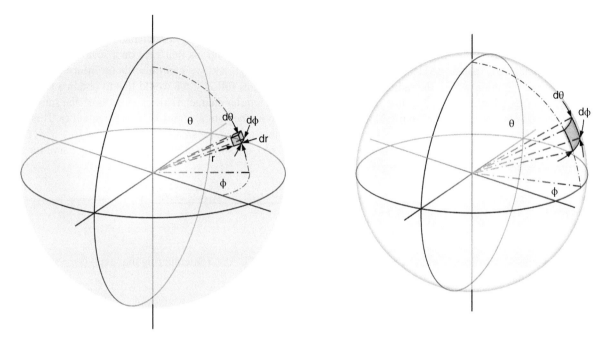

Figure 6.89 Geometries used in a comparison of the moments of inertia of a solid sphere and a spherical shell. Left: a volume element $dV = r^2\, dr\, \sin\theta\, d\theta\, d\phi$ at a distance r from the centre in a solid sphere of radius r_s and homogeneous density $\rho_V = 3M/(4\pi r_s^3)$. Right: a surface element $dS = r_g^2 \sin\theta\, d\theta\, d\phi$ of a hollow spherical shell of radius r_g, infinitesimal wall thickness, and areal density $\rho_S = M/(4\pi r_g^2)$.

as the object under investigation (or, more accurately, the same number of electrons), all packed within the shell wall. It is a measure of the object's average spatial extent.

The radius of gyration of a solid sphere can be determined by equating its moment of inertia with that of hollow sphere with the same total mass M and infinitesimal shell thickness (see Figure 6.89). The volume density of the solid sphere is $\rho_V = 3M/(4\pi r_s^3)$, while the areal density of the shell is $\rho_S = M/(4\pi r_g^2)$. We integrate the moments of inertia of all volume elements $dI_V = \rho_V dV r^2$ in the solid sphere, and those of the surface elements $dI_S = \rho_S dS\, r_g^2$ of the shell:

$$\rho_V \int_0^{2\pi} \int_0^{\pi} \int_0^{r_s} r^4 dr \sin\theta d\theta\, d\phi = \rho_S r_g^4 \int_0^{2\pi} \int_0^{\pi} \sin\theta d\theta\, d\phi$$

$$\Rightarrow \rho_V \frac{4\pi r_s^5}{5} = 4\rho_S \pi r_g^4$$

$$\Rightarrow r_g^2 = \frac{3r_s^2}{5}. \tag{6.74}$$

One of the most important radii of gyration used in SAXS is indeed that of a hollow spherical shell, as this can often satisfactorily approximate the shape of biological cells in fluid media. It is given by

$$r_g^2 = \frac{3}{5} \frac{r_2^5 - r_1^5}{r_2^3 - r_1^3}, \tag{6.75}$$

where r_g is the radius of gyration, and r_1 and r_2 are the inner and outer shell radii, respectively[32].

Let us now consider the scattering-vector regime where $Q \ll 2\pi/a$, a being the characteristic length-scale of the object. Because, in this region, we can only resolve density fluctuations that vary on a scale $D = 2\pi/Q$, which is much larger than a, all details of the particle shape are lost, and we can only obtain information on the overall size, in other words, the radius of gyration. We proceed as follows: we would like to obtain a simple expression for $I(Q)$ in the small-Q regime. Because all objects of similar characteristic sizes a 'look' the same in this regime, we choose one with an expression that lends itself to mathematical simplification for small Q. This is Equation (6.65) for a solid sphere. It can be simplified by remembering that, for small x, $\sin x \approx x - x^3/6 + x^5/120$ and $\cos x \approx 1 - x^2/2 + x^4/24$. Inserting these into Equation (6.65), the small-Q limit of the scattering curve for a sphere is accurately approximated by

$$I(x \equiv Qr \ll 1) \approx 3I_0 \left(\frac{x - x^3/6 + x^5/120 \cdots - x + x^3/2 - x^5/24}{x^3} \right)^2$$

$$= I_0(1 - x^2/10)^2 \approx I_0(1 - x^2/5).$$

But we have just determined that, for a sphere of radius r, $r^2 = 5r_g^2/3$. Also remembering that $\exp(-x) \approx (1 - x)$, we therefore finally obtain

$$I(Qr \ll 1) \approx I_0 \exp(-Q^2 r_g^2/3). \tag{6.76}$$

Remember, in this regime, we don't actually know what the shape is – we used the equation for a sphere only because it was mathematically convenient. Equation (6.76), called Guinier's law, is therefore approximately valid for all objects for which $Qr_g < \sqrt{3}$. A so-called 'Guinier plot' of $\ln[I(Q)]$ against Q^2 thus exhibits a linear descent,

$$\ln[I(Q)] = \ln I_0 - Q^2 r_g^2/3, \tag{6.77}$$

directly yielding the radius of gyration.

The radii of gyration of some commonly used geometrical shapes to describe systems in SAXS are given in Table 6.2.

An extended, or intermediate-Q Guinier range becomes available if two of the characteristic lengths of an object differ significantly in magnitude, such as for elongated rods or lamella. In the case of elongated rods, the intermediate region can be approximated by

$$I(Q) = \frac{AI_0}{Q} \exp\left(-\frac{Q^2 r_{gr}^2}{2} \right), \tag{6.78}$$

where A is a fitting constant and $r_{gr} = r/2^{1/2}$ is the radius of gyration for the rod in this range. To extract the rod radius r, $\ln[QI(Q)]$ should be plotted against Q^2 between $Q = \sqrt{2}/r_g$ and $\sqrt{3}/r$. The gradient is $-r^2/4$.

For lamellae, the intermediate-Q Guinier approximation is

$$I(Q) = \frac{AI_0}{Q^2} \exp(-Q^2 r_{gl}^2), \tag{6.79}$$

---

[32] We can check the validity of this by expanding the numerator and denominator, so that

$$r_g^2 = \frac{3}{5} \frac{(r_2 - r_1)(r_2^4 + r_2^3 r_1 + r_2^2 r_1^2 + r_2 r_1^3 + r_1^4)}{(r_2 - r_1)(r_2^2 + r_2 r_1 + r_1^2)}$$

$$= \frac{3}{5} \frac{(r_2^4 + r_2^3 r_1 + r_2^2 r_1^2 + r_2 r_1^3 + r_1^4)}{(r_2^2 + r_2 r_1 + r_1^2)}.$$

If we then set $r_2 = r_1 = r$, as in the case of a hollow shell with infinitesimal shell width, we immediately obtain $r_g = r$, as asserted above.

Table 6.2 Formulae describing the radius of gyration r_g for some commonly used homogeneous triaxial bodies as models in SAXS.

Object	r_g^2
Solid sphere radius r	$\dfrac{3}{5}r^2$
Hollow sphere radii r_1 and $r_2 > r_1$	$\dfrac{3}{5}\dfrac{r_2^5 - r_1^5}{r_2^3 - r_1^3}$
Solid cylindrical rod radius r, height L	$\dfrac{r^2}{2} + \dfrac{L^2}{12}$
Solid rectangular beam width W, height H, length L	$\dfrac{W^2 + H^2 + L^2}{12}$
Hollow tube radii r_1, r_2, height L	$\dfrac{r_1^2 + r_2^2}{2} + \dfrac{L^2}{12}$
Solid ellipsoid semi-axes a, b, c	$\dfrac{a^2 + b^2 + c^2}{5}$
Hollow ellipsoid outer semi-axes a, b, c, inner semi-axes αa, βb, γc	$\dfrac{(1 - \alpha^3\beta\gamma)a^2 + (1 - \alpha\beta^3\gamma)b^2 + (1 - \alpha\beta\gamma^3)c^2}{5(1 - \alpha\beta\gamma)}$
Solid elliptical cylinder semi-axes a, b, height L	$\dfrac{a^2 + b^2}{4} + \dfrac{L^2}{12}$
Hollow elliptical cylinder outer semi-axes a, b, outer height L, inner semi-axes αa, βb, inner height γL	$\dfrac{3(1 - \alpha^3\beta\gamma)a^2 + 3(1 - \alpha\beta^3\gamma)b^2 + (1 - \alpha\beta\gamma^3)L^2}{12(1 - \alpha\beta\gamma)}$
Randomly folded polymer chain, n monomers of length a	$\dfrac{a^2 n}{6}$

where $r_{gl} = L/(12)^{1/2}$ is the radius of gyration for the lamellae. Here, $\ln[Q^2 I(Q)]$ should be plotted against Q^2, whereby the valid range extends from $Q = \sqrt{3}/r_g$ to $\sqrt{12}/L$ (see Figure 6.90).

Importantly, Equations (6.78) and (6.79) depend only on the size of the rod diameter and lamella thickness, respectively. Hence, ensembles of rods with equal diameter but varying lengths (as long as they are all much larger than the radius) will still provide information on r; equally, intermediate Guinier plots of an ensemble of lamellae of varying diameter will deliver values on their common thickness.

The varied scattering behaviours of different systems in the intermediate-Q regime according to their dimensionality, exhibited in Figure 6.90, can be further highlighted in a semiquantitative manner by so-called 'Kratky plots'. These are simply a plot of $Q^2 I(Q)$ against Q (or, alternatively, $\ln[Q^2 I(Q)]$ vs. Q). Kratky plots of the 'Gaussian-chain' polymers described by Equation (6.73) are particularly revealing. At high-Q, their Kratky plots converge asymptotically to $2/r_g^2 = 12/a^2 n$. This provides important information on the average polymer length (see Figure 6.91).

It lies beyond the scope of this text to detail all the structural subtleties that can be extracted in Guinier, Kratky, and other plots (including 'Zimm' plots and Porod plots), other than to say these include fractality (structure within structure), polymer configurations, and protein conformations. The reader is referred to the literature for more detailed descriptions [99, 100, 102].

6.15.2.4   *Porod's Regime and Porod's Law*

The scattering amplitude is proportional to the contrast, or difference in the electron density Δn_e between the scatterer and its surroundings, and therefore the scattering intensity is proportional to $(\Delta n_e)^2$. Moreover, the intensity

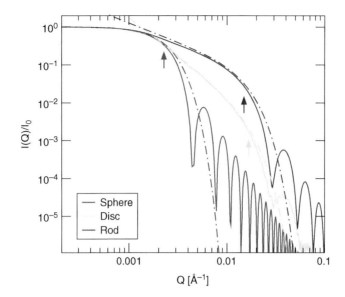

Figure 6.90 Plot of scattering curves for a 100 nm radius sphere, a disc with $r = 5L$, and a rod with $L = 20r$. All three objects have the same magnitude radius of gyration ($r_g^2 = 600\,000$ Å^2). The three dot-dashed curves are the 'normal' Guinier approximation [red, Equation (6.76)], the intermediate approximation for lamellae [yellow, Equation (6.79)], and the intermediate approximation for rods [blue, Equation (6.78)]. Fitting the latter two to the data between $Q = \sqrt{3}/r_g$ (red arrow, upper limit of the 'normal' Guinier regime) and $\sqrt{2}/r$ (rods, blue arrow) or $\sqrt{12}/L$ (lamellae, yellow arrow) yields, through the gradient, the rod radius and lamella thickness, respectively. Beyond this intermediate region, one enters the Porod regime of high-Q, for which the scattering curves exhibit a Q^{-4} dependence.

is proportional to the total number of scatterers in the irradiated volume, N_p, assuming they are separated from one another by more than the coherence length of the x-rays being used. In other words

$$I(Q) \propto N_p\, n_e^2. \tag{6.80}$$

At higher Q-values, towards the tail of the scattering curve, the signal comes from differences in contrast (electron density) over correspondingly small distances, given by a characteristic length $2\pi/Q$ (see the high-Q region shown in Figure 6.84).

For particles within which the average electron density varies significantly less than the difference in electron density as one crosses the particle's surface, scattering will occur primarily at the interface between the particle and the medium in which it is suspended[33]. Consider Figure 6.92. We divide up the surface of a particle into spherical elements having this characteristic size. The number N_p of such spheres is equal to the surface area of the particle divided by the cross-sectional area of the scattering spheres, and is therefore proportional to Q^2. On the other hand, the number of electrons per scattering sphere (equivalent to Δn_e above) is proportional to the cube of its characteristic size, or $1/Q^3$. From Equation (6.80), we thus obtain for large Q

$$I(Q) \propto Q^{-4}. \tag{6.81}$$

A more rigorous derivation results in

$$I(Q) = 2\pi(\Delta\rho)^2 S_p Q^{-4}, \tag{6.82}$$

---

[33] For example, the electron densities of water and a typical protein are 0.33 and 0.44 e Å^{-3}, respectively

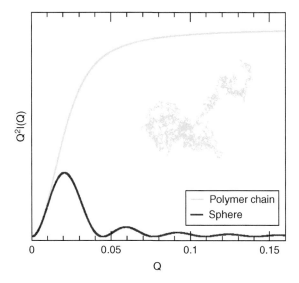

Figure 6.91 Globular, or compactly folded, clusters (blue curve), modelled by a sphere or ellipsoid, show a distinctive peak when their scattering curves are weighted by Q^2 and plotted against Q. In contrast, the similarly weighted scattering from the more loosely configured Gaussian chain (yellow curve) converges asymptotically at high-Q to a value of $12/(a^2n)$. A typical Gaussian chain containing $n = 4000$ monomer units, created using a random-number generator, is also shown.

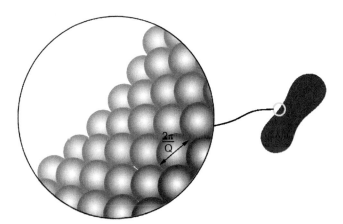

Figure 6.92 At high Q, the observation window, of size $2\pi/Q$, is much smaller than the scattering particles, and contrast is only obtained across the interface of the particles and the medium in which they reside. Hence, the interface can be thought of as an array of spherical scatterers of diameter $2\pi/Q$. This leads to Porod's law.

where S_p is the total probed surface area of the particle ensemble. This is Porod's law, and can be used to measure the surface area of nanoscale structures, if $\Delta\rho$ is known.

One can also arrive at the same dependence if it is assumed that at sufficiently high Q, the probed interface accurately approximates a flat surface. We have already derived the inverse fourth-power dependence of the scattering intensity on the incident angle α for reflectivity from surfaces, given by Equation (6.50). We reasoned that in SAXS, α is proportional to Q, hence in this manner, we obtain again Porod's law.

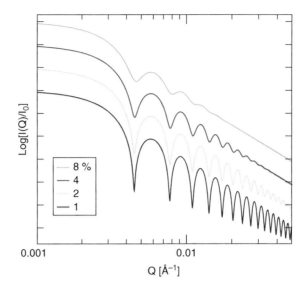

Figure 6.93   Theoretical scattering curves for ensembles of spheres with varying polydispersity. A Gaussian distribution was used; the legends refer to the standard deviation σ in percent of the size distribution. The curves have been shifted vertically for purposes of clarity.

6.15.2.5   *Polydispersity and Sample Concentration*

Until now, we have considered only ensembles of identical particles (except in the case of investigating the magnitudes of the smaller dimensional parameter using the intermediate Guinier regime, described above). Such ensembles are 'monodisperse'. Most real systems, however, deviate from this to a greater or lesser extent – that is, they exhibit a degree of 'polydispersity'. Polydispersity may arise because of a range of scales of otherwise identical particles, but might also be described by variations in one or two structural parameters only.

Depending on the type, dispersity can be accurately modelled, thereby providing information on the statistical distribution of sizes and shapes. A simple example is illustrated in Figure 6.93.

When the particle concentration is no longer dilute (defined as the average neighbour separation being significantly larger than the coherence length of the radiation[34]), the scattering cross-section contains contributions from interparticle scattering [$S(Q)$ in Equation (6.63)]. Note that, as Q increases, the probing length decreases. In general, therefore, the Porod region remains unaffected by particle concentration.

A general mathematical description of polydispersity and concentration effects lies well beyond the remit of this introductory text; the reader is referred to the literature for details [99, 100, 102].

6.15.3   **Practical Considerations**

One of the most stringent practical requirements for the success of SAXS experiments is the minimization of vibrations and long-term drifts. Vibrations on the Hz scale or higher can originate from passing traffic, air-conditioning fans, mechanical vacuum pumps, or humming from high-power electrical components, for example. Long-term drifts occur if the temperature of the experimental hutch is allowed to fluctuate by more than approximately $\pm 1°$. In general, lateral shifts between the incident beam and sample should be significantly smaller than the sample

---

[34] If we assume the largest of the coherence lengths to be approximately 100 μm in a DLSR-SAXS beamline, the dilute upper limit requires concentrations of less than 10^6 cm^{-3}.

Figure 6.94   The 7 m flight tube installed at the cSAXS beamline, Swiss Light Source. Courtesy Oliver Bunk and Andreas Menzel, Paul Scherrer Institute.

dimensions, while the detector should remain stable to within less than the linear pixel size, typically measured in microns to a few tens of microns.

The pixel size itself determines the largest scale which can be probed in the SAXS experiment, while the smallest sample dimensions are determined in principle by the size of the detector. In most instances, however, because the scattering intensity drops as $1/Q^4$ (Porod's law), at large Q-values close to the detector edge, the signal is so weak that the dynamic range of the detector precludes reliable measurements. The very high brilliance of DLSRs will, however, push back this limitation, and further significant gains are expected.

Air scatter should also be avoided, as this can swamp low-intensity SAXS signal. Modern SAXS stations are equipped with monstrous vacuum flight tubes pumped down to the mbar regime or below (Figure 6.94), between the sample at one end and the detector at the other.

Lastly, the beam divergence should be compatible with the pixel resolution. This is normally achieved by focussing the beam at the detector and, if necessary, slitting down the beam near the horizontal focussing element. Again, the two orders of magnitude improvement in horizontal emittance at DLSRs will further extend the experimental limits.

6.15.4   Grazing Incidence SAXS

Grazing incidence SAXS, or GISAXS, is a powerful nondestructive tool to study nanoscale structures on surfaces, interfaces, and thin films [103].

One can think of GISAXS as marrying reflectivity with 'normal' SAXS. By allowing the x-ray beam to impinge on a sample at an angle that is grazing with respect to its average surface (i.e. close to the critical angle for total external reflection), only the surface region is probed, along with any structures that might have formed on it. Because the grazing beam samples a large fraction of the surface, measured in several hundred or even thousands of

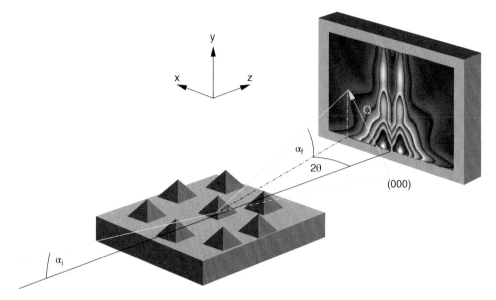

Figure 6.95   A grazing incidence x-ray beam is scattered by nanostructures on a surface. The small-angle component is recorded on an area detector. The simulated GISAXS image has been adapted from [104] with permission of AAAS.

square microns, averaged statistical information over the sample surface is obtained. Typical applications include investigations of quantum-dot arrays, kinetic or time-resolved studies (such as changes in morphology of nanostructures during thin film growth [104]), as well as the observation of spontaneous formation of self-organized nanostructures in films of polymers. Indeed, as an *in situ* technique for a wide range of environmental parameters (including, most importantly, pressure and temperature) it is invaluable, because direct-imaging methods such as atomic-force microscopy or scanning tunnelling microscopy are essentially impossible to use under these conditions. Moreover, GISAXS can be used to investigate buried and internal structures. Increasingly, GISAXS is being applied to catalytic and biological systems, such as proteins attached to the surfaces of lipid layers.

A typical GISAXS setup is shown in Figure 6.95. The component of the scattering vector $\mathbf{Q}$ in plane (the *x*-direction in Figure 6.95), is given by

$$Q_x = |k| \sin 2\theta \, \cos \alpha_f \tag{6.83}$$

and yields information on the widths of the objects in the surface plane, while the out-of-plane component

$$Q_y = |k|(\sin \alpha_i + \sin \alpha_f) \tag{6.84}$$

does likewise for the object heights. Any correlations in the distribution of neighbour-to-neighbour distance D of the nanoparticles will also be highlighted by interference maxima in-plane, separated by $2\pi/D$.

6.16   Concluding Remarks

The periodic nature of crystalline systems lends them physical and electronic properties that can only be understood in terms of that periodicity and the configuration of the atoms within each period. At the same time, this periodicity provides the necessary scattering conditions for diffraction phenomena. X-ray diffraction is a nondestructive structural tool with unsurpassed resolution, capable of determining dimensions with femtometre accuracy.

The primary obstacle to reconstructing structural information from diffraction data is the ubiquitous phase problem. The approach to its solution essentially encapsulates the whole art of crystallography. An impressive arsenal of tools has been developed over the last century to crack this problem, and in no more a breathtaking manner than that used in macromolecular crystallography, which nowadays can determine the structure and provide insights into the functionality of biological systems consisting of tens to hundreds of thousands of atoms – the progress from solving the structures of diamond and diopside by the Braggs in the first decades of the twentieth century to determining the form of enzymes, ribosomes, and other biological systems to a resolution of a few angstroms eighty years later must count as one of the most remarkable in the history of modern science.

Diffraction and elastic-scattering techniques continue to play leading rôles in modern synchrotron science. The reason is clear – which other methods are capable of obtaining atomic-scale images of objects consisting of as many as several hundred thousand atoms? Indeed, with the advent of the free-electron laser, scattering methods will undoubtedly enjoy a renewed prominence, particularly in the field of ultrafast time-resolved studies.

Problems

1. CuO is a monoclinic crystal with lattice parameters $a = 4.684$ Å, $b = 3.423$ Å, $c = 5.129$ Å, $\alpha = 90°$, $\beta = 99.54°$, and $\gamma = 90°$. Calculate the interplanar spacing of the (211) planes. (2P)

2. Bragg's law is normally written as

$$\lambda = 2d_{hkl} \sin \theta,$$

whereby d_{hkl} is the separation of the (hkl)-planes in a crystal and θ is the Bragg angle. Prove that, when refraction effects are considered, Bragg's law is modified to

$$\lambda = 2d_{hkl} \sin \theta \left[1 - \frac{2\rho r_0 d_{hkl}^2}{\pi} \right],$$

whereby ρ is the electron density and $r_0 = 2.82 \times 10^{-5}$ Å is the Thomson scattering length. Make an order-of-magnitude estimate as to how much this fractionally changes the Bragg angle for low-index Bragg reflections. (5P)

3. The extinction depth of a single crystal is given by Equation (6.20). Determine the extinction depth of diamond(111), given that the atomic form factor of carbon f_C is equal to 2.584 e at the Bragg condition. The lattice constant of diamond is $a = 3.5672$ Å. [Hint: use also Equation (6.17).] (4P)

4. A single crystal is mounted on a diffractometer to be investigated using the selected-Bragg-peak method (see Section 6.8.2). The detector is moved to a Bragg peak via rotation by a two-circle stage – the first circle rotates the detector horizontally by an amount γ; the second vertically by an amount δ (see Figure 6.96). Derive an expression for 2θ in terms of γ and δ. (2P)

5. Prove that the theoretical maximum number of reflections N that can be recorded for a single crystal of unit-cell volume V_c at an x-ray wavelength λ [Equation (6.10)] is

$$N = \frac{32\pi}{3} \frac{V_c}{\lambda^3} = 33.5 \frac{V_c}{\lambda^3}.$$

(3P)

6. The atomic form factors of Sr^{2+}, Ti^{4+}, and O^{2-} are plotted as a function of Q in Figure 6.97(a). Using the method shown in Figure 6.14 and the structure of the centrosymmetric simple-cubic crystal $SrTiO_3$ given in Figure 6.97(b), determine the ratio of the structure factors of the (001) and (002) Bragg peaks, assuming kinematical diffraction and no absorption. The lattice constant of $SrTiO_3$ is 3.905 Å. (5P)

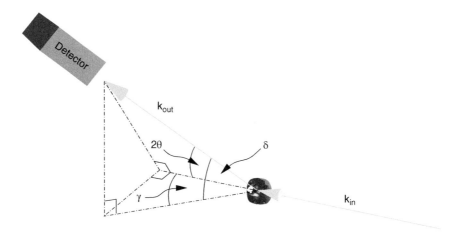

Figure 6.96 A Bragg peak from a single crystal is recorded by rotating a detector arm in the horizontal plane by an amount γ then vertically by an amount δ. The angular difference between the incoming and scattered radiation is 2θ.

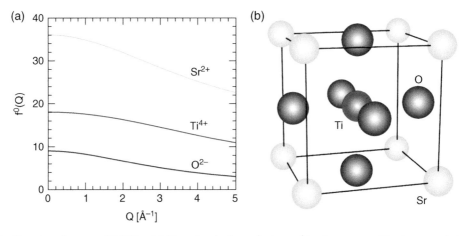

Figure 6.97 Structure factors of $SrTiO_3$. (a) The atomic form factors of ionic oxygen, titanium, and strontium as a function of scattering vector Q. (b) The centrosymmetric structure of simple-cubic strontium titanate. The Sr ions occupy the cube vertices, the oxygen atoms the face centres, and titanium the body centre.

7. Given that the set of complex structure factors $\{F_{hkl}\}$ is the Fourier transform of the electron-density distribution $\rho(\mathbf{r})$ within a unit cell of a crystalline material, and that any one measured diffraction signal $I_{hkl} = |F_{hkl}|^2$, demonstrate using the convolution theorem that the Patterson function $\mathscr{F}(\{I_{hkl}\})$ is indeed $\rho(\mathbf{r}) \otimes \rho(-\mathbf{r})$, the convolution of the electron-density distribution with its inverse. (3P)

8. Four related aromatic compounds are shown in Figures 6.98(a) to (d). Determine their respective Patterson functions from Figures 6.98(e) to (h). (2P)

9. Ferroelectricity demands noncentrosymmetric crystals. The unit cell of the classic ferroelectric material $PbTiO_3$ (PTO) is shown in Figure 6.99: the octahedral oxygen sublattice is symmetric with regards to the unit-cell dimensions, but the Pb and Ti atoms are displaced along the c-axis from the high-symmetry locations by the amounts shown in the figure.

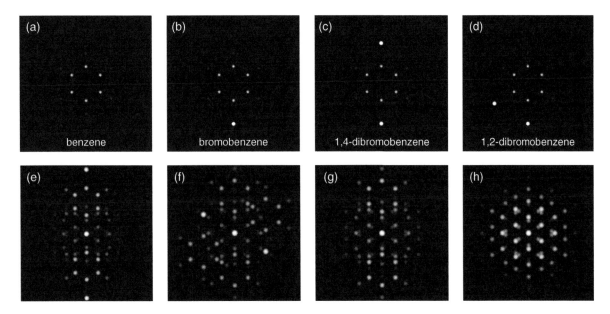

Figure 6.98 (a) to (d) Four aromatic compounds. (e) to (h) Their Patterson functions, shown in a different order.

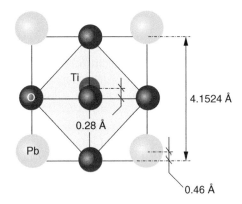

Figure 6.99 The unit cell of PbTiO$_3$.

Accurately sketch the Argand diagrams for the (001) and (002) Bragg reflections for a photon energy of 4 keV. The information required for this is given in Table 6.3. From this, determine the ratios $I_{001}/I_{00\bar{1}}$ and $I_{002}/I_{00\bar{2}}$. Lastly, without the need to sketch the Argand diagram, predict the ratio $I_{100}/I_{\bar{1}00}$ and give your reasonings behind this. (7P)

10. Derive Equation (6.28). Lysozyme has an attenuation coefficient $\mu = 2.0 \times 10^{-4}$ μm^{-1} at a photon energy of 12.4 keV. It is investigated using a focussed beam with a cross-section of 5×5 μm^2 and flux of 5×10^{12} photon per second. Calculate the time until the illuminated volume has received a dose of 10^7 Gy. Lysozyme crystals have a mass density of 1.37 g cm^{-3}. (4P)

11. A $200 \times 200 \times 200$ μm^3 protein crystal with a mass density of 1.35 g cm^{-3} and heat capacity of $C_p = 500$ J (K kg)$^{-1}$ is irradiated with 10^{13} 1 Å photons per second. The attenuation length at this photon

Table 6.3   Energy- and Q-dependent structure factors of Pb, Ti, and O at 4 keV for the (001) and (002) Bragg reflections at $Q = 1.513$ Å^{-1} and 3.026 Å^{-1}, respectively.

	O	Ti	Pb
$f_1^{(001),\ 4\ keV}$	7.425	16.939	67.189
$f_1^{(002),\ 4\ keV}$	4.994	14.389	57.945
$f_2^{(4\ keV)}$	0.143	0.680	25.450

      energy is 5000 μm. Determine the heating rate of the crystal, assuming no thermal conduction to its surroundings.   (2P)

12. A protein crystal is illuminated with x-rays at three energies in the neighbourhood of 12.66 keV and the resulting diffraction patterns are recorded on an area detector of diameter 440 mm, positioned 140 mm downstream of the crystal. The centre of the detector is coincident with the direct beam (but protected with a tungsten beamstop). Calculate the maximum resolution of the structure of the protein achievable in this configuration. Can you suggest a reason why this particular range of photon energies was chosen?   (2P)

13. Demonstrate that the angular separation δ of two adjacent Bragg peaks at $m(h, k, l)$ and $(m + 1)(h, k, l)$ is given by

$$\delta = \frac{\lambda}{d \cos \theta}.$$

Hint: differentiate m with respect to θ.

      Two adjacent Bragg peaks $(h + 1, 0, 0)$ and $(h, 0, 0)$ from a cubic protein crystal lie close to the direct beam and are separated by a distance of 0.91 mm. The detector lies 90 mm downstream from the crystal, and 12.66 keV radiation is being used. Determine the unit-cell size.   (3P)

14. In the main text, it is shown in the Argand diagram of Figure 6.14 that the structure factors of centrosymmetric crystals are real, that is, they lie along the real axis. Close to an absorption edge, however, this is no longer true. Demonstrate, also using an Argand diagram, that despite this, centrosymmetric crystals nevertheless *always* obey Friedel's law, that is, $|F_{hkl}| = |F_{\overline{hkl}}|$.   (2P)

15. Using the arguments detailed in Section 6.11.4 and summarized in Table 6.1 for a chemically 'average' protein molecule, demonstrate that the anomalous signal from sulfur in crambin does indeed dominate. You may ignore the contribution from hydrogen.   (4P)

16. In a surface x-ray diffraction (SXRD) experiment, the structure factors of a given crystal-truncation rod (hk) are often recorded by fixing the grazing incident angle of the incoming beam ($\mathbf{k}_{in}$, see Figure 6.100), rotating the sample about its normal, and simultaneously tracking the position where the CTR penetrates the Ewald sphere, that is, the value $Q_\perp = l$ of the out-of-plane component of Q, which satisfies the Bragg condition. By applying the cosine rule, demonstrate that the l-value of the rod that intersects the Ewald sphere for a given ϕ is given by

$$l = (2kQ_\parallel \cos \phi - Q_\parallel^2)^{1/2},$$

whereby $Q_\parallel$ is the in-plane component of Q, has a magnitude equal to the radius of the dashed circle in Figure 6.100, and is smaller than $|k|$. You can assume that $\mathbf{k}_{in}$ is parallel to the substrate surface.

      What is the maximum l-value on the (21)-CTR which can be accessed via this recording mode, for a cubic crystal having a lattice constant of 4 Å and using 16 keV photons?   (5P)

17. The reflectivity curve of an atomically flat silicon surface is recorded at 10 keV. Determine the critical angle α_c of Si at this energy. What is the reflectivity R at 2°? Next, a germanium surface with a surface of 8 Å root-mean-square roughness is also measured at the same energy. What is its reflectivity at the same angle of 2°? The lattice constants of Si and Ge are 5.431 Å and 5.646 Å, respectively. $Z_{Si} = 14$ and $Z_{Ge} = 32$.(4P)

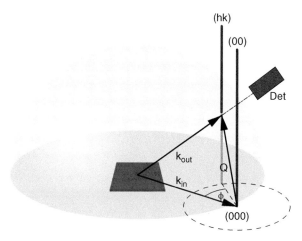

Figure 6.100 A typical scattering geometry for recording a crystal-truncation rod in SXRD. The incident beam $\mathbf{k}_{in}$ is fixed at a shallow angle relative to the surface (shown in red) and the sample is rotated about its normal axis. This causes a given CTR (*hk*) to rotate around the (00)-CTR. Depending on its position ϕ on the dashed circle, the CTR intersects the Ewald sphere (shown as the semitransparent hemisphere) at different values of *l*. The structure factor at each *l* can then be recorded by navigating a detector to be positioned along the vector $\mathbf{k}_{out}$.

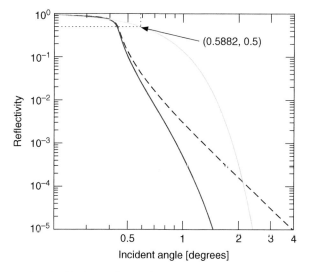

Figure 6.101 Shown in blue in the double logarithmic plot is the measured reflectivity of a gold surface recorded at 10 keV. The ratio of this curve to that of a perfectly flat gold surface (black dashed line) is shown by the yellow curve. This assumes a value of 50% (0.5) at an angle of 0.5882°.

18. δ, the refractive-index decrement of a material, is 1.6×10^{-6}. Determine the reflectivity R at an incident angle $\alpha = 0.5°$ on an atomically smooth surface of this material. (1P)

19. The x-ray reflectivity of a gold crystal surface is investigated using 10 keV photons, shown in Figure 6.101. From the curve showing the ratio of the actual reflectivity to that of a perfectly flat surface, determine the surface roughness. (2P)

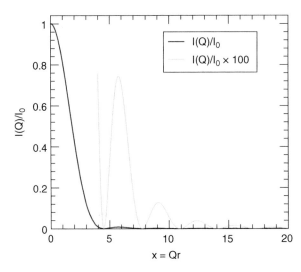

Figure 6.102   Plot of $I(Q)/I_0$ for dilute monodisperse spheres as a function of $x = Qr$. The curve in yellow is expanded in the ordinate by a factor of 100 in order to better distinguish the oscillations.

20. The SAXS form factor of a hollow sphere with outer radius r_2 and inner radius r_1 is given by

$$F(Q, r_1, r_2) = \frac{3}{r_2^3 - r_1^3} \left[\frac{r_2^3}{Qr_2} j_1(Qr_2) - \frac{r_1^3}{Qr_1} j_1(Qr_1) \right],$$

whereby $j_1(x)$ is the first-order spherical Bessel function, given by

$$j_1(x) = \frac{\sin x}{x^2} - \frac{\cos x}{x}.$$

Beginning with this, derive an expression for $F(Q, r_1, r_2)$ for $Qr_{1,2} \ll 1$, that is, in the Guinier regime, and, by using Equation (6.76), demonstrate that this directly leads to the radius of gyration for a hollow sphere given in Table 6.2.   (6P)

21. Derive the expression given in Table 6.2 for the radius of gyration for a solid cylinder by integrating the moment of inertia in cylindrical coordinates and equating this to Mr_g^2, the moment of inertia of a thin spherical shell of radius r_g.   (4P)

22. Demonstrate that the low-Q approximation of the scattering curve of an ensemble of randomly oriented slender rods of length L, Equation (6.72), leads to $r_g = L^2/12$, as expected.   (4P)

23. The form factor for an ensemble of monodisperse and dilute spheres $I(Q)$ in SAXS is given by Equation (6.65). The term within the curly brackets is shown graphically in Figure 6.102. Provide from this an expression for $x = Qr$ for the mth minimum in the curve. As m increases, what value does x asymptotically approach? Demonstrate that for $x = 0$, $I(Q = 0)/I_0 = 1$. Hint: each minimum has an intensity of zero.   (5P)

24. Apoferritin is a protein with a structure that is a near-perfectly spherical shell. From TEM images, the outer diameter, $2r_2$, of apoferritin is known to be 130 Å; however, an accurate determination of the inner diameter using the same technique is much more difficult to ascertain using TEM. To determine this more easily, apoferritin was investigated using SAXS, and is shown in Figure 6.103. From the plot, determine the inner diameter $2r_1$ to the nearest angstrom.   (1P)

25. In terms of SAXS experiments, the bacterium *Escherichia coli* can be approximated as being a hollow prolate ellipsoid of length 2 μm and 0.5 μm diameter. Its cell wall is 40 nm thick. Calculate the radius of gyration of

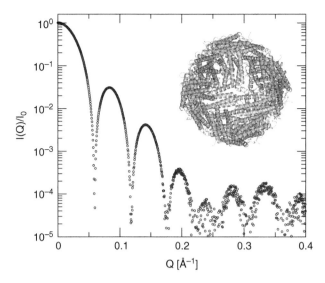

Figure 6.103 SAXS plot of apoferritin, shown in the inset.

E. coli. A SAXS experiment using 2 Å radiation illuminating a dilute collection of *E. coli* uses an Eiger area detector with a 75 μm pixel size. How far away from the sample must the detector be in order for the radius of gyration to be determined experimentally? Assume that at least ten pixels are required to obtain a reliable value for r_g and that the maximum value for Q in the Guinier regime is $\pi/2r_g$. (4P)

26. Beginning with Equation (6.73), demonstrate that the asymptotic value of a Kratky plot of a Gaussian chain is indeed $12/(a^2 n)$. (2P)

References

[1] B. D. Cullity, *Elements of x-ray diffraction.* Addison-Wesley, 1978.

[2] B. E. Warren, *X-ray diffraction.* Dover, 1990.

[3] C. Giacovazzo, ed., *Fundamentals of crystallography.* Oxford University Press, 2011.

[4] A. M. Glazer, *Crystallography: a very short introduction.* Oxford Science, 2016.

[5] B. Warren and W. L. Bragg, "The structure of diopside, $CaMg(SiO_3)_2$,". *Z. Kristallogr.*, vol. 69, pp. 168–193, 1928.

[6] M. F. Perutz, "How W. L. Bragg invented x-ray analysis," *Acta Crystallogr. A*, vol. 46, pp. 633–643, 1990.

[7] P. Strickland, ed., *International Tables for X-ray Crystallography.* International Union of Crystallography, 2006.

[8] A. L. Patterson, "A direct method for the determination of the components of interatomic distances in crystals," *Z. Kristallogr.*, vol. 90, pp. 517–542, 1935.

[9] H. M. Rietveld, "A profile refinement method for nuclear and magnetic structures," *J. Appl. Crystallogr.*, vol. 2, pp. 65–71, 1969.

[10] P. Thompson, D. E. Cox, and J. B. Hastings, "Rietveld refinement of Debye-Scherrer synchrotron x-ray data from Al_2O_3," *J. Appl. Crystallogr*, vol. 20, pp. 79–83, 1987.

[11] S. Marchesini, "A unified evaluation of iterative projection algorithms for phase retrieval," *Rev. Sci. Instrum.*, vol. 78, p. 011301, 2007.

[12]   R. W. Gerchberg and W. O. Saxton, "Practical algorithm for determination of phase from image and diffraction plane pictures," *Optik*, vol. 35, pp. 237–246, 1972.

[13]   J. R. Fienup, "Phase retrieval algorithms - a comparison," *Appl. Optics*, vol. 21, pp. 2758–2769, 1982.

[14]   V. Elser, "Phase retrieval by iterated projections," *J. Opt. Soc. Am. A*, vol. 20, pp. 40–55, 2003.

[15]   G. Oszlanyi and A. Suto, "Ab initio structure solution by charge flipping," *Acta Crystallogr. A*, vol. 60, pp. 134–141, 2004.

[16]   K. Moffat, D. Szebenyi, and D. Bilderback, "X-ray Laue diffraction from protein crystals," *Science*, vol. 223, pp. 1423–1425, 1984.

[17]   J. W. Pflugrath, "The finer things in X-ray diffraction data collection," *Acta Crystallogr. D*, vol. 55, pp. 1718–1725, 1999.

[18]   M. Mueller, M. Wang, and C. Schulze-Briese, "Optimal fine ϕ-slicing for single-photon-counting pixel detectors," *Acta Crystallogr. D*, vol. 68, pp. 42–56, 2012.

[19]   J. M. Harp, D. E. Timm, and G. J. Bunick, "Macromolecular crystal annealing: overcoming increased mosaicity associated with cryocrystallography," *Acta Crystallogr. D*, vol. 54, pp. 622–628, 1998.

[20]   B. Heras and J. L. Martin, "Post-crystallization treatments for improving diffraction quality of protein crystals," *Acta Crystallogr. D*, vol. 61, pp. 1173–1180, 2005.

[21]   J. D. Emery, C. M. Schlepütz, P. Guo, et al., "Atomic layer deposition of metastable β-Fe_2O_3 via isomorphic epitaxy for photoassisted water oxidation," *ACS Appl. Mater. Interf.*, vol. 6, pp. 21894–21900, 2014.

[22]   S. Gaudet, K. De Keyser, S. Lambert-Milot, et al., "Three dimensional reciprocal space measurement by x-ray diffraction using linear and area detectors: Applications to texture and defects determination in oriented thin films and nanoprecipitates," *J. Vac. Sci. Technol. A*, vol. 31, p. 021505, 2013.

[23]   C. Mocuta, M.-I. Richard, J. Fouet, et al., "Fast pole figure acquisition using area detectors at the DiffAbs beamline –Synchrotron SOLEIL," *J. Appl. Crystallogr.*, vol. 46, pp. 1842–1853, 2013.

[24]   K. Fadenberger, I. E. Gunduz, C. Tsotsos, et al., "In situ observation of rapid reactions in nanoscale Ni-Al multilayer foils using synchrotron radiation," *Appl. Phys. Lett.*, vol. 97, p. 144101, 2010.

[25]   "International Centre for Diffraction Data website." http://www.icdd.com/.

[26]   S. J. L. Billinge and M. G. Kanatzidis, "Beyond crystallography: the study of disorder, nanocrystallinity and crystallographically challenged materials with pair distribution functions," *Chem. Commun.*, vol. 7, pp. 749–760, 2004.

[27]   D. A. Dimitrov, A. L. Ankudinov, A. R. Bishop, and S. D. Conradson, "Pair distribution function and x-ray absorption signatures of rotational and radial distortions in a model system with average long-range order," *Phys. Rev. B*, vol. 58, pp. 14227–14237, 1998.

[28]   A. Ben-Shem, L. Jenner, G. Yusupova, and M. Yusupov, "Crystal structure of the eukaryotic ribosome," *Science*, vol. 330, pp. 1203–1209, 2010.

[29]   H. P. Austin, M. D. Allen, B. S. Donohoe, et al., "Characterization and engineering of a plastic-degrading aromatic polyesterase," *Proc. Natl. Acad. Sci. USA*, vol. 115, pp. E4350–E4357, 2018.

[30]   A. Bhattacharya, "Protein structures: Structures of desire," *Nature*, vol. 459, pp. 24–27, 2009.

[31]   H. M. Berman, "The protein data bank: a historical perspective," *Acta Crystallogr. A*, vol. 64, pp. 88–95, 2008.

[32]   M. Levitt, "Growth of novel protein structural data," *Proc. Natl. Acad. Sci. USA*, vol. 104, pp. 3183–3188, 2007.

[33]   J. D. Bernal and D. M. Crowfoot, "X-ray photographs of crystalline pepsin," *Nature*, vol. 133, pp. 794–795, 1934.

[34]   C. H. Carlisle and D. Crowfoot, "The crystal structure of cholesterol iodide," *Proc. Royal Soc. A*, vol. 184, pp. 64–83, 1945.

[35]   J. C. Phillips, A. Wlodawer, M. M. Yevitz, and K. O. Hodgson, "Applications of synchrotron radiation to protein crystallography: preliminary results," *Proc. Natl. Acad. Sci. USA*, vol. 73, pp. 128–132, 1976.

[36]   H. Hope, "Cryocrystallography of biological macromolecules: a generally applicable method," *Acta Crystallogr. B*, vol. 44, pp. 22–26, 1988.

[37]   T.-Y. Teng, "Mounting of crystals for macromolecular crystallography in a free-standing thin film," *J. Appl. Crystallogr.*, vol. 23, pp. 387–391, 1990.

[38]   T.-Y. Teng and K. Moffat, "Primary radiation damage of protein crystals by an intense synchrotron x-ray beam," *J. Synchrotron Rad.*, vol. 7, pp. 313–317, 2000.

[39]   J. W. Murray, E. F. Garman, and R. B. G. Ravelli, "X-ray absorption by macromolecular crystals: the effects of wavelength and crystal composition on absorbed dose," *J. Appl. Crystallogr.*, vol. 37, pp. 513–522, 2004.

[40]   R. L. Owen, E. Rudino-Pinera, and E. F. Garman, "Experimental determination of the radiation dose limit for cryocooled protein crystals," *Proc. Natl. Acad. Sci. USA*, vol. 103, pp. 4912–4917, 2006.

[41]   E. F. Garman, "Radiation damage in macromolecular crystallography: what is it and why should we care?," *Acta Crystallogr. D*, vol. 66, pp. 339–351, 2010.

[42]   M. Weik, R. B. G. Ravelli, G. Kryger, et al., "Specific chemical and structural damage to proteins produced by synchrotron radiation," *Proc. Natl. Acad. Sci. USA*, vol. 97, pp. 623–628, 2000.

[43]   H. Fischer, I. Polikarpov, and A. F. Craievich, "Average protein density is a molecular-weight-dependent function," *Prot. Sci.*, vol. 13, pp. 2825–2828, 2004.

[44]   E. T. White, W. H. Tan, J. M. Ang, S. Tait, and J. D. Litster, "The density of a protein crystal," *Powd. Technol.*, vol. 179, pp. 55–58, 2007.

[45]   W. A. Hendrickson, "Atomic-level analysis of membrane-protein structure," *Nat. Struct. Mol. Biol.*, vol. 23, pp. 464–467, 2016.

[46]   E. M. Landau and J. P. Rosenbusch, "Lipidic cubic phases: a novel concept for the crystallization of membrane proteins," *Proc. Natl. Acad. Sci. USA*, vol. 93, pp. 14532–14535, 1996.

[47]   J. L. Smith, R. F. Fischetti, and M. Yamamoto, "Micro-crystallography comes of age," *Curr. Op. Struct. Biol.*, vol. 22, pp. 602–612, 2012.

[48]   S. Botha, K. Nass, T. R. M. Barends, et al., "Room-temperature serial crystallography at synchrotron x-ray sources using slowly flowing free-standing high-viscosity microstreams," *Acta Crystallogr. D*, vol. 71, pp. 387–397, 2015.

[49]   P. Nogly, D. James, D. Wang, et al., "Lipidic cubic phase serial millisecond crystallography using synchrotron radiation," *IUCrJ*, vol. 2, pp. 168–176, 2015.

[50]   F. Stellato, D. Oberthür, M. Liang, et al., "Room-temperature macromolecular serial crystallography using synchrotron radiation," *IUCrJ*, vol. 1, pp. 204–212, 2014.

[51]   C. Gati, G. Bourenkov, M. Klinge, et al., "Serial crystallography on in vivo grown microcrystals using synchrotron radiation," *IUCrJ*, vol. 1, pp. 87–94, 2014.

[52]   C.-Y. Huang, V. Olieric, P. Ma, et al., "In meso in situ serial x-ray crystallography of soluble and membrane proteins," *Acta Crystallogr. D*, vol. 71, pp. 1238–1256, 2015.

[53]   K. Diederichs and M. Wang, "Serial synchrotron x-ray crystallography (SSX)," *Methods Mol. Biol.*, vol. 1607, pp. 239–272, 2017.

[54]   J. S. Fraser, H. van den Bedem, A. J. Samelson, et al., "Accessing protein conformational ensembles using room-temperature x-ray crystallography," *Proc. Natl. Acad. Sci. USA*, vol. 108, pp. 16247–16252, 2011.

[55]   M. Warkentin, J. B. Hopkins, R. Badeau, et al., "Global radiation damage: temperature dependence, time dependence and how to outrun it," *J. Synchrotron Rad.*, vol. 20, pp. 7–13, 2013.

[56]   R. L. Owen, N. Paterson, D. Axford, et al., "Exploiting fast detectors to enter a new dimension in room-temperature crystallography," *Acta Crystallogr. D*, vol. 70, pp. 1248–1256, 2014.

[57]   C.-Y. Huang, V. Olieric, P. Ma, et al., "In meso in situ serial x-ray crystallography of soluble and membrane proteins at cryogenic temperatures," *Acta Crystallogr. D*, vol. 72, pp. 93–112, 2016.

[58]   C. Nave and M. A. Hill, "Will reduced radiation damage occur with very small crystals?," *J. Synchrotron Rad.*, vol. 12, pp. 299–303, 2005.

[59] R. Sanishvili, D. W. Yoder, S. B. Pothineni, et al., "Radiation damage in protein crystals is reduced with a micron-sized x-ray beam," *Proc. Natl. Acad. Sci. USA*, vol. 108, pp. 6127–6132, 2011.

[60] J. M. Holton and K. A. Frankel, "The minimum crystal size needed for a complete diffraction data set," *Acta Crystallogr. D*, vol. 66, pp. 393–408, 2010.

[61] Q. Liu and W. A. Hendrickson, "Crystallographic phasing from weak anomalous signal," *Curr. Op. Struct. Biol.*, vol. 34, pp. 99–107, 2015.

[62] A. Wagner, R. Duman, K. Henderson, and V. Mykhaylyk, "In-vacuum long-wavelength macromolecular crystallography," *Acta Crystallogr. D*, vol. 72, pp. 430–439, 2016.

[63] A. F. Bent, G. Mann, W. E. Houssen, et al., "Structure of the cyanobactin oxidase ThcOx from Cyanothece sp. PCC 7425, the first structure to be solved at Diamond Light Source beamline I23 by means of S-SAD," *Acta Crystallogr. D*, vol. 72, pp. 1174–1180, 2016.

[64] W. A. Hendrickson, "Evolution of diffraction methods for solving crystal structures," *Acta Crystallogr. A*, vol. 69, pp. 51–59, 2013.

[65] W. A. Hendrickson, "Anomalous diffraction in crystallographic phase evaluation," *Quart. Rev. Biophys.*, vol. 47, pp. 49–93, 2014.

[66] M. G. Rossmann and D. M. Blow, "The detection of sub-units within the crystallographic asymmetric unit," *Acta Crystallogr.*, vol. 15, pp. 24–31, 1962.

[67] G. Scapin, "Molecular replacement then and now," *Acta Crystallogr. D*, vol. 69, pp. 2266–2275, 2013.

[68] J. M. Cork, "The crystal structure of some of the alums," *Philos. Mag.*, vol. 4, pp. 688–698, 1927.

[69] M. F. Perutz, M. G. Rossmann, A. F. Cullis, et al., "Structure of haemoglobin: A three-dimensional Fourier synthesis at 5.5 Å resolution, obtained by x-ray analysis," *Nature*, vol. 185, pp. 416–422, 1960.

[70] E. Garman and J. W. Murray, "Heavy-atom derivatization," *Acta Crystallogr. D*, vol. 59, pp. 1903–1913, 2003.

[71] D. Harker, "The determination of the phases of the structure factors of non-centrosymmetric crystals by the method of double isomorphous replacement," *Acta Crystallogr.*, vol. 9, pp. 1–9, 1956.

[72] W. A. Hendrickson, "Analysis of protein structure from diffraction measurements at multiple wavelengths," *Trans. Am. Crystallogr. Assoc.*, vol. 21, pp. 11–21, 1985.

[73] W. A. Hendrickson, "Determination of macromolecular structures from anomalous diffraction of synchrotron radiation," *Science*, vol. 254, pp. 51–58, 1991.

[74] W. A. Hendrickson and M. M. Teeter, "Structure of the hydrophobic protein crambin determined directly from the anomalous scattering of sulfur," *Nature*, vol. 290, pp. 107–113, 1981.

[75] B.-C. Wang, "Resolution of phase ambiguity in macromolecular crystallography," *Meth. Enzymol.*, vol. 115, pp. 90–112, 1985.

[76] I. P. Rose, B. C. Wang, and M. S. Weiss, "Native SAD is maturing," *IUCrJ*, vol. 2, pp. 131–140, 2015.

[77] T. Weinert, V. Olieric, S. Waltersperger, et al., "Fast native-SAD phasing for routine macromolecular structure determination," *Nature Meth.*, vol. 12, pp. 131–133, 2015.

[78] J. P. Abrahams and A. G. W. Leslie, "Methods used in the structure determination of bovine mitochondrial F_1 ATPase," *Acta Crystallogr. D*, vol. 52, pp. 30–42, 1996.

[79] H. N. Chapman, P. Fromme, A. Barty, et al., "Femtosecond X-ray protein nanocrystallography," *Nature*, vol. 470, pp. 73–77, 2011.

[80] R. Neutze, R. Wouts, D. van der Spoel, E. Weckert, and J. Hajdu, "Potential for biomolecular imaging with femtosecond x-ray pulses," *Nature*, vol. 406, pp. 752–757, 2000.

[81] J. Tenboer, S. Basu, N. Zatsepin, et al., "Time-resolved serial crystallography captures high-resolution intermediates of photoactive yellow protein," *Science*, vol. 346, pp. 1242–1246, 2014.

[82] M. Suga, F. Akita, M. Sugahara, et al., "Light-induced structural changes and the site of O=O bond formation in PSII caught by XFEL," *Nature*, vol. 543, pp. 131–135, 2017.

[83] D. Wang, U. Weierstall, L. Pollack, and J. Spence, "Double-focusing mixing jet for XFEL study of chemical kinetics," *J. Synchrotron Rad.*, vol. 21, pp. 1364–1366, 2014.

[84] V. Srajer, T.-Y. Teng, T. Ursby, et al., "Photolysis of the carbon monoxide complex of myoglobin: nanosecond time-resolved crystallography," *Science*, vol. 274, pp. 1726–1729, 1996.

[85] A. Batyuk, L. Galli, A. Ishchenko, et al., "Native phasing of x-ray free-electron laser data for a G protein-coupled receptor," *Science Adv.*, vol. 2, p. e1600292, 2016.

[86] M. S. Hunter, C. H. Yoon, H. DeMirci, et al., "Selenium single-wavelength anomalous diffraction de novo phasing using an x-ray-free electron laser," *Nature Comms.*, vol. 7, p. 13388, 2016.

[87] R. Feidenhans'l, "Surface-structure determination by x-ray diffraction," *Surf. Sci. Rep.*, vol. 10, pp. 105–188, 1989.

[88] I. K. Robinson and D. J. Tweet, "Surface x-ray-diffraction," *Rep. Prog. Phys.*, vol. 55, pp. 599–651, 1992.

[89] D. Martoccia, M. Björck, C. M. Schlepütz, et al., "Graphene on Ru(0001) – a corrugated and chiral structure," *New J. Phys.*, vol. 12, p. 043028, 2010.

[90] C. M. Schlepütz, M. Björck, E. Koller, et al., "Structure of ultrathin heteroepitaxial superconducting $YBa_2Cu_3O_{7-x}$ films," *Phys. Rev. B*, vol. 81, p. 174520, 2010.

[91] R. Herger, P. R. Willmott, O. Bunk, et al., "Surface of strontium titanate," *Phys. Rev. Lett.*, vol. 98, p. 076102, 2007.

[92] Y. Murakami, J. P. Hill, D. Gibbs, et al., "Resonant x-ray scattering from orbital ordering in $LaMnO_3$," *Phys. Rev. Lett.*, vol. 81, pp. 582–585, 1998.

[93] P. D. Hatton, S. B. Wilkins, T. A. W. Beale, et al., "Resonant soft x-ray diffraction – in extremis," *J. Synchrotron Rad.*, vol. 12, pp. 434–441, 2005.

[94] Y. Murakami, H. Kawada, H. Kawata, et al., "Direct observation of charge and orbital ordering in $La_{0.5}Sr_{1.5}MnO_4$," *Phys. Rev. Lett.*, vol. 80, pp. 1932–1935, 1998.

[95] L. G. Parratt, "Surface studies of solids by total reflection of x-rays," *Phys. Rev.*, vol. 95, pp. 359–369, 1954.

[96] H. Kiessig, "Untersuchung zur Totalreflexion von Röntgenstrahlen," *Ann. der Phys.*, vol. 10, pp. 715–768, 1931.

[97] P. R. Willmott, C. M. Schlepütz, B. D. Patterson, et al., "In situ studies of complex PLD-grown films using hard X-ray surface diffraction," *Appl. Surf. Sci.*, vol. 247, pp. 188–196, 2005.

[98] P. R. Willmott, R. Herger, C. M. Schlepütz, D. Martoccia, and B. D. Patterson, "Energetic surface smoothing of complex metal-oxide thin films," *Phys. Rev. Lett.*, vol. 96, p. 176102, 2006.

[99] O. Glatter and O. Kratky, eds., *Small-angle x-ray scattering*. Academic Press, 1982.

[100] P. Lindner and T. Zemb, eds., *Neutrons, x-rays, and light: scattering methods applied to soft condensed matter*. North-Holland Elsevier, 2002.

[101] T. D. Grant, J. R. Luft, J. R. Wolfley, et al., "Small angle x-ray scattering as a complementary tool for high-throughput structural studies," *Biopolymers*, vol. 95, pp. 517–530, 2011.

[102] L. A. Feigin and D. I. Svergun, *Structure analysis by small-angle x-ray and neutron scattering*. Springer, 1987.

[103] G. Renaud, R. Lazzari, and F. Leroy, "Probing surface and interface morphology with grazing incidence small angle x-ray scattering," *Surf. Sci. Rep.*, vol. 64, pp. 255–380, 2009.

[104] G. Renaud, R. Lazzari, C. Revenant, et al., "Real-time monitoring of growing nanoparticles," *Science*, vol. 300, pp. 1416–1419, 2003.

7

Spectroscopic Techniques

7.1 Introduction

In this chapter we will discuss the experimental methods which probe, on the one hand, the absorption of ultraviolet and x-ray light by matter and, on the other, the subsequent mechanisms, such as fluorescence and photoelectron emission.

Spectroscopic methods in general measure the response of a system as a function of energy. The energy that is scanned may be that of the incident beam (whatever that beam might be – photons, electrons, neutrons, etc.), or the energy of the outgoing particles (for example, photons in x-ray fluorescence or electrons in x-ray photoelectron spectroscopy). Some techniques also consider the *direction* of the incoming or outgoing beam, such as in angle-resolved photoelectron spectroscopy, in which case one is interested in the momentum as well as the energy. Images are obtained by spatially resolving the spectroscopic response of heterogeneous samples: spectromicroscopies record the changes of full-field images as a function of polarization or energy (be it that of the incident photon energy or that of the subsequent Auger electron, photoelectron, or fluorescence photon), while microspectroscopies record energy spectra from a scanned, tightly focussed incident beam (Figure 7.1).

All the spectroscopic techniques described in this chapter use synchrotron light as the incident probe beam. Absorption of this results, to a greater or lesser extent, in excitation of one or more electrons. The different techniques discussed here are broadly divided into three categories. X-ray absorption spectroscopies (XAS) are concerned with the change in response of a system as a function of the incident photon energy and as such can only be performed at synchrotrons. In general, an electron is excited to either an unoccupied, bound, valence state, or is completely ejected from the electrostatic influence of its parent atom. This state can then relax via emission of a photon, be scattered elastically or inelastically by neighbouring atoms, or may be radiationless, leading to the ejection of photoelectrons, Auger electrons, and a cascade of low-energy secondary electrons. As such, x-ray absorption spectroscopies probe the unoccupied density of states of the system [see Figure 7.2(a)].

In contrast, x-ray photoelectron spectroscopies (XPS) involve the collection and measurement of the energies of directly ejected electrons, yielding information about the core or valence state from which the electrons originated. Hence, each point in an XAS spectrum includes all the possible processes occurring subsequent to absorption of a photon of energy $h\nu$, and can thus be thought of as being the energy-integrated intensity of an XPS spectrum recorded using the photon energy associated with that point in the XAS spectrum [Figure 7.2(b)], that is

$$I_{\mathrm{XAS}} = \int I_{\mathrm{XPS}}(\mathscr{E}_e, h\nu)d\mathscr{E}_e. \qquad (7.1)$$

An Introduction to Synchrotron Radiation: Techniques and Applications, Second Edition. Philip Willmott.
© 2019 John Wiley & Sons Ltd. Published 2019 by John Wiley & Sons Ltd.

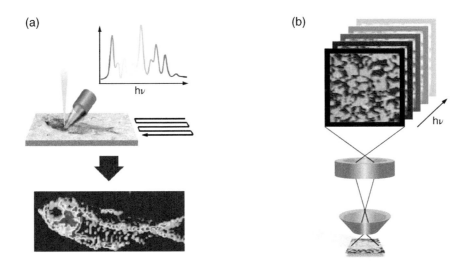

Figure 7.1 Examples of microspectroscopy and spectromicroscopy. (a) Scanning x-ray fluorescence. A tightly focussed beam produces fluorescence in a sample; the spectrum is recorded using an energy-dispersive detector. A set of images for different fluorescent-photon energies are created by scanning the sample in two transverse directions. Lower image adapted from [1] and courtesy of Troy Rasbury, Stony Brook University, with permission of Elsevier. (b) Photoemission electron microscopy. A sample is bathed in x-rays and the resultant secondary electron emission is imaged using electron microscope optics on to a detector. The image can change by altering either the polarization or photon energy.

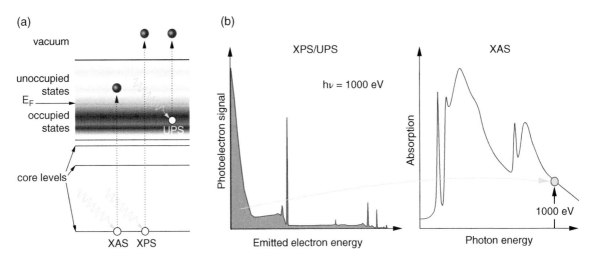

Figure 7.2 The relationship between absorption and photoelectron spectroscopies. (a) XAS promotes core-level electrons to unoccupied valence-state levels and thereby probes these upper states by varying the incident photon energy. In contrast, in XPS and UPS, the electrons are directly promoted into the vacuum and their signal intensity as a function of emitted electron energy, $\mathscr{E}_e$ (or binding energy, E_B) is recorded for a fixed incident photon energy. (b) Each point in an XAS spectrum can be thought of as the integrated XPS signal for the photon energy of interest.

X-ray fluorescence (XRF) and resonant inelastic x-ray scattering (RIXS), on the other hand, record the spectral response of the emitted photons produced by relaxation of excited states formed by photoabsorption at a given incident photon energy[1].

The information obtained by the broad available palette of synchrotron-based spectroscopies is to a greater or lesser extent interlinked – because both x-ray fluorescence and photoelectron spectroscopy rely on the absorption of x-rays, there is a direct correlation between them and absorption spectroscopy. We therefore begin this chapter with x-ray absorption.

7.2 X-ray Absorption Processes

As we have discussed in Section 2.6.3, the absorbing power of a material is given by the absorption coefficient μ, which describes the exponential drop in intensity of an incident beam passing through a medium. μ depends on the types of atoms constituting the medium, how they are distributed, the nature of their bonding, magnetism, the light polarization, and the wavelength of the x-rays. In the simplest case, μ is determined by comparing the intensities of the beam entering and after it emerges from a sample of known thickness d (see Figure 7.3), such that

$$\frac{I}{I_0} = e^{-\mu d}. \tag{7.2}$$

This is the Beer–Lambert law for linear absorption.

The relative absorption spectrum can also be indirectly acquired by detecting the yield of secondary electrons. In the case of liquid or solid samples, the measurements must be corrected for reflections from the boundaries of the solid, although reflectivities are normally very low, except for ultraviolet photons up to approximately 100 eV and for x-ray photons incident at shallow angles [i.e. near or below the critical angle, see Equation (2.28)].

Conservation of energy demands that the energy of a system after absorbing a photon is increased by the photon energy $h\nu$, i.e.

$$E_f = E_i + h\nu. \tag{7.3}$$

Secondly, Pauli's exclusion principle requires that the final quantum state E_f was unoccupied before absorption took place. Quantum mechanics shows us that the probability of an optical transition by the absorption of a photon from an initial state (labelled as the *state vector* or *ket* $|\psi_i\rangle$ in the commonly used Dirac formalism) to a final state ($|\psi_f\rangle$) is proportional to the square of the transition-dipole matrix element

$$P_{if} \propto |\langle \psi_f |e\mathbf{r}|\psi_i\rangle|^2, \tag{7.4}$$

which is shorthand for

$$P_{if} \propto \left| \int \psi_f^* \, e\mathbf{r} \, \psi_i dV \right|^2, \tag{7.5}$$

where $e\mathbf{r}$ is the transition-dipole operator, ψ^* is the complex conjugate of the electron wavefunction ψ, and integration is over all space (Figure 7.4). When a photon has an energy equal to the energy difference between two states involved in an allowed electronic transition, it is said to be *resonant* with that transition.

There may be other isoenergetic transitions involving other quantum states, and so the total probability for absorption of a photon of energy $h\nu$ is the sum of the probabilities for all such transitions. Hence the study of the variation of μ with photon energy provides extremely valuable information on the electronic structure of the material and can only be performed using tuneable synchrotron radiation.

---

[1] Strictly speaking, XRF is also possible and is often carried out using an incident beam of quasi-monoenergetic electrons in an electron microscope, in which case the technique is called energy-dispersive x-ray spectroscopy, EDX.

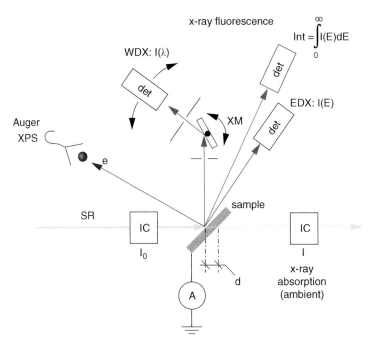

Figure 7.3 X-ray absorption and x-ray fluorescence experiments. Monochromatic synchrotron radiation (SR) is allowed to impinge on a sample. X-ray absorption spectra can be recorded by measuring the amount of light that passes through a thin sample. The x-ray intensities before entering the sample (I_0) and after (I) are measured using ionization chambers (IC), or other beam-intensity monitors, and are particularly suited for samples that cannot be placed in vacuum, such as in biological or catalytic experiments. The total electron current (A) can also be used to indirectly determine the absorption spectrum. In this case, the sample and detectors must be in vacuum. X-ray fluorescence spectra can be recorded, either using a crystal monochromator (XM) in wavelength-dispersive spectra (WDX), or by using a dispersive solid-state semiconductor device (EDX). The integrated fluorescence yield can also be used as a measure of absorption strength. Unwanted detection of elastically scattered x-rays is best achieved by placing the detector on the polarization axis of the synchrotron radiation (see also Figure 2.8).

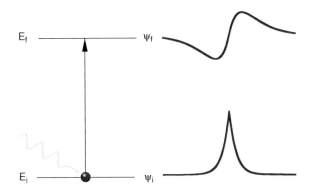

Figure 7.4   Schematic of a dipole transition via absorption of a photon of energy $h\nu = E_f - E_i$ from a lower initial state with wavefunction ψ_i and energy E_i to an unoccupied upper state ψ_f at energy E_f.

7.2.1   Energy-level Schemes of Atoms, Molecules, and Solids

Now we have discussed the basic ideas behind electronic transitions induced by absorption of a photon (optical transitions), it is instructive to consider the general features of electronic states in different states of matter. We begin with the simplest system, the isolated atom, of which the simplest of these is the hydrogen atom.

The energy required to remove the single electron from the lowest energy level of hydrogen (i.e. the $1s$ state, principal quantum number $n = 1$) is $E_1 = 13.6$ eV. Normally, the energy scale is so defined that the electron is said to have zero energy exactly at the ionization threshold, and therefore the discrete energy states below this have negative energies. The energy of a level n in a hydrogen atom is given by

$$E_n = -\frac{R_H}{n^2}, \tag{7.6}$$

where $R_H = 13.6$ eV is the Rydberg constant and is equal to the binding energy of an electron in the $1s$ state. The energy of a photon resonant with a transition from the $1s$ state of hydrogen to an excited state n is therefore

$$R_H \left(1 - \frac{1}{n^2}\right). \tag{7.7}$$

As n becomes large, the energy of the state approaches the ionization threshold asymptotically. States with large n are called Rydberg states [see Figure 7.5(a)] and are closely spaced in energy. Electrons excited to above the ionization threshold (IT) are completely free from the constraints of the system and can therefore assume any kinetic energy (i.e. their energies are no longer quantized) Above the IT, there is a true continuum of electronic states.

For atoms heavier than hydrogen, the binding energy of the $1s$ state increases approximately as Z^2, where Z is the nuclear charge, as was shown in Figure 2.3. This is because the Coulombic potential at a given radius is proportional to the product of the two charges involved (i.e. $Ze \times e$), while the radius of the $1s$ state is proportional to $1/Z$, (the Coulombic potential being itself inversely proportional to the radius). The reason this dependence is only approximate is due to the presence of the other $Z - 1$ electrons of the atom, which will interact with the photoelectron as it is ejected from the atom. Z ranges in magnitude from 1 to approximately 100, hence optical transitions in heavy atoms can result in the absorption of photons with energies as high as approximately 100 keV. For example, the binding energy of the $1s$ (K state) of uranium ($Z = 92$) is 115.6 keV. Synchrotron sources cover this range of photon energies.

The electronic state of an atom partaking in chemical bonding is different from that of the same atom in an isolated state, due to the redistribution of (primarily) the outermost electrons in the valence or conduction band in chemical bonding [Figure 7.5(b)]. Absorption spectra associated with the excitation of a core-level electron (which is normally shielded by outer electrons from the effect of chemical bonding) are only marginally affected by the nature of the chemical bond, as this latter involves the more weakly bound valence electrons. Nonetheless, we will see in Section 7.9 that the small changes in the binding energy (so-called 'chemical shifts') of core-electrons due to differences in the distribution of the valence electrons caused by chemical bonding can provide valuable information via the energy spectra of ejected photoelectrons about the chemical nature of the sample under investigation.

Molecules also show absorption features called 'shape resonances'. Although these are quasi-unbound (and are hence fairly diffuse), they still 'feel' the influence of the electrostatic field of the molecule and are associated with electron states that are physically close to the molecule.

The absorption spectra of condensed matter in the x-ray region associated with the ejection of core electrons exhibit well-defined and (to a greater or lesser extent) sharp absorption edges. However, as we will see in Sections 7.4 and 7.5, subtle (and sometimes not so subtle) differences compared to absorption spectra from the same elements found in isolated atoms and molecules can be evident. These are caused by interactions of the excited electron with scattering atoms within the local surroundings of the absorbing atom.

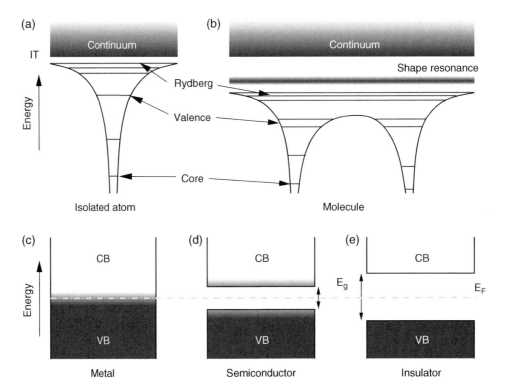

Figure 7.5 Schematic electron-energy-level schemes for different systems. (a) Isolated atoms contain core-level-state (most tightly bound) electrons, valence-state electrons, and excited-state electrons, of which those with the highest-energy bound levels occupy closely lying Rydberg states. (b) In molecules, the energy of the system is lowered by sharing or exchange of valence electrons. The valence states are therefore generally lower in energy than for the individual atoms. Condensed matter (c)–(e) generally exhibits continuous bands of electron-state *density* due to the interactions between many mutually interacting valence electrons (see text). The boundary between the bound-state valence band (VB) and the unbound-state conduction band (CB) at absolute zero temperature is the Fermi energy, E_F. A metal has no gap between the valence and conduction bands and will be electrically conducting. Semiconductors and insulators exhibit energy gaps (E_g) around the Fermi level, in which there are no electronic states. The magnitude of E_g is determined by the strength of the bonding of the valence electrons. The gap in semiconductors is sufficiently small that a small fraction of the valence electrons are thermally promoted to the conduction band. In insulators, E_g is so large that the conduction band remains entirely unoccupied.

The energy-level schemes of condensed-matter samples differ most notably from those of isolated systems (atoms and small molecules) with regards to the valence states. In condensed matter, the valence electrons can physically extend over large distances, thereby interacting with other electrons in a complex manner. Consider the interaction of two electrons which, when they are in isolated atoms, have similar binding energies. When these two electrons approach one another, they interact to form two new 'eigenstates'[2], which are a combined configuration of these electrons, one with a lower energy than the original isolated state, and another with a higher energy. In general, the lower state is associated with a bonding configuration, while the upper state is unbound, or 'antibonding'.

Let us extend this concept to a large number of such electrons (of the order of Avogadro's number, 6×10^{23}), as one might find in a macroscopic solid or liquid. Now, instead of there being one upper and one lower energy state,

---

[2] 'Eigen' is German for 'actual', or 'real'.

there are of the order of 10^{23} lower and upper states. Instead of being well-separated as in isolated atoms, these states are extremely densely packed (with separations of the order of 10^{-23} eV) and therefore form continuous bands of electrons.

In the case of a metal, the valence band and the conduction band butt up against each other. At absolute zero temperature, the valence band is fully occupied and the conduction band is empty – the boundary between the valence and conduction band is known as the Fermi energy, E_F [Figure 7.5(c)]. At finite temperatures, those electrons in the valence band that lie within about kT of E_F can be thermally promoted into the conduction band. kT is approximately 25 meV at room temperature.

Metals are therefore characterized by containing a 'gas' of delocalized and nearly free valence electrons. However, for systems in which the attraction of the valence electron to the ionic core increases, the valence and conduction bands separate in energy, leaving a 'gap' in between, in which there are no allowed states. For small energy gaps (E_g in Figure 7.5) up to a few eV the system is a semiconductor and a small fraction of the valence electrons can be thermally promoted to the conduction band. There is no specific lower limit to the bandgap which results in a material being an insulator. Rather, an insulator is better defined as a material which melts or decomposes before a detectable fraction of its valence electrons can be promoted into the conduction band.

Remembering Pauli's exclusion principle, we immediately recognize that metals, semiconductors, and insulators must have different absorption signatures. Electrons residing in metals can be excited by electromagnetic radiation of any energy and this is the reason why metals reflect light so well in the visible and infrared regions. Semiconductors above absolute zero have a certain (though often very small) fraction of electrons thermally promoted to the conduction band, which is why semiconductors do conduct, albeit poorly. Interband transitions are allowed, but require that $hv \geq E_g$. In insulators, in contrast, no intraband transitions are possible, while the minimum photon energy required to induce an interband transition is larger than that in semiconductors.

7.2.2 Absorption Features

Absorption spectra have several characteristic features. As mentioned in Chapter 2, μ varies approximately as the inverse third power of the photon energy. This trend is interrupted by step-like increases in absorption as the photon energy matches the ionization potential of an occupied electron state in the atom.

7.2.2.1 *Core-level Absorption Edges*

The K-edge results from the increase in absorption due to resonance of the photon energy with the ionization threshold of the K or $1s$ state (see Figure 2.3), while the L- and M-edges are those associated with ionization of the $n = 2$ and $n = 3$ states, respectively. All absorption edges except the K-edges exhibit three or more energetically closely lying 'sub-edges' associated with nondegenerate quantum states. These arise from different (quantized) possibilities of coupling the nonzero orbital- and spin-angular-momentum components of the electronic states (see Section 2.6.3). The reason these are not seen in K edges is that in this case, the orbital angular momentum is zero, as $l = 0$ (see Figure 7.6) and there can therefore be no coupling.

There can be particularly strong fluctuations in intensity around the absorption edge, known as the x-ray absorption near-edge structure (XANES), discussed in Section 7.4. Starting approximately 50 eV above the edges, the absorption profile exhibits structure that can extend several hundred eV or more. This 'extended x-ray absorption fine structure' (EXAFS) is discussed in detail in Section 7.5.

7.2.2.2 *Delayed Onset*

The shape of the absorption edges depends on the initial state being excited. As a rule of thumb, those states with a low angular-momentum quantum number l (i.e. s states with $l = 0$ and p states with $l = 1$) have sharp rising edges across only a few electron volts. States with larger l values have broader absorption features, with the maximum several tens of eV higher in energy than the onset of the absorption edge. This so-called 'delayed onset' can be explained qualitatively as follows. Quantum states with high principal quantum number tend to remain away from

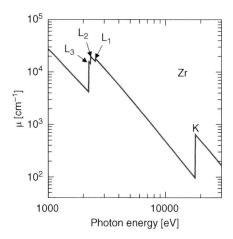

Figure 7.6   The absorption coefficient of Zr across its K- and L-edges. Note that the L-edge has three components, corresponding to ionization from the $2s$, $2p_{1/2}$, and $2p_{3/2}$ levels, denoted as L_1, L_2, and L_3, respectively.

the nucleus – classically, one would expect them to have a large circular orbit, while s and p orbitals spend some of their time very close to the nucleus.

Referring back to Equation (7.5), it can be seen that the maximum probability for absorption occurs if the integrand of the transition matrix element is as symmetric as possible, as we integrate over all space. The dipole transition operator r is, however, antisymmetric, so for the integrand in Equation (7.5) to be symmetric, we require the overlap integral $\langle \psi_i | \psi_f \rangle$ to be as antisymmetric as possible, that is, the two involved states should be dissimilar in symmetry[3]. Vacuum states just above the ionization threshold also have circular-like orbitals that keep the electron away from the nucleus, and hence the transition probability between initial states with high n and final vacuum states just above threshold is low. At still higher excess energies, however, the electron behaves more like a plane wave, and the transition probability therefore increases. An example of this delayed onset is shown for gold in Figure 7.7.

7.3   Photoelectron Energies, Wavelengths, and Absorption Regions

The kinetic energy $\mathscr{E}_e$ of a photoelectron[4] ejected from an atom by absorption of a photon is equal to the photon energy $h\nu$ minus the binding energy of the electron E_B

$$\mathscr{E}_e = h\nu - E_B. \tag{7.8}$$

But the de Broglie relationship describing particle–wave duality states that a particle with momentum p has an associated wavelength $h/p = \hbar k$, where h is Planck's constant and k is the wavevector $2\pi/\lambda$.

Moreover, the kinetic energy and momentum of a nonrelativistic particle of mass m are related by

$$E = \frac{p^2}{2m}. \tag{7.9}$$

---

[3] Note, for example, that because they have the same angular symmetry, the probability of a dipole transition from the $1s$ to the $2s$ level in the hydrogen atom is zero.

[4] Note the distinction in nomenclature between the kinetic energy of the electrons within the synchrotron storage ring $\mathscr{E}$ used in Chapter 3 and that of photoelectrons $\mathscr{E}_e$ described in this chapter.

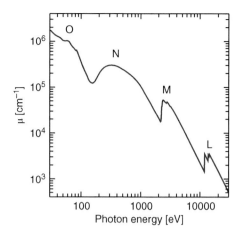

Figure 7.7 The 'delayed onset' can be clearly seen in the absorption spectrum of gold. The L-edge and, to a lesser extent, the M-edge, are sharp, while the N- and O-edges at lower excitation energies are broader and less well-defined.

Hence the photoelectron's linear momentum $\hbar k_e$ is related to its energy $\mathscr{E}_e$ by

$$\mathscr{E}_e = h\nu - E_B = \frac{\hbar^2 k_e^2}{2m_e},\tag{7.10}$$

and so

$$\lambda_e = h[2m_e(h\nu - E_B)]^{-1/2},\tag{7.11}$$

or, in practical units

$$\lambda_e[\text{Å}] = \frac{12.28}{\sqrt{\mathscr{E}_e\ [eV]}},\tag{7.12}$$

or

$$k_e[\text{Å}^{-1}] = 0.512\ \sqrt{\mathscr{E}_e\ [eV]}.\tag{7.13}$$

Thus, the wavelengths of photoelectrons with energies below ~ 100 eV are comparable to or larger than inter-atomic distances. We will see that this energy provides a natural divide between the two related but distinct methods XANES (also called NEXAFS) and EXAFS, shown in Figure 7.8. A word of caution: the dividing energy between XANES and EXAFS cannot be universally defined, as the transition is gradual and in any case shifts according to the typical nearest-neighbour distances in the system under investigation.

The abscissa in Figure 7.8 is the photon energy relative to that for the first point of inflection in the absorption edge [2], which can be accurately determined from the derivative of the spectrum.

7.3.1 The Universal Curve

The inelastic mean free path (IMFP) Λ_e of an electron in condensed matter is the distance it will travel on average before being inelastically scattered with a probability $1 - 1/e = 63\%$. Λ_e depends strongly on the electron's kinetic energy $\mathscr{E}_e$, but only weakly on the nature of the type of matter it is travelling through. Particularly when they have energies between approximately 20 and 1000 eV, electrons can be strongly inelastically scattered by excitation of plasmons (collective movements of bound electrons relative to the nuclear cores), especially in the case of

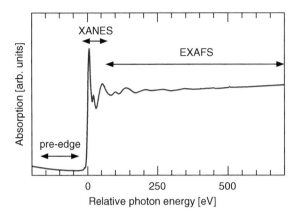

Figure 7.8 The important regions of a typical x-ray absorption spectrum around a core-electron absorption edge. The energy scale is relative to the first point of inflection of the absorption edge. Some features may be found in the pre-edge region due to transitions of core electrons to empty bound states. In the region around the absorption edge, one obtains x-ray absorption near-edge structure (XANES), dominated by transitions to unoccupied bound states. The absorption spectrum at energies from about 50 eV above the absorption edge often shows a series of oscillations up to as much as 1000 eV. This extended x-ray absorption fine structure (EXAFS) contains information about the local structure around the absorbing atom.

metals. Their mean free path, and therefore their escape depth in this energy region is consequently very small, of the order of a few angstroms. At lower energies, the electrons have insufficient energy to excite the plasmons, and the IMFP increases. Conversely, at higher energies, the cross-section for inelastic processes drops off because of the decreasing cross-section associated with their de Broglie wavelengths.

These inelastic scattering processes for a given electron kinetic energy depend primarily on the average separation between electrons, which in turn is proportional to the cube-root of the electron density ρ. But, as we have already seen for the elements in Figure 2.17, ρ rarely varies by more than a factor of 30 across the large majority of materials. Consequently, the scatter in Λ_e for a given electron energy for almost all condensed matter lies within a band of $\pm\sqrt[3]{30/2} \approx \pm 2.5$. When displayed as a double logarithmic plot, the universality of the dependence of Λ_e on $\mathscr{E}_e$ becomes apparent, and is hence referred to as the 'universal curve' (see Figure 7.9).

The photoabsorption depth of UV photons and x-rays in most materials is normally at least an order of magnitude greater than even the large IMFP for electrons with $\mathscr{E}_e \sim 1$ eV. Therefore, the majority of photoelectrons produced by the initial absorption process will be inelastically scattered multiple times before they can escape the sample surface. As they slow down to energies of the order of 10 eV, Λ_e begins to increase. The result is that so-called 'secondary electrons' with energies of the order of an electronvolt or less dominate the total electron yield from a sample (Figure 7.10).

Conversely, and as we will discuss in detail in Section 7.9, direct photoelectrons that have not undergone any inelastic scattering can only be detected with any reasonable intensity if they originated in the first one or two IMFPs from the surface. Methods such as ARPES, XPS, and XPD, which detect such electrons, are therefore highly surface sensitive.

7.3.2 σ- and π-polarizations

The orientation of the electric field of a linearly polarized electromagnetic wave incident on a surface relative to that surface can play a crucial rôle in how the surface, or any molecules sitting on it, will react. The electric field is responsible for electronic dipole transitions, hence the orientation of any electric dipoles due, for example, to

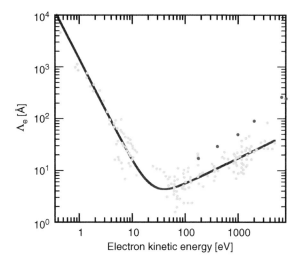

Figure 7.9   The universal curve. Plot of the inelastic mean free path (IMFP) of electrons in condensed matter, as a function of electron kinetic energy $\mathcal{E}_e$. The solid blue line describes the general expression $\Lambda_e = A/\mathcal{E}_e^2 + B\sqrt{\mathcal{E}_e}$, encapsulating both the low-energy (below 15 eV) and high-energy (above 150 eV) limiting physical cases. The best least-squares fit results in $A = 1430$ and $B = 0.54$ if $\mathcal{E}_e$ is expressed in eV. The yellow points are experimentally determined values, mainly from elemental samples. The red points are for water, important for solid–liquid-interface data in high-pressure XPS experiments. Adapted from [3] with permission from John Wiley.

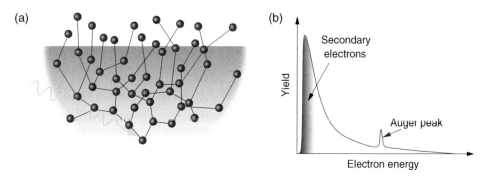

Figure 7.10   Secondary-electron production. (a) Secondary electrons are produced by a cascade of inelastic collisions seeded by the production of an initial photoelectron. As the electrons' kinetic energies decrease to below $\approx$ 10 eV, the mean free path Λ_e increases dramatically (note the longer paths between collisions further along any given chain), resulting in (b) there being an intense peak in the electron yield of the lowest-energy electrons compared to Auger electron intensities at higher energies.

bonding relative to the electromagnetic radiation will contribute to the extent to which they are coupled (that is, the degree of absorption).

Consider Figure 7.11, showing light impinging on a surface. If the plane defined by the incident propagation direction and the surface normal contains the electric field, the radiation is π- (or p)-polarized; if the electric field is perpendicular to the scattering plane it is σ- (or s)-polarized. For an incident angle of $\pi/2$ (that is, normal to the surface), the light polarization is undefined.

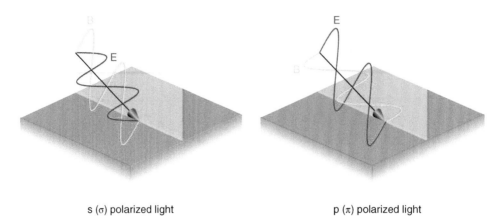

s (σ) polarized light   p (π) polarized light

Figure 7.11   σ- (or s-) and π- (or p)-polarizations. When the electric field of an x-ray beam incident on a surface is perpendicular to the plane defined by the propagation direction and the surface normal, it is said to be σ- or s-polarized (from the German for perpendicular, 'senkrecht'). When the electric field lies in the plane, the radiation is π- or p-polarized (from 'parallel').

7.4   X-ray Absorption Near-edge Structure, XANES

7.4.1   Introduction

The absorption cross-section for photoexcitation of a core-electron to vacuum is not a smooth function, but exhibits features in the neighbourhood of the absorption edge (see Figure 7.8). These are caused by excitation to low-energy bound states and also the quasi-continuum just above the ionization threshold that exhibit large fluctuations in their density of states. This signal is referred to as near-edge x-ray absorption fine structure (NEXAFS), or x-ray absorption near-edge structure (XANES). Photoabsorption creates excited states that can be described by electronic configurations in which a core electron is promoted to an energy level (orbital or band) that is empty in the ground state. and images the transition to unoccupied orbitals. NEXAFS and XANES are synonymous and here we will make no distinction (and for consistency's sake, plump for XANES[5]).

For the low photoelectron energies typical of XANES, the mean free path becomes significantly larger than at higher energies (see Figure 7.9). XANES thus 'sees' a large range of absorber–scatterer distances. This results in multiple scattering of the photoelectron, which means theoretical modelling of XANES still remains a significant challenge. XANES spectra are therefore mostly *interpreted* and not quantitatively analysed, although progress is being made in theoretical modelling and quantitative interpretation [4, 5].

There is an important aspect underlying the detection of a XANES signal. After the absorption of an x-ray photon by a core level, a photoelectron is generated. Importantly, this might not be ejected into the vacuum continuum above the ionization threshold, but only have enough energy to be promoted to an unoccupied but *bound* level. The core hole resulting from the absorption process can be filled either via an Auger process or by capture of an electron from another shell followed by emission of a fluorescent photon. The difference between XANES and traditional photoemission experiments is that in the latter, the initial photoelectron itself is measured (see

---

[5] Also, the orthographical similarity of NEXAFS to EXAFS can be a source of confusion.

Section 7.9), while in XANES the absorption signal is recorded via attenuation, fluorescence, Auger electrons, or secondary electrons.

This distinction may sound trivial but it is actually significant: in photoemission spectroscopy the final state of the directly emitted electron captured in the detector is an extended, free-electron state. As such, only detection of excitations beyond the ionization threshold is possible. In XANES, one can measure the final state of bound photoelectrons, since the photoelectron itself need not be detected. Hence, the effect of measuring fluorescent photons, Auger electrons, and directly emitted electrons is to sum over all possible final states of the photoelectrons, meaning that what XANES measures is the total density of states of the initial core level with all final states, consistent with conservation rules. The distinction is critical, because in spectroscopy final states are more susceptible to many-body effects than initial states, meaning that XANES spectra are more easily calculable than photoemission spectra.

7.4.2   The XANES Signal

What is the source for these large fluctuations in absorption intensity in the XANES regime? There are two major factors, namely the intrinsic probability of a transition from one state to another [Equation (7.5)]; and the density of states in which the excited state is embedded. These can be encapsulated in the famous Fermi's golden rule[6]

$$P_{if} = \frac{2\pi}{\hbar} |\langle f|H'|i\rangle|^2 \rho, \tag{7.14}$$

where $\langle f|H'|i\rangle$ is the transition matrix element that we have already met in Equation (7.4), H' is the perturbation Hamiltonian responsible for the transition (which in the simplest case of a dipole transition to a continuum state in the vacuum is the dipole operator $e\mathbf{r}$), and ρ is the density of states in which the final state is embedded. Let us now discuss how these factors affect XANES features.

The lowest-energy XANES signals actually occur *below* the absorption edge, and are referred to pre-edge transitions. Pre-edge features are strongly influenced by the symmetry of the local environment around the absorbing atom. To understand this, one must first appreciate that orbitals in molecules and solid-state systems can assume the character of a mixture of atomic orbitals and are related to the symmetry of the local environment, such as the simple example the famous tetragonally symmetric sp^3-orbitals of diamond formed by hybridization of one 2s- and three 2p-atomic orbitals.

Those orbitals around an atom to which a transition is allowed are determined by the symmetry of the local environment. In general, orbitals exhibit a so-called 'parity' of $(-1)^l$. An orbital has positive parity and is symmetric if it is invariant under the operation of moving each volume element of the orbital amplitude wavefunction from $\mathbf{r}$ to $-\mathbf{r}$. It is said to be inversion symmetric. Hence, the spherically symmetric s states have positive parity (as do the d-orbitals), while the p and f states are antisymmetric and therefore have negative parity (Figure 7.12). If the value of the integral given in Equation (7.4) is zero, the transition is forbidden. In practice, the integral itself does not need to be calculated to determine a selection rule; it is sufficient to know the symmetries of the initial and final states and that of the transition-moment operator, which in the case of a dipole moment $e\mathbf{r}$, is antisymmetric. This leads to the conclusion that the initial and final states must have opposite symmetries.

One can also have local environments which are inversion symmetric, such as any of the atom sites in a rocksalt crystal. In such environments, atomic orbitals with different parities cannot hybridize, as the symmetry of such hybridized orbitals does not match that of the environment. So, for example, an sp^3-orbital is incompatible with an inversion-symmetric environment.

Consider as an example a material for which a transition between a K state (1s, $l = 0$, symmetric) and a bound excited state of mostly (symmetric) d character is observed with high intensity. We have already argued that in

---

[6] Although popularized by Fermi, who coined the phrase 'golden rule', most of the theory behind the rule was worked out by P. A. M. Dirac.

Figure 7.12   Parity and symmetry. The parity of an atomic electron orbital is either positive or negative, depending on how it is transformed when moving all the elements j of the orbital's amplitude wavefunction from r_j to $-r_j$. So, for example, p orbitals are antisymmetric with negative parity, while d orbitals are symmetric and have positive parity.

the dipole approximation, only p states (antisymmetric) are accessible from s states. Hence the upper bound state must have both p-type (antisymmetric) and d-type character. From this we learn that the environment around the absorbing atom cannot be inversion symmetric, because of the partial p character. The presence of bound-state, resonant absorption peaks therefore provides important information on the local symmetry of the system. Qualitative information regarding bond lengths can also be relatively easily obtained, as the intensities of pre-edge peaks drop sharply with increasing bond length, due to the exponential-decay character of the orbital wavefunctions.

Depending on the interaction strength with neighbouring atoms and molecules, excited electronic states immediately above the highest occupied state of a system can maintain, to a large degree, the same structure they have in isolated molecules (see Figure 7.13). This is particularly true for molecules adsorbed with low coverage on surfaces, where 'crosstalk' or 'scrambling' of their individual character caused by interaction with neighbouring atoms or molecules is less pronounced and hence the density of states is sparser. The large majority of XANES experiments are performed on organic compounds, where, more often than not, any interaction between the molecule of interest and the 'substrate' on which it is absorbed is small.

In contrast to 'normal' solid-state orbitals within quasi-continuous energy bands, the HOMO (highest occupied molecular orbital) and LUMO (lowest unoccupied molecular orbital) orbitals may therefore be fairly discrete. Hence, illumination with photons of the appropriate energy can promote a core electron to an unoccupied but isolated state. Importantly, because these unoccupied states maintain much of their 'isolated' character, the selection rules governing such discrete-to-discrete transitions are far stricter than normally encountered by solid-state physicists.

The intensity of transitions to bound-state, unoccupied, orbitals can be high, depending on their density and selection rules. In early experiments for which the spectra were recorded on photographic strips, these bound, excited-state features were easily identifiable as sharp and strong 'white lines', and this is how they were named. An illustrative example of how the intensity of white lines can change from material to material is shown in Figure 7.14, for the XANES spectra of the five $5d$ metals Re, Os, Ir, Pt, and Au [6]. The ground-state electronic configuration of Re is [Xe] $6s^2\,4f^{14}\,5d^5$. As one moves to the right in the periodic table from Re, the remaining five available $5d$ states are filled. In gold, these states are all occupied. In the elements Re to Pt, the bound but unoccupied $5d$ states, which are below the ionization threshold, are seen as intense white lines. As gold has no such unoccupied states, excitation is directly into the unbound continuum.

The probability of a photon being absorbed depends on the orientation of the electric-field component (the linear polarization, **P**) of the electromagnetic field of the x-rays relative to the polarization direction of the orbital

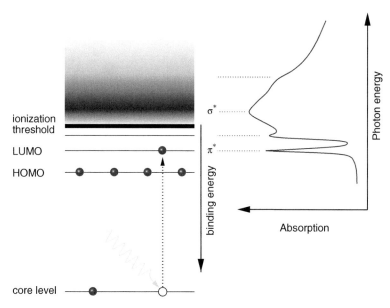

Figure 7.13 The XANES signal of a core-level electron may consist of sharp, bound features, corresponding to unoccupied bound states marginally below the ionization threshold (I.T.), of which excitons and white lines are examples, and broader features up to 30 or 40 eV above the I.T. In the case of unsaturated organic compounds (i.e. those with double or triple carbon bonds), the lowest unoccupied molecular orbital (LUMO) immediately above the highest occupied molecular orbital, (HOMO) is invariably a π^* orbital, while the lowest quasi-unbound state above the ionization threshold is σ^*. Their energies in carbon XANES spectra are approximately 286 and 295 eV, respectively, though crucially, these can shift by as much as two or three eV, according to the local chemical environment, a feature exploited to identify organic compounds using XANES.

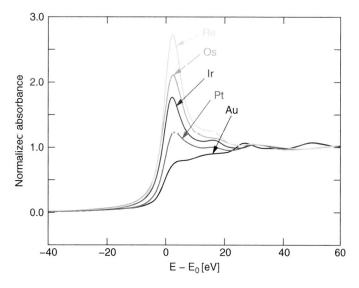

Figure 7.14 XANES spectra comparing the L_3 absorption edges of five of the 5d metals, rhenium to gold. Adapted from [6] with permission of the American Chemical Society.

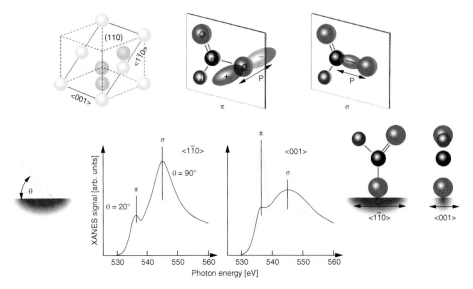

Figure 7.15   XANES spectra of the formate radical absorbed on the Cu(110) surface change according to the orientation of the x-ray polarization vector (*P*) relative to the molecule. The yellow curves are for grazing-incidence radiation, the blue curves for normal incidence. From these spectra, the orientation of the radical on the surface could be unambiguously identified. Adapted from [7] with permission of the American Physical Society.

being excited[7]. So, for example, the π-orbital of the formate radical shown in Figure 7.15 lies perpendicular to the molecular plane, and is most efficiently excited if **P** has the same orientation. Because the electric field is always perpendicular to the propagation direction of the x-ray beam, this means that the latter must lie in the molecular plane. The formate radical binds strongly to the surface of Cu(110). The orientation of the molecule was unambiguously determined by the change in the oxygen-XANES signal with x-ray-polarization orientation and incident angle on the surface [7]. Importantly, when the x-rays strike the surface at a glancing angle (20°) and **P** is parallel to the $\langle 1\bar{1}0 \rangle$ crystal axis of the copper, excitation of the π-orbital is suppressed. In contrast, it is strongly enhanced if the sample is rotated so that **P** is parallel to the $\langle 001 \rangle$ direction. This leads to the conclusion that the formate radical is oriented with the molecular plane perpendicular to the surface and parallel to the $\langle 1\bar{1}0 \rangle$ direction (Figure 7.15).

Hence, for systems with well-defined structural orientations (i.e. not polycrystalline material or powders), XANES can provide direct and simply interpretable information on the configuration via the dependence of the transition probability on the polarization of the exciting photons.

7.5   Extended X-ray Absorption Fine Structure, EXAFS

7.5.1   Introduction

Extended x-ray absorption fine structure (EXAFS) signal refers to oscillations in the x-ray absorption coefficient starting at approximately 50 eV above an absorption edge and extending up to several hundred eV. The interpretation and analysis of EXAFS signal was first presented by Sayers, Stern, and Lytle in their seminal paper in 1971 [8].

---

[7] This dependence of the absorption spectral response on the orientation of the linear polarization is called x-ray linear dichroism (XLD). The observation of XLD generally indicates an asymmetry in the electron charge density.

EXAFS partially relies on extensive libraries of reference-compound spectra of well-characterized materials, which provide a comparison for new investigations [9]. Nonetheless, because (in contrast to XANES) multiple scattering is insignificant in EXAFS, the theory of EXAFS is well described and quantitative comparisons with experimental results can be drawn [8, 10, 11].

The photoelectrons associated with EXAFS have mean-free paths of the order of a few angstroms to a nm (see Figure 7.9) and hence EXAFS only probes the immediate neighbourhood of the absorbing atom. EXAFS is therefore an important technique in noncrystalline solids and liquids, as it yields element-specific information of the short-range structure of materials. It is much used for investigating liquids, clusters, nanoparticles, biological materials, low-concentration impurities, and catalysts, although it can also be used for well-ordered structures, and is thus complementary to diffraction techniques. The sensitivity of EXAFS is typically below 100 parts per million (ppm) and can be as low as 10 ppm, depending on the element under investigation.

Surface regions, involving perhaps the first five or so monolayers, are exceedingly interesting for materials scientists, as it is normally here that reactions and physical changes take place. Chemisorption and physisorption, corrosion, catalytic processes, reconstructions, and surface alloying and segregation are typical examples of mechanisms that take place in the surface region of a condensed-matter sample. However, EXAFS is normally a bulk technique. To obtain surface sensitivity, surface-EXAFS (SEXAFS) uses surface-sensitive detection techniques, the most common of which is Auger-electron spectroscopy (AES), or one impinges on the sample with grazing incidence.

7.5.2 The EXAFS Signal

EXAFS is caused by single-backscattering events by nearest-neighbour atoms of the outgoing photoelectron produced through x-ray absorption. The scattered waves interfere with the original photoelectron wave, which results in a modulation of the absorption probability and hence also the absorption strength $\mu(E)$ as the photon energy is scanned (see Figure 7.16).

The theory of EXAFS is, in principle, straightforward [11]. Here, we cover the basics, explaining the quintessential features. We begin by looking more closely at the interference between the outgoing photoelectron wave and a wave produced by its backscattering by a neighbouring atom at a distance R, as shown schematically in Figure 7.17.

The most important parameters are therefore the material properties (in particular the atomic absorption coefficient μ), the electron wavelength λ_e [Equations (7.11) and (7.12)], the number and type of nearest neighbours, and the distance R between the absorbing atom and neighbouring atoms.

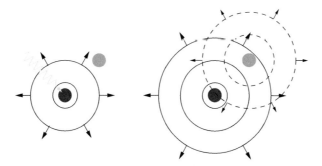

Figure 7.16 The origin of EXAFS. An x-ray photon is absorbed by an atom, resulting in promotion of a core-level electron to an unoccupied continuum state. As the electron wave propagates out from the excited atom (solid circles), it can be scattered by neighbouring atoms. The scattered waves (dashed circles) interfere with the outgoing waves, thereby modulating the absorption cross-section as a function of photon energy.

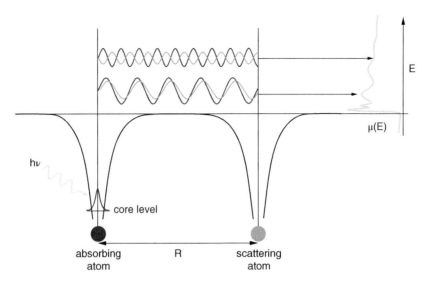

Figure 7.17   Creation of the EXAFS signal. Photoelectrons promoted to the vacuum continuum by absorption of x-rays can be partially scattered by neighbouring atoms. Depending on the electron wavevector k_e, and the phase of the scattered wave, interference between the outgoing wave and scattered wave can be constructive or destructive, resulting in a modulation of the absorption spectrum.

Consider first the absorption coefficient of an isolated atom, $\mu_0(k_e)$, which varies monotonically and smoothly with electron wavevector (or energy) above the absorption edge. In condensed matter, however, the absorption coefficient $\mu(k_e)$ is modulated by the so-called EXAFS function $\chi(k_e)$ because of interference with backscattered waves, and so

$$\mu(k_e) = \mu_0(k_e)[1 + \chi(k_e)], \tag{7.15}$$

which we can rearrange to obtain

$$\chi(k_e) = \frac{\mu(k_e) - \mu_0(k_e)}{\mu_0(k_e)}. \tag{7.16}$$

$\chi(k_e)$ is therefore an expression of the effect on an isolated atom's absorption characteristics caused by it no longer being isolated. An illustrative example of the effect of neighbouring atoms is shown in Figure 7.18 for the K-edge of krypton in its gas, liquid, and solid phases [12].

Clearly, because the fine-structure oscillations are produced by interference between the outgoing and backscattered waves, which are in turn determined by the local atomic structure, we should be able to extract this structural information from the EXAFS signal.

We now derive an expression for $\chi(k_e)$, based on the concept of 'shells'. We define a set of concentric shells centred around the absorbing atom, whereby the jth shell passes through N_j identical atoms [see Figure 7.19(a)]. First, we should recognize that the EXAFS signal is angle-dependent, as formulated by Stern and Heald [13]

$$\chi_j(k_e, \theta) = \sum_{N_j} 3\langle \cos^2\theta_j \rangle \chi_j^{\text{iso}}(k_e), \tag{7.17}$$

where θ_j is the angle between the electric field vector and the vector connecting the absorbing atom to the backscattering atom, and $\chi_j^{\text{iso}}(k_e)$ is the isotropic contribution of the jth shell. In the majority of cases where there is no long-range order and the sample is amorphous or polycrystalline, there is no macroscopic preferred orientation in the sample and the term $\langle \cos^2\theta_j \rangle$ averages to $1/3$, in which case, $\chi_j(k_e)$ reduces to

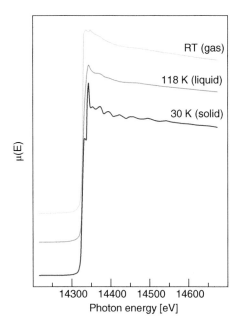

Figure 7.18 The change in $\mu(E)$ for the K-edge absorption spectra of krypton in its gas, liquid, and solid phases. The gas-phase spectrum can be considered to be very close to that of a truly isolated krypton atom. Adapted from [12] with permission of the American Physical Society.

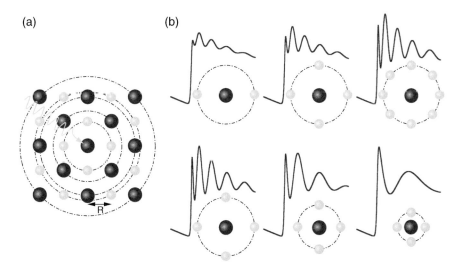

Figure 7.19 Rules of thumb in EXAFS. (a) A two-dimensional representation of the first five shells in a regular square array of two different atom types. Here, $N_j = 4, 4, 4, 8,$ and 4; $R_j = R, \sqrt{2}R, 2R, \sqrt{5}R,$ and $2\sqrt{2}R$, for $j = 1, 2, 3, 4,$ and 5, respectively. (b) The EXAFS oscillatory signal increases with the number of neighbours in a given shell; while the oscillation period dilates with decreasing shell radius.

$$\chi_j(k_e) = \sum_{N_j} \chi_j^{\text{iso}}(k_e). \tag{7.18}$$

We now present the standard expression used for $\chi^{\text{iso}}(k_e)$ and then justify its content. The EXAFS equation is

$$\chi^{\text{iso}}(k_e) = \sum_j N_j \frac{t_j(k_e) \sin[2k_e R_j + \delta_j(k_e)]}{R_j^2}$$

$$\times \exp(-2k_e^2 \sigma_j^2) \, \exp(-2R_j/\Lambda_e). \tag{7.19}$$

The summation is over all the j considered shells, each shell containing N_j atoms. Shells with radii much larger than a few angstroms do not normally need to be considered, due to the local nature of EXAFS (which is essentially determined by the small photoelectron mean-free path, the Debye–Waller factor, and the R_j^{-2} term, discussed below).

The factor $t_j(k_e)$ is the scattering amplitude for the scatterers in the jth shell and has dimensions of area. It depends on the type of backscattering atom and is so small for the first four or five elements in the periodic table that these atoms cannot normally be detected by EXAFS. t_j increases significantly and exhibits progressively complex oscillatory behaviour as a function of k_e with increasing atomic number Z [10, 14].

The denominator of R_j^2 takes into account the inverse-square decrease in the electron density of the spherically expanding electron wave from the absorbing atom.

The heart of the equation is the part that describes the oscillations [Figure 7.19(b)]. Let us assume to begin with that there is no phase shift associated with the backscattering of the outwards propagating photoelectron wave. In this case, the backscattered wave interferes with the outgoing wave at the origin constructively if $R = n\lambda_e/4$, where n is an even integer. Conversely, if n is odd, interference is destructive. But $\lambda_e = 2\pi/k_e$, hence the conditions for constructive and destructive interference can be re-expressed as

$$2k_e R_j = n\pi \tag{7.20}$$

for n even and odd, respectively. Until now, we have ignored any phase shifts in the backscattering process. Coulomb interactions between the electron and the cores of the emitting and scattering atoms will, however, induce a phase shift $\delta_j(k_e)$, which must be considered [10, 15]. Hence the oscillatory term is given by

$$\sin[2k_e R_j + \delta_j(k_e)], \tag{7.21}$$

as in Equation (7.19).

The first exponential term in Equation (7.19) accounts for the fact that the neighbouring atoms are not stationary, but vibrating with an amplitude σ_j parallel to $\mathbf{k}_e$. This is the well-known Debye–Waller factor, and is normally only marginally below unity for samples at room temperature. The final exponential term accounts for inelastic scattering of the electron wave (mainly by plasmons and phonons) on its round trip of $2R_j$, where Λ_e is the electron's mean-free path length, shown earlier as a function of energy in Figure 7.9.

Data analysis of EXAFS spectra proceeds as follows (Figure 7.20). The oscillatory function $\chi(k_e)$ is extracted from the raw data $\mu(E)$ by removal of the background signal fit to the pre-edge signal and the monotonically varying signal $\mu_0(E)$ above the absorption edge. The EXAFS function is then re-expressed as a function of k_e instead of $\mathscr{E}_e$ using Equation (7.13) and weighted to enhance the weak oscillations at higher k_e by multiplying $\chi(k_e)$ by k_e^2 or k_e^3. This function is then Fourier transformed, which converts the data from being expressed in terms of frequencies (proportional to k_e) to a plot which reveals characteristic lengths, associated with the interatomic distances R_j.

Generally, the phase shifts $\delta_j(k_e)$ are not incorporated into the Fourier transform, the result of which being that the peaks that emerge tend to be shifted towards lower R-values than the actual bond lengths.

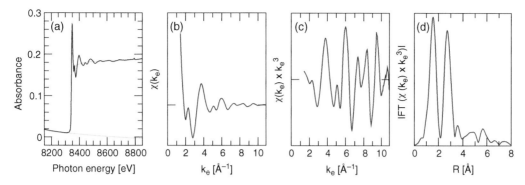

Figure 7.20 Data analysis of EXAFS spectra. A background using a low-order polynomial is fit to the pre-edge region of the absorbance spectrum $\mu(E)$ in (a), followed by a spline fit to the EXAFS region to obtain $\mu_0(E)$ (both shown here in yellow). The resulting $\chi(k_e)$ function in (b) is then weighted by multiplication by k_e^2 or k_e^3, shown in (c), which is then Fourier-transformed (d) to obtain the interatomic spacings. The example here is for the nickel K-edge for Ni-absorbates in the mineral montmorillonite.

The quality of the EXAFS data and the degree of possible interpretation depend very much on the number of oscillations, which in turn depend on the degree of local ordering and on the absorption strength of the partaking atoms [i.e. $t_j(k_e)$ and Λ_e in Equation (7.19)], hence comparison with well-characterized reference samples, preferably of similar chemical nature, can be invaluable in obtaining meaningful information.

When using a DCM or CCM in an EXAFS experiment (which is invariably the case), it is important to first collimate the incident beam on the monochromator, in order to minimize the spatial dispersion of the radiation incident on the sample, as shown in Figure 5.29, which would otherwise affect the spectral resolution. The lower limit is given by the Darwin width of the crystal reflection used in the monochromator [Equation (6.21)]. The relative bandwidth of $\Delta E/E = 1.31 \times 10^{-4}$ for Si(111) may be insufficiently narrow for some experiments, in which case, the Si(311) reflection is often chosen, for which $\Delta E/E = 2.78 \times 10^{-5}$.

An example of using EXAFS fingerprinting to identify chemical processes is shown in Figure 7.21. Some lichens have developed the ability to tolerate high concentrations of metals and thereby prosper in heavily contaminated areas unsuitable to other plant species. The biochemical mechanisms responsible for this resilience are largely unknown. The high sensitivity of EXAFS makes it ideally suited to study such pollutants in biological (noncrystalline) systems [16].

Two species of lichen were investigated – *Diploschistes muscorum*, harvested near a zinc and lead smelter, and *Xanthoria parietina*, sampled on cement poles near a tetraethyl and tetramethyl lead factory. Chemical tests on *D. muscorum* showed it to contain concentrations of Zn and Pb that are approximately three times higher than those recognized as being harmful to 'normal' plants; *D. muscorum* is thus 'classified as a 'hyperaccumulator'. *X. parietina* accumulates lead somewhat less voraciously, and is said to be 'tolerant'.

It is known that metals collect as inactive complexes of carboxylic groups on the inside walls of *Penicillin chrysogenum*, which has a very similar cell-wall structure to *D. muscorum* and *X. parietina*. It was therefore thought that the two lichens under investigation might use this same mechanism, and so a sample of *P. chrysogenum* was contaminated with Zn and Pb, and used as a reference check.

EXAFS experiments were carried out around the Zn K-edge (9659 eV) for *D. muscorum* and at the Pb L_{III}-edge (13 035 eV) for both lichen species. The results were compared to a selection of other recorded lead and zinc metallorganic compounds, as summarized in Figure 7.21(b). From these, two distinct adaptation strategies to these highly adverse environments were identified. In *D. muscorum*, Pb and Zn are accumulated through an as yet unidentified mechanism in which the synthesis of oxalate $[(C_2O_4)^{2-}]$ is enhanced, allowing the precipitation of the toxic metals as insoluble oxalate salts. *X. parietina*, on the other hand, complexed Pb to carboxylic groups of the fungal-wall cells in the same manner as does *P. chrysogenum*.

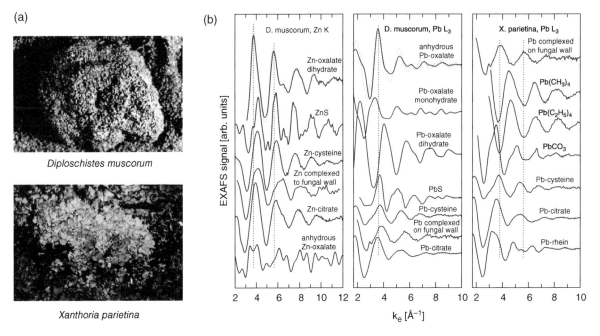

Figure 7.21 EXAFS of metallic pollutants in lichens. (a) The two lichens *Diploschistes muscorum* and *Xanthoria parietina*. Courtesy Leif Stridvall and Anita Stridvall. (b) Left panel: EXAFS spectra taken at the Zn K-edge of *D. muscorum* and other Zn-containing compounds. Centre panel: Comparison of the Pb L_{III}-edge EXAFS spectrum of *D. muscorum* and other Pb-containing compounds. Right panel: Comparison of the Pb L_{III}-edge EXAFS spectrum of *X. parietina* and other Pb-containing compounds. The spectra for the two lichens are shown in yellow, the reference spectra in blue. Adapted from [16] with permission of the American Chemical Society.

7.5.3 Time-resolved Absorption Spectroscopy

Conventional EXAFS spectra, in which energy scans of the order of 1000 eV with 1 eV resolution are recorded, may require acquisition times of several minutes to a few tens of minutes, depending on the sample type and recording mode. A good fraction of the time is taken up not with data acquisition, but with movements of the monochromator motors between data points in the energy scan, resulting in dead times required to allow transient vibrations and settling oscillations to decay to an acceptable level.

Chemical kinetics studied using x-ray spectroscopy, especially of catalytic systems, require an improvement in the temporal resolution of at least two orders of magnitude. With this in mind, two distinct approaches to improve the time resolution of XAS have been developed in the last two to three decades, namely energy-dispersive x-ray absorption spectroscopy (EDXAS) and quick-EXAFS (QEXAFS); both are briefly described below.

7.5.3.1 EDXAS

Energy-dispersive x-ray absorption spectroscopy (EDXAS) is based on the simple idea of spatially dispersing the component parts of a polychromatic beam, then bringing these to a common focus where a sample is placed. The transmitted signal is then recorded using a position-sensitive detector (Figure 7.22). The time-resolution is limited by the readout time of the detector, which for modern devices can be substantially smaller than a millisecond.

The optics used to disperse the polychromatic incident beam and focus the components onto a common micron-sized spot have developed considerably since the first device was reported in 1981 [17, 18]. In general, they consist of bending a single crystal (commonly made from silicon) into a cylindrical or elliptical shape. This 'polychromator crystal' disperses the incident beam by picking out the different wavelengths according to Bragg's law. Overtones can be suppressed by bouncing the dispersed beam off an x-ray mirror, as discussed in Section 5.4.2.

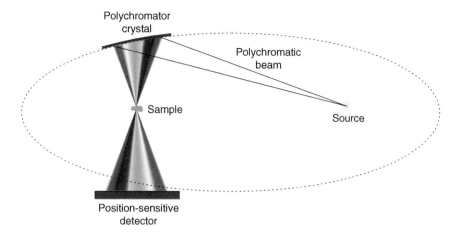

Figure 7.22   The principle of EDXAS. A polychromatic beam is dispersed into its monochromatic components via a polychromator crystal. The elliptical configuration in which the source and sample lie at the ellipse's foci and the polychromator crystal on its perimeter is chosen so that the dispersed x-rays are tightly focussed onto the sample and the transmitted signal is spread across a position-sensitive area- or strip detector.

The ideal focus is achieved using an elliptical curvature of the polychromator crystal and placing the sample and source at the foci of that ellipse (see Figure 7.22). In this manner, submicron spot sizes can be achieved. Because there are no moving parts, the beam stability can be excellent, allowing for spatial mapping. However, even very small distortions of the polychromator crystal due to thermal drifts, sample inhomogeneities, or other effects will cause the dispersed signal to drift spatially. This sets exceedingly stringent specifications in EDXAS for mechanical and thermal stability. Moreover, samples which produce strong small-angle scattering signal are intrinsically unsuitable for EDXAS. Lastly, detection using fluorescence, Auger signal, or total electron yield is excluded, due to the parallel nature of data acquisition, hence samples need to be thin enough to allow sufficient transmission.

These limitations mean that EDXAS is somewhat a niche technique. It is often used for studies at extreme conditions of pressure using diamond-anvil cells (which have small volumes and limited acceptance angles) and 'one shot' pump-probe experiments in which the sample is destroyed.

7.5.3.2   QEXAFS

Quick-EXAFS (QEXAFS) was first presented by Ronald Frahm in 1988 [19, 20]. Essentially, the idea is to minimize the number of motor movements, and make these as smooth as possible. In modern QEXAFS stations, a single motor continuously rotates a channel-cut monochromator crystal with lattice spacing d (often using piezo drives, which are notably free of vibrations), while modern encoder systems are able to monitor the crystal's Bragg angle with sufficient resolution and 'on-the-fly', i.e. without stopping the motor. Consider, for example, a Si(111) crystal scanning the energy across the iron K-edge from 7000 to 8000 eV. This requires the Bragg angle to change by marginally more than 2°. The encoder must be therefore able to rapidly read out the angular position with an accuracy measured in thousandths of a degree and gate the detector within the time needed to rotate the crystal through an angle corresponding to the required energy resolution. For shallow Bragg angles θ, the relationship between angular step size $\Delta\theta$ and energy step ΔE is almost perfectly linear and approximately given by

$$\Delta\theta \approx \frac{hc}{2d}\frac{\Delta E}{E^2} = \frac{355.188}{d[\text{Å}]\ E^2[\text{keV}^2]}\Delta E[\text{keV}]. \tag{7.22}$$

whereby, in the practical expression on the right, $\Delta\theta$ is in degrees.

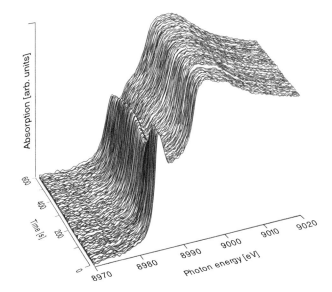

Figure 7.23 QEXAFS cascade plot at the Cu-K-edge of copper dispersed in the zeolite catalyst Cu-SSZ-13 at 225°C, dosed with NO and NH_3, the latter being cut off after 60 s. The full data set recorded spectra from 8900 to 9500 eV with 0.14 eV step sizes, every 0.5 s. Only a subset of this, with temporal separations of adjacent spectra of 5 s, is shown here, in order to highlight the critical changes in the XANES region. Note the rapid drop in the pre-edge peak at 8983 eV as soon as the NH_3 flow is stopped, and the more gradual appearance of a shoulder at approximately 8987 eV. Adapted from [25] with permission from Macmillan Publishers Ltd.

Modern QEXAFS setups [21, 22] are capable of measuring complete EXAFS spectra within a few tens of milliseconds, meaning that individual data points on any given spectrum must be recorded with up to MHz frequencies, depending on the signal-to-noise ratio.

Because in QEXAFS one limits the monochromator movements to a single motor rotation of a channel-cut crystal, the beam will drift vertically during the scan by an amount proportional to the channel height [see Equation (5.26)], unless a profiled channel is milled [23]. Hence the channel-cut crystals used in QEXAFS are normally made to be as compact as possible. The beam drift can be further reduced by using a vertically focussing mirror downstream of the monochromator. If these small movements are unacceptable for certain experiments, a truly fixed-exit monochromator using only one channel-cut crystal is possible, if the two reflecting surfaces are milled to a special shape [24].

An example is shown in Figure 7.23 of the dynamics and optimization of NO_x reduction to N_2 and water on a copper/zeolite catalyst using careful, temperature-dependent doses of ammonia, investigated at the SuperXAS beamline of the Swiss Light Source [25].

7.5.3.3   HEROS

High energy-resolution off-resonant spectroscopy (HEROS) is an emerging technique for XANES spectra in a single shot, making it eminently suitable for time-resolved measurements, particularly using XFELs. HEROS relies on the excitation of a virtual (off-resonant) state above the Fermi edge but below the vacuum level. In this second-order photon–atom process, a decay via an electronic transition from an inner- or outer-electronic state into the core hole occurs simultaneously (not subsequently) in a coherent manner, resulting in the emission of a photon. The energy of this photon depends on the incident photon energy, the difference in binding energies of the two states involved in the relaxation process, and the energy of the virtual state. Importantly, the electronic states above the Fermi level can thus be probed simultaneously by dispersing the emitted photon energies using a von Hamos spectrometer (see also Section 5.7.10). The recorded spectrum mirrors the XANES spectrum [26–28].

Because the process is second order, the signal intensity is weak. On the other hand, because in HEROS the absorption spectra can be recorded in a single shot, it may find significant interest at XFEL facilities.

7.6   Fluorescence Spectroscopies

7.6.1   Introduction

Fluorescence spectroscopies are in general photon-in–photon-out techniques and can therefore be used for both conducting and insulating samples and can often be applied under ambient or indeed *in vivo* environments, depending on the energy range of interest.

In the following, we will consider three types of experiment, namely 'conventional' x-ray fluorescence (XRF), resonant inelastic x-ray scattering (RIXS), and x-ray standing wave spectroscopy (XSW). RIXS can be considered as a special case of XRF in which the photon has an energy resonant with a transition from a core level to an unoccupied conduction or valence band state instead of having a higher energy which promotes the core electron to the vacuum.

XSW, on the other hand, measures the depth-dependent fluorescence[8] from atoms bathed in standing waves produced by interference between incoming and elastically scattered x-ray wavefronts and as such is a structural technique capable of yielding sub-angstrom information for two-dimensional systems like crystal surfaces, absorbates, or films.

7.6.2   X-ray Fluorescence

X-ray fluorescence (XRF) is a powerful method for determining the chemical composition of objects that are either too thick for transmission-absorption measurements, or are insulating and therefore preclude photoemission techniques. XRF is used widely in geochemistry, archaeology, forensic science, and investigations of rare or ancient artefacts. It exploits the 'fingerprint' characteristic radiation of the elements for chemical and elemental analysis.

The basic experimental setup for x-ray fluorescence measurements has been shown in Figure 7.3. Fluorescence can be used as a measure of absorption, as we have already mentioned, if the integrated fluorescence yield is recorded, in other words, if the fluorescence signal is not dispersively probed. X-ray fluorescence *spectroscopy* applies to methods which disperse the fluorescence spectrum. We have already discussed in Section 5.7.10 the relative merits of different dispersive photon detector types. In scanning x-ray fluorescence, the dispersed XRF signal is mapped out spatially. This is only practical with regards to the recording time if one uses an EDX detector, which, however, limits the spectral resolution and the detection-sensitivity limit.

XRF signal is strongly absorbed by air for emission below approximately 3500 eV, which corresponds to Ca or thereabouts ($Z = 20$, $K\alpha$ emission line at 3692 eV, see also Figure 5.14). For emission signal below this photon energy, the path to the detector should be evacuated to avoid absorption by the residual gas, such that the product of sample–detector distance and pressure is lower than approximately 1 Pa m.

Examples of scanning XRF spectromicroscopy are shown in Figure 7.24.

7.6.3   Resonant Inelastic X-ray Scattering

Resonant inelastic x-ray scattering (RIXS) is a rapidly developing 'photon-in–photon-out' technique in which photons resonant with electronic transitions are inelastically scattered from matter [32–34]. The energy loss is transferred to the sample in the form of low-energy excitations. RIXS has many analogies with Raman spectroscopy in the visible and infrared regime.

---

[8] XSW can be performed with detection via electron emission, although this is less common.

Figure 7.24   Examples of culturally significant discoveries using scanning XRF. Lost portraits by (a) Vincent van Gogh and (b) Edgar Degas hidden beneath more recent works were revealed by monitoring the fluorescence from elements found in the paints of the earlier pieces but (largely) absent in the later paintings. From [29] with permission of the American Chemical Society and [30] with permission from Macmillan Publishers Ltd. (c) The elemental contents of fossilized bones of *Archaeopteryx thermopolis* indicate that, contrary to previous assumptions, a significant fraction of the original elements in the bones remains sequestered in the fossil and thus mineralization does not entirely obliterate these. This provides invaluable potential information about previously unknown compounds critical to understanding biological structures, such as pigmentation. From [31] with permission of the National Academy of Sciences. In each case, the investigated region using scanning XRF is highlighted with a yellow rectangle on the original.

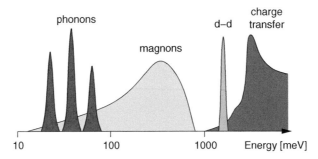

Figure 7.25   Some of the most important excitation processes probed by RIXS and their typical energy regimes. Adapted from [34] with permission from the American Physical Society.

The energy difference between the inelastically scattered photon and the incident photon can range between meV and a few eV (see Figure 7.25). This energy loss, plus the momentum transfer and change in polarization between the incoming and outgoing photon, can be directly associated with important processes related to electronic and superconducting properties such as phonon, magnon, or low-energy electron–hole pair excitations.

One can consider RIXS as being the marriage of conventional x-ray absorption and x-ray fluorescence in a single experiment. RIXS signal can be directly generated following resonant absorption of a photon by an atom

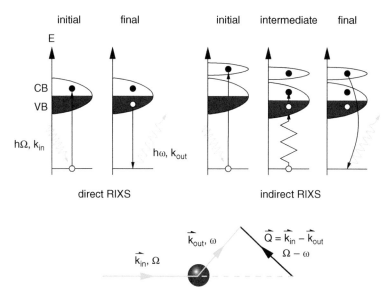

Figure 7.26   RIXS processes. In direct RIXS, a previously unoccupied conduction-band state is accessed by resonant absorption of a photon, and the core-level hole is subsequently filled by relaxation of an electron from a previously occupied valence-band level, resulting in the production of an electron–hole pair. As this is the only difference between the initial state before excitation and the final state, their momentum and energy differences describe the electron–hole excitation. In indirect RIXS, an electron is promoted to an empty state then relaxes back to its original core state. Importantly, the intermediate state before relaxation can impose a strong electrostatic screening potential, which shakes up electrons in the valence band, thereby exciting an electron–hole pair. In both cases, the energy loss and momentum transfer of the system is transferred to the generated electron–hole pair and is the reason why the valence and conduction bands are shown to be at lower energies after the initial excitation of the core level.

promoting an electron to an unoccupied conduction-band state followed by relaxation of a lower-lying state to the core, resulting in the formation of an electron–hole pair. Indirect RIXS signal occurs when absorption and relaxation are between the same states, but when the atom is in the excited intermediate state, a strong electrostatic screening potential around the core hole 'shakes' the atom, resulting in the promotion of an electron from the valence band to the conduction band (see Figure 7.26).

Importantly, while XAS and XRF alone yield no information regarding the momenta of the electrons, excluding insights into electronic band structure[9], RIXS does indeed provide this information, at least for wide-bandgap materials, and, as such, is an important adjunct to ARPES, described in Section 7.9.2, as a probe for the electronic band structure of a wide variety of materials. Another important advantage of RIXS over conventional XRF is that it 'bypasses' the intermediate state containing the core hole. As a result, RIXS spectra are not limited in resolution by the natural-linewidth broadening associated with the very short core-hole lifetime. Finally, because excitations are second-order and involve the *virtual* absorption of a photon, certain excitations (such as the *d-d* excitation between 3*d* states in transition-metal oxides) are allowed in RIXS, but are forbidden in direct optical processes.

An important distinction between ARPES and RIXS is that the former can only probe occupied states of a material (by measuring the energy and momentum of an ejected electron formally in an occupied state), while the

---

[9] The reason why both XAS and XRF cannot probe the electron momenta is that a core hole is involved – in the case of XAS it is in the final state and in XRF it is in the initial state. This results in a breakdown of the crystal's translational symmetry at exactly the location where the measurement is being made, which has the consequence that k, the momentum, is no longer a good quantum number. In RIXS, neither the initial nor the final states contain a core vacancy and momentum remains a good quantum number.

latter involves normally unoccupied states (as part of the electron–hole pair) and can therefore be used to probe the transitions between occupied and unoccupied bands[10].

RIXS is both bulk sensitive and site (i.e. chemically) selective. Because the technique is photon-in–photon-out, one can apply external electric or magnetic fields, experimental 'knobs' excluded from photoemission techniques such as ARPES.

Conservation of momentum and energy (see Figure 7.26) requires that the energy transfer and momentum transferred to the electron–hole pair are given by

$$E = \hbar\Omega - \hbar\omega \tag{7.23}$$

and

$$\mathbf{Q} = \mathbf{k}_{\text{in}} - \mathbf{k}_{\text{out}}, \tag{7.24}$$

respectively. Note that in the special case of $\omega = \Omega$, there is no net energy loss, and one speaks of 'Rayleigh scattering' or 'resonant elastic x-ray scattering'.

What sort of excitations does RIXS probe? The most modern apparatuses at DLSRs promise access to phonon excitations (collective vibrations of the crystal lattice) with energies as small as a few meV. One of the most important classes of excitations within the scientific framework of condensed-matter physics, particularly with regards to high-temperature superconductivity, is that of so-called 'magnons', which describe a collective response of magnetic spins in systems which may show some magnetic order (be it antiferromagnetic, ferromagnetic, or ferrimagnetic). Magnon excitations typically cost between a few meV and a few tenths of an eV, depending on the system. Another very important process in condensed matter, in particular in many metal oxides, is charge-transfer excitation. These require energies of the order of a few eV.

The RIXS intensity is a second-order process and is in general weak, due to the normally low quantum efficiency of x-ray emission, especially for low-Z elements and for L-edge and lower-energy x-ray absorption transitions (see Figure 2.28). One of the main drivers for the increased interest in RIXS in the last decade has been the availability of high-brilliance undulator sources at third- and fourth-generation synchrotron facilities, allowing one to record the dispersed, weak RIXS signal in reasonable times.

A unique feature of RIXS is the very large range of energies and momenta that can be transferred to the system under investigation in the inelastic scattering event. This is because photons instead of electrons (used in photoelectron studies) are used as probes. For a given wavelength λ, or more conveniently, wavenumber k (or k_e), the ratio of the photon energy to electron energy is

$$\begin{aligned}
\frac{h\nu}{\mathscr{E}_e} &= \frac{\hbar c k}{\hbar^2 k_e^2 / 2m_e} \\
&= \frac{2m_e c}{\hbar k} = 517.3/k, \tag{7.25}
\end{aligned}$$

whereby $k = k_e$ is given in reciprocal angstroms. So, for typical wavevectors of the order of 1 Å^{-1}, the range of energies and scattering vectors $\mathbf{Q}$ available to RIXS surpass other x-ray methods by approximately two orders of magnitude.

The second reason why RIXS has only relatively recently gained so much popularity is due to marked technical improvements in the obtainable resolution. If energy losses well under an eV are to be distinguished, a resolving power $E/\Delta E \geq 10\,000$ is required. This has only become a feasible prospect in the first decade of the twenty-first century, such as at the SAXES instrument at the ADRESS beamline of the Swiss Light Source, which, depending on the photon energy used, can provide an energy resolution of up to 33 000 [36], an improvement of well over an order of magnitude in less than a decade.

---

[10] The exception to this is the resonant 'normal' fluorescence part of RIXS, which probes the occupied density of states – the initial state in the photon-emitting relaxation process must be occupied.

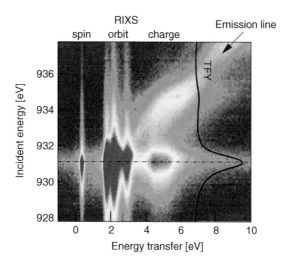

Figure 7.27 Two-dimensional plot showing RIXS features in Sr_2CuO_3 at the L_3-edge of copper. The superimposed black curve shows the total fluorescence yield (TFY), while the dot-dashed line marks the energy of the maximum of the Cu L_3 absorption resonance. The three signals which exhibit constant, nonzero energy transfer are RIXS resonant Raman peaks. The diagonal feature is a 'normal' fluorescence line associated with emitted photons with constant energy. Adapted from [35] with permission of Macmillan Publishers Ltd.

One distinguishes two types of energy-loss features in RIXS, according to whether they vary as a function of the incident photon energy $\hbar\Omega$ or not. Features which are independent of $\hbar\Omega$ appear as vertical lines in a two-dimensional plot of energy transfer versus incident energy (Figure 7.27). They represent pair excitations associated with two well-defined states and are therefore products of RIXS processes. Their analogy in the optical regime are Stokes phonons in Raman spectroscopy, from which they obtain their name of Raman peaks. In contrast, those features which exhibit a constant emission photon energy are called x-ray fluorescence or emission lines.

According to whether one is probing absorption edges in the soft or hard x-ray regime, the equipment used for RIXS differs considerably (see Figure 7.28). Soft x-rays are used to investigate the K-edges of low-Z atoms (up to silicon or thereabouts at 1840 eV, and the L- and M-edges of the transition metals and heavier elements. Tender and hard x-rays are needed for K-edge RIXS for the elements above sulfur. Soft RIXS instruments use diffraction gratings for both monochromatization and signal dispersion, with line spacings around a few thousand per millimetre, operating at grazing incidence [36, 37]. The dispersed signal propagates in ultra-high vacuum to an area detector such as a CCD. The larger the sample–detector distance, the higher the line density of the grating, and the smaller the pixel size, the higher is the resolution.

Hard RIXS equipment uses crystal monochromators and Johann or von Hamos geometry analysers in much the same manner as that described for WDX experiments in conventional x-ray fluorescence, described in Section 5.7.10 [38].

The procedure for RIXS experiments is first to record a conventional x-ray absorption spectrum at the absorption edge of interest. Next, x-ray emission spectra are recorded for different incident photon energies. In order to obtain the dispersion of the RIXS signal as a function of momentum transfer, the sample and detector orientations must be scanned in a concerted manner.

7.6.4 X-ray Standing Waves

The method of structural determination by x-ray standing waves (XSW) depends intimately on details of dynamical diffraction around Bragg peaks [39–43].

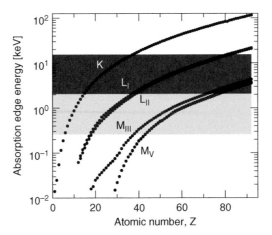

Figure 7.28   Absorption-edge energies of the elements between hydrogen and uranium. Soft x-ray RIXS machines cover approximately the energy regime highlighted in yellow, hard x-ray RIXS equipment the range in blue.

Interference between the incident beam and the elastically scattered beam at the Bragg condition produces a standing wave with a periodicity λ_s equal to the interplanar distance d_{hkl} associated with that Bragg peak, that is

$$\lambda_s = d_{hkl} = \lambda/(2\sin\theta). \tag{7.26}$$

Depending on the length scale under investigation, different Bragg maxima can be selected. The smallest length scale for λ_s is given by the energy (wavelength) of the x-ray beam and for $\theta = \pi/2$, in which case $\lambda_{s,\min} = \lambda/2$. In diffraction experiments, such a normal geometry cannot be accessed, as this requires that the x-ray detector lies in the path of the incident beam. This poses no problem for XSW, however, as here we are interested in the absorption changes of the standing wave by the atoms in the crystal and not in the properties of the Bragg peak. This begs the question: how do the absorption properties change across a Bragg peak?

According to the dynamical theory of diffraction, the phase of the standing wave produced by interference between the incident and elastically scattered waves changes relative to the scattering centres (the atoms) as one scans across a Bragg peak. Consider Figure 7.29. At the leading edge (lower angles), the phase v is π radians and the atoms sit in the nodal regions of the standing wave. As one increases the incident angle, the phase of the standing wave shifts vertically until at the opposite edge, the maxima lie exactly above the atoms.

As well as scattering elastically, the atoms absorb. Photoabsorption is stronger when the standing-wave maxima lie on top of the atoms and the phase is zero and is the reason why the reflectivity curve is marginally less intense on the high-angle side compared to the low-angle side. The degree of absorption can be measured using fluorescence detectors (such as the lithium-doped semiconductor EDX detectors described in Section 5.7.10) or Auger-electron detectors. The strength of this signal depends also on the choice of photon energy and the absorption edges of the atoms within the crystal.

In the above, we made the simplification that the distance between atomic planes is $\lambda_s = d_{hkl}$ and that there are no other atoms lying between them. This is of course not normally the case – for example, the (002) spacing of GaAs contains two planes of atoms (see Figure 6.22). So, according to the positions of different atoms within a unit cell, their absorption response (and fluorescence yield) will vary in different manners as one scans the Bragg peak.

Note that, according to the photon energy and the material under investigation, the width of the Bragg peak (the Darwin width) can be very small, of the order of 50 μrad (3×10^{-3} degrees). This sets strict constraints on the accuracy and resolution of the diffractometer, which should be able to move reliably in steps of approximately 10^{-4} degrees. The beam divergence in the scattering plane should also be considerably smaller than the Darwin width, which usually means using beam defining slits and a loss of beam intensity.

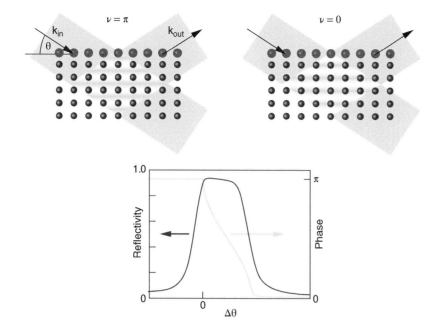

Figure 7.29   The principle of XSW. As one scans across a Bragg peak, the phase v between the standing-wave maxima and the scattering atoms flips from being π radians to zero. The sample absorbs differently according to the phase, and can be used to measure the positions of the atoms within the unit cell. Chemical sensitivity is provided by the fluorescence signal, hence the type and positions of absorbates or other surface species (shown here in red) can be determined.

XSW is also an important technique for investigating the configuration of surface absorbates, particularly if they contain atoms not found elsewhere in the bulk crystal [44].

7.7   Scanning Transmission X-ray Microscopy, STXM

7.7.1   Introduction

Since the introduction of high-brilliance undulator radiation in third-generation synchrotron facilities, x-ray focal spot sizes of the order of a micron or even smaller have become routinely achievable. With this, scanning microspectroscopy methods were developed, in which the chemistry of heterogeneous samples could be mapped out with high spatial resolution. One of the most important of these to emerge was scanning transmission x-ray microscopy (STXM), particularly in the field of polymer chemistry and physics, where phenomena such as segregation and heterogeneous morphology must be understood at a submicron scale. STXM can be thought of as scanning XANES operated in the transmission mode.

7.7.2   The Water Window

There are many problems in biology, organic chemistry, and polymer physics that require detailed chemical analysis at a submicron scale, but also mapping over macroscopic areas. Although traditional methods such as infrared spectroscopy and nuclear magnetic resonance can differentiate chemical species (by observing subtle differences in bond strengths caused by the local chemical environment), their spatial resolution is limited to the millimetre-scale.

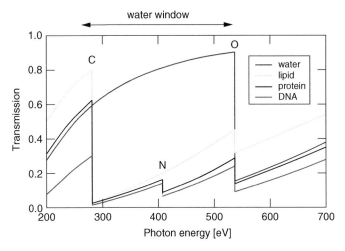

Figure 7.30 Transmission curves of 1 μm-thick biological materials in the water window between approximately 284 and 540 eV.

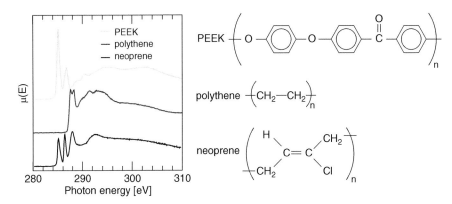

Figure 7.31 Examples of XANES spectra for three common polymers: the aromatic polymer PEEK, the saturated polymer polythene, and the unsaturated polymer neoprene. The spectra were downloaded from the database described in [45].

STXM is a technique for spatially mapping with submicron resolution the chemical contents of organic samples. This is particularly important for experiments performed within the energy range defined by the so-called 'water window', which extends from about 260 eV, below the K-edge of carbon at 284 eV, to 543 eV, the K-edge of oxygen (Figure 7.30). In this energy range, biological and *in vivo* specimens absorb approximately an order of magnitude more strongly than water and can therefore reside in an aqueous environment and need not be stained with heavy metals or sliced (both of which are common procedures in the preparation of organic specimens in transmission electron microscopy). Note also that the L-edges of several biologically relevant elements, in particular potassium, sulfur, and calcium, lie within the water window, while phosphorus's L_1-edge is sometimes also accessible at 189 eV. STXM is frequently used to characterize polymer films [45], because of near-edge resonances that correspond to antibonding molecular orbitals, allowing identification of the structure of organic molecules via their very rich and informative C 1*s* XANES spectra (see Figure 7.31).

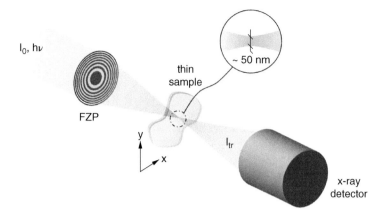

Figure 7.32   A schematic diagram of a scanning transmission x-ray microscope (STXM). The x-rays are focussed to as little as a few tens of nanometres using a Fresnel zone plate. To generate an image, the energy-dispersed transmitted x-ray signal is recorded as the thin specimen is rastered in the *x*- and *y*-directions.

7.7.3   Modes in STXM

A typical STXM setup is shown in Figure 7.32. STXM is performed in one of three modes. In the mapping mode, the sample is rastered across a focussed monochromatic x-ray beam. This yields a transmission map of the sample which, depending on the photon energy relative to absorption edges, provides the distribution of a given element or, if the energy resolution is sufficient, even of a chemical group. The rastering resolution and reproducibility should be finer than the size of the x-ray focus, which can set stringent specifications on the translation table and suppression of unwanted sources of vibration (see also Section 5.5).

In the spectrum mode, the sample remains fixed, the photon energy is scanned, and the transmitted signal I_{tr} is recorded, to yield the absorption spectrum, also known as the optical density. In the so-called E-stack mode, a sequence of maps at different energies are recorded. These images can then be aligned with one another using reference 'anchor points', either small grains of material that absorb strongly over the entire energy range of interest, or holes where no material exists. Spectrum-mode energy scans can thereby be retrieved by 'looking down' at a particular coordinate through the aligned stack (Figure 7.33).

Finally, STXM can be operated in different detection modes. In the normal mode of operation, the transmitted signal is measured using a photon detector and as such STXM is usually considered to be a bulk technique, though limited to samples thin enough to provide a detectable transmission signal. However, one can instead measure the photoelectron yield using detectors such as channeltrons. Typical escape depths of photoelectrons lie in the range of one or two nanometres, and hence STXM operating in this mode becomes surface sensitive [46]. Moreover, one can also use a fluorescence detector in STXM to simultaneously perform XRF mapping. Experiments can also exploit dichroism–linear dichroism to quantitatively map the orientation of organic molecules and XMCD to measure the magnetization of thin metal films (see below).

7.8   Photoemission Electron Microscopy, PEEM

7.8.1   Basics of PEEM

Photoemission electron microscopy (PEEM) is a technique that images the spatial distribution of electrons emitted from a sample by x-ray absorption in the XANES region [47]. Although the technique uses photoelectrons for its signal, it is not an electron-spectroscopy technique, as the energies of the electrons are not distinguished. Instead,

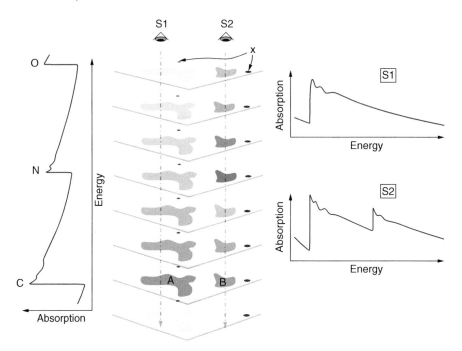

Figure 7.33   Schematic diagram of STXM in the E-stack mode. A specimen consists of two features, A and B. A is carbon-rich while B also contains high concentrations of nitrogen-containing compounds. Spatially resolved absorption maps of the sample are recorded over a range of energies. Two small features, labelled 'x', absorb strongly over the entire energy range of interest (they may, for example, be high-Z inclusions such as grains of silicate rock) and can be used as reference points to accurately align the images on top of one another and thereby circumvent problems associated with sample drift. Spectra can therefore be extracted from this 'E-stack' by viewing any particular coordinate down through the stack (S1 and S2 in the figure).

the absorption coefficient is indirectly measured via the yield of *secondary* electrons – not the photoelectrons directly emitted after photoabsorption but the electrons released by the system after multiple scattering events in a cascade process originating with the directly produced photoelectron from the absorbing atom (see Figure 7.10). This initial photoelectron may derive from an atom relatively deep in the material, as the x-rays can penetrate to depths of several tens of nanometres. The initial photoelectron can only travel a few angstroms, however, before interacting inelastically with its surroundings via one of several scattering phenomena, which produces a cascade of 'daughter' electrons with ever decreasing energies. As the electrons' kinetic energy approaches only a few eV, their mean free path increases rapidly (see Figure 7.9) and they can escape the surface with higher probability – a typical graph of yield versus energy shows a maximum just above the work function, which then drops off with energy [Figure 7.10(b)]. The majority of emitted secondary electrons therefore have energies of the order of an electronvolt. Although their detection does not yield absolute results for the absorption coefficient, the spectral dependence is adequately reflected (see Section 7.2).

A PEEM experiment essentially consists of tuning synchrotron radiation illuminating an entire sample in vacuum and imaging the spatial variation of the subsequent secondary-electron yield using an electrostatic and/or electromagnet lens system very similar to that used in electron microscopes (Figure 7.34). As such, PEEM is a form of 'spectromicroscopy' – full-field images are recorded at different photon energies, in contrast to STXM, a XANES microspectroscopy technique in which a focussed x-ray beam is rastered across the sample. The secondary electrons are imaged using a 2D detector. The magnified image of the surface can therefore be observed directly and in real time.

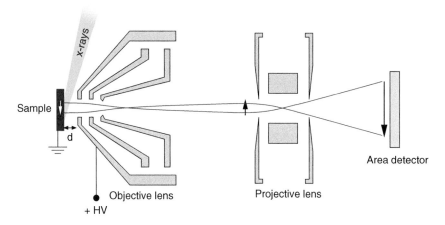

Figure 7.34   Secondary electrons are imaged in PEEM using electron optics very similar to those used in conventional electron microscopy.

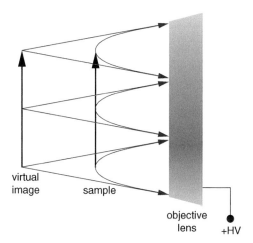

Figure 7.35   Secondary electrons emitted from the sample with nonzero lateral kinetic energy execute a parabolic trajectory as they are accelerated to the objective lens. Extrapolation of the tangent to these paths where the electrons enter the objective for a virtual image with the same size. The accelerating field therefore acts as a lens with unitary magnification.

Since electrons are used for imaging in PEEM, the resolution is not limited by the wavelength of the x-ray photon beam. Instead, a high electrostatic field between the sample and the objective lens accelerates the released electrons to energies of typically $eV_L = 10$ to 20 keV across a distance d of the order of 2 mm. This accelerating field acts as a lens – the trajectories of the electrons as they leave the surface form a set of parabolas. The tangents to the parabolas at the point where the electrons enter the objective lens extrapolate back to form a virtual image with unity lateral magnification (see Figure 7.35).

The accelerating field is critical in determining the ultimate resolution. Because the electrons have a range of energies and emission angles, the virtual image will become blurred to a greater or lesser extent. The range of electron energies $\Delta\mathcal{E}_e$ can be reduced by introducing an aperture between the sample and objective. This limits the largest parabola width that can enter the electron microscope and therefore acts as a low-pass energy filter, though obviously at the expense of signal intensity.

To a first approximation, the resolution Δx is given by

$$\Delta x = \frac{d\,\Delta\mathcal{E}_e}{eV_L}. \tag{7.27}$$

The best achievable lateral resolution is therefore about 20 nm.

PEEM thus combines the techniques of x-ray absorption spectroscopy and electron microscopy. Chemical contrast is obtained by tuning the synchrotron radiation to a strong absorption edge and recording an image. Image 'stacks' can be recorded as a function of photon energy.

7.8.2 PEEM and Magnetic Dichroism

7.8.2.1 Introduction

Dichroism, meaning 'two-coloured', is the phenomenon of a material having an absorption spectrum which changes according to the polarization of the electromagnetic radiation used. Magnetic dichroism describes the dependence of the absorption of a magnetic material on the polarization *and* the relative orientation of an applied external magnetic field. The discovery of magnetic dichroism in core-level x-ray spectroscopy has opened the route to magnetic-domain imaging [48].

Magnetic dichroism using tuneable and polarized soft or hard x-rays at synchrotron sources offers the unique feature of chemical specificity. Magnetic features down to the 100 nm scale can be imaged by exploiting the dependence of the absorption on the polarization of x-rays in ferromagnetic and antiferromagnetic materials. The methods are called x-ray magnetic circular dichroism (XMCD) and x-ray magnetic linear dichroism (XMLD), respectively. Imaging down to the nanometre range is very important for magnetic structures, as at this scale, the influence of domain boundaries between one magnetic direction and another becomes significant and new phenomena can occur which would be negligible in larger structures. An understanding of the energetics of nanomagnetism is therefore essential in the drive to further miniaturize magnetic-memory-storage devices.

7.8.2.2 Spin and Magnetic Materials

Here we first describe the influence of spin on the energetics of core levels and valence-band electrons in magnetic materials, then describe how dichroism in such systems can arise.

The intrinsic magnetic moment of an electron is produced by its spin and can be thought of as being a tiny bar magnet[11]. Note that the direction of the magnetic moment is antiparallel to that of the spin. Electrons always possess spin angular momentum of quantum number $s = 1/2$ and magnitude $\sqrt{3/4}\hbar$. The orientation of the spin relative to a magnetic field (either $+1/2$, 'up', or $-1/2$, 'down') determines its only two possible orientations.

Bound electrons in atoms which have nonspherically symmetric orbitals (i.e. not *s*-type) also have an orbital angular momentum, *l*, and therefore generate a second magnetic field that can be thought of classically as being produced by the 'current' of the electron as it orbits the atom. These magnetic moments of the spin and the orbital angular momenta couple with each other just as two magnets are influenced by each others' fields, in so-called 'spin-orbit coupling'. Depending on whether the spin is oriented up or down relative to the axis of the orbital magnetic moment, the energy of the electron is higher or lower (see Figure 7.36).

A magnetic moment in ferromagnetic materials is produced by there being an imbalance between spin-up and spin-down electrons. This normally arises because the spin-up and spin-down sub-bands have different

---

[11] This analogy should be taken with a pinch of salt – spin is one of only very few quantum-mechanical phenomena that has no classical analogy.

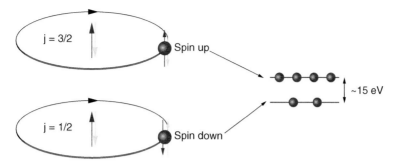

Figure 7.36 An example of spin-orbit coupling. The $2p$ orbital has an orbital angular momentum $l = 1$ and an associated magnetic moment which couples to that of the spin of the electron, causing the level to split into two distinct eigenstates $j = 1 + 1/2 = 3/2$ and $j = 1 - 1/2 = 1/2$. When these angular momenta are parallel, the magnetic moments are aligned unfavourably compared to when they lie antiparallel to each other. Hence $j = 3/2$ has a higher energy than $j = 1/2$. The maximum allowed occupation of each state is determined by the number of quantum-mechanically allowed projections m_j of j relative to the magnetic axis and is equal to $2j + 1$. The magnetic moments, shown in yellow, are antiparallel to their associated angular momenta, shown in blue.

energies, caused by the so-called 'exchange interaction'. We can qualitatively explain exchange interaction by considering Pauli's exclusion principle – each electron within an atom must have its own unique set of quantum numbers. The part of the quantum-mechanical wavefunction that describes the spin is special insofar that it does not affect the spatial distribution of the electron. Hence, those electrons that only differ in their spin quantum numbers overlap significantly and are therefore repelled from each other by Coulomb forces. This is the cause of the splitting both of the core levels, but also of the valence band levels, which, despite being extended wavefunctions, locally maintain much of their properties as isolated atoms. The splitting of the $3d$-band caused by the exchange interaction is of the order of an electronvolt, while the bandwidth is typically a few eV.

7.8.2.3   XMCD

Consider Figure 7.37(a). A ferromagnetic material such as cobalt with $3d$ valence electrons forms domains with different magnetic directions. In region 1, there are more occupied $3d$ states with their spin down than with spin up, hence the magnetization of this domain is upwards (i.e. the opposite direction to the imbalance in spin states because, remember, the spin direction is opposite to the magnetization direction). In contrast, the number and density of *unoccupied* spin-up states above the Fermi level E_F in the $3d$ band is greater than spin-down states for region 1 [Figure 7.37(b)] and the transition probability for absorption of a photon will be greater. In addition, however, one must take a further selection rule into consideration, which states that, for dipole transitions, spin flips are forbidden. This rule is not entirely strict for condensed matter, but is sufficiently observed to induce dichroism. To obtain this so-called x-ray magnetic circular dichroism (XMCD), one uses circularly polarized light. So in the case of region 1, an incident RCP photon has its orbital angular momentum pointing forwards (in the direction of propagation, see Section 3.7.4) and will preferentially excite the $2p_{3/2}$ (L$_3$) levels over the $2p_{1/2}$ (L$_2$) levels. The opposite case arises for LCP photons, whereby the $2p_{1/2}$ states are favoured [Figure 7.37(c)]. Lastly, if we now focus on region 2, the whole story reverses, as the spin-down $3d$-subband is now shifted to higher energies than the spin-up subband.

By recording PEEM images of ferromagnetic domains with RCP and LCP x-rays and dividing the intensities of one image, pixel-for-pixel, by those of the other, the domain structure is revealed with maximum contrast. An example is given in Figure 7.38 for magnetic structures fabricated from thin films of Ni-Fe [49].

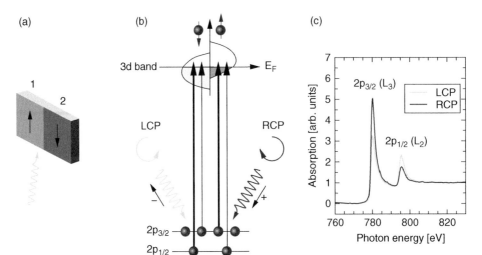

Figure 7.37   The principle of x-ray magnetic circular dichroism. (a) An experiment where circularly polarized x-rays tuned to the L-edge of a magnetic material are shone at glancing incidence on a sample with different magnetic domain directions. (b) Absorption of RCP photons mainly excites spin-up electrons, while LCP light mainly excites spin-down electrons, resulting in a dichroic absorption spectrum, shown in (c) for the L-edge of cobalt. Courtesy Frithjof Nolting, Paul Scherrer Institute. Adapted from [48] with permission of Elsevier for (a) and (b).

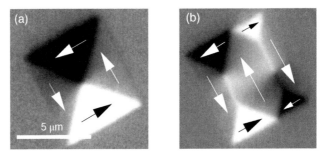

Figure 7.38   (a) The orientation of microscopic domains in Ni-Fe thin films minimize the stray field energy. (b) Sometimes, such domains can break up into higher-energy metastable configurations with an activation barrier to lower-energy states. Recorded using the PEEM apparatus at the Surfaces and Interfaces Microscopy Beamline at the Swiss Light Source. Courtesy Christoph Quittmann, Paul Scherrer Institute.

7.8.2.4   XMLD

X-ray magnetic linear dichroism (XMLD) is a technique used to investigate the domain structure of *anti*ferromagnetic (AFM) materials [50]. AFM materials are characterized by having antiparallel orientations of neighbouring magnetic moments and thus no net spin but a defined magnetic axis below a critical temperature (the Néel temperature). A simple schematic example is shown in Figure 7.39 for the antiferromagnet nickel oxide, NiO.

XMLD signal arises from the fact that, due to spin-orbit coupling, the spatial distribution of the electron density is marginally distorted, providing the necessary dichroism depending on whether the linear polarization vector is parallel to or perpendicular to the magnetic axis. The effect is usually small and detecting XMLD is more

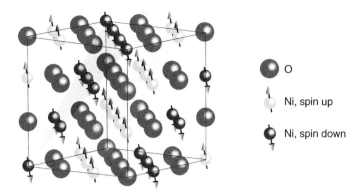

Figure 7.39 NiO is an antiferromagnetic cubic crystal with a face-centred cubic rocksalt structure. Below the Néel temperature, the spins of the nickel ions in the (111) crystallographic planes (shown here in blue and yellow) alternate between being all spin-up and all spin-down. There is therefore no net magnetic moment, but there does exist a magnetic axis, given by the spin orientations.

challenging than XMCD. It has a $\cos^2\theta$ dependence, where θ is the angle between the polarization vector of the x-ray light and the magnetic axis of the domain. The maximum XMLD effect therefore occurs when the signal for the polarization parallel to the magnetic axis is compared to that perpendicular to the axis.

7.9 Photoemission Spectroscopy

7.9.1 Introduction

The goal of photoemission spectroscopy is to determine the kinetic energy and, in the case of angle-resolved studies, also the three orthogonal momentum components of the wavevector of electrons emitted by the photoelectric effect. Armed with this information, one may draw important conclusions about the material's chemical nature or its electronic band structure.

In Section 7.2.2 it was argued that the probability was very high for the ejection of a core electron into an unbound state if a photon with an energy higher than the binding energy of that electron was absorbed. The kinetic energies of these directly produced photoelectrons can be measured, which yields information on the elemental and chemical composition of materials. Photoemission spectroscopy (PES) exploits Einstein's explanation for the photoelectric effect to draw quantitative conclusions about the electronic and/or chemical nature of materials over a broad range of excitation x-ray energies. The reader is recommended a review of photoemission spectroscopies in all their varied guises as an excellent general introduction [51].

The valence electrons of atoms, molecules, and condensed matter determine their electronic, optical, mechanical, and magnetic properties. Techniques that probe the spatial distribution, energy, and momentum of valence electrons are, therefore, indispensable tools in understanding the properties of matter. These include on the one hand optical techniques in the visible or near-visible regime such as ellipsometry, reflectivity, and transmission spectroscopy, in which the response of electrons which remain bound to the ionic cores is investigated. In the ultraviolet and soft x-ray range, on the other hand, the photon energy is sufficient to eject valence electrons from bound states and thus allow them to be detected and analysed for their energy and momentum. This information can then be interpolated back to the electrons' original properties within the bound system.

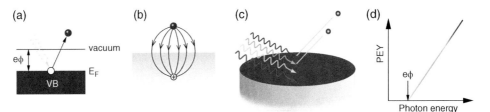

Figure 7.40   The photoelectric effect and the work function. (a) An electron in the highest-energy occupied state of the valence band (VB) still requires an energy $e\phi$ in order to reach the vacuum level and free itself of the system. (b) The reason for this is that, as it leaves the surface of a material, the electron is electrostatically attracted to its positively charged mirror image produced in the near-surface region caused by screening effects due to the electron's removal. (c) Only photons with energies above $e\phi$ are thus capable of ejecting an electron. Below $e\phi$, no photoelectrons are produced, no matter what the beam intensity is (or rate of impinging photons), a result of the corpuscular (quantum) nature of light. (d) The photoelectric yield (PEY) therefore increases with photon energy above the threshold energy $e\phi$.

7.9.1.1   The Work Function

The minimum photon energy required to produce direct photoelectrons is given by the work function $e\phi$ of a given material, as famously described by Einstein in his Nobel-Prize winning explanation of the photoelectric effect in 1905. This lies around 5 eV, hence photoelectron spectroscopy (PES) requires photons in the ultraviolet and x-ray regions. The work function arises because, as an electron originally residing in a state at the top of the valence band (the Fermi level) is removed from the surface of a material, it is attracted by its positively charged mirror image produced by a rearrangement of the surface charge resulting from its departure (see Figure 7.40).

Only at distances of more than about 100 Å does this electrostatic-image force become negligible. This image potential is the reason why electricity does not 'leak' out from wires and electrons do not shoot out of kinks in current-carrying wires; for typical work functions of 5 eV, one requires electric field strengths of the order of 100 MeV m^{-1} to electrostatically remove electrons from metal surfaces. On the other hand, the phenomenon of thermionic emission results from a certain fraction of valence electrons in a resistively heated metal obtaining sufficient thermal energy that they can overcome the work-function barrier and 'boil off'.

We now include the work function $e\phi$ in describing the kinetic energy of an emitted photoelectron, by modifying our definition of the binding energy E_B of electrons in isolated atoms that was first given in Equation (7.8). From our explanation of the origin of the work function, it should be clear that it is a manifestation of condensed matter. Hence, now, E_B is the energy needed to promote the electron to the continuum (or 'vacuum level') but is insufficient to pull it away from its positively charged mirror image, for which we still need the work function $e\phi$, that is

$$\mathcal{E}_e = h\nu - E_B - e\phi. \tag{7.28}$$

It is noted that the binding energies of electrons in condensed matter are conventionally measured with respect to the Fermi level, rather than the vacuum level.

7.9.1.2   Energy Regimes of Photoelectron Spectroscopy

Here, we discuss only photoelectron spectroscopy of condensed matter. One can divide PES into three broad categories according to the energy range of the photons that are used (see Figure 7.41).

Techniques such as ultraviolet photoelectron spectroscopy (UPS) and the more sophisticated angular-resolved photoelectron spectroscopy (ARPES) probe the electronic structure of matter by exciting valence states with low-energy photons from a few eV to a few hundred eV. More bulk-like electronic properties can be gleaned by using higher photon energies of the order of a keV in soft-x-ray ARPES (SX-ARPES), but at the cost of signal intensity (see Section 7.9.3). Elemental and chemical analysis is primarily carried out in x-ray photoelectron

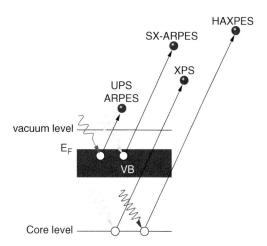

Figure 7.41   Low-energy ultraviolet and soft x-rays are used in UPS and ARPES to investigate the electronic structure of matter by probing the valence electrons. The bulk electronic properties of condensed matter are probed in soft-x-ray ARPES (SX-ARPES) with x-rays of the order of a keV or more, for which the electron kinetic energies are associated with larger inelastic mean free paths Λ_e (Figure 7.9). In XPS, elemental and chemical information is gleaned from the energy and signal intensity of photoelectrons originating from core levels. HAXPES also provides similar information, but from deeper within the sample using hard x-rays as high as several keV.

spectroscopy (XPS, also called electron spectroscopy for chemical analysis, ESCA) by probing photoelectrons originating from core levels, and typically spans the energy range of a few hundred to one or two thousand electronvolts. Finally, hard x-ray photoelectron spectroscopy (HAXPES) is a variant of XPS in which the incident photon energy far exceeds the electron binding energy, hence the photoelectron energy is also high. This has the advantage that the escape depth of high-energy electrons is large (see Figure 7.9) and so the properties of the bulk material or of buried interfaces far below the sample surface can be investigated.

We now discuss each of these techniques in turn.

7.9.2   Ultraviolet Photoemission Spectroscopy

Ultraviolet photoemission spectroscopy (UPS) uses photons in the vacuum ultraviolet and soft x-ray regimes from approximately 10 to 200 eV and utilizes energy-dispersive analysis of the emitted photoelectrons to study the electronic states of the near-surface region of a sample, down to a few nanometres.

In laboratory-based UPS equipments, the choice of radiation is limited to fixed-energy line sources such as He I (21.1 eV) and He II radiation (40.8 eV). Importantly, UPS performed at synchrotron sources has the added experimental degrees of freedom that the photon energy can be tuned and the polarization changed. Hence, by using a synchrotron, the relative change in photoemission cross-section for various electron states can be used to determine the partial density of states (PDOS).

In addition to analysing the energies of the photoelectrons, it is also possible to probe their angular distribution. This enables the energy versus wavevector of the electronic states to be measured in so-called 'energy dispersion curves' (EDCs, described below). When the angular distribution is measured, UPS is termed angle-resolved photoelectron spectroscopy (ARPES), otherwise the UPS measurement is angle-integrated.

We now proceed first by introducing concepts related to valence bands in a qualitative manner, after which we discuss how these features can be probed in a UPS setup. The purpose here is not to provide a rigorous description of band theory, but to convey the relevant physical phenomena that can be probed using UPS. For rigorous derivations, there are many classic texts (see for example [52, 53]), or specialist reviews [54, 55].

7.9.2.1   Valence-band Structure

As we have already discussed at the beginning of this chapter, the valence electrons in a solid occupy very closely packed levels within bands. We begin with the simplest case of a good metal. To a first approximation, one can assume that electrons in metals are essentially free to move and, because of efficient screening by the other surrounding electrons (the 'electron gas'), are not influenced by the periodic electrostatic field of the ionic cores. In this case, the allowed electron states are determined by the permitted de Broglie wavelengths of the electrons, which in turn are determined only by the boundary condition that the electric field must be zero at the surface of the metal.

The energy of an electron in these metallic bands depends on its momentum $\hbar k_e = h/\lambda_e$. More precisely,

$$E_{k_e} = \frac{p^2}{2m_{eff}} = \frac{\hbar^2 k_e^2}{2m_{eff}}, \tag{7.29}$$

where $k_e = 2\pi/\lambda_e$ is the wavevector of the electron in the crystalline solid and m_{eff} is its effective mass. This last quantity describes the apparent mass of an electron if one were to assume it were completely free – a large effective mass indicates a high degree of localization. It is different from the rest mass of a free electron because, in a crystal, an electron interacts with the surrounding ions and other electrons. Hence our assumption that an electron in a metal is free is only approximately true – it is perhaps better to focus on the statement that electrons in a metal do not feel the *periodic* potential of the crystal field. This potential is, as we will shortly see, responsible for the formation of bandgaps – energy regions which contain no electron states. In spite of these, the bands may still be thought of as being *locally* parabolic (that is, their energy is proportional to the square of their momentum). Therefore, from Equation (7.29), the parabola describing energy versus wavevector is shallow and wide for an electron with a large electron mass, while a low effective mass results in a steep and narrow parabola.

The energy of the electron within a solid thus increases parabolically with its wavevector. Note that E_{k_e} is the energy of the electron *within* the material and should not be confused with $\mathscr{E}_e$, the kinetic energy of an escaped photoelectron, although we will see that by measuring the latter, the former can be deduced.

How closely packed *are* adjacent valence-band levels within a macroscopic solid? Let us first consider the simplest case of the allowed wavelengths along a linear object of length L. Each level has its own wavelength and is associated with two states with opposite spins. The electron wavelength can only assume values $2L/n$, where n is a positive integer, because the boundary conditions dictate that the amplitude must be zero at the ends (or surfaces, in 3D). Hence

$$k_e = n\pi/L. \tag{7.30}$$

The difference Δk_e between adjacent states n and $n + 1$ is π/L and so the density of states is therefore linear with respect to k_e. From Equation (7.29), we obtain

$$E_{k_e}^{(n)} = \frac{\hbar^2}{2m_{eff}} \left(\frac{n\pi}{L} \right)^2. \tag{7.31}$$

and the energy separation ΔE_{k_e} between adjacent states is for $n \gg 1$ accurately approximated by

$$\Delta E_{k_e} = \frac{n}{m_{eff}} \left(\frac{\hbar\pi}{L} \right)^2. \tag{7.32}$$

Depending on n and considering macroscopically sized objects, the highest occupied level, ΔE_{k_e}, is many orders of magnitude smaller than an electronvolt, and is much smaller still for three-dimensional objects.

What values do the wavevector and energy take on at the highest occupied level at absolute zero, the Fermi level? Let us assume that in the linear model introduced above that each unit 'cell' of length a contributes a single valence electron to the band structure. The total number of electrons is therefore $N = L/a$. At the Fermi energy,

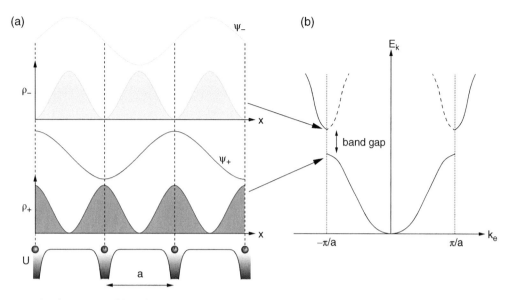

Figure 7.42   The formation of band gaps. (a) Two standing waves ψ_+ and ψ_- with wavevectors $n\pi/a$ are possible solutions of electrons moving back and forth within a material of periodicity a. The probability ρ of finding an electron at any one position is proportional to the square of the wavefunction. ρ_+ has maxima that lie exactly above the positively charged ionic cores and is therefore more strongly bound than ρ_-, which exhibits minima at the ion cores. The result of this is shown in the plot of energy versus wavevector in (b) whereby at these special wavevector values, an energy band gap is produced. One can also represent the band structure in only one zone between $-\pi/a$ and $+\pi/a$ by translation by a reciprocal lattice vector $2\pi/a$ (shown here as dashed lines).

the total number of valence electrons N equals $2n = 2n_F$, as each level can accommodate a spin-up and spin-down state. Hence

$$k_F = \frac{N\pi}{2L},\tag{7.33}$$

$$E_F = E_{k_e}^N = \frac{\hbar^2}{2m_{\mathrm{eff}}}\left(\frac{N\pi}{2L}\right)^2.\tag{7.34}$$

As already mentioned, this 'free-electron model' ignores the periodic electrostatic potential of the core ions. In reality, the valence electrons are affected by this potential energy field, U, and interact with it most strongly when

$$k_e = \frac{m\pi}{a},\tag{7.35}$$

where m is an integer. This can happen either in an attractive or repulsive manner, depending on whether the electron density (which is proportional to the square of the electron wavefunction) is highest or lowest above the ion cores, respectively (see Figure 7.42). The result of this interaction is to lower or raise the energy of those electrons, depending on the phase of their wavefunctions relative to the ion core array.

At which energy does the band gap appear? If we equate Equations (7.30) and (7.35), we see that

$$\frac{n\pi}{L} = \frac{m\pi}{a}.\tag{7.36}$$

Let us consider the first band gap, for $m = 1$, for which therefore $L = na$, which we have argued above is true for $n = N$. In other words, the band gap opens up around the Fermi energy.

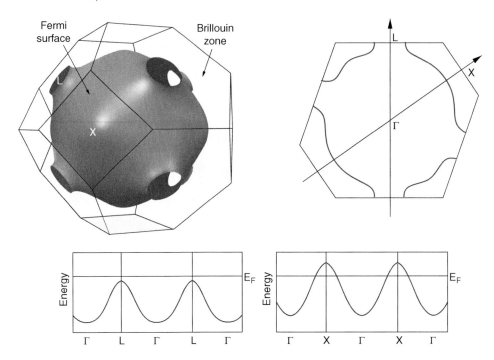

Figure 7.43　The Fermi surface of copper within the Brillouin zone, which, for the face-centred cubic Cu-lattice assumes the form of a truncated octahedron. A slice through the Fermi surface and Brillouin zone is shown on the right. The centre of the Brillouin zone is called the Γ-point, while the two crystallographic directions (111) (denoted 'L') and (100) ('X') are also shown. Note that the Fermi surface never crosses the Γ–L line and in this direction all the states are below the Fermi energy and therefore a gap opens up. In contrast, in the Γ–X direction, the Fermi surface lies within the Brillouin zone and copper is conducting in this direction.

Lastly, because of the periodic nature of crystals, the reciprocal space is also periodic, with the periodic unit in one dimension being $2\pi/a$. One can therefore 'fold' the band structure within the central region bounded by $\pm\pi/a$ by translation by a reciprocal lattice vector. This is shown in Figure 7.42 by the dashed lines reflecting the bandstructure above the band gap.

7.9.2.2　*The Fermi Surface*

In our one-dimensional model, the highest occupied state is a point along the k_e-axis with energy E_F and wavevector k_F, which we call the 'Fermi point'. In three-dimensional k-space, we speak of the 'Fermi surface', which describes the wavevectors in every possible direction below which the electron states are filled at absolute zero. In the case of a metal, the Fermi level lies within a continuous band, *at least along a crystallographic axis*. In some arbitrary direction, however, the magnitude of the wavevector might assume some other value, depending on the interaction between the electron and the ionic electrostatic potential in that direction, and a band gap might be formed. In other words, the form of the Fermi surface is a function of the crystallographic directions, and even for the most simple elemental metals such as copper, the Fermi surface can assume complex forms, as shown in Figure 7.43.

In this figure, the Fermi surface is seen to be encased within a polyhedron called the 'Brillouin zone'. Due to invariance of the band structure with respect to translations by integral multiples of the reciprocal lattice vector (a consequence of the periodicity of the crystal), the entire band structure is contained within the Brillouin zone. We see from Figure 7.42 that the boundaries of the Fermi surface (or 'Fermi point' in one dimension) were at

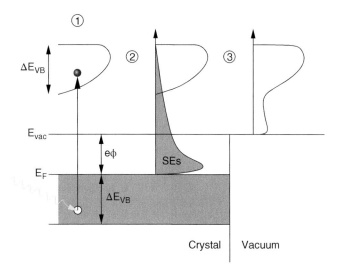

Figure 7.44 The three-step model. First, a photoelectron is produced by absorption of an x-ray photon at some finite depth in the crystal. Next, the photoelectron travels to the surface. Because the escape depth is in general significantly smaller than the x-ray absorption depth, the photoelectron can be scattered, producing secondary electrons (SEs). The electrons then penetrate the surface. The spectral form consists of the direct (unscattered) photoelectrons, plus a low-energy tail of those secondary electrons which have energies above $e\phi$, the work function.

$\pm\pi/a$, in other words exactly halfway in between adjacent reciprocal lattice points at $\pm2\pi/a$. If we extend this concept to three dimensions in k-space, what we obtain is the Brillouin zone. This is therefore the polyhedron defined by the smallest volume to be enclosed by planes that bisect at right angles lines connecting adjacent reciprocal lattice points. We have already come across this construct in our discussion of Wigner–Seitz cells in Section 6.3.1. Indeed, the Brillouin zone is the Wigner–Seitz cell of the reciprocal lattice of the crystal type being investigated. In the example shown in Figure 7.43, the reciprocal lattice of the face-centred cubic lattice of copper is a body-centred cubic lattice, which has the shown truncated octahedral Brillouin zone.

Hence, the Fermi surface of a material reflects the interaction of the valence electrons with their crystallographic environment, which in turn determines the electronic properties (conductivity) and optical properties (reflectivity and colour). A detailed knowledge of the Fermi surface is therefore of great importance. This, and related information, are the primary goals of UPS and, in particular, ARPES, experiments.

7.9.2.3   *The Three-step Model*

It is convenient to discuss the physical processes in photoemission by breaking down the ejection of a photoelectron from matter into successive events in the so-called 'three-step model' (see Figure 7.44):

- photo-induced excitation of an electron from a filled state to an empty conduction band state within the crystal;
- ballistic (i.e. without scattering) transport of the electron to the surface – this might also be contaminated by signal originating from inelastically scattered secondary electrons;
- transmission of the electron through the surface, resulting in its emission from the solid.

We begin with the first process. We have already discussed the conservation of energy [Equation (7.28)] with regards to photoemission. Let us next consider momentum conservation. The momentum of the electron is $\hbar\mathbf{k_e}$ with $|k_e| \sim 2$ Å^{-1}. On the other hand, the magnitude of the wavevector of a 50 eV photon is only approximately 0.025 Å^{-1}. Hence, photoabsorption imparts insignificant momentum transfer and the transition from the initial to final electronic state is 'vertical', that is, they both have essentially the same wavevector $\mathbf{k_e}$. This is an important

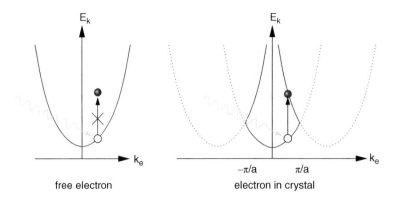

Figure 7.45 A truly free electron cannot absorb a photon, as the gain in energy would put it into a state of higher energy, but almost equal momentum. Such a state does not exist for free electrons, where $E(k_e)$ has unique solutions. Only in bound or periodic structures can states arise with different energies but the same wavevectors, allowed by the folding of higher-energy band structures back into the first Brillouin zone.

result – a truly free electron has an open, parabolic energy-versus-momentum curve and for a given energy, the wavevector k_e is uniquely determined by Equation (7.29). Changing the kinetic energy of a free electron is associated with a corresponding change in k_e, but this cannot occur, as the photon momentum is insignificant. Therefore incompatibility between energy conservation and momentum conservation means that photon-stimulated transitions are forbidden for truly free electrons. On the other hand, photoemission transitions *are* allowed in periodic structures, thanks to the fact that the band structure can be folded back on itself within the Brillouin zone (see Figure 7.45). The folding process means that the momentum of the final (upper) state differs from that of the initial state by an integer multiple of $2\pi/a$.

The depth to which photoelectrons are generated is determined by the absorption depth of the x-rays being used in the material under investigation, and is of the order of a few tens of nanometres for soft x-rays between 50 and 200 eV. However, the escape depth of the photoelectrons, given by the universal curve (Figure 7.9) is an order of magnitude smaller, hence valence-band structure measurements are sensitive only to the first few atomic layers. In addition, spectral peaks have 'loss tails' towards low kinetic energies, which consist of photoelectrons that have lost some energy due to inelastic scattering while passing through the material.

Lastly, we must consider what happens when the electron passes through the surface and into vacuum. Because it needs to overcome the work function, the electron slows down as it passes through the surface [Figure 7.46(a)]. Hence in this third process, only the component of $\mathbf{k_e}$ parallel to the surface is preserved, that is

$$k_{\text{out},\parallel} = k_{\text{in},\parallel} \equiv k_\parallel.$$

The potential step at the surface breaks the translational invariance perpendicular to that surface and so, unlike $k_\parallel$, $k_\perp$ is not conserved. From Equation (7.29),

$$k_{\text{out}} = \sqrt{\frac{2m_{\text{eff}}}{\hbar^2}\mathscr{E}_e},$$

$$k_{\text{in}} = \sqrt{\frac{2m_{\text{eff}}}{\hbar^2}(\mathscr{E}_e + V_0)},$$

where V_0 is the so-called 'inner potential', the energy required to remove an electron from the bottom of the parabolic potential within the solid to vacuum. From Figure 7.46, we see that

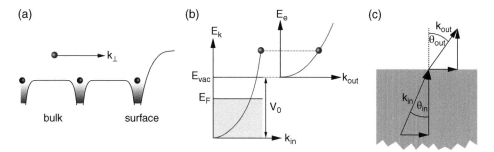

Figure 7.46 Transmission of photoelectrons through a surface. (a) Because of the work function, there is a potential barrier for the photoelectron leaving the surface, which slows it down in the direction normal to the surface. (b) In order to determine the wavevector $k_e = k_{in}$ of the photoelectron within the solid, the free-electron parabolas must be matched, which requires a measurement of $\mathscr{E}_e$, the kinetic energy of the photoelectron after leaving the surface and thereby $k_e = k_{out}$, its wavevector, plus a knowledge of V_0, the inner potential. (c) For a given kinetic energy of the photoelectron $\mathscr{E}_e$, there is a maximum angle θ_{in}^{max} that produces a photoemission signal. For larger angles, the component of the energy of the photoelectron within the solid perpendicular to the surface is insufficient to overcome V_0.

$$\sin\theta_{in} = \frac{k_\parallel}{k_{in}}$$

$$= \frac{\sin\theta_{out}\sqrt{\frac{2m_{eff}}{\hbar^2}\mathscr{E}_e}}{\sqrt{\frac{2m_{eff}}{\hbar^2}(\mathscr{E}_e + V_0)}},$$

which assumes its maximum value when $\theta_{out} = 90°$, that is

$$\sin\theta_{in}\big|_{max} = \sqrt{\frac{\mathscr{E}_e}{\mathscr{E}_e + V_0}}. \tag{7.37}$$

Hence, by detecting photoelectrons emitted from a surface at different emission angles, the energy of the electrons as a function of the momentum vector may be determined. This process is known as 'band mapping' and is a powerful probe of the electronic structure of crystalline materials. The measurements can usually be compared with theoretical predictions.

7.9.2.4   Presentation of ARPES Data

One can map out the density of occupied states in the valence and conduction bands of a material using ultraviolet photons. It is tempting to assume that the energy distribution of the electrons measured by UPS is a direct map of the band density of states. That is only true if the photoemission probability is the same for all valence electrons. From our discussions of photoabsorption in Section 7.2, in particular Equations (7.4) and (7.14), it was argued that both the initial and final states are of importance in the photoemission process and the photoemission probability may vary across the range of binding energies of the valence band. Therefore, the intensities of the UPS spectra are weighted by the transition-matrix element representing the coupling of the initial and final states.

If we assume a constant transition probability for all valence-band states, the resulting intensity distribution for a fixed incident photon energy as a function of photoelectron energy is therefore equal to the valence-band density of occupied states multiplied by the Fermi–Dirac function (which accounts for thermal promotion of valence-band

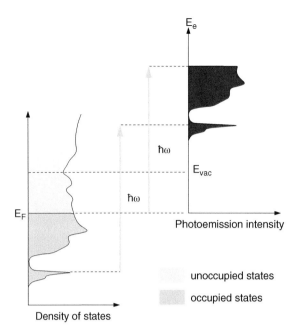

Figure 7.47 Energy-dispersion curves. EDCs are generated by recording the photoelectron intensity as a function of photoelectron kinetic energy for a fixed photon energy and detector angle. Assuming a constant transition probability, the intensity at a kinetic energy $\mathscr{E}_e = h\nu - e\phi - E_B$ is proportional to the local density of states at E_B having the in-plane wavevector $k_\parallel$ selected by the detector orientation. In this schematic, the photoemission intensity has been plotted as a function of $\mathscr{E}_e$ for didactical reasons. More commonly, however, it is plotted as a function of E_B.

electrons into the conduction band) and is shown schematically in Figure 7.47 for a UPS experiment at a fixed emission angle. Such plots are referred to as 'energy dispersion curves' (EDCs).

We argued above that a knowledge of the Fermi surface of a crystalline material is indispensable in predicting the optical and electronic properties of that material. The binding energy E_B at the Fermi surface is zero. We can therefore map out the Fermi surface by setting the kinetic energy of our detector (the CHA, Section 5.7.10) to $\mathscr{E}_e = h\nu - e\phi$ and varying the probed exit angles of the photoelectron. In this manner, the Fermi surface can be plotted as a function of the orthogonal in-plane wavevectors k_x and k_y, whereby $k_\parallel^2 = k_x^2 + k_y^2$. An example of the Fermi-surface map (FSM) of Ni(110) is shown in Figure 7.48.

The FSM of Ni(110) shown in Figure 7.48 is for $E_B = 0$, that is, at the Fermi energy. Similar maps can be generated for electrons originating from occupied states below the Fermi energy by adjusting the CHA. Vertical slices through the resulting three-dimensional intensity distribution, shown on the right-hand side of Figure 7.48, expose energy distribution maps (EDMs), which are also shown. These reveal some of the most important information on the electronic properties of crystalline materials, in particular on the presence and directions of band gaps, and details of the overall band structure.

7.9.2.5 Surface States

Surface states are associated with phenomena which are localized normal to the surface. These can be due to a rearrangement of bonds and change in bond strengths at a surface, often seen in semiconductors, but also occur in quasi two-dimensional systems such as in the conducting planes of graphite or in the superconducting CuO_2 sheets of high-temperature superconductors.

How can one distinguish between 'normal' bulk states and surface states? To answer this, the most important aspect to recognize is the *localization* of surface states to a two-dimensional plane. If a state is physically localized

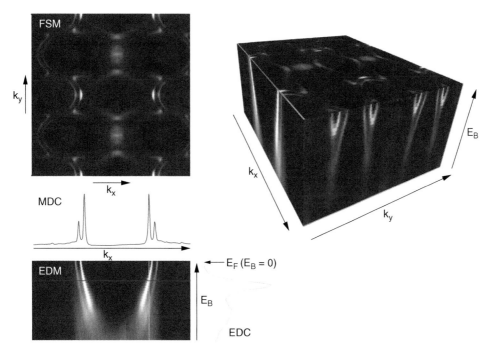

Figure 7.48 Representations of ARPES data. A Fermi-surface map (FSM) is generated by recording the intensity at $\mathcal{E}_e = h\nu - e\phi$ for different k_x and k_y. Momentum-distribution curves (MDCs) track the change in intensity of the ARPES signal for a fixed binding energy as one navigates in a fixed in-plane direction (e.g. along k_x). Energy-distribution maps can either be generated by stacking MDCs for a range of binding energies, or by stacking EDCs for different positions along a certain in-plane momentum axis. Finally, a full three-dimensional representation of ARPES data is possible by plotting the intensity as a function of E_B and the in-plane wavevectors k_x and k_y. The results shown here are for Ni(110). Courtesy Luc Patthey, Paul Scherrer Institute, adapted from [56] with permission of the American Physical Society.

in one direction, the corresponding periodicity is lost, which can be mathematically formulated as an *infinite* periodic wavelength in this direction. In contrast, as the wavevector is inversely proportional to the periodicity, this becomes completely *de*localized perpendicular to the two-dimensional feature, and can take on any value. This is shown schematically in Figure 7.49.

In reciprocal space, therefore, surface states are present for all possible values of $k_\perp$, while the in-plane wavevectors are well-defined and fixed. Surface states hence have a fixed binding energy E_R, independent of $k_\perp$.

Now let us imagine an experiment in which we probe only those electrons emerging perpendicularly from the surface, for which $k_{\text{in},\parallel} = k_{\text{out},\parallel} = 0$. From Figure 7.50, it becomes immediately apparent that, in contrast to bulk states, the position of surface states does not disperse if one changes the photon energy.

Surface states are in general located in bandgap regions for which there are no bulk-like states. The reason for this is that if a surface state were degenerate with a bulk state, they could couple and the surface state would take on some of the character of the bulk state, thereby losing its surface localized character.

A surface state will be strongly affected by adsorption; the introduction of contaminants to the surface of the sample under investigation can kill surface-state signal. Because, however, the influence of some contaminants can extend several monolayers into a sample, states which are normally considered to be bulk-like can be affected, and this test is less stringent than the first.

An example of a surface state in the Fermi surface map of Cu(111) is shown in Figure 7.51. Normal to the surface, the bulk density of states at the Fermi energy is zero and copper has a bandgap in this direction. However, a localized signal is indeed found in this direction, associated with a surface state [54].

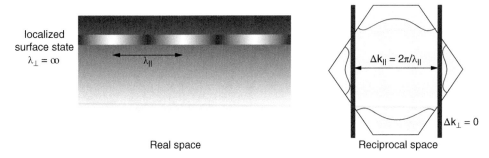

Real space   Reciprocal space

Figure 7.49   A localized, or isolated, two-dimensional electronic feature contains so-called 'surface states', in which there is a well-defined periodicity in a plane parallel to the surface, but not one out of that plane. One can therefore assign an out-of-plane wavelength to such states of $\lambda_\perp = \infty$, as the states are never repeated in this direction. In reciprocal space, this translates to the out-of-plane wavevector $\Delta k_\perp = 2\pi/\lambda_\perp = 0$ and in this direction, the surface state is entirely *de*localized.

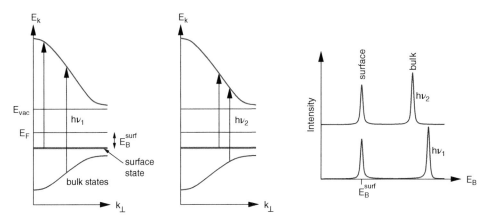

Figure 7.50   Identification of surface states. Consider two binding-energy spectra recorded with two different photon energies $h\nu_1$ and $h\nu_2$ and for normal emission of the photoelectrons. Different photon energies probe different bulk states according to which $k_\perp$ corresponds to the energy difference between the occupied and unoccupied states. Because surface states have a constant binding energy E_B^{surf} independent of $k_\perp$, their positions in the binding-energy spectra remain invariant and can thus be distinguished from bulk states.

7.9.2.6   *Experimental Considerations*

From the above description of the three-step model, the basic idea of ARPES should be clear – by determining both the kinetic energy and three orthogonal momentum components of the wavevector of the emitted electrons, one can infer the energy as a function of momentum of the occupied states within the solid, in other words, the material's electronic band structure.

How does one achieve angular resolution in ARPES? For electron-energy analysers with small entrance apertures (shown as S in Figure 5.65), this is achieved either by moving the orientation of the sample relative to the stationary analyser (normally a concentric hemispherical analyser, CHA, see Section 5.7.10), or alternatively by mounting the detector on a two-circle goniometer.

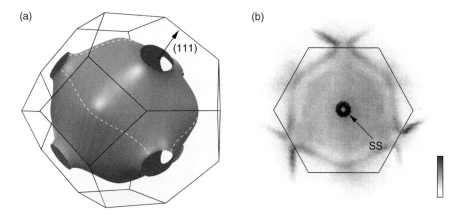

Figure 7.51 Surface states in Cu(111). (a) The (111) crystallographic direction penetrates the Brillouin zone but not the Fermi surface – in this direction, copper has a band gap. (b) A Fermi-surface map of the Cu(111) surface. The projection of the Brillouin zone on the (111) surface is shown partially as the hexagon. A high partial density of states at the Fermi energy can be recognized by the arcs of darker bands, also highlighted by the dashed white lines on the Fermi surface in (a). In the normal direction ($k_x = k_y = 0$) a localized high-intensity signal is detected, associated with a surface state. Adapted from [54] with permission of Elsevier.

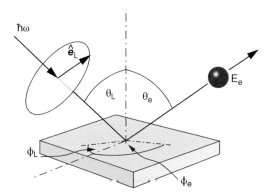

Figure 7.52 Schematic diagram of an angular-resolved photoelectron experiment, including the possible experimental variables. Those relating to the incident beam are suffixed with L, while those relating to the electron are suffixed with e. The unit vector $\hat{\mathbf{e}}_L$ is a polarization vector. The angles are normally referenced to a high-symmetry axis of the crystal.

The second possibility of keeping the sample fixed and moving the electron analyser poses the problem of maintaining ultra-high vacuum conditions when repositioning the analyser. The basic geometry of an angular-resolved photoelectron experiment is shown in Figure 7.52.

In modern CHAs, data acquisition is accelerated by using area electron detectors such as multichannel plates (shown in Figure 5.66), such that a range of angles (and thereby momenta) can be recorded simultaneously.

7.9.3 Soft X-ray ARPES

The very shallow escape depth Λ_e of conventional ARPES in the range of 20 to 200 eV means that (a) sample preparation is critical, requiring the production and maintenance of atomically well-defined surfaces, which in turn demands ultra-high-vacuum conditions (that is, pressures at or below 10^{-10} mbar); and (b) any results gleaned from the data may not represent bulk-like properties of the material.

Moreover, the momentum-resolution of ARPES for three-dimensional systems is limited by the fundamental difficulty in resolving $k_\perp$, the component of the momentum vector k_e perpendicular to the surface. At the photoelectron-escape stage of the three-step model, $k_\perp$ is altered by the momentum absorbed by the surface barrier, and is thus not directly measurable. It can only be recovered if one knows the dispersion of the final state at the initial photoexcitation stage (i.e. the first step in the three-step model). The free-electron approximation commonly applied at this point is in principle invalid, because, for many materials, the final states can feature complicated quasi-bound and excited-state self-energy effects. Over and above this, the final state is confined within the photoelectron escape depth. By applying Heisenberg's uncertainty principle ($\Delta p \Delta x = \hbar$; $\Delta p = \hbar \Delta k_\perp$), one immediately obtains an intrinsic broadening $\Delta k_\perp = 1/\Lambda_e$, which can be a significant fraction of $k_\perp$.

By performing ARPES, not in the conventional VUV regime, but instead extending it to soft-x-ray photon energies of the order of 1 keV, these problems are, to a large extent, resolved [57, 58]. The photoelectron energies become much larger than the crystal potential modulations, which makes the final states truly free-electron-like; in addition, the increase of Λ_e with electron energy results in a sharpening of $\Delta k_\perp$ by up to an order of magnitude. Further advantages of soft x-ray ARPES (SX-ARPES) include greater bulk sensitivity and associated simplified matrix elements that essentially reduce to atomic cross-sections and thus vary smoothly and predictably with binding energy and momentum vector. One further potential advantage is that SX-ARPES covers important absorption edges, such as the L-edges of the Group-4 transition metals and the M-edges of many of the rare-earths. This, in turn, permits resonant excitation of their valence d and f states (see below), which play crucial rôles in the physics of strongly correlated electron systems, such as cuprate superconductors, colossal magnetoresistive materials, and Mott insulators.

On the other hand, SX-ARPES suffers from a reduction of the photoexcitation cross-section of valence states by a few orders of magnitude compared to the VUV-energy range, due to its $1/(h\nu)^3$-dependence. This can only be overcome by the use of the most brilliant synchrotron and XFEL radiation.

Photoelectrons with energies of the order of 1 keV have de Broglie wavelengths of approximately 0.3 Å. Although this is still an order of magnitude larger than typical thermal-motion amplitudes, the electrons begin to interact significantly with, and be inelastically scattered by, phonons, resulting in spectral broadening and an increased background signal.

Lastly, the absolute resolving power $\Delta \mathscr{E}_e$ is directly proportional to $\mathscr{E}_e$, which means that, even with state-of-the-art CHAs, features narrower than approximately 10 meV are unresolvable.

An example of recent results obtained by SX-ARPES is summarized in Figure 7.53. This is the first direct visualization of the Fermi surface and k-resolved electronic structure of a buried heterostructure. The device was a quantum well formed by band-bending at the interface between AlN and GaN. The two-dimensional electron gas resides in pure, undoped GaN. The lack of dopant atoms dramatically increases the electrons' mobility, thereby boosting such devices' high-frequency performance in applications such as receivers in mobile phones [59].

7.9.3.1   Resonant Photoelectron Spectroscopy

Direct measurements of valence states close to the Fermi edge can be difficult if they originate from a material buried at a depth beneath the surface that is large compared to Λ_e. In order to increase the escape depth, larger photon energies are required, which have a low absorption cross-section with the valence state of interest. This problem can be mitigated by selecting a photon energy that resonates with unoccupied states associated with the valence band, but just above the Fermi energy (see Figure 7.54) – relaxation from the valence band back to the core level resonantly enhances Auger emission from the same valence band. The kinetic energy of this Auger electron is close to the binding energy of the initially excited core electron. In this manner, unusual electronic phenomena at buried interfaces can be probed [60].

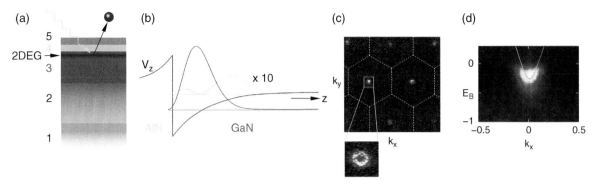

Figure 7.53 SX-ARPES of a quantum-well high-electron mobility transistor. (a) A quantum-well device was fabricated on a sapphire substrate (1) by first laying down 700 nm of a graded concentration of $Al_xGa_{1-x}N$ (2) in order to remove misfit dislocations. On this, 500 nm of pure, undoped, GaN (3) was deposited, followed by a thin 2 nm layer of AlN (4) and a 1 nm cap of $Al_{0.5}Ga_{0.5}N$ (5). (b) The interface between the GaN and AlN layers induce band-bending and the formation of a quantum well and trapping of a two-dimensional electron-gas (2DEG). The electron-density-probability functions of the first two discrete energy levels of the quantum well are sketched. (c) A Fermi-surface map exhibits the sixfold crystallographic symmetry of the system. Closer inspection of the signals around the $\overline{\Gamma}$-points reveals them to be open circles, a manifestation of cuts through the parabolic form of the conduction band, shown in detail in the energy-density map in (d). The units for k_x and k_y are Å^{-1} and, for E_B, electronvolts. Adapted from [59] with permission from Macmillan Publishers Ltd.

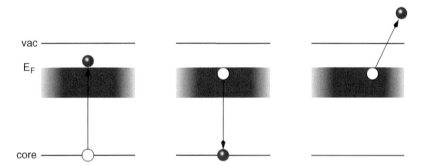

Figure 7.54 Resonant photoelectron spectroscopy. The photon energy is tuned to promote a core level to just above the Fermi edge. Relaxation of an electron from the valence band back to the core level resonantly enhances the ejection of an Auger electron from the same band.

7.9.4 X-ray Photoelectron Spectroscopy

The chemical composition of the surface region of a material can be determined down to a fraction of a percent using x-ray photoelectron spectroscopy (XPS). In addition to providing information on the elemental constituents, XPS also yields information on the type of chemical bonds in the material. Indeed, XPS is alternatively called ESCA, or electron spectroscopy for chemical analysis.

7.9.4.1 The Chemical Shift

The chemical bond between atoms or ions arises due to the rearrangement of the spatial distribution of one or more of the valence electrons – for example, in covalent bonds pairs of electrons between atoms are shared, while

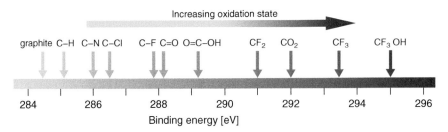

Figure 7.55   A plot of the shift in binding energies of the C $1s$ core electron in different carbon-containing compounds. Values are approximate and can vary by up to an eV from compound to compound.

in ionic bonds there is a transfer of one or more electrons from one atom to another, resulting in the formation of oppositely charged and Coulomb-attracted ions.

As an example, a $1s$ electron from oxygen ejected by an x-ray photon has a binding energy, which, depending on the local chemical environment, can differ by as much as 10 eV around approximately 543 eV. As a rule of thumb, elements in a higher oxidation state have electrons with higher binding energies, and vice versa. This is because atoms (or ions) in higher oxidation states have fewer electrons within their immediate neighbourhood, hence the electrostatic repulsion between the ejected electron and those electrons remaining is smaller[12]. The $1s$ binding energy of carbon for different oxidation states is shown in Figure 7.55. Chemical shifts can be detected with a resolution of 50 meV or better, using modern synchrotrons.

Although XPS setups at synchrotrons can be used for high-resolution chemical analysis, this is nowadays normally considered to be a somewhat wasteful use of beamtime. One clear advantage of synchrotron sources, other than the increased brightness and sensitivity, is the fact that the beam can be focussed down to approximately a few tens of nm. This allows one to perform two-dimensional maps in 'photoelectron microscopy' [61].

7.9.4.2   *High-pressure XPS*

An exception to the rule of thumb that XPS at synchrotrons is not in general used for high-resolution chemical analysis is the case of high-pressure XPS [HPXPS, or ambient-pressure PES (APPES), or high-pressure PES, (HPPES)]. Although photoemission is a highly powerful tool for investigating clean surfaces under ultra-high-vacuum conditions, many interesting processes occur on surfaces under reaction conditions, which can be at several orders of magnitude higher pressure than those typically used for traditional XPS experiments. Such systems are excluded from investigation using standard PES spectrometers, and have been the spur to develop the HPXPS technique [62–64]. A typical application of HPXPS is the investigation of the surface chemistry and its change under different ambient conditions (in particular gas pressure and type) in heterogeneous catalysis [65, 66].

The inelastic mean free path Λ_e of an electron travelling through a gas depends on the gas pressure P and the electron's kinetic energy $\mathscr{E}_e$. For a given gas type and electron energy, the product of Λ_e and P is a constant. An example of the variation of $P\Lambda_e$ with $\mathscr{E}_e$ is shown in Figure 7.56 for electrons travelling through oxygen gas. For instance, a photoelectron with a kinetic energy $\mathscr{E}_e$ of 400 eV would have a mean free path in 100 Pa (1 mbar) oxygen of approximately 4 mm.

XPS measurements at high pressures are made possible by employing one or more differentially pumped stages between the sample surface and the electron analyser (see Figure 7.57). The diameter D_0 of the primary aperture is decisive for the fundamental scales of the experiment with regards to observable photoelectron energies and maximum possible pressures. There is a fine balance between placing the aperture too close to the sample, thereby perturbing the chemical environment, and positioning it too far away, when scattering of the primary photoelectron

---

[12] A useful mnemonic is OIL RIG: *Oxidation Is Loss, Reduction Is Gain* (of electrons).

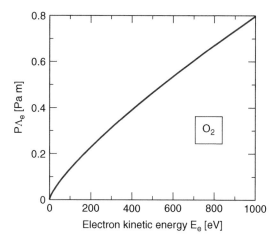

Figure 7.56 The product of gas pressure and electron inelastic mean-free path in oxygen gas for electron kinetic energies up to 1 keV. Adapted from [64] with permission from Elsevier.

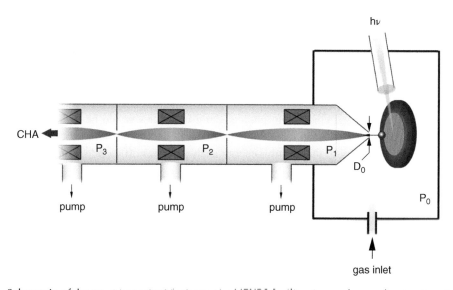

Figure 7.57 Schematic of the most important features of a HPXPS facility. A sample on a heater stage in the reaction environment of pressure P_0 is irradiated with synchrotron radiation incident via a re-entrant window. The photoemission signal is allowed to pass through a skimmer with nozzle diameter D_0 into a first differentially pumped stage at pressure P_1, where it is focussed using electron optics; and then two subsequent differentially pumped stages. Finally, the electrons' energies are analysed using a concentric hemispherical analyser (CHA, see also Figure 5.65).

by the gas becomes highly probable, and the solid angle subtended by the aperture becomes so small that the captured photoelectron signal becomes unacceptably weak. In practice, the ideal separation between the sample surface and the primary aperture is a few times D_0. The maximum allowed ambient gas pressure is inversely proportional to D_0, hence reducing D_0 allows a higher ambient pressure, but at the cost of photoemission signal, which scales with D_0^2. Typical modern HPXPS chambers have apertures in the range of 0.5 mm, sample–aperture distances of the order of a few mm, and hence maximum working pressures approximately 1000 Pa (10 mbar).

The introduction of electrostatic lenses in the differentially pumped stages in order to collect a larger fraction of the photoemission signal represented a breakthrough in HPXPS technology [63] (Figure 7.57).

In addition to reactions at the interface between solids and gas, many chemical processes occur at liquid–gas interfaces and, even more importantly, solid–liquid interfaces. With this in mind, liquid microjets [67] and droplet trains [68] can be employed for the former, and a 'dip and pull' meniscus approach for the latter [69]. The liquid layer needs to be thinner than a few tens of nm and x-rays in the region of 2 to 10 keV should be employed in order to obtain a sufficient escape depth through the liquid layer. The IMFP of water as a function of electron energy is highlighted as red data points in Figure 7.9. This is approximately an order of magnitude larger than for most condensed matter, with values in excess of 25 nm in the hard x-ray regime.

7.9.5   Hard X-ray Photoelectron Spectroscopy

Curiously, the first x-ray spectroscopic measurements, by Siegbahn and co-workers in the 1950s, were carried out using photon energies in the range of 5 to 8 keV [70, 71]. However, a desire to improve the spectroscopic resolution [see Equation (5.50)] and scientific interest in the surface properties of materials rapidly led to a shift to lower energies, where the surface sensitivity is highest, of the order of 1 to 2 nm (see Figure 7.9).

As we have already argued in Section 6.12, the surface region of a material, i.e. that which structurally or electronically differs from the bulk material, can extend down to several monolayers, or in other words, the same depth as the photoelectron escape depth between approximately 30 to 2000 eV. Hence, conclusions drawn about the chemistry or electronic structure of a sample using photoelectron spectroscopy in this energy range may not in fact represent the bulk properties.

Therefore, those experiments which are designed to investigate bulk properties, or embedded regions deeper than a few monolayers, need to use photoelectrons with larger escape depths. This implies a move to higher x-ray energies, as suggested by Lindau *et al.* in 1974 [72]. For a long time this idea was not seriously pursued, as the absorption cross-sections and consequent photoelectron intensities would be too low – if we refer back to Figure 2.23, showing the absorption coefficient as a function of energy, we remember that this drops off as $1/(h\nu)^3$. In addition, the sensitivity of the electron-energy analyser drops off as $\mathscr{E}_e^{-1}$. Hence, the photoemission signal intensity for experiments using photons with energies far in excess of the binding energy of the quantum state of interest is very small.

With the development of highly brilliant, modern synchrotron sources, however, the idea of using hard x-rays (between 4 and 15 keV) in photoelectron spectroscopy (HAXPES) has been recently revived, and now several groups are developing this technique [73]. In this energy range, the escape depth of the photoelectrons increases to approximately 10 nm at 5 keV and to some 25 nm at 15 keV, far larger than the extent of typical surface regions.

In conventional XPS carried out with soft x-rays at synchrotron facilities, the spectral resolution is typically $\Delta E = 0.1$ to 0.2 eV, which is necessary in order to unambiguously distinguish chemical shifts corresponding to different chemical or valence states (see Figure 7.55). The same absolute energy resolution is also desirable in HAXPES, but, as we see from Equation (5.50), this sets stringent specifications on the detector geometry and on the accuracy of setting the pass energy, which in HAXPES must be much smaller than the original kinetic energy of the electrons as they emerge from the sample surface. Most commercial electron analysers designed for conventional XPS are unable to detect energies much above 5 keV, although some do extend as far as 12 keV. At the time of writing, electron energies as high as 15 keV could only be detected using custom-made equipment.

In addition, the degree of monochromatization of the synchrotron radiation must be of the order of $\Delta\lambda/\lambda = 10^{-5}$. The intrinsic Darwin width of the most common monochromator crystal, Si(111), is, at around 10^{-4}, one order of magnitude larger. The width can be decreased to approximately 2×10^{-5} by using the (333) or (400) reflections, or by using a dispersive geometry in a double-crystal monochromator [see Figure 5.28(d)], although this is accompanied by a reduction in flux. In this case, the resolution at 10 keV is approximately 0.2 eV.

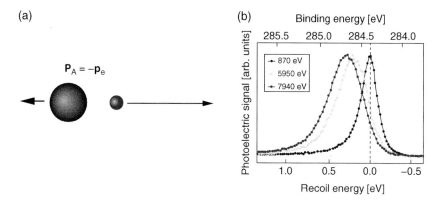

Figure 7.58 The recoil effect. (a) Because of the conservation of momentum, some of the kinetic energy of the ejected photoelectron is lost to the recoil of the absorbing atom. (b) The recoil effect in the 1s signal of carbon. The mass ratio of a carbon atom to a photoelectron is 21 874. Adapted from [74] with permission of the American Physical Society.

Despite all these technical difficulties, there are important advantages to HAXPES. First, because the bulk is being probed, the need for ultrahigh vacuum conditions, required in normal XPS in order to maintain surface cleanliness, is obviated. Moreover, because the photon wavelengths at these high energies are generally significantly smaller than typical interatomic spacings, the photoemission matrix elements reduce to atomic cross-sections, thus simplifying data analysis.

The momentum of photoelectrons in HAXPES can be significant, causing the emitting atom to recoil [74]. According to the conservation of momentum [see Figure 7.58(a)],

$$\mathbf{P}_A = -\mathbf{p}_e,$$

whereby $\mathbf{P}_A$ and $\mathbf{p}_e$ are the momenta of the emitting atom and the photoelectron, respectively. Thus, the photoelectron loses a fraction of its kinetic energy $\Delta\mathscr{E}_e$ given by

$$\Delta\mathscr{E}_e = \frac{P_A^2}{2M_A}$$

$$= \frac{p_e^2}{2m_e}\frac{m_e}{M_A}$$

$$= \mathscr{E}_e\frac{m_e}{M_A},$$

whereby M_A is the atom's mass. Thus, particularly for light atoms, this energy loss [and apparent increase in binding energy, see Figure 7.58(b)] can be of the same order of magnitude as chemical shifts, and must therefore be considered in analysis.

7.10   Concluding Remarks

Spectroscopic techniques that use synchrotron radiation have the unique advantage that the incoming x-rays can be tuned and/or scanned in energy. This adds an invaluable new dimension to experiments excluded to those based on laboratory x-ray sources and has been exploited to full advantage in many of the above-described methods, not least in XAS, ARPES, and RIXS.

The high brilliance of storage rings, coupled with the ability to focus down to the nanometre scale, has opened up new vistas in synchrotron spectroscopy. Enormous progress has also been made in the performance of fast, energy-dispersive detectors, and it is expected that the range of experiments available will continue to burgeon as rapidly as has been experienced in the last decade.

Spectroscopy using x-rays will remain among the most prominent and sought-after general synchrotron techniques because of the detailed information that can be gleaned about the chemical and electronic properties of matter. In an era when ever more novel and unexpected materials are being tailored and synthesized, synchrotron spectroscopy will continue to play a central rôle in materials research.

Problems

1. Derive the practical expression

$$f'' = f_2 = 1.431 \times 10^{-8} E[eV] \sigma_a [\text{barn}],$$

   where f_2 is the anomalous component to the atomic form factor, E is the photon energy, and σ_a is the cross-section for photoabsorption. (1P)

2. Prove Equation (5.26). Determine the change in Δy in a channel-cut monochromator using Si(111) for a channel width $D = 10$ mm, as the photon energy is changed from 8 to 11 keV. The Si(111) plane separation is $5.431/\sqrt{3} = 3.1356$ Å. (2P)

3. A pellet of $SrTiO_3$ (mass density $= 5.12$ g cm^{-3}) is to be prepared for a XAS measurement around the K-edge of Sr at 16.105 keV. The pellet is made from $SrTiO_3$ powder mixed with transparent polymer powder to produce a pellet of sufficient thickness to make it mechanically robust. The pellet is formed by compressing the powder mixture in a pressing die (13 mm diameter) using a hydraulic press. What mass of $SrTiO_3$ powder should be used to obtain a ratio of transmissions just below and above the Sr K-edge of $I_{\text{below}}/I_{\text{above}} = 10$? The absorption coefficients of $SrTiO_3$ directly below and above the Sr K-edge are $\mu_- = 7.999 \times 10^{-3}$ μm^{-1} and $\mu_+ = 0.03146$ μm^{-1}, respectively. (2P)

4. The flux at a beamline needs to be measured. The beamline scientist finds an old silicon photodiode in a drawer. It has a protective aluminium coating. Neither the thickness of the silicon nor that of the aluminium coating are known. However, the beamline scientist is clever, so she measures the transmission through the diode using another diode, and obtains values of 2.05×10^{-5} at 6 keV and 0.0868 at 10 keV. What are the thicknesses of the silicon diode and the aluminium coating? Use the cxro website to obtain the attenuation coefficients of Si and Al. (3P)

5. Calculate the velocity and de Broglie wavelength of a helium atom with a kinetic energy of 100 meV. (2P)

6. If relativistic effects need to be considered for particles with mass when $v \geq c/10$, at approximately which kinetic energy do electrons start to become relativistic?
   The relativistically corrected equation for the de Broglie wavelength of an electron is given by

$$\lambda_e = \frac{hc}{[eV(2m_e c^2 + eV)]^{1/2}},$$

   whereby eV is the kinetic energy of the electron, and $m_e c^2$ is its rest-mass energy. Show that for the limit of $eV \ll m_e c^2$, λ_e reduces to the classical expression, while for $eV \gg m_e c^2$, the electron wavelength asymptotically approaches that of a photon with the same energy. Show also that the 'turning point' in energy $\mathscr{E}_e$ in between these two limiting cases is centred at $2m_e c^2$. (3P)

7. Demonstrate that the ratio of the wavelength of electrons with nonrelativistic energies to those of photons of the same energy is given by

$$\frac{\lambda_e}{\lambda_p} = \frac{v}{2c}.$$

(2P)

8. The inelastic mean free path (IMFP) of electrons in condensed matter is only weakly dependent on the specific material type, leading to the curve describing the IMFP as a function of electron kinetic energy being called the 'universal curve' (see Figure 7.9). A close fit to the universal curve is given by $\Lambda_e = A/\mathscr{E}_e^2 + B\sqrt{\mathscr{E}_e}$, where Λ_e is in angstroms and $\mathscr{E}_e$ in eV, and $A = 1430$ and $B = 0.54$. Determine the theoretical minimum of this function. (2P)

9. Using the same best-fit equation for the universal curve, determine the ratio of probing depths of (a) an SX-ARPES experiment using 2000 eV photons and (b) a standard ARPES experiment using 100 eV photons investigating the Se $3d$ electrons in the layered semiconductor $TiSe_2$. The work function of $TiSe_2$ is 5.4 eV. (2P)

10. A beamline operating in the UV regime is used to excite Rydberg states of hydrogen. The monochromator used at the beamline has a resolving power of $\Delta E/E = 10\,000$. Make an estimate of the highest Rydberg state n which can be resolved by the beamline. (1P)

11. Derive Equation (7.22). A QEXAFS experiment is carried out on a palladium catalyst, centred at the palladium K-edge and covering ± 500 eV at a rate of 50 spectra per second. Spectra are recorded in both the forward and reverse directions of the Si(111)-monochromator rotation and need to be resolved with a resolution of 1 eV. Determine the rotation rate in degrees per second and the minimum recording rate of the gridded ionization chamber used to record the data. (5P)

12. A 2 mW x-ray beam of 1.24 Å photons is incident at a grazing angle of 0.5° on a 2-μm-thick foil made of a material with an attenuation length of 5 μm at this photon energy. The fluorescence efficiency is 0.8. How many fluorescent photons per second are produced? (2P)

13. A 90 nm thin film of ZnO is deposited on silica glass. This is probed using monochromatized x-radiation a little above the K-edge of Zn, at 9900 eV, with an incident flux of 10^{13} photons per second. A photon-counting fluorescence detector with a 4 cm-diameter sensitive area is placed in front of the film (that is, in the backscattering direction compared to the incident beam) at a distance of 20 cm. Calculate the count rate of fluorescent photons, assuming the silica substrate contributes to this only insignificantly. The absorption length of ZnO at 9800 eV is 9 μm. Hint: don't forget Figure 2.28! (3P)

14. In an x-ray photoemission spectroscopy experiment investigating metallic TiN using 500 eV photons, the de Broglie wavelength of an electron ejected due to a transition from the $2p$ to $3d$ state in Ti ($\Delta E = 460$ eV) is determined to be $\lambda_e = 2.05$ Å. Calculate the work function $e\phi$. (1P)

15. The binding energies of the titanium $2p_{3/2}$ state are measured using x-ray photoelectron spectroscopy for titanium in various chemical combinations, namely TiF_3, Ti_2O_3, TiO, Ti metal, TiO_2, and TiN. Allocate the binding energies, shown in Figure 7.59, to the given compounds. (1P)

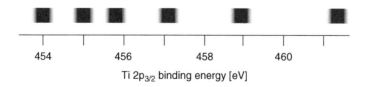

Figure 7.59 Six experimentally determined positions for the Ti $2p_{3/2}$ state for titanium bound in different chemical combinations.

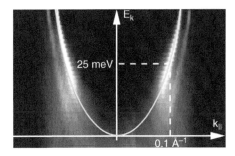

Figure 7.60   The EDM of metallic copper. At a parallel wavevector of $k_\parallel = 0.1$ Å^{-1}, the energy of the electron within the copper is 25 meV.

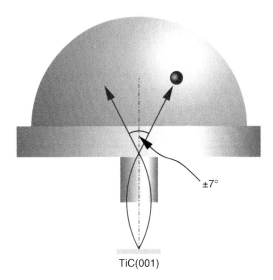

TiC(001)

Figure 7.61   Setup of a HAXPES experiment investigating the Fermi surface of TiC.

16. The energy-distribution map for metallic copper shown in Figure 7.60 has been fit to a parabolic distribution. At a parallel wavevector of $k_\parallel = 0.1$ Å^{-1}, the energy of the electron within the copper is $E_k = 25$ meV. Calculate the electron's effective mass. By what factor does this differ from m_e, the mass of a free electron? (2P)

17. A concentric hemispherical analyser (CHA) has an acceptance angle of $\pm 8°$. The Fermi surface of cubic TiC(001) buried under an ultrathin layer of zirconium nitride is investigated using 460 eV photons in a HAXPES experiment. The sample normal points to the centre of the entrance of the CHA (see Figure 7.61). Determine the largest (nth) Brillouin zone boundary which can be recorded without changing the relative orientation of the sample to the CHA. Calculate at which angles the boundaries of the first to nth Brillouin zones would be expected. The work function is 5 eV, and the lattice constant of cubic TiC is 4.4 Å. (2P)

References

[1] J. M. Cole, J. Nienstedt, G. Spataro, et al., "Phosphor imaging as a tool for in situ mapping of ppm levels of uranium and thorium in rocks and minerals," *Chem. Geology*, vol. 193, pp. 127–136, 2003.

[2] J. A. Bearden and A. F. Burr, "Reevaluation of atomic x-ray energy levels," *Rev. Mod. Phys.*, vol. 39, pp. 125–142, 1967.

[3] M. P. Seah and W. A. Dench, "Quantitative electron spectroscopy of surfaces: a standard data base for electron inelastic mean free paths in solids," *Surf. Interface Anal.*, vol. 1, pp. 2–11, 1979.

[4] J. J. Rehr and R. C. Albers, "Theoretical approaches to x-ray absorption fine structure," *Rev. Mod. Phys.*, vol. 72, pp. 621–654, 2000.

[5] J. J. Rehr and A. L. Ankudinov, "Progress in the theory and interpretation of XANES," *Coord. Chem. Rev.*, vol. 249, pp. 131–140, 2005.

[6] G. Meitzner, G. H. Via, F. W. Lytle, and J. H. Sinfelt, "Analysis of x-ray absorption-edge data on metal-catalysts," *J. Phys. Chem.*, vol. 96, pp. 4960–4964, 1992.

[7] A. Puschmann, J. Haase, M. D. Crapper, C. E. Riley, and D. P. Woodruff, "Structure determination of the formate intermediate on Cu(110) by use of x-ray-absorption fine-structure measurements," *Phys. Rev. Lett.*, vol. 54, pp. 2250–2252, 1985.

[8] D. E. Sayers, E. A. Stern, and F. W. Lytle, "New technique for investigating noncrystalline structures – Fourier analysis of extended x-ray absorption fine structure," *Phys. Rev. Lett.*, vol. 27, pp. 1204–1207, 1971.

[9] M. Newville, S. A. Carroll, P. A. O'Day, G. A. Waychunas, and M. Ebert, "A web-based library of XAFS data on model compounds," *J. Synchrotron Rad.*, vol. 6, pp. 276–277, 1999.

[10] B. K. Teo and P. A. Lee, "Ab-initio calculations of amplitude and phase functions for extended x-ray absorption fine-structure spectroscopy," *J. Am. Chem. Soc.*, vol. 101, pp. 2815–2832, 1979.

[11] P. A. Lee, P. H. Citrin, P. Eisenberger, and B. M. Kincaid, "Extended x-ray absorption fine-structure – its strengths and limitations as a structural tool," *Rev. Mod. Phys.*, vol. 53, pp. 769–806, 1981.

[12] Q. Wang, C. X. Li, X. J. Niu, et al., "Structure of liquid krypton under atmospheric pressure: An EXAFS and reverse Monte Carlo study," *Phys. Rev. B*, vol. 72, p. 092202, 2005.

[13] E. A. Stern and S. M. Heald, *Handbook on Synchrotron Radiation*, vol. 1b. Amsterdam: North-Holland, 1983.

[14] B. K. Teo, P. A. Lee, A. L. Simons, P. Eisenberger, and B. M. Kincaid, "EXAFS - approximation, parameterization, and chemical transferability of amplitude functions," *J. Am. Chem. Soc.*, vol. 99, pp. 3854–3856, 1977.

[15] P. A. Lee, B. K. Teo, and A. L. Simons, "EXAFS – new parameterization of phase-shifts," *J. Am. Chem. Soc.*, vol. 99, pp. 3856–3859, 1977.

[16] G. Sarret, A. Manceau, D. Cuny, et al., "Mechanisms of lichen resistance to metallic pollution," *Environ. Sci. Technol.*, vol. 32, pp. 3325–3330, 1998.

[17] T. Matsushita and R. P. Phizackerley, "A fast x-ray absorption spectrometer for use with synchrotron radiation," *Jap. J. Appl. Phys.*, vol. 20, pp. 2223–2228, 1981.

[18] P. G. Allen, S. D. Conradson, and J. E. Pennerhahn, "A 4-point crystal bender for dispersive x-ray absorption spectroscopy," *J. Appl. Crystallogr.*, vol. 26, pp. 172–179, 1993.

[19] R. Frahm, "Quick scanning EXAFS: first experiments," *Nucl. Instrum. Methods A*, vol. 270, pp. 578–581, 1988.

[20] R. Frahm, "QEXAFS – x-ray absorption studies in seconds," *Physica B*, vol. 158, pp. 342–343, 1989.

[21] O. Müller, D. Lützenkirchen-Hecht, and R. Frahm, "Quick-scanning monochromator for millisecond in situ and in operando x-ray absorption spectroscopy," *Rev. Sci. Instrum.*, vol. 86, p. 093905, 2015.

[22]   O. Müller, M. Nachtegaal, J. Just, D. Lützenkirchen-Hecht, and R. Frahm, "Quick-EXAFS setup at the SuperXAS beamline for *in situ* x-ray absorption spectroscopy with 10 ms time resolution," *J. Synchrotron Rad.*, vol. 23, pp. 260–266, 2016.

[23]   S. Oestreich, B. Kaulich, and J. Susini, "Single-movement fixed-exit channelcut x-ray monochromator based on profiled surfaces," *Rev. Sci. Instrum.*, vol. 70, pp. 1921–1925, 1999.

[24]   P. Spieker, M. Ando, and N. Kamiya, "A monolithic x-ray monochromator with fixed exit beam position," *Nucl. Instrum. Methods A*, vol. 222, pp. 196–201, 1984.

[25]   A. Marberger, A. W. Petrov, P. Steiger, et al., "Time-resolved copper speciation during selective catalytic reduction of NO on Cu-SSZ-13," *Nature Catalysis*, vol. 1, pp. 221–227, 2018.

[26]   H. Hayashi, R. Takeda, Y. Udagawa, et al., "Lifetime-broadening-suppressed/free XANES spectroscopy by high-resolution resonant inelastic x-ray scattering," *Phys. Rev. B*, vol. 68, p. 045122, 2003.

[27]   W. Blachucki, J. Szlachetko, J. Hoszowska, et al., "High energy resolution off-resonant spectroscopy for x-ray absorption spectra free of self-absorption effects," *Phys. Rev. Lett.*, vol. 112, p. 173003, 2014.

[28]   J. Szlachetko, C. J. Milne, J. Hoszowska, et al., "Communication: the electronic structure of matter probed with a single femtosecond hard x-ray pulse," *Struct. Dyn.*, vol. 1, p. 021101, 2014.

[29]   J. Dik, K. Janssens, G. Van der Snickt, et al., "Visualization of a lost painting by Vincent van Gogh using synchrotron radiation based X-ray fluorescence elemental mapping," *Anal. Chem.*, vol. 80, pp. 6436–6442, 2008.

[30]   D. Thurrowgood, D. Paterson, M. D. de Jonge, et al., "A hidden portrait by Edgar Degas," *Sci. Reports*, vol. 6, p. 29594, 2016.

[31]   U. Bergmann, R. W. Morton, P. L. Manning, et al., "Archaeopteryx feathers and bone chemistry fully revealed via synchrotron imaging," *Proc. Natl. Acad. Sci. USA*, vol. 107, pp. 9060–9065, 2010.

[32]   S. Eisebitt, J. Lüning, J. E. Rubensson, and W. Eberhardt, "Resonant inelastic soft X-ray scattering as a bandstructure probe: A primer," *Phys. Stat. Sol. B*, vol. 215, pp. 803–808, 1999.

[33]   A. Kotani and S. Shin, "Resonant inelastic x-ray scattering spectra for electrons in solids," *Rev. Mod. Phys.*, vol. 73, pp. 203–246, 2001.

[34]   L. J. P. Ament, M. van Veenendaal, T. P. Devereaux, J. P. Hill, and J. van den Brink, "Resonant inelastic x-ray studies of elementary excitations," *Rev. Mod. Phys.*, vol. 83, pp. 705–767, 2011.

[35]   J. Schlappa, K. Wohlfeld, K. J. Zhou, et al., "Spin-orbital separation in the quasi-one-dimensional Mott insulator Sr_2CuO_3," *Nature*, vol. 485, pp. 82–86, 2012.

[36]   G. Ghiringhelli, A. Piazzalunga, C. Dallera, et al., "SAXES, a high resolution spectrometer for resonant x-ray emission in the 400-1600 eV energy range," *Rev. Sci. Instrum.*, vol. 77, p. 113108, 2006.

[37]   V. N. Strocov, T. Schmitt, U. Flechsig, et al., "High-resolution soft x-ray beamline ADRESS at the Swiss Light Source for resonant inelastic x-ray scattering and angle-resolved photoelectron spectroscopies," *J. Synchrotron Rad.*, vol. 17, pp. 631–643, 2010.

[38]   J. P. Hill, D. S. Coburn, Y.-J. Kim, et al., "A 2 m inelastic X-ray scattering spectrometer at CMC-XOR, advanced photon source," *J. Synchrotron Rad.*, vol. 14, pp. 361–365, 2007.

[39]   B. W. Batterman, "Effect of dynamical diffraction in x-ray fluorescence scattering," *Phys. Rev.*, vol. 133, pp. A759–A764, 1964.

[40]   B. W. Batterman and H. Cole, "Dynamical diffraction of x-rays by perfect crystals," *Rev. Mod. Phys.*, vol. 36, pp. 681–717, 1964.

[41]   S. K. Andersen, J. A. Golovchenko, and G. Mair, "New applications of x-ray standing-wave fields to solid-state physics," *Phys. Rev. Lett.*, vol. 37, pp. 1141–1145, 1976.

[42]   J. Zegenhagen, "Surface-structure determination with x-ray standing waves," *Surf. Sci. Rep.*, vol. 18, pp. 199–271, 1993.

[43]   J. Zegenhagen, B. Detlefs, T.-L. Lee, et al., "X-ray standing waves and hard X-ray photoelectron spectroscopy at the insertion device beamline ID32," *J. Electron Spectrosc. Rel. Phenom.*, vol. 178, pp. 258–267, 2010.

[44] M. J. Bedzyk, D. H. Bilderback, G. M. Bommarito, M. Caffrey, and J. S. Schildkraut, "X-ray standing waves – a molecular yardstick for biological membranes," *Science*, vol. 241, pp. 1788–1791, 1988.

[45] O. Dhez, H. Ade, and S. G. Urquhart, "Calibrated NEXAFS spectra of some common polymers," *J. Electron Spectrosc. Rel. Phenom.*, vol. 128, pp. 85–96, 2003.

[46] B. Watts and C. R. McNeill, "Simultaneous surface and bulk imaging of polymer blends with x-ray spectromicroscopy," *Macromol. Rapid Commun.*, vol. 31, pp. 1706–1712, 2010.

[47] S. Anders, H. A. Padmore, R. M. Duarte, et al., "Photoemission electron microscope for the study of magnetic materials," *Rev. Sci. Instrum.*, vol. 70, pp. 3973–3981, 1999.

[48] J. Stöhr, "Exploring the microscopic origin of magnetic anisotropies with X-ray magnetic circular dichroism (XMCD) spectroscopy," *J. Magn. Magn. Mater.*, vol. 200, pp. 470–497, 1999.

[49] J. Raabe, C. Quitmann, C. H. Back, et al., "Quantitative analysis of magnetic excitations in Landau flux-closure structures using synchrotron-radiation microscopy," *Phys. Rev. Lett.*, vol. 94, p. 217204, 2005.

[50] G. van der Laan, B. T. Thole, G. A. Sawatzky, et al., "Experimental proof of magnetic x-ray dichroism," *Phys. Rev. B*, vol. 34, pp. 6529–6531, 1986.

[51] C. S. Fadley, "X-ray photoelectron spectroscopy: Progress and perspectives," *J. Electron Spectrosc. Rel. Phenom.*, vol. 178, pp. 2–32, 2010.

[52] C. Kittel, *Introduction to Solid State Physics*. John Wiley, 8th edn., 2005.

[53] N. W. Ashcroft and N. D. Mermin, *Solid State Physics*. Brooks Cole, 1st edn., 1976.

[54] P. Aebi, J. Osterwalder, R. Fasel, D. Naumovic, and L. Schlapbach, "Fermi-surface mapping with photoelectrons at UV energies," *Surf. Sci.*, vol. 307, pp. 917–921, 1994.

[55] A. Damascelli, Z. Hussain, and Z. X. Shen, "Angle-resolved photoemission studies of the cuprate superconductors," *Rev. Mod. Phys.*, vol. 75, pp. 473–541, 2003.

[56] X. Y. Cui, E. E. Krasovskii, V. N. Strocov, et al., "Final-state effects in high-resolution angle-resolved photoemission from Ni(110)," *Phys. Rev. B*, vol. 81, p. 245118, 2010.

[57] A. Sekiyama, S. Kasai, M. Tsunekawa, et al., "Technique for bulk Fermiology by photoemission applied to layered ruthenates," *Phys. Rev. B*, vol. 70, p. 060506(R), 2004.

[58] V. N. Strocov, M. Kobayashi, X. Wang, et al., "Soft x-ray-ARPES at the Swiss Light Source: from 3D materials to buried interfaces and impurities," *Synch. Rad. News*, vol. 27, pp. 31–40, 2014.

[59] L. L. Lev, I. O. Maiboroda, M.-A. Husanu, et al., "k-space imaging of anisotropic 2D electron gas in GaN/GaAlN high-electron-mobility transistor heterostructures," *Nature Comms.*, vol. 9, p. 2653, 2018.

[60] C. Cancellieri, M. L. Reinle-Schmitt, M. Kobayashi, et al., "Interface Fermi states of $LaAlO_3$/$SrTiO_3$ and related heterostructures," *Phys. Rev. Lett.*, vol. 110, p. 137601, 2013.

[61] G. Margaritondo, "Photoelectron spectromicroscopy and spectronanoscopy at synchrotrons: Growing impact on life sciences and materials science," *J. Electron Spectrosc. Rel. Phenom.*, vol. 178, pp. 273–291, 2010.

[62] R. W. Joyner, M. W. Roberts, and K. Yates, "A "high-pressure" electron spectrometer for surface studies," *Surf. Sci.*, vol. 87, pp. 501–509, 1979.

[63] D. F. Ogletree, H. Bluhm, G. Lebedev, et al., "A differentially pumped electrostatic lens system for photoemission studies in the millibar range," *Rev. Sci. Instrum.*, vol. 73, pp. 3872–3877, 2002.

[64] M. Salmeron and R. Schlögl, "Ambient pressure photoelectron spectroscopy: A new tool for surface science and nanotechnology," *Surf. Sci. Rep.*, vol. 63, pp. 169–199, 2008.

[65] F. Tao, M. E. Grass, Y. Zhang, et al., "Reaction-driven restructuring of Rh-Pd and Pt-Pd core-shell nanoparticles," *Science*, vol. 322, pp. 932–934, 2008.

[66] S. Blomberg, M. J. Hoffmann, J. Gustafson, et al., "In situ x-ray photoelectron spectroscopy of model catalysts: at the edge of the gap," *Phys. Rev. Lett.*, vol. 110, p. 117601, 2013.

[67] M. A. Brown, A. B. Redondo, I. Jordan, et al., "A new endstation at the Swiss Light Source for ultraviolet photoelectron spectroscopy, x-ray photoelectron spectroscopy, and x-ray absorption spectroscopy measurements of liquid solutions," *Rev. Sci. Instrum.*, vol. 84, p. 073904, 2013.

[68] D. E. Starr, E. K. Wong, D. R. Worsnop, K. R. Wilson, and H. Bluhm, "A combined droplet train and ambient pressure photoemission spectrometer for the investigation of liquid/vapor interfaces," *Phys. Chem. Chem. Phys.*, vol. 10, pp. 3093–3098, 2008.

[69] S. Axnanda, E. J. Crumlin, B. Mao, et al., "Using "tender" x-ray ambient pressure x-ray photoelectron spectroscopy as a direct probe of solid-liquid interface," *Sci. Reports*, vol. 5, p. 09788, 2015.

[70] C. Nordling, E. Sokolowski, and K. Siegbahn, "Evidence of chemical shifts of the inner electronic levels in a metal relative to its oxides (Cu, Cu_2O, CuO)," *Arkiv f. Physik*, vol. 13, pp. 483–506, 1958.

[71] E. Sokolowski, C. Nordling, and K. Siegbahn, "Chemical shift effect in inner electronic levels of Cu due to oxidation," *Phys. Rev.*, vol. 110, pp. 776–776, 1958.

[72] I. Lindau, P. Pianetta, S. Doniach, and W. E. Spicer, "X-ray photoemission spectroscopy," *Nature*, vol. 250, pp. 214–215, 1974.

[73] L. Kover, "X-ray photoelectron spectroscopy using hard X-rays," *J. Electron Spectrosc. Rel. Phenom.*, vol. 178, pp. 241–257, 2010.

[74] Y. Takata, Y. Kayanuma, M. Yabashi, et al., "Recoil effects of photoelectrons in a solid," *Phys. Rev. B*, vol. 75, p. 233404, 2007.

8

Imaging Techniques

8.1 Introduction

Synchrotrons are often described to the general public as being very large and powerful microscopes. With some notable exceptions, this generalization holds true (see Figure 8.1). For example, coherent x-ray diffractive imaging (CXDI), described in this chapter, is based on the same physical phenomena as x-ray diffraction (elastic scattering); recovery of the real-space structure is, like XRD, realized through phasing. Although most crystallographers would not necessarily regard XRD as being first and foremost an imaging technique, in the majority of studies, it really is – through Fourier analysis, the three-dimensional arrangement of the constituent atoms in crystals is recovered. Yet XRD has already been covered in a separate chapter.

It is thus apparent that imaging and microscopy techniques using x-rays extend across most x-ray disciplines. Many of these have already been discussed within the context of other phenomena – for example, scanning transmission x-ray microscopy (STXM) and photoemission electron-microscopy (PEEM) were introduced as two distinct (scanning microspectroscopic and spectromicroscopic, respectively) imaging modes of x-ray absorption near-edge structure (XANES) in Sections 7.7 and 7.8.

It seems, therefore, that x-ray imaging using synchrotron radiation is ubiquitous in modern facilities; what disciplines, then, qualify to be included in this chapter?

The choices are, to a large extent, arbitrary. Certainly, it would have been difficult to include classical absorption contrast x-ray microtomography (XTM) in either Chapters 6 or 7. Because XTM and ptychographic nanotomography are so closely linked, the latter is also covered here. For the sake of completeness, CXDI, which can be thought of as being the necessary 'predecessor' to ptychography, is likewise handled in this chapter.

Many materials, including biological structures and biomimetic systems, are characterized by their nested, or hierarchical, ordering over a large range of length scales. In nature, the evolutionary drive for this has been to maximally exploit a material's strength while simultaneously minimizing mass and energy cost. A classic example of such biological 'ultrastructures' is the nested architecture of bone (Figure 8.2). A challenge in modern imaging in fields as diverse as molecular biology and chemical engineering is to investigate the full range of scales; this challenge is being met by recent advances in x-ray full-field tomographies and scanning lensless techniques.

We begin with XTM, a technique with its roots in the oldest and most widespread x-ray application, radiographic imaging.

An Introduction to Synchrotron Radiation: Techniques and Applications, Second Edition. Philip Willmott.
© 2019 John Wiley & Sons Ltd. Published 2019 by John Wiley & Sons Ltd.

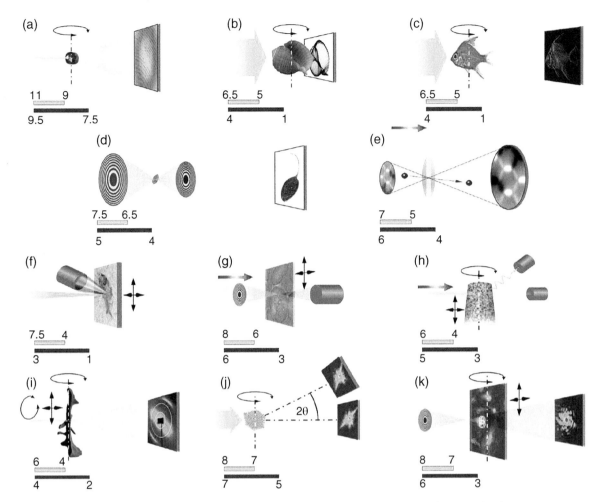

Figure 8.1 A gallery of imaging modes using x-rays covered in this book. (a) X-ray diffraction; (b) absorption-contrast XTM; (c) phase-contrast XTM; (d) full-field hard-x-ray (including Zernike) microscopy; (e) PEEM/XMCD spectromicroscopy; (f) scanning XRF microspectroscopy; (g) STXM; (h) chemical line tomography; (i) scanning SAXS; (j) CXDI and Bragg CXDI; (k) ptychography/ptychographic tomography. The rainbow-coloured arrows indicate that the photon energy is tuned as a necessary part of the experiment. The numbers at the ends of the yellow bars show the exponent X (in 10^{-X}) of the typical lower and upper limits of the resolution of the technique in metres, while the blue bar likewise indicates typical linear sample dimensions (in the case of XRD, this is the unit-cell size).

8.2   X-ray Computed Microtomography

8.2.1   Introduction

It is the relative transparency of condensed matter to hard x-rays that makes them such a powerful tool for investigating the internal structures of objects in a nondestructive manner. Indeed, this benefit of x-rays was the first to be recognized. In an age where medical radiographs are so commonplace, it is perhaps difficult for us to appreciate just how sensational x-ray radiographs must have been immediately after the discovery of x-rays more than a century ago (Figure 1.7).

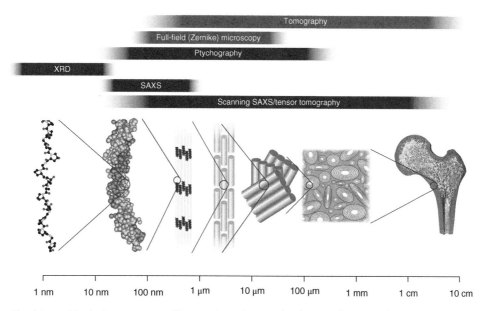

Figure 8.2   The hierarchical ultrastructure of bone. The polypeptide-chain collagen molecule combines in triplets to form helical tropocollagen. This is bunched and cross-linked in staggered units about 67 nm long interleaved with mineral crystals to form microfibrils, which in turn form first fibrils, then lamellae of fibres. These combine in larger structures called osteons on the 10 μm to 1 mm scale, the largest features of cortical bone. The bars at the top represent the typical length scales covered by different imaging techniques. Red bars correspond to real-space techniques, blue to scattering (reciprocal-space) methods.

The word 'tomography' is derived from the Greek words 'tomos' meaning 'to slice' and 'graph' meaning 'image'. Tomography in its most general sense is the generation of a three-dimensional image by analysing several transmission radiographic projections taken of a specimen at different angles. Absorption-contrast tomography requires radiation which has an absorption length of the same order of magnitude as the size of the specimen under investigation. Nowadays, computed tomography (CT) with a resolution of approximately 500 μm is a standard diagnostic medical imaging tool using x-ray tubes in hospitals around the world (Figure 8.3). It is estimated that approximately 100 million CT scans per year are performed in the United States alone.

High-resolution x-ray tomographic microscopy (XTM) is a synchrotron-based, nondestructive technique for creating quantitative, three-dimensional images of solid objects, including their detailed internal architecture, with a spatial resolution which can extend below a micron [1–3]. It is useful for a wide range of materials, including inorganic substances such as rock, ceramic, or metal, and also for objects made from bone and soft tissue.

The elements of a tomographic reconstruction, the so-called *voxels* (short for 'volumetric pixel'), provide a quantitative three-dimensional structure, which one can then view as one pleases, for example, sliced across any desired cross-section, or rendered to highlight boundaries between materials of different density, to name just two possibilities. The contrast between voxels is in general higher than that between neighbouring pixels in a standard radiographic projection, as in the latter, one records for any given pixel the integrated contribution of all the voxels that lie in path of the x-rays impinging on that pixel (Figure 8.4).

The tuneability of synchrotron radiation can be exploited to record tomograms above and below absorption edges of certain elements, thereby enhancing three-dimensional maps of their concentrations. An example of the diffusion of the contaminant caesium in a geological medium is shown in Figure 8.5.

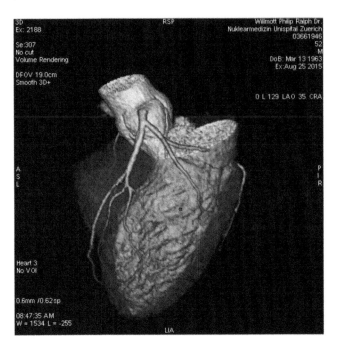

Figure 8.3 Medical x-ray computed tomogram of a human heart.

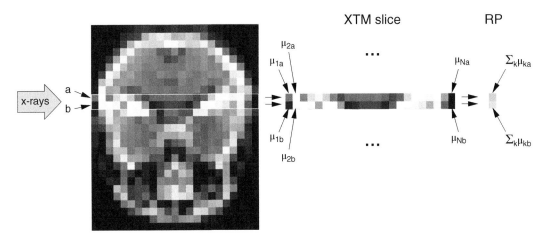

Figure 8.4 Tomographic versus radiographic imaging. The contrast in attenuation strength between adjacent voxels μ_{ka} and μ_{kb} can be high in a reconstructed tomogram slice. Along the x-ray propagation direction, however, the contrast between the integrated attenuation strength of neighbouring pixels in the radiographic projection (RP) will be small.

8.2.2 General Concepts

A typical experimental setup for recording XTM data is shown schematically in Figure 8.6. A sample is rotated through 180° around an axis perpendicular to the quasi-parallel incident x-ray beam. The transmitted signal is recorded at regular intervals during the rotation using a scintillator, magnification optics, and CCD camera. The

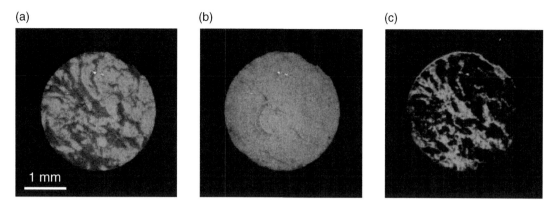

Figure 8.5   Diffusion of a 'contaminant plume' of caesium in a geological sample. Contrast in the elemental distribution of Cs was enhanced by recording radiographic projections below and above the Cs K-edge at 35.985 keV. (a) Tomogram at 37 keV, (b) the same tomogram recorded at 35 keV, and (c) the difference (a)–(b), clearly revealing the Cs distribution. Courtesy Daniel Grolimund, Paul Scherrer Institute.

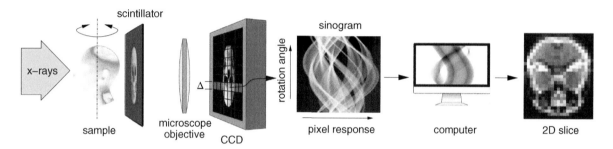

Figure 8.6   The principle of x-ray computed microtomography. A parallel beam of x-rays is allowed to pass through a sample, and the transmitted signal is detected by a scintillator, which is magnified using a microscope objective lens onto a CCD array. The response from each row of pixels (size Δ) as a function of sample-rotation angle, the so-called 'sinograms', are fed into a computer, from which two-dimensional tomograms are generated. Finally, the 3D reconstruction is created from the stack of 2D tomograms.

spatial resolution of the image is therefore determined by five factors, namely the point-spread function of the scintillator, the magnification factor of the microscope objective optics M, the signal-to-noise ratio of the CCD image, and the linear pixel size Δ and point-spread function of the CCD camera. The magnification can vary typically between $1 \leq M \leq 50$, depending on the desired resolution and size of the illuminated object.

The number of projections N of the sample which must be recorded within the angular range is given by the so-called 'sampling theorem' and is equal to

$$N = N_p \times \pi/2, \tag{8.1}$$

whereby N_p is the number of resolution elements in the row that provide transmission data.

From each row in the CCD array, a tomogram is generated. To obtain this, the response of the N_p pixels in the row as a function of rotation angle, known as a *sinogram*, is fed into a computer. An example of a sinogram of a simple system is shown in Figure 8.7.

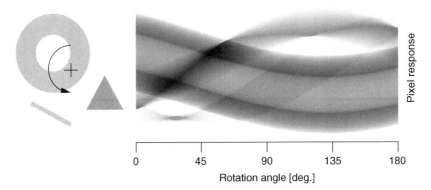

Figure 8.7   The generation of sinograms. The response of a line of pixels in the plane of three partially absorbing, simple geometrical two-dimensional objects as a function of their rotation around the crosshair between 0° and 180° results in the corresponding sinogram.

The response of a detector pixel is proportional to the intensity of the transmitted x-rays that impinges upon it after travelling along a path $L = M\Delta z$ through an object, that is

$$I = I_0 e^{-\mu_1 \Delta z}\, e^{-\mu_2 \Delta z}\, e^{-\mu_3 \Delta z} \cdots e^{-\mu_M \Delta z} \tag{8.2}$$

$$= I_0 \exp\left(-\sum_{k=1}^{M} \mu_k \Delta z\right). \tag{8.3}$$

We define the *projection measurement p* as

$$p = -\ln\left(\frac{I}{I_0}\right) = \sum_{k=1}^{M} \mu_k \Delta z \approx \int_L \mu(z)dz. \tag{8.4}$$

In other words, p for a given pixel at a given angle is the line integral of the object's attenuation coefficient, related to the imaginary part of the refractive index, along the line L. Given these sets of line integrals for different angles, the task is to reconstruct the attenuation distribution throughout the entire object.

Each tomogram is reconstructed from its respective sinogram using one of several algorithms based on the 'Radon transform', developed by Johann Radon in 1917 [4], called filtered back-projection (FBP). The basic idea behind FBP is simple to understand. Consider the simple 2D objects shown in Figure 8.8. In one direction, the radiographic absorption profile is recorded. This projection is then run back through the image, whereby the projection intensity is evenly distributed among all the pixels along each ray path. This is then repeated for a set of projection angles spanning 180°, whereby for each angle, the back projection is added to the image. The smaller the angular shift, the more closely the final overlapping set of back-projections resembles the original object (Figure 8.8).

Although the general features of the original object can be retrieved in this manner, there is significant distortion due to overlapping projections producing star-like artefacts and a blurring, which becomes increasingly evident near the centre of the reconstruction. These artefacts can be suppressed by careful filtering of the absorption profiles. This is carried out by Fourier analysis – each profile is Fourier-transformed and then treated with a filter, which suppresses certain ranges of the Fourier components. Lower frequency components tend to be responsible for blurring, while sharper, high-frequency, features might be removed in order to smooth statistical noise. A common type of filter is shown in Figure 8.9. In real-space, this filter function, or 'kernel', has the appearance

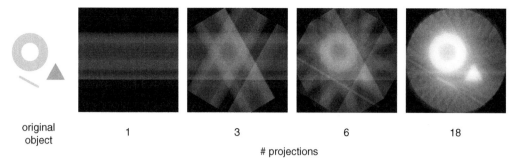

original
object

1 3 6 18

projections

Figure 8.8 Illustration of filtered back-projection in XTM. The transmission profile of an object is recorded in several direction. For each direction, the profile is run back through the image space. Those regions in the space where the back projections overlap add up. The more projections are added, the closer their summation resembles the original object. Without filtering, the reconstructed image suffers from blurring and star-like artefacts.

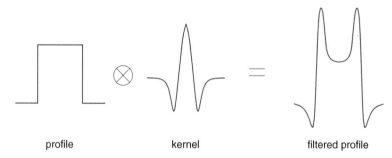

profile kernel filtered profile

Figure 8.9 Filters for back projections. A 'kernel' is convoluted with the original profile to produce a filtered profile which compensates the blurring due to overlapping of adjacent back projections. The detailed shape of the kernel determines the spatial frequencies most affected.

of a distorted 'w', which is convoluted with the back projection profile. In Fourier space, the FT of the profile is therefore *multiplied* by the FT of the kernel (according to the convolution theorem, see Section 6.5.1). The detailed shape of the kernel's FT determines which frequencies are filtered out.

Advances in processing power of modern computers now allow tomogram slices to be rapidly and more directly constructed using fast-Fourier-transform algorithms and exploiting the so-called 'Fourier-slice theorem'. Shown schematically in Figure 8.10, this states that the one dimensional Fourier transform of the projection of a two-dimensional object (i.e. a slice) onto a line (i.e. the transmission profile at the height of the slice) is equal to the one-dimensional slice through the centre of the two-dimensional Fourier transform of the same object. So, by Fourier transforming in one dimension a given projection, we obtain a line in the 2D-Fourier transform of the object. By collecting all the projections between 0 and 180° (the sinogram), we fill the entire Fourier-space representation of the object. It is then a simple procedure to first interpolate this data from its polar grid to a Cartesian coordinate system and then recover the object (i.e. the tomogram) by inverse Fourier transforming the data.

The final three-dimensional reconstruction consists of a stack of all the reconstructed tomograms. The resolution of the reconstruction is determined by the dimensions of voxels, or resolution elements. Tomographic images comprise an array of these voxels on a regular three-dimensional grid, whereby each voxel is associated with a certain physical value, such as the average linear absorption coefficient within that voxel, or the phase shift induced by the voxel.

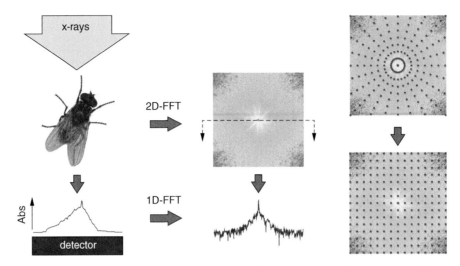

Figure 8.10 The Fourier slice theorem. The one-dimensional (fast) Fourier transform (1D-FFT) of a projection of an object yields the same result as the central slice of the 2D-FFT of the same object. After rotating the sample through 180°, the radially distributed FT data must be interpolated to an evenly distributed Cartesian data set.

8.2.3 Practical Considerations

Each radiographic projection of a data set must be corrected for two factors: 'dark' noise generated by the detector electronics of the CCD camera in the absence of an x-ray beam; and inhomogeneities in the incident x-ray beam profile. The first factor is corrected by recording a 'dark-field' image in the absence of x-rays, while the second 'flat-field' image records the direct beam with the sample removed.

The sample sits on a stage, which rotates through 180°. While rotating, the rotation axis can drift laterally (the *eccentricity*), and can also *wobble*. State-of-the-art rotation stages offer eccentricities and wobbles of the order of a few nanometres and less than a microradian, respectively [5, 6].

XTM images sometimes contain features which are artefacts of the technique. Regions in the centre of the object appear to have lower attenuation than identical regions near its periphery, if significantly polychromatic radiation ('pink beam') is used. The radial distribution of the photon energy of a polychromatic x-ray beam emanating from a bending magnet or wiggler at a synchrotron facility is such that the high-energy photons are more concentrated in the centre of the beam than the soft x-rays, which become more dominant at the peripheries. Because the lower-energy x-rays in the beam are more strongly absorbed than are the higher-energy x-rays, the energy-distribution spectrum of the beam changes as it passes through the object. The emergent beam therefore contains a higher proportion of hard x-rays; this phenomenon is known as 'beam-hardening'. This shift in the distribution of photon energies in the x-ray beam manifests itself during image reconstruction as an apparent decrease in attenuation in the centre of the object[1].

In many XTM experiments, the highest degree of monochromacity is not needed, while the demand for high photon rates is paramount. Multilayer monochromators, described in Section 5.4.2, provide bandwidths of the order of $\Delta E/E = 0.01$, some hundred times larger than those using single-crystal monochromators. With the advent of the much cleaner undulator spectra at DLSRs, a similar increase can be obtained by selecting the entirety of any given harmonic, described in detail in Section 3.9.4.

---

[1] Note that CT experiments that use broadband laboratory x-ray sources also suffer from beam hardening, although in this case, border regions tend to seem anomalously bright (in other words, they seem to have a *low* attenuation). The x-ray beam from a lab source is spatially homogeneous with regards to its spectral content, but because borders tend to present a shorter absorption path for the x-rays, here, a disproportionately high fraction of soft x-rays are transmitted.

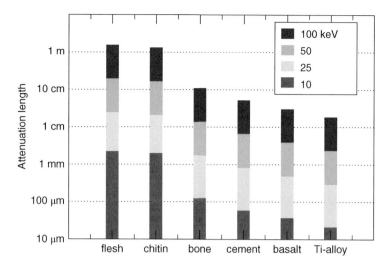

Figure 8.11 The attenuation lengths $1/\mu$ of six materials commonly investigated in x-ray imaging techniques for different photon energies. The following chemical compositions and mass densities were used. Flesh: $C_{12}H_{53}O_{18}N$, $\rho = 0.985$ g cm^{-3}; chitin: $C_{16}H_{28}O_{11}N_2$, $\rho = 1.4$ g cm^{-3}; bone (hydroxylapatite): $Ca_5P_3O_{13}H$, $\rho = 1.75$ g cm^{-3}; Portland cement: Ca_3SiO_5, $\rho = 3.15$ g cm^{-3}; basalt: $Si_{50}OK_4O_2Na_6O_3Al_{32}O_{48}Mg_7O_7Ca_{10}O_{10}Fe_{10}O_{10}Ti_2O_4$, $\rho = 2.9$ g cm^{-3}; aeronautical Ti alloy: $Ti_{89}Fe_{0.2}Al_6V_4$, $\rho = 4.43$ g cm^{-3}.

8.2.4 Phase-contrast Tomography

8.2.4.1 Introduction

Soft matter presents two problems in imaging experiments using x-rays. A primary goal of x-ray imaging of soft matter is to distinguish structural features within a sample. The photoabsorption attenuation lengths of representative materials are shown in Figure 8.11. When soft tissue or other samples containing components that differ in their absorption lengths by a small fraction, and/or whose absorption lengths are significantly larger than the size of the sample (that is, $\mu z \ll 1$), absorption contrast becomes very weak.

The contrast in an absorption radiograph between two points A and B is defined as being

$$C_a = \frac{I_A - I_B}{I_A},\tag{8.5}$$

where $I_A > I_B$. For an equal distance z passed through the sample to produce the intensities at A and B, we have

$$C_a = \frac{\exp(-\mu_A z) - \exp(-\mu_B z)}{\exp(-\mu_A z)}$$
$$= 1 - \exp[(\mu_A - \mu_B)z].$$

For weakly absorbing materials with similar attenuation coefficients ($\mu_A z \sim \mu_B z \ll 1$), this is approximated by

$$C_a = (\mu_B - \mu_A)z.\tag{8.6}$$

Let us consider the illustrative example of a sample consisting of distinct volumes of fat (mass density 0.9 g cm^{-3}) and muscle (1.06 g cm^{-3}), probed using 20 keV (0.62 Å) radiation. At this energy, the absorption coefficients of fat and muscle are, respectively, $\mu_f = 5.3 \times 10^{-5}$ μm^{-1} and $\mu_m = 6.2 \times 10^{-5}$ μm^{-1}. The contrast is therefore $9 \times 10^{-6}z$, whereby z is in microns. So, for example, for $z = 1$ mm, the contrast remains less than 1% [7].

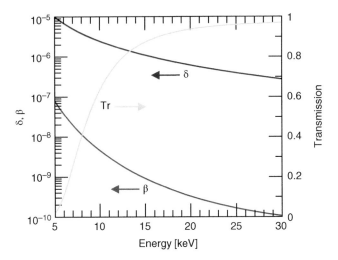

Figure 8.12   The refractive index decrement δ and absorption index β for flesh from 5 to 30 keV, plus the transmission through a 1 mm sample. To calculate these curves, the chemical formula for flesh was taken as $C_{12}H_{53}O_{18}N$ and the density as 0.985 g cm^{-3}.

One solution is to lower the photon energy, increase the photoabsorption coefficient, and thereby obtain higher contrast [8, 9]. Although such approaches have been pursued, they suffer from (a) a dramatic increase in absorbed dose and rapid degradation of the sample (particularly true for soft organic matter), (b) a need to evacuate the sample environment and flight tube to the sample in order to avoid absorption by air, and (c) the need to cryocool the sample, a consequence of both (a) and (b).

We have already discussed the variation with photon energy of the refractive-index decrement δ and the absorption index β in Chapter 2 [Equation (2.45)]. δ scales as $(h\nu)^{-2}$, β as $(h\nu)^{-4}$. In other words, absorption falls off much more steeply with energy than do refraction effects. This is most pronounced for lighter elements, as β scales as Z^4, while δ is approximately proportional to Z.

Exploiting fluctuations in δ can thus be used to advantage in so-called phase-contrast imaging techniques.

8.2.4.2   *Edge Diffraction, Refraction, and Propagation-based Tomography*

The phenomena of refraction and diffraction are both used to enhance image quality in tomography. Refraction effects have no requirements on temporal coherence (monochromacity) and phase shifts are produced simply by the propagation of the x-rays through different types of material.

The effect of δ alone (ignoring absorption) is to cause the wavefronts within the medium to expand by an amount $(1-\delta)^{-1}$ as the phase velocity of the x-rays increases from c in vacuum to c/n in the medium. Hence two parallel x-ray beams which are initially in phase become out of phase by an amount

$$\phi = \frac{2\pi L\delta}{\lambda} \tag{8.7}$$

if one of these passes through a refracting sample of length L and refractive index n (Figure 8.13). Hence, the part of the beam propagating through the medium a distance $L = \lambda/\delta \sim 100\,\mu$m for 0.5 Å x-rays in flesh (Figure 8.12), will have advanced by a phase shift of approximately π relative to the unperturbed part of the beam. If these two beams are allowed to overlap, they will produce a fluctuation in the intensity in the region propagated downstream of the heterogeneous interface.

Here, we discuss edge-enhanced imaging due to the two phenomena of refraction and Fresnel edge diffraction. Under the correct conditions, these effects cause the borders of objects to be highlighted, somewhat like a line

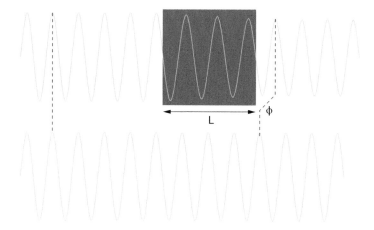

Figure 8.13 The part of a beam which travels through a medium of refractive index n has a phase velocity $v_p = c/n > c$ – the peaks and troughs speed up. The result is that the emerging beam is phase shifted by an amount ϕ relative to the part of the beam travelling through vacuum.

drawing, even for object with very low absorption contrast. First, we should identify under which geometrical arrangement one observes and exploits edge enhancement and diffraction phenomena.

When a coherent wave is scattered by an object, interference between the differently scattered components produces diffraction features. This requires a certain propagation distance, however, for the different parts of the scattered wave to overlap. Observed close to the object, therefore, the scattered pattern still resembles the object, albeit with increasing evidence of diffraction phenomena with distance z from the object. The so-called Fresnel-diffraction regime extends until the diffraction features (fringes) are separated by a distance similar to the linear dimensions of the scattering object itself (a in Figure 8.14), that is, when

$$F_0 = \frac{a^2}{z\lambda} \sim 1,$$

where F_0 is known as the Fresnel number. At very large distances $F_0 \ll a^2/z\lambda$, in the so-called far-field, or Fraunhofer-diffraction, regime, the object profile is completely obscured by the diffraction effects, and the pattern is in fact the square of the Fourier transform of the object (assuming the object to be smaller than the spatial coherence of the incident beam). This is further discussed in Section 8.4.

Fresnel edge diffraction is observed in both fully opaque and transparent objects. In order to understand its origin, we begin by introducing the concept of the Huygens–Fresnel construction. This imagines that every point along a wavefront (let us say, for convenience's sake, the wave maximum) is itself a point emitter of radiation of the same wavelength. In vacuum, interference between the spherically propagating wavelets emanating from the row of point sources causes the amplitude in all directions other than that perpendicular to the row to cancel out, thereby generating the next wavefront parallel to the previous [Figure 8.15(a)]. If some sort of object lies in the path of part of the beam, the wavelets passing through that object assume a wavelength $\lambda/n > \lambda$ and this fraction of the beam is refracted and will subsequently interfere with other parts of the beam that travelled along a different path.

So, for example, the pattern on a detector at a distance z from a parallel beam of x-rays partially blocked by an opaque screen is not a step function at z, but instead exhibits interference fringes in the region close to the projection of the edge of the screen [Figure 8.15(b)]. Although it goes beyond the scope of this book to derive an

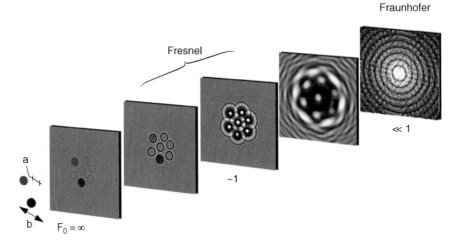

Figure 8.14 The regimes of Fresnel and Fraunhofer diffraction. The transmission function of an object (here, composed of five transparent 'phase' discs and two partially absorbing discs of diameter a, distributed over an area of characteristic length b) changes with propagation distance z. As $F_0 = a^2/z\lambda$, the dimensionless Fresnel number is decreased from infinity at $z = 0$, the projected image of the object develops fringes at the boundaries between the discs and the background, where the refractive-index decrement makes a step-like change. In the Fresnel regime at $F_0 \sim 1$, the object remains recognizable, but becomes increasingly fuzzy at the boundaries with increasing z, due to the formation of fringes. At very large distances $F_0 \ll 1$, the detector records the far-field diffraction pattern of the object in the so-called Fraunhofer-diffraction regime, the condition for 'normal' diffraction and SAXS experiments.

expression for this pattern[2], we can gain a physical insight using the Huygens–Fresnel principle. If the propagation of a planar wavefront can be considered to be formed by a row of point emitters as argued above, then removing those emitters below the opaque screen means that they cannot interfere with those not blocked by the screen and the constructed wavefront must therefore deviate from a perfect plane wave, particularly in the neighbourhood of the screen edge. In contrast, the missing contribution to the planar wave for the constructed wave far above the screen is negligible, as here the amplitudes of the missing wavelets, which drop off inversely with the propagation radius, are very small.

Where can we expect the first interference minimum to occur? Consider Figure 8.15(c). We estimate x_1, the lateral displacement of the first dark fringe relative to the geometric projection of the screen edge, by assuming that here a ray A associated with the forward-propagating planar wavefront interferes destructively with ray B originating from the last point source before the plane wave is occluded by the screen edge. We next make the assumption that x_1 is small compared to the screen–detector distance z, in other words θ is small and $\sin\theta \approx \tan\theta$, from which

$$\frac{x_1}{z} \approx \frac{\text{OPD}}{x_1}, \tag{8.8}$$

where OPD $= \lambda/2$ is the optical path difference. From this, we obtain

$$x_1 = \sqrt{\frac{z\lambda}{2}}. \tag{8.9}$$

---

[2] Its calculation involves the so-called 'Fresnel integrals', $C(v)$ and $S(v)$, and is equal to $\{[1/2 - C(v)]^2 + [1/2 - S(v)]^2\}/2$.

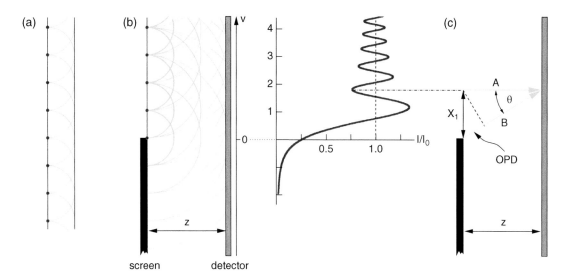

Figure 8.15   Edge diffraction and enhancement. (a) A propagating wavefront can be constructed by assuming each point on that wavefront acts as a source emitting spherical waves. Only in the direction of propagation of the plane wave is interference between the wavelets constructive, resulting in a new wavefront. (b) If the wavefront is partially blocked by an opaque screen, the wavelets from this part are missing, and the new wavefront can no longer form a perfect plane wave, as a fraction of the contributing components has been blocked – as a result, interference fringes occur. (c) Simplified geometric explanation for the occurrence of the first dark interference fringe, whereby the optical path difference (OPD) between the forward propagating plane wave and a wavelet originating from the screen edge is $\lambda/2$.

Hence, the further away the detector is from the screen, the larger the fringe separations. We therefore introduce the dimensionless parameter

$$v = x\sqrt{\frac{2}{z\lambda}}. \tag{8.10}$$

Fringe patterns from an opaque edge are thus invariant if plotted as a function of v, as in Figure 8.15(b). Here, we see that in fact $v \approx 1.8$ for the first fringe minimum, that is, we have underestimated x_1 by a factor of 1.8. This is explained by realizing that we have ignored all the other wavelets between rays A and B which still contribute partially constructively. Note that because the positions of the fringes are proportional to $\sqrt{\lambda}$, one still observes edge diffraction (albeit more smeared out to a greater or lesser extent) for relatively broadband x-ray sources.

Directly opposite the screen edge, the fringe intensity equals $I_0/4$, where I_0 is the beam intensity far away from the screen. This is simply explained by the fact that half the wavefront is occluded here and hence the amplitude in the forward direction is halved. But the intensity is proportional to the square of the amplitude, thus here it drops to one quarter. Note also that the constructive-interference fringes have higher intensities than I_0 – the wavelets blocked by the screen that would otherwise have interfered destructively with the rest of the wavefront are missing and so cannot contribute – the amplitude (and hence intensity) is therefore enhanced. This effect is termed *edge enhancement*.

Lastly, not only do the fringes become weaker away from the projection of the screen edge, they also become narrower. From Equation (8.8), we see that for an increase in the OPD of 2π radians between adjacent fringes Δx, the separation between fringes on the detector, is approximately $2\pi/\sin\theta$.

What might be typical dimensions for observing edge enhancement using hard x-rays? Edge enhancement is used in tomography to highlight internal and external boundaries of heterogeneous samples for which the absorption coefficients of the component parts are similar, hence the first fringe maximum should be resolved, but not

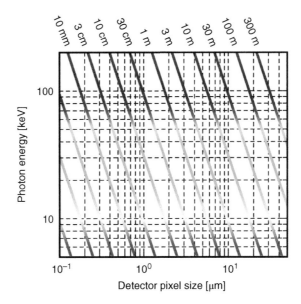

Figure 8.16   The relationship in propagation tomography between the detector-pixel size, photon energy, and optimal sample–detector distance. The three unlabelled contours on the left are the natural progression of $Z_0 = 3$, 1, and 0.3 mm. Adapted from [10], with permission from IUCr].

necessarily any of the subsequent, more narrowly spaced, fringes. The sample–detector distance Z_0 should be in the Fresnel regime at $F_0 \sim 1$ (Figure 8.14) adjusted so that the lateral resolution (pixel size, Δ) matches this fringe separation. As this requires that at least two pixels differentiate the fringe, we arrive at a value of $Z_0 = (2\Delta)^2/\lambda$. A plot of the relationship between Δ, Z_0, and $h\nu$ is shown in Figure 8.16 [10].

Although in the above discussion, we assumed that the screen was totally opaque to x-rays, it can be easily shown that similar effects can be observed for semi- or indeed completely transparent screens, as long as the screen has a different refractive index than the medium through which the x-rays are otherwise travelling.

In summary, if edge-enhanced-optimized radiographs are used in propagation-based tomographic reconstructions, the three-dimensional data contains edge detection superimposed on the absorption contrast. This can be crucial for resolving features, most notably in biological specimens, for which the absorption contrast is very low.

Propagation-based x-ray imaging (PBI) and tomography (PBT) [11, 12] require a degree of spatial coherence best provided by synchrotrons – the large source size of laboratory-based sources will convolute with the fringe patterns and smear them out unacceptably. As mentioned above, the specifications on the longitudinal coherence (in other words, monochromacity) are less strict, due to the $\sqrt{\lambda}$-dependence on the fringe separations. Because PBI uses no lenses and functions merely by allowing the transmitted wavefront to propagate, the stability requirements are very modest. In PBT, the sample rotation stage should have a circle of confusion smaller than the features which are to be resolved, a requirement that nowadays is easily met using modern instrumentation [5, 6].

Tomograms based on phase maps can be extracted from edge-enhanced images as follows. Consider Figure 8.17(a). Refraction and interference result in a modulation in the intensity in the Fresnel regime at the geometric projection of the boundary between two parts of a phase object (that is, an object with insignificant absorption) which induce different relative phases – this fringe profile can be approximated by a function of the form $x \exp(-x^2/2\sigma^2)$, that is, the derivative of a Gaussian [the red curve in Figure 8.17(a)]. Doubly integrating this produces a sigmoid function with a difference between the limiting values at $\pm\infty$ proportional to the phase difference $\Delta\phi$ of the adjacent parts (the blue curve).

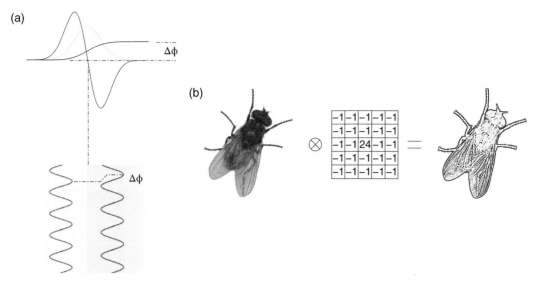

Figure 8.17 From edge diffraction to phase contrast. (a) The dominant fringe feature in the Fresnel regime resulting from refraction and interference between x-rays passing through parts of a phase object with different refractive indices can be approximated by the red curve. Doubly integrating this results in the blue curve, whereby the difference in intensity at the limits is proportional to the phase difference between the two transmitted waves, $\Delta\phi$. (b) Edge enhancement of an image can be generated by several means. Here, the original image is convoluted with the shown 5×5 kernel. For any pixel in the original, the central strongly positive (+24) element is exactly cancelled by the other 24 singly negative elements, assuming the surrounding pixels of the original have the same (or very similar) value. Where the intensity in the original changes rapidly, however, the convolution with the kernel will produce a nonzero value, resulting in edges being highlighted. The edge-enhanced image on the right has been inverted to mimic the Fresnel transmission image of the original.

There are several kernels one can employ in image manipulation to enhance edge features. Shown in Figure 8.17(b) is a kernel called the Laplace edge detector (which we mathematically label here as L, that produces an edge-enhanced image E when convoluted with the original O. An approach therefore to produce a phase-contrast image (equivalent to O) from the recorded edge-enhanced Fresnel-diffraction image (equivalent to E) is to employ the convolution theorem, that is,

$$O = \mathscr{F}^{-1}\left\{\frac{\mathscr{F}(E)}{\mathscr{F}(L)}\right\}. \tag{8.11}$$

An example of edge enhancement and phase contrast is shown in Figure 8.18.

Because the achieved spatial resolution in PBI/PBT is mainly limited by the degree of spatial coherence of the beam, the new generation of DLSR synchrotrons, with up to two orders of magnitude higher coherent fraction in the hard x-ray regime, promise a quantum leap in both recording rates and reduced sample dosage – consider, for example, Figure 3.20. For a given desired coherent fraction, the photon energy at which this is achieved for DLSRs is, in the hard x-ray regime, approximately equal to the energy exhibiting that coherent fraction at a third-generation facility multiplied by the increase in brilliance (between one and two orders of magnitude). Such large shifts will require detectors that are more sensitive to high-energy photons, such as provided by sensors based on CdTe or other high-Z materials.

The resolution of PBT can be enhanced in so-called zoom tomography [14], in which the sample is placed downstream from the focal point of a focussed beam (see Figure 8.19). Depending on the relative positions of the sample and detector, resolutions of a few tens of nm are routinely achievable.

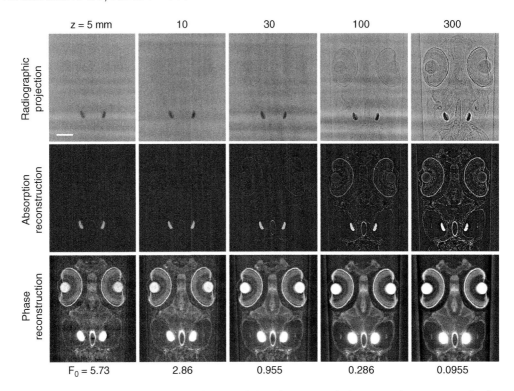

Figure 8.18 Propagation-based tomography of a zebrafish embryo recorded using 21 keV x-rays and a detector with a pixel size of $\Delta = 0.65$ μm. Zebrafish embryos are popular vertebrate models for developmental biology, toxicology, and pharmacokinetics; importantly, they exhibit good biofunctional correlations to higher vertebrates [13]. The pair of absorbing features seen in all the images are calcium carbonate-based inner-ear otoliths (Greek for 'ear stones'), which are sensitive to gravity and acceleration and are required by zebrafish for maintaining equilibrium. As the fish–detector distance z is increased, edge enhancement becomes progressively evident. At $F_0 = (2\Delta)^2/z\lambda \sim 1$, edge enhancement is clearly visible, but not so pronounced that intensity-oscillation artefacts (such as seen at 300 mm) begin to dominate. Even for very short distances, the phase shifts at the boundaries between distinct features are sufficient to produce clear phase reconstructions, in contrast to both the radiographic projections and absorption reconstructions, which can only identify the otoliths. Scale bar in top-left image = 100 μm. Courtesy Christian Schlepütz, Paul Scherrer Institute, and Jörg Huwyler and Emre Cörek, Institute of Pharmaceutical Sciences, University of Basel.

In zoom tomography, the sample is positioned a distance z_1 downstream of the focus and the detector a distance z_2 from the sample. Because the act of focussing generates a spherical-wave illumination of the sample, the propagation distance differs from the plane-wave case. The equivalent propagation distance for the magnified Fresnel image is

$$z = \frac{z_1 z_2}{z_1 + z_2}, \tag{8.12}$$

while the degree of magnification is

$$M = \frac{z_1 + z_2}{z_1}. \tag{8.13}$$

For large magnifications ($z_1 \ll z_2$), $z \approx z_1$, while for near-unity magnification ($z_1 \gg z_2$), it is approximately z_2, as in 'standard' PBT.

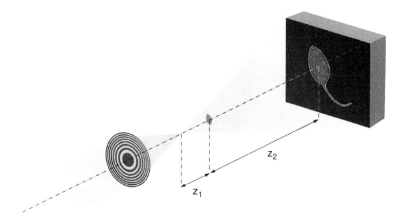

Figure 8.19   Zoom tomography. A sample is placed a distance z_1 downstream of a focussed x-ray beam. The propagated image is recorded at a distance z_2 downstream from the sample. The magnification factor is $(z_1 + z_2)/z_1$.

8.2.5   Fast XTM

'Fast' XTM is defined as pertaining to experiments for which resolutions are on the micron scale and tomographic acquisition rates are above a few Hz (translating to individual radiograph exposure times of the order of a few milliseconds, or even substantially shorter).

The speed of data acquisition and computer power are nowadays so high that XTM can be used to probe dynamical processes which exhibit changes down to the few millisecond time-scale. Three interlinked factors determine the feasibility of ultrafast XTM, namely the brilliance of the x-ray beam, the detector performance (sensitivity and readout time), and accurate synchronization of sample position and exposure time [15]. Instead of rotating the sample in a step-like manner and recording the radiographic projections between steps, the sample is continuously rotated and its angular position must be recorded accurately 'on the fly' by an encoder triggered by the detector.

Because the readout times for CCD chips containing millions of pixels are relatively long, and certainly not in the desired millisecond range, CMOS (complementary metal-oxide semiconductor) technology is an obvious alternative for fast XTM. The advantage of CMOS sensors is their rapid and parallel readout times. This, however, requires more electronic architecture associated with the light-sensitive pixels, which therefore tend to be smaller for a given pixel repetition length and hence their light-gathering power is lower than that of CCDs. Lastly, the effects of centrifugal distortion of the sample must be considered. The outwards-acting acceleration of a sample rotating at 10 Hz (allowing a tomographic acquisition rate of 20 Hz) around a radius of 5 mm is 19.73 m s^{-2}, that is, twice the earth's gravitational acceleration. This can result in unacceptable 'equatorial bulging' in soft-matter samples.

Phase contrast is more commonly used than absorption contrast, as the contrast of the former is higher for weak signal, due to the weaker dependence of the refractive-index decrement δ on photon energy than that of the absorption index β.

Several recording modes are possible for fast XTM. The most obvious is simply to record images at a rate such that a full tomogram can be recorded within the time interval of interest. This can lead to frame rates well in excess of 10 kHz, setting a premium on photon arrival rates, detector sensitivity, and data-transfer hardware [16], which can be several GBytes per second. Indeed, nowadays, it is not uncommon for a tomography experiment to accumulate hundreds of Tbyte over a beamtime of a few days.

If a process is repeatable with a spatial reproducibility equal to or superior than the desired resolution, a stroboscopic approach is possible. Processes that are only approximately cyclic can still be recorded by pooling radiographs into sets corresponding to the different phases of the cycle, from a redundantly large data set. A

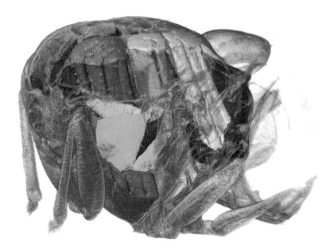

Figure 8.20   Cutaway visualization of the thorax of a blowfly, showing the five small steering muscles (green to blue) and the larger power muscles (yellow to red). Adapted from [17] with permission from the Public Library of Science.

Figure 8.21   X-ray laminography. Tomography has problems with samples in which the extent of an object in one of the three dimensions is much smaller than in the other two. For such 'plate-like' samples, laminography is used, whereby the rotation axis, perpendicular to the 'plate' surface, is set to an angle less than 90° to the beam axis.

beautiful example of this approach has been a study of the mechanics of blowfly flight mechanisms [17] (see Figure 8.20)[3] using phase-contrast techniques. The fly was rotated at a little under 1 Hz and its wingbeat frequency of approximately 150 Hz was captured and resolved using radiographic frame rates of 2500 Hz.

8.2.6   Laminography

Laminography is a special configuration of tomography used for large flat samples [18–20]. Although the sample might have a reasonable transmission if viewed face on, in standard tomography, this will drop drastically as the sample is rotated to being edge on (Figure 8.21). In laminography, the sample is tilted so its rotation axis (set to be perpendicular to the large, flat, facet of the sample) is neither parallel nor perpendicular to the incident beam; the choice of angle should, however, be as close to being perpendicular to the beam that still allows a reasonable transmission.

---

[3] Although this image is quite arresting, it does not do the study full justice – I would strongly recommend you to look at the movies linked to this article to see it in its full splendour.

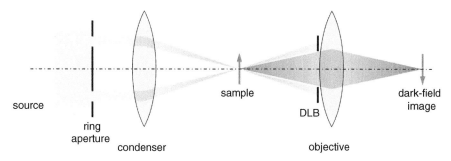

Figure 8.22 Dark-field imaging. In this full-field microscopy example, a sample is illuminated by focussing an annular source using a condenser lens (typically a FZP or PZP). The directly transmitted radiation (yellow rays) is blocked using an annular direct-light blocker (DLB). That radiation which is scattered (shown here as the red rays) is refocussed onto a detector using an objective lens to produce a dark-field image.

8.3 Full-field Microscopy

In the introductory comments to Chapter 6, I argued that full-field x-ray microscopy with subatomic resolution is impossible, due to limitations of lens fabrication and magnification factors. At more modest resolutions, of the order of tens of nanometres, full-field microscopy is nowadays feasible, employing state-of-the-art diffractive lenses [21]. Magnifications can vary from close to unity up to approximately 100, leading to ultimate resolutions of approximately 50 nm.

An important variant of both full-field microscopy and tomography [22, 23] is dark-field imaging. In this technique, the direct, undeviated beam after sample illumination is blocked, and only the scattered signal is collected (Figure 8.22). Dark-field imaging is especially useful in highlighting small, dense features (which scatter strongly) embedded in a large object. Because dark-field imaging collects only the scattered radiation, it is particularly sensitive to abrupt changes in refractive index and small objects that produce significant small-angle scattering signal. Note, however, that it is insensitive to phase variations, as there is no reference wave with which these can be compared. This aspect is addressed in Zernike microscopy.

Full-field microscopies in the hard x-ray regime have the great advantage over similar techniques using soft x-rays in that they have a much larger depth of field [which is, for a desired resolution, directly proportional to the photon energy, see Equation (5.17)]. So, for example, while full-field microscopies operating in the water window at approximately 300 eV might have a depth of field of a micron or less, hard x-ray microscopes have depths of field measured in several tens of microns.

8.3.1 Zernike X-ray Microscopy

Frits Zernike received the Nobel Prize in Physics in 1953 for his invention of phase-contrast microscopy. The idea behind this technique is to convert phase shifts in radiation scattered from a sample to intensity changes, by vectorially adding the scattered wavevectors to a reference vector. The amplitude (and hence also intensity, proportional to the square of the amplitude) of this summed vector changes, according to the relative phase between the reference and scattered waves.

Consider Figure 8.23(a). A weakly absorbing object, composed of distinct elements of differing composition but very similar absorption coefficients, will exhibit almost no contrast regarding its internal architecture in a conventional full-field x-ray microscope using hard x-rays with energies of the order of 10 keV or higher. This is because the scattering components have both relatively small amplitudes and phases that, for elements of the order of a few microns or smaller, are substantially less than $\pi/2$ [see Equation (8.7)]. In other words, the unperturbed fraction of the beam dominates the signal intensity.

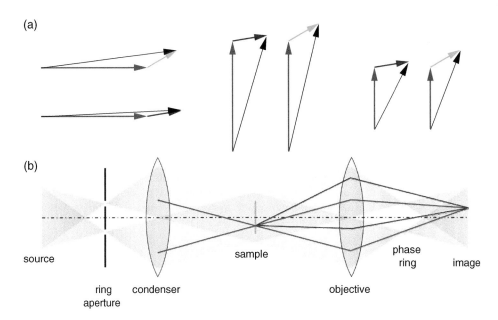

Figure 8.23 Zernike x-ray full-field microscopy. (a) A weakly absorbing sample will exhibit very small variations in the amplitude of the x-rays it scatters. On the other hand, the phase shift induced by different scattering elements of the sample can vary much more significantly (the blue and green vectors), although in general they are much less than $\pi/2$ (see text). Vector addition of these scattering amplitudes to a 'reference' wave (shown in red) that is nearly parallel to them results in vector sums (black arrows) that are almost indistinguishable in magnitude. If, however, one is able to rotate the reference vector by $\pi/2$, the fractional difference between the vector sum of this with the blue and with the green sample-scattered vectors is much larger. If the magnitude of the reference vector is reduced by absorption, the relative contrast can be made to be larger still. (b) In Zernike full-field microscopy, a ring aperture is placed in the back focal plane of a condenser lens (normally a FZP). The ring illumination is collimated with the condenser (shown in red) and floods a weakly absorbing sample. Positioned symmetrically downstream from the sample is an objective lens, which images the ring-illumination aperture onto a so-called 'phase ring' – a (partially) transparent ring with a thickness tailored to advance the phase of the illumination (or reference) rays by $\pi/2$. X-rays scattered by the sample (shown in blue) are imaged using the objective lens onto the image plane, where they converge with the $\pi/2$-phase-shifted reference signal. Most of the scattered signal passes by the phase ring and is therefore not phase-shifted by $\pi/2$. The only phase shift the scattered x-rays experience is that produced by the sample itself, which, for hard x-rays and weakly absorbing and small samples, is much smaller than $\pi/2$.

However, by illuminating the sample with an annular source in the configuration shown in Figure 8.23(b), one can manipulate the phase of the unperturbed radiation using a so-called 'phase ring' without affecting the radiation scattered by the sample (except for the very small fraction that is precisely forward scattered). The thickness of the phase ring is chosen [using Equation (8.7)] so that the phase of the otherwise unperturbed 'reference' radiation is shifted by $\pi/2$. Now, the vector addition of the scattered radiation with the reference wave produces significantly greater differences with relative phase angle. Indeed, the contrast can be increased still further by choosing a material for the phase ring that not only induces the desired phase shift in the reference wave, but is also partially absorbing, reducing thereby in a controlled manner the magnitude of the reference component [the red arrows on the right-hand side of Figure 8.23(a)] in what can be thought of as a combination of Zernike and partial dark-field microscopy. A recent example of Zernike microscopy is shown in Figure 8.24.

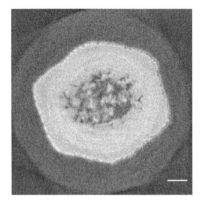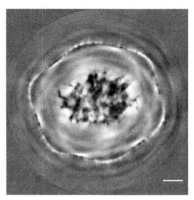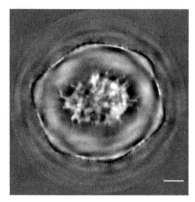

Figure 8.24   Zernike microscopy of a Nb$_3$Sn superconductor filament. Wires containing hundreds of such 50 μm diameter wires are being developed for the bending magnets for the upgrade of the Large Hadron Collider at CERN. Quality control of their integrity is a crucial aspect, including the identification of porosities and individual grains. Left: Zernike microscopy image in absorption mode. Centre: the same sample in positive-phase ($\phi = \pi/2$) contrast. Right: in negative-phase ($\pi = 3\pi/2$) contrast. The scale bars are 10 μm. Courtesy Anne Bonnin, Paul Scherrer Institute, and Christian Scheuerlein, CERN [24].

8.4   Lensless Imaging

8.4.1   Introduction

Full-field x-ray microscopy using lenses works by a beam of light or x-rays being scattered by a sample (which in the far field is the Fourier transform of that sample), and this scattered radiation being inverted back to an image of the sample by the action of the lens. As such, it has limitations regarding lens quality, restricting the resolution of such techniques to a few tens of nanometre. For example, Fresnel zone plates, the most commonly used type of submicron-focussing x-ray lenses, have resolutions limited by the width of the outermost zone, and become exceedingly difficult to fabricate for desired resolutions of a few nanometres [see Equations (5.40) and (5.41)]. Imaging based on scattering alone, so-called lensless, or coherent-diffraction, imaging dispenses with lenses and their associated aberrations (see Figure 8.25).

One can think of lensless imaging as being 'normal' diffraction writ big. Whereby in standard crystallography, one is interested in determining the atomic components and their positions within a unit cell to a resolution often substantially below an angstrom, lensless imaging attempts to determine structures on the scale of a few nanometres to tens of nanometres. From Bragg's law, this implies for hard x-rays scattering angles typically smaller than a degree and often down to a few tens of microradian. Hence, while in diffraction, scattering is induced by the electron clouds surrounding atoms, in lensless imaging, it occurs due to fluctuations in the electron density on the scale of nanometres.

Other than the elimination of lens aberrations, what advantages does lensless imaging confer? Δx, the resolving power of a lens, is inversely proportional to the lens diameter (Figure 8.25). X-ray lenses, especially 'big' ones, are difficult to manufacture such that they produce no aberrations, hence dispensing with them solves this problem. The resolution of lensless imaging is given by

$$\Delta x = 4\lambda \frac{Z_0}{D}, \tag{8.14}$$

where Z_0 is the sample–detector distance and D is the detector size. Hence, it is in principle easier to obtain high resolution in lensless imaging, simply by using area detectors which subtend a large angle with the sample. As we saw in Section 6.15.2, the scattering power drops off with the fourth power of the scattering vector, hence the

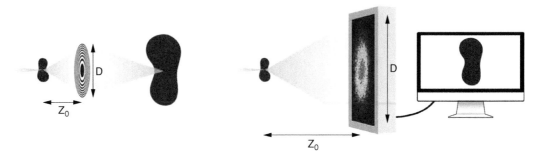

Figure 8.25   Focussing optics versus lensless imaging. Left: The resolution of imaging using lenses is limited by the size of the lens, D and the sample–lens distance Z_0. Because x-ray lenses are difficult to make aberration-free, especially for large D, lensless imaging (right) becomes attractive for resolutions below approximately 100 nm. The resolution improves linearly with detector size D. Drawbacks include the need to inverse-Fourier transform the far-field scattering image of the pattern recorded by the detector, which is hampered by the phase problem, and the low scattered intensity at large angles.

signal intensity drops off sharply away from the direct beam. Lensless imaging is therefore limited by a tradeoff between resolution and signal statistics.

Lensless imaging neatly bridges the resolution gap between full-field x-ray microscopies (of the order of 100 nm) and electron microscopies (nanometre or subnanometre). Electron microscopies also suffer from the fact that the penetration depth of the electrons is very small, typically measured in small fractions of a micron. Because of the weak interaction of x-rays with matter, one can apply the kinematical approximation, thereby minimizing the computational effort in reconstructing the image. This brings us to the main drawback of the technique, namely the ubiquitous phase problem. As we will see later in this section, sophisticated methods have been developed to overcome this.

Because standard x-ray crystallography is concerned with the arrangement of atoms within the unit cell, the spatial coherence of the x-ray beam in principle needs only to be large compared to the linear dimensions of the unit cell. However, the intensity of diffraction maxima increases with the square of N, the number of coherently illuminated unit cells in any given direction, while the narrowness of the signal is inversely proportional to N. From Equation (3.46), we can calculate that for beamlines not immediately concerned with high coherence, the upper limit to the transverse coherence lengths is of the order of 50 µm for hard x-rays of approximately 1 Å wavelength, and can be substantially smaller than this, due to imperfections in the beamline optics, resulting in typical coherence lengths an order of magnitude smaller than the theoretical limits.

In order to obtain useful results in lensless imaging (also referred to as coherent x-ray diffraction imaging, CXDI), two conditions must be met: first, the sample must be entirely bathed in the coherent part of the beam (though this is not necessary in ptychography, see below); and second, the maximum path-length difference must be smaller than the longitudinal coherence length [25].

Fourth-generation DLSR synchrotrons exhibit up to a two orders of magnitude higher coherent fraction than third-generation sources, offering several percent in the 1 Å hard x-ray regime. Beamlines that utilize the coherent part of the synchrotron radiation are characterized by a small source size, minimization of optical components that can distort the wavefront, suppression of sources of vibration, and a large source-to-sample distance. A beamline optimized for coherence can have transverse coherence lengths of the order of 100 μm or more for 1 Å radiation, even in the horizontal plane. Normally samples have dimensions considerably smaller than this, in which case the need for coherence is somewhat relaxed and one can partially focus the incident beam to increase the signal intensity.

One can divide up samples into two types – noncrystalline and single crystal. In the first case, one is interested in the signal close to the forward scattering direction around the direct beam. In the case of crystalline samples,

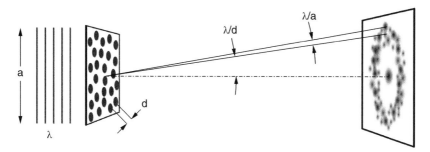

Figure 8.26 Scattering of coherent illumination by an object produces a speckle pattern with an angular extent of the order of λ/d, and 'speckle' features within the pattern separated from each other by λ/a. If the transverse coherent lengths are much smaller than the sample extent a, the scattering signal reverts to a standard SAXS pattern.

one can investigate the details of the scattering around the Bragg peaks associated with the crystal, which can lend some important advantages, described below.

To understand the physics of CXDI, we next turn to the concept of speckle.

8.4.2 Speckle

Speckle is the result of interference between wavelets produced by scattering of x-rays caused by spatial variations in the electron density within an object (see Figure 8.26) [25]. It therefore has its origins in the same phenomena responsible for diffraction patterns, and is the square of the Fourier transform of the electron-density distribution of the object. To obtain a speckle pattern, the object must be coherently illuminated – if this is not the case, the speckles become averaged out due to the fluctuation of the form of the wavefront incident on the sample; in the forward direction around the direct beam, this incoherent signal is referred to as small-angle x-ray scattering (SAXS), covered in Section 6.15.

There are three characteristic length scales related to speckle: the wavelength λ of the x-rays, measured in angstroms; the typical linear dimensions of illumination of the sample a, which should be smaller than the transverse coherent length (of the order of 100 µm); and d, the characteristic length describing the variation in electron density (or distribution of 'particles') within the illuminated object, which can range between 10 and 1000 nm. The x-rays are on average scattered through an angle of the order of λ/d, which can be as small as 100 µrad (less than 0.01°). The speckles arise through interference between all the scattered wavelets across the coherently illuminated area and therefore have angular separations that can be as small as $\lambda/a \sim 1$ µrad. In order to avoid 'aliasing', the linear detector pixel size Δ should be no larger than half the separation between speckle maxima (that is, the speckle pattern has been at least 'Nyquist sampled', see Section 8.4.4). From an argument of similar triangles, a minimum distance z from sample to detector is given by

$$z \geq \frac{2a\Delta}{\lambda}. \tag{8.15}$$

So, for example, in order to resolve speckle features, a detector having $\Delta = 75$ µm-sized pixels would need to be at least 7.5 m downstream of the sample, for 1 Å x-radiation and a coherent illumination extent of $a = 5$ µm. Alternatively, softer x-radiation can be used, although the increased absorption at lower photon energies sets an upper limit to the thickness of objects that can be studied.

Note that because the speckles arise from interference between the scattered wavefronts from all the particles in the sample, the entire speckle pattern will change even if only one particle moves (see Figure 8.27). The speckle pattern therefore contains information on the positions of all the particles, which can be obtained by inverting the

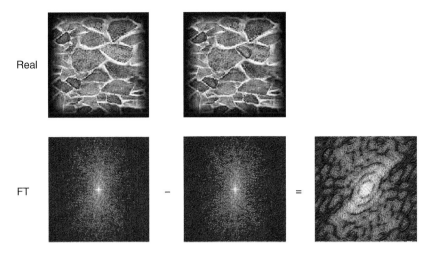

Figure 8.27   The change in speckle due to changes in the sample architecture. Top: two simulated grain distributions. The one on the right differs from that on the left only by the inclusion of a small grain, whose perimeter is highlighted in red. Both distributions have borders grading to black to suppress edge effects in their Fourier transforms, shown below. To their right is the logarithm of the absolute value of their difference. Note that this affects the entire speckle pattern and not just a limited part. This is important in XPCS, discussed later in this chapter.

pattern, involving solving the phase problem. This is nowadays commonly achieved by applying phase-retrieval techniques such as the so-called 'difference-map' algorithm [26–28].

8.4.3   Noncrystalline and Crystalline Samples

In order to obtain the maximum amount of information about the three-dimensional internal structure of a noncrystalline object, one must rotate it through at least 180° in order to record the entire three-dimensional scattering pattern. The reason for this is that, in reciprocal space, only that part of the pattern that lies on the surface of the Ewald sphere (see Figure 6.11) will be recorded, hence in order to obtain the whole speckle pattern centred around the forward-scattered beam [the (000) point], it must be rotated by at least 180°, and with sufficiently small steps so that information is not lost in between.

In contrast to noncrystalline objects, crystalline samples also have strong Bragg peak signals far away from the direct beam. As we have already discussed in Section 6.12.2, the diffraction pattern of finite-sized crystals is the convolution of the diffraction pattern of an infinitely large crystal with the Fourier transform of the function describing the boundary of the actual crystal (the 'shape function'), hence each Bragg peak has the form of the shape function (see Figure 8.28).

The exact shape of the Bragg peaks can be further modified by other phenomena such as internal strain [29]. Importantly, however, it reflects the entire internal structure of the crystal, and the shape function is just the speckle for ordered systems.

This has three important consequences. First, so-called 'Bragg coherent x-ray diffractive imaging' (Bragg-CXDI) can be carried out far away from the direct beam. This means one can dispense with the beamstop (which excludes features in the reconstruction above a certain size) and also avoid diffuse scattering in the forward direction produced by other sources than the sample. Secondly, because the Bragg-peak shape is only affected by the electron density of the crystalline part of the object, it is highly sensitive to crystalline defects and strain fields [29, 30]. Lastly, one must rotate the sample through a much smaller angular range than in the case of

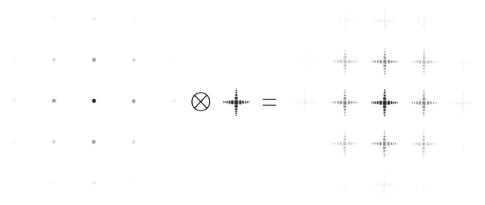

Figure 8.28 Each Bragg peak of the diffraction pattern of a nanocrystalline sample reflects, via the convolution theorem, the 'shape function' of the object.

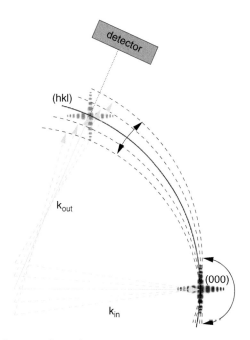

Figure 8.29 The speckle pattern for crystalline objects can be recorded around Bragg peaks far away from the direct beam. The angular range one must rotate the sample in order for the Ewald sphere to pass across the scattering pattern and thereby record the entire pattern, is much reduced compared to that required for noncrystalline samples, which only scatter significantly in the forward direction.

noncrystalline samples in order for all the features to pass through the Ewald sphere (Figure 8.29), significantly relaxing the technical specifications of the rotation stage[4].

---

[4] The scattering pattern of a square function in one dimension is a sinc^2 function with its first minima at $\pm 2\pi/na$, where n is an integer, a is the unit cell size in that direction, and na is the width of the signal. Extension to two and three dimensions (a 'nanocube') is trivial. The angle one must rotate to record these first minima is approximately $\pm 1/n$. Normally one would record to $\pm 5/n$, but even this, for $n \sim 100$, equates to rotation angles of the order of $5°$.

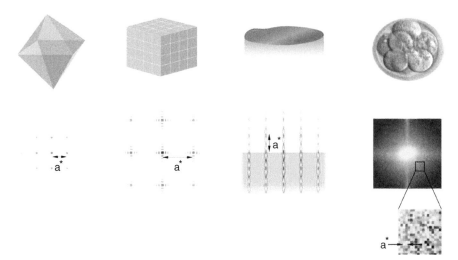

Figure 8.30 Four systems, their scattering patterns, and oversampling. Large bulk crystals have diffraction patterns that consist of equally separated sharp Bragg peaks in a sea of zero intensity. The oversampling factor $\beta_O = 1/2$. In contrast, nanocrystals with linear dimensions of only a few unit cells in any one direction exhibit modulations perpendicular to their surfaces; atomically flat single crystals also have measurable intensities in the out-of-plane direction. Both these systems can therefore have $\beta_O > 1$. Noncrystalline objects exhibit a continuous scattering pattern (the speckle, highlighted in the inset) which can also be oversampled. In each case, the Nyquist sampling rate would require recording data separated in reciprocal space by half the size of the shown repeat dimensions a^*.

8.4.4   Oversampling and Redundancy

Now we have discussed the basic concepts and practices of lensless imaging, and before we move on to scanning CXDI, or ptychography, it behoves us to consider the quality and quantity of the scattering data required to obtain reliable real-space reconstructions.

We begin by introducing the concept of the 'Nyquist frequency', f_N. This is the minimum rate at which a signal can be sampled without introducing errors, and is equal to two times the highest frequency associated with that signal. As an example, consider a pure sinusoidal sound wave at middle-C (261.62 Hz). The bare minimum sampling frequency required to faithfully record this is 523.24 Hz. Even then, artefacts will most probably arise: first, if the recording instrument assumes a straight line variation between sampling points, it will generate a triangular wave, which contains odd overtones of the Nyquist frequency. Secondly, if the wave is not sampled exactly at the peaks and troughs, the recorded amplitude will be lower.

The oversampling factor β_O is the ratio of the sampling frequency f_S to the Nyquist frequency, that is,

$$\beta_O = \frac{f_S}{f_N}. \tag{8.16}$$

What, then, is oversampling in the context of diffraction, or scattering, patterns? Consider Figure 8.30. In the case of diffraction patterns from bulk, macroscopically sized, crystals, the goal is to determine the atomic structure within the repeat unit of the crystal, that is, the unit cell. The diffraction signal consists of sharp, equally spaced, maxima, whereby their separation in any given direction in reciprocal space is inversely proportional to the corresponding lattice parameter in real space. The Nyquist sampling frequency would have a periodicity half of the peaks' separation. But, because there is no diffraction intensity between the Bragg peaks, *oversampling of bulk diffraction patterns is impossible.*

Nanosized crystals, such as investigated with Bragg-CXDI, often exhibit extended and modulated features, such as the sinc^2 function for a cubic nanocrystal. Samples with atomically flat surfaces, likewise, have a continuous scattering signal in the out-of-plane direction (crystal truncation rods, see Section 6.12.2). Such systems can therefore be oversampled ($\beta_O > 1$).

Oversampling can be used to solve the phase problem in noncrystalline objects. As already explained, the measured scattering pattern provides only the amplitudes; half the information, namely the phases, is missing. This is the phase problem. The 'oversampling method' can be applied to solve the phase problem by padding the real-space object with a zero-density volume. If this zero-density region is larger than that of the sample, the phase problem can, in principle, be solved. The method proceeds as follows: the larger real-space object, including the zero volume and the guessed sample structure, is Fourier-transformed; the resulting FT is sampled in a given direction at a frequency that is increased by the ratio of the size of the larger, zero-padded object to the unpadded version in the corresponding real-space axis. The amplitudes of the FT are exchanged for the measured values (that is, the square-root of the measured intensities), and this is then inverse-Fourier transformed. The real-space model resulting from this IFT has its electron density set to zero outside the known extent of the real-space sample. This procedure is iterated until self-consistency is achieved.

Note that the oversampling method depends on the physical extent of the sample being known. This might be only approximately true, or indeed not at all true. In more general cases, other approaches must be adopted. These are becoming increasingly sophisticated and can often be used in combination with each other [31–33].

Redundancy, on the other hand, determines how ambiguous any one solution to the phase problem might be. If the experimental data provides M independent values, and the object under investigation has N unknown variables (for example, N 'voxels' of electron density, the linear size of each voxel corresponding to the minimum spatial fluctuation of the density of the sample), then the redundancy is given by M/N. The larger the redundancy, the more likely one is to obtain the correct solution. Note also that equating M to the number of pixels in the detector is the absolute upper limit. In nearly all cases, the intensities towards the outside of the detector are so weak that they cannot be used. Also, if the pixel size is significantly smaller than the typical speckle separations (relating back to the concept of oversampling) then one cannot assume that adjacent pixels record 'independent' values.

8.4.5 Ptychography

At the beginning of this section, we argued that, for lensless imaging of a given object, a detector placed in the Fraunhofer regime must have a pixel size inversely proportional to the object size (Figure 8.26). Because microscopists normally want to place an object within a larger spatial context, this can set stringent and often unrealistic specifications on the detector. Imagine, for example, one would like to image a 100 μm sample with a resolution of 20 nm using 2 Å radiation. Each pixel should, according to the Nyquist condition, subtend an angle smaller than $\lambda/2a = 10^{-6}$, or 10 μm at a sample-to-detector distance of 10 m. The desired resolution demands that the detector needs to have a radius of at least 10 cm (10 000 pixels). X-ray ptychography overcomes this limitation while also providing a robust and rapid algorithm to solve the phase problem [34].

Ptychography is an experimental method originally developed in the 1970s for electron microscopy [35]. In the first decade of the twenty-first century it began to be applied to x-rays, with dramatic results [36–38]. It allows the investigation of macroscopic structures, many times larger than the coherence volume of the beam, to be investigated with the same resolution as offered by CXDI. The upper limit to the size of the sample under investigation is given by technical issues, in particular the volume of data generated, the computing power needed to solve the structure, and the accuracy and reproducibility of the scanning stage of the sample.

Modern x-ray ptychography can be thought of as a marriage of coherent x-ray diffraction imaging (CXDI) and scanning transmission x-ray microscopy (STXM, see Section 7.7). The major (but crucial) differences between STXM and ptychography are that in the latter, coherent x-radiation is used and the transmitted signal is spatially resolved using an area detector, rather than recording just the integrated signal (as in STXM). Because of this, ptychography overcomes both the limited sample size of CXDI and the limited resolution of STXM.

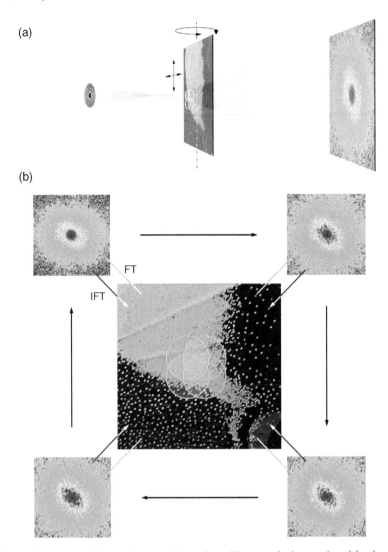

Figure 8.31 Ptychography setup and algorithm. (a) The setup. The sample is translated in the *x*- and *y*-directions, each step being sufficiently small to guarantee overlap between adjacent illuminated regions. In ptychographic nanotomography, the sample is also rotated to probe the sample's three-dimensional structure. (b) The basic scheme of the ptychographic iterative engine cycles over the scan positions via IFTs and FTs. Looping continues until the difference between the measured intensities and the Fourier transform of the illuminated areas fails to decrease any more. Central image courtesy Anita Kironde.

The configuration of ptychography and the principle behind it are shown in Figure 8.31(a). A portion of a sample is illuminated with an x-ray beam over an area of a few square microns, and the speckle pattern is recorded. The sample is then shifted to illuminate a new region, for which a different scattering pattern is recorded. Importantly, this region should overlap with the previously illuminated area. Logically, this must mean that any correct real-space structural solution for both of these speckle patterns contains a common region regarding their detailed structures where they overlapped. This imposes a strong constraint to the solution and provides added redundancy to the algorithm used to solve the structure. If the overlap is too large, little added information is gained and the

Figure 8.32   Three-dimensional ptychographic nanotomography of an Intel processor with 15 nm resolution. The bar in the inset is 500 nm. Adapted from [42] with permission from Macmillan Publishers Ltd.

scanning procedure is slowed down, while if there is too little overlap, the degree of ambiguity of the solution increases. An overlap of approximately 60% has been found experimentally to be optimal [39].

The phase-retrieval algorithm is sketched in Figure 8.31(b). It was introduced in 2004 as the 'ptychographic iterative engine' (PIE) [40]. The power of the algorithm was increased in 2008 by including a nested iterative algorithm to determine precisely the illumination function, which is very difficult to know *a priori* [37, 41]. This development also permits the use of focussed x-rays without the need to model the wavefront on the sample, as it is reconstructed anyway as part of the algorithm.

Three-dimensional objects can be studied in ptychographic nanotomography by also rotating the sample. Using a state-of-the-art sample manipulator, 15 nm isotropic resolution was demonstrated [42] (see Figure 8.32).

8.4.6   Scanning SAXS and Small-angle Scattering Tensor Tomography

The physical properties of many macroscopic objects depend on the arrangement and orientation of hierarchical structures over several distinct length scales, such as shown in Figure 8.2.

Scanning SAXS can provide information on the shapes and sizes of components making up a larger object down to the nanometre scale. In many cases, certain components can be anisotropic and also exhibit nonrandom orientations within the larger object [44, 45]. Tensor-scanning SAXS provides information within a sample's volume on the orientations of the different components, thus yielding six-dimensional information – the cartesian coordinates of the components within the sample and the angular orientation of each component (that is, its tensor). In order to obtain the overall structure, therefore, six independent experimental data sets are required. These are the translational scanning coordinates x and y, the two-dimensional SAXS image, and *two* mutually perpendicular rotational motions. A beautiful example of the anisotropic distribution of mineralized collagen fibrils in a human trabecula bone, investigated using small-angle scattering tensor tomography, is shown in Figure 8.33 [43].

8.4.7   X-ray Photon Correlation Spectroscopy

X-ray photon correlation spectroscopy (XPCS) sits somewhat uncomfortably in a chapter on imaging, as it does not produce any real-space images. However, because it uses the phenomenon of speckle to extract information about dynamical processes, I decided to include this subsection as an adjunct to the section on lensless imaging.

We have already argued in Section 8.4.2 that because speckle arises from interference between scattered wavelets from all objects illuminated by the coherent x-ray beam, any movement of even one of these objects will alter the entire speckle pattern (Figure 8.27). Photon correlation spectroscopy (PCS) in the visible, using lasers,

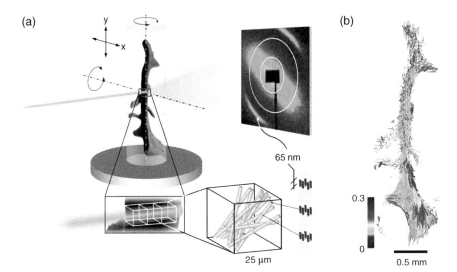

Figure 8.33 Small-angle scattering tensor tomography. (a) Experimental setup, showing the required six degrees of freedom. (b) The orientation and its degree of nanoscale mineralized collagen fibrils in a human trabecula bone sample, recorded with a spatial (step or voxel) resolution of 25 microns and structural features within each voxel to a resolution as high as a few nm. The rod orientations equate to those of the collagen fibrils in the corresponding voxels, while their size and colour indicate the degree of anisotropy, from blue (low) to red (high), provided by the circumferential extent of the SAXS features at 65 nm indicated in (a). Adapted from [43], with permission from Macmillan Publishers Ltd.

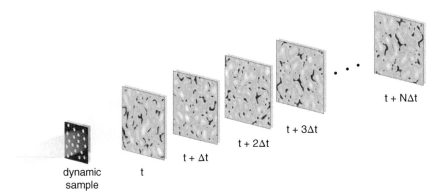

Figure 8.34 X-ray photon correlation spectroscopy probes dynamics of systems by observing the temporal changes in the speckle pattern they produce.

has been an established technique for studying dynamics in optically transparent systems for several decades. However, PCS is unable to probe length scales smaller than about 200 nm, and is excluded from systems that are opaque, or are embedded in an opaque medium. By extending PCS to the hard x-ray regime, it was recognized in the landmark paper by Sutton *et al.* in 1991 [46] that XPCS can be used as a powerful method to investigate dynamics on length scales given by the speckle pattern; and to time-scales that are limited either by the repetition rate of the x-ray detector (which nowadays can have frame rates above 10 kHz for area detectors, and orders of magnitude more still using scintillator point detectors), or more commonly, by the limited photon flux and scattering power of the sample.

With the advent of hard x-ray free-electron lasers and, more recently, DLSRs, XPCS is likely to experience a renaissance in the field of atomic-resolution dynamics and has been demonstrated to be able to probe nanoscale fluctuations up to frequencies in the kHz range [47] (see Figure 8.34). Depending on the sample, XPCS is sensitive to structural fluctuations ranging between several hundred nanometres and a few angstroms, while the largest system that can be probed is given by the transverse coherence length, of the order of ten microns in the vertical direction, and an order of magnitude smaller in the horizontal direction.

A serious current limitation of XPCS is its modest temporal resolution. XPCS measures the correlation *between* two photons rather than interference of a photon with itself. This means that, in contrast to 'normal' diffraction phenomena, the signal-to-noise ratio scales with intensity rather than the scattering amplitude. The temporal resolution can be pushed towards the microsecond regime if point detectors or avalanche-photodiode arrays such as the AGIPD detector [48] are employed rather than standard 2D detectors, which have readout times presently limited to the kHz range. Extension of this to tens or even hundreds of MHz would allow experiments to be performed on many systems with characteristic time-scales that are presently inaccessible, such as magnetic spin reorientation (10^9 Hz), ferroelectric-domain switching ($\sim 10^8$ Hz), or structural phase transitions (10^5 to 10^9 Hz).

The signal-to-noise ratio S/N for the autocorrelation function between two speckle patterns separated by a time Δt is given by

$$S/N = F_{\mathrm{coh}}(T\,\Delta t\,n_x n_y)^{1/2}, \tag{8.17}$$

where T is the total accumulated recording time, and $n_x \times n_y$ is the number of pixels in the detector that record 'usable' signal. So, for a given acceptable lower limit to S/N, an X times increase in the coherent flux would permit an X^2 reduction in the minimum time separation Δt. For a typical 50 times increase in F_{coh} for DLSRs, this therefore corresponds to an improvement in temporal resolution by over three orders of magnitude. A probable bottleneck to fully exploiting this substantial improvement will be detector technology, especially for area detectors. Another important aspect to consider is the fact that synchrotron radiation has a time structure, with pulses arriving at hundreds of MHz, and pulse lengths expected to be approximately 200 ps. As soon as one begins to probe processes on time-scales of the order of a train of a few pulses (of the order of 10 ns), this time structure must be considered. It has been suggested that one could perform XPCS within the duration of a single x-ray pulse [47], though the detector technologies required to achieve this (i.e. with readout times significantly shorter than the pulse width, of the order of 10 ps) means that such experiments lie much further in the future.

Provided that these technological challenges are successfully addressed, radiation-damage issues are likely to remain. For example, in the case of XPCS performed in the forward-scattering (SAXS) geometry, the scattered intensity drops as the fourth power of the scattering vector Q. Studies of fast fluctuations and at short length scales (large Q) are inherently impaired by the low intensity, requiring often prohibitively intense incident-beam intensities to compensate for this. The applicability of XPCS is likely therefore to be limited to radiation-tolerant systems and either slow processes or phenomena that include intensity enhancements (i.e. above the Q^{-4}-dependence of 'standard' XPCS), such as through the use of Bragg peaks or crystal-truncation rods, further limiting the scope of sample types that lend themselves to this method.

8.5 Concluding Remarks

Imaging using x-rays is an enormously profitable enterprise, in which insights on the sub-micron scale can often be rapidly obtained – indeed, the first demonstration of the power of x-rays was made by Röntgen with his radiograms of human hands. The impact these must have had on society can hardly be appreciated in a modern world in which medical radiographic imaging is so commonplace. However, the detailed architectural information provided by tomography on a microscopic scale can still both surprise and delight. The immediacy of the feedback and the relative ease of at least qualitative interpretation supplied by x-ray imaging makes it one of the most attractive synchrotron techniques.

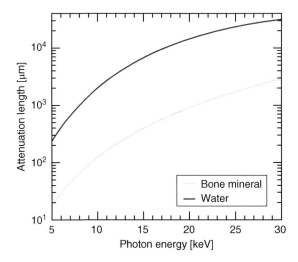

Figure 8.35   The attenuation lengths in microns of water and bone mineral between 5 and 30 keV.

As still more DLSRs and XFELs begin to come online, their associated high power densities mean that any optical components such as focussing lenses will be circumvented if they are not absolutely necessary. Lensless imaging is emerging not just as an attractive alternative to conventional x-ray microscopy, but may in many cases be the only realistic approach in investigating complex and large-scale nanoscaled structures, aided by high coherent fractions. This drive has already resulted in several advances in scattering and diffraction analysis [17, 37, 42, 49] and still more exciting developments can surely be expected in the near future.

Problems

1. The Hounsfield unit (HU) is used in radiology and tomography to define the absorption of a given material, and is defined as

$$\text{HU} = 1000 \times \frac{\mu_X - \mu_w}{\mu_w},$$

   whereby μ_w and μ_X are the linear attenuation coefficients of water and the material of interest, respectively. In a tomography experiment at a hard x-ray beamline of dense bone mineral (chemical formula $Ca_{10}(PO_4)_6(OH)_2$, density of 1.75 g cm^{-3}), the bone is found to have HU = 11 500. From the graph shown in Figure 8.35, determine the photon energy at which this experiment was performed. Hint: for an accurate determination, download the curves in Figure 8.35 by using the cxro website for x-ray attenuation lengths.   (2P)
2. Prove the Fourier-slice theorem operating on a two-dimensional function $f(x, y)$.   (4P)
3. Phase-contrast tomography experiments are performed on the wing of a drosophila fly and on the retinal cone cells of human eyes. In the former, the structure of the veins down to a resolution of 5 μm is required, while in the latter, the internal structure of the cell is to be investigated with 0.25 μm resolution. Estimate the ideal sample–detector distances to obtain optimal edge enhancement in both experiments, if 25 keV radiation is used.   (2P)
4. Flesh has an absorption coefficient $\mu = 6.48 \times 10^{-5}$ μm^{-1} at 20 keV. Calculate the transmission of a 2 mm thick flesh biopsy at 50 keV.   (1P)
5. The phase shift induced by a block at a certain photon energy is 40° relative to the same plane wave travelling through vacuum. If the photon energy is doubled, what value does the phase shift assume? (You can

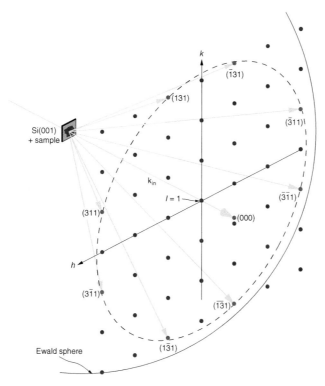

Figure 8.36 Setup of a CXDI experiment in which nine speckle patterns from a sample are observed simultaneously by adjusting the orientation of a Si(001) crystal so that the (001) axis lies perfectly parallel to the incident beam k_{in} and tuning the photon energy so that the eight equivalent (311) Bragg reflections (highlighted in red) all lie on the Ewald sphere.

presume both experiments use photon energies far above any absorption edges associated with the sample material.) (1P)

6. Consider Figure 8.23. If the distance along the central axis from the sample to the objective lens is p, and that from the objective to the image is q, what is the objective-to-phase-ring distance in terms of p and q. For a setup for which the magnification is 50, and the axial sample–objective distance is 50 mm, determine the positions of the phase ring and the image plane. (2P)

7. Consider again Figure 8.23(a) and the case where the reference wavevector amplitude (in red) is 40 times larger than those from the two scattering elements (in green and blue). Without a phase ring, the green and blue vectors have a phase of $\phi_g = 15°$, and $\phi_b = 45°$, respectively, relative to the reference wavevector. Determine I_g/I_b, the ratio of the intensities from these two elements in absorption mode.

   Next, the phase ring is added to induce a $\pi/2$ phase shift in the reference wave. It is made of a material with a refractive-index decrement $\delta = 7 \times 10^{-7}$ and an attenuation length of 11 μm for 0.5 Å radiation. What is I_g/I_b now? (5P)

8. A speckle pattern is detectable out to a radius on the detector of 7 cm, using a 172-μm-pixel Pilatus detector. The object is $10 \times 10 \times 5$ μm^3 in volume and should be resolved with 200 nm voxel accuracy. Determine the redundancy of the speckle image. You can assume that each pixel provides a useable independent data point. (1P)

9. A novel Bragg-CXDI experiment is able to simultaneously record nine speckle patterns by tuning the photon energy so that, in addition to the forward-scattered speckle pattern around the direct beam, the eight equivalent (311) reflections from a silicon single crystal placed immediately upstream of the sample all lie on the Ewald

sphere (see the reciprocal-lattice points highlighted in red in Figure 8.36). Calculate the required photon energy and Bragg angle θ. The fcc-unit-cell size of silicon is 5.431 Å.   (4P)

10. Given that the horizontal and vertical transverse coherence lengths, $\xi_{t,h}$ and $\xi_{t,v}$, respectively, at a distance R from an x-ray source of lateral size $w_h \times w_v$ are given by

$$\xi_{t,h} = \frac{\lambda R}{w_h},$$

$$\xi_{t,v} = \frac{\lambda R}{w_v},$$

prove Equation (3.51), that the coherent flux F_{coh} in photons per second is

$$F_{\mathrm{coh}} = \mathscr{B} \cdot \lambda^2 \cdot \frac{\Delta\lambda}{\lambda},$$

whereby $\mathscr{B}$ is the brilliance, λ is the x-ray wavelength, and $\Delta\lambda/\lambda$ is the fractional bandwidth of the x-ray source.

A beamline optimized for coherent x-ray diffraction studies has a brilliance at 7 keV of $\mathscr{B} = 10^{21}$ photons/s/mm^2/mrad^2/0.1% BW. The quasi-monochromatic x-ray beam has a bandwidth determined by the Si(111) crystals in the monochromator of 1.3×10^{-4}. The source size of the x-ray beam is $w_h \times w_v = 200 \times 50$ μm^2. Determine the horizontal and vertical transverse coherence lengths $\xi_{t,h}$ and $\xi_{t,v}$, respectively, at a distance from the source of $R = 40$ m, plus the coherent flux.   (6P)

References

[1]   J. H. Kinney and M. C. Nichols, "X-ray tomographic microscopy (XTM) using synchrotron radiation," *Ann. Rev. Mater. Sci.*, vol. 22, pp. 121–152, 1992.

[2]   U. Bonse and F. Busch, "X-ray computed microtomography (μ-CT) using synchrotron radiation (SR)," *Prog. Biophys. Mol. Biol.*, vol. 65, pp. 133–169, 1996.

[3]   A. C. Kak and M. Slaney, *Principles of Computerized Tomographic Imaging*. Soc. Indust. Appl. Math., 2001.

[4]   J. Radon, "On the determination of functions from their integral values along certain manifolds," *IEEE Trans. Med. Imaging*, vol. 5, pp. 170–176, 1986.

[5]   M. Holler, J. Raabe, A. Diaz, et al., "An instrument for 3D x-ray nano-imaging," *Rev. Sci. Instrum.*, vol. 83, p. 073703, 2012.

[6]   M. Holler, A. Diaz, M. Guizar-Sicairos, et al., "X-ray ptychographic computed tomography at 16 nm isotropic 3D resolution," *Sci. Reports*, vol. 4, p. 3857, 2014.

[7]   K. A. Nugent, T. E. Gureyev, D. F. Cookson, D. Paganin, and Z. Barnea, "Quantitative phase imaging using hard x-rays," *Phys. Rev. Lett.*, vol. 77, pp. 2961–2964, 1996.

[8]   G. Schneider, E. Anderson, S. Vogt, et al., "Computed tomography of cryogenic cells," *Surf. Rev. Lett.*, vol. 9, pp. 177–183, 2002.

[9]   D. Y. Parkinson, G. McDermott, L. D. Etkin, M. A. Le Gros, and C. A. Larabell, "Quantitative 3-D imaging of eukaryotic cells using soft x-ray tomography," *J. Struct. Biol.*, vol. 162, pp. 380–386, 2008.

[10]   T. Weitkamp, D. Haas, D. Wegrzynek, and A. Rack, "ANKAphase: software for single-distance phase-retrieval from inline X-ray phase-contrast radiographs," *J. Synchrotron Rad.*, vol. 18, pp. 617–629, 2011.

[11]   A. Snigirev, I. Snigireva, V. Kohn, S. Kuznetsov, and I. Schelokov, "On the possibilities of x-ray phase contrast microimaging by coherent high-energy synchrotron radiation," *Rev. Sci. Instrum.*, vol. 66, pp. 5486–5492, 1995.

[12] P. Cloetens, R. Barrett, J. Baruchel, J. P. Guigay, and M. Schlenker, "Phase objects in synchrotron radiation hard x-ray imaging," *J. Phys. D*, vol. 29, pp. 133–146, 1996.

[13] S. Sieber, P. Grossen, P. Detampel, et al., "Zebrafish as an early stage screening tool to study the systematic circulation of nanoparticulate drug delivery systems in vivo," *J. Controll. Release*, vol. 264, pp. 180–191, 2017.

[14] R. Mokso, P. Cloetens, E. Maire, W. Ludwig, and J.-Y. Buffière, "Nanoscale zoom tomography with hard x rays using Kirkpatrick-Baez optics," *Appl. Phys. Lett.*, vol. 90, p. 144104, 2007.

[15] R. Mokso, F. Marone, and M. Stampanoni, "Real time tomography at the Swiss Light Source," *AIP Conf. Proc.*, vol. 1234, pp. 87–90, 2010.

[16] R. Mokso, C. M. Schlepütz, G. Theidel, et al., "GigaFRoST: the gigabit fast readout system for tomography," *J. Synchrotron Rad.*, vol. 24, pp. 1250–1259, 2017.

[17] S. M. Walker, D. A. Schwyn, R. Mokso, et al., "In vivo time-resolved tomography reveals the mechanics of the blowfly flight motor," *PLOS Bio.*, vol. 12, p. e1001823, 2014.

[18] L. Helfen, T. Baumbach, P. Mikulik, et al., "High-resolution three-dimensional imaging of flat objects by synchrotron-radiation computed laminography," *Appl. Phys. Lett.*, vol. 86, p. 071915, 2005.

[19] L. Helfen, T. Baumbach, P. Cloetens, and J. Baruchel, "Phase-contrast and holographic computed laminography," *Appl. Phys. Lett.*, vol. 94, p. 104103, 2009.

[20] L. Helfen, A. Myagotin, P. Mikulik, et al., "On the implementation of computed laminography using synchrotron radiation," *Rev. Sci. Instrum.*, vol. 82, p. 063702, 2011.

[21] M. Stampanoni, R. Mokso, F. Marone, et al., "Phase-contrast tomography at the nanoscale using hard x-rays," *Phys. Rev. B*, vol. 81, p. 140105, 2010.

[22] F. Pfeiffer, M. Bech, O. Bunk, et al., "Hard-x-ray dark-field imaging using a grating interferometer," *Nature Mater.*, vol. 7, pp. 134–137, 2008.

[23] M. Bech, O. Bunk, T. Donath, et al., "Quantitative x-ray dark-field computed tomography," *Phys. Med. Biol.*, vol. 55, pp. 5529–5539, 2010.

[24] C. Scheuerlein, B. Fedelich, P. Alknes, et al., "Elastic anisotropy in multifilament Nb_3Sn superconducting wires," *IEEE Trans. Appl. Supercond.*, vol. 25, p. 8400605, 2015.

[25] F. van der Veen and F. Pfeiffer, "Coherent x-ray scattering," *J. Phys.: Condens. Matter*, vol. 16, pp. 5003–5030, 2004.

[26] V. Elser, "Phase retrieval by iterated projections," *J. Opt. Soc. Am. A*, vol. 20, pp. 40–55, 2003.

[27] D. Shapiro, P. Thibault, T. Beetz, et al., "Biological imaging by soft x-ray diffraction microscopy," *Proc. Natl. Acad. Sci. USA*, vol. 102, pp. 15343–15346, 2005.

[28] P. Thibault, V. Elser, C. Jacobsen, D. Shapiro, and D. Sayre, "Reconstruction of a yeast cell from x-ray diffraction data," *Acta Crystallogr. A*, vol. 62, pp. 248–261, 2006.

[29] M. A. Pfeifer, G. J. Williams, I. A. Vartanyants, R. Harder, and I. K. Robinson, "Three-dimensional mapping of a deformation field inside a nanocrystal," *Nature*, vol. 442, pp. 63–66, 2006.

[30] I. Robinson and R. Harder, "Coherent x-ray diffraction imaging of strain at the nanoscale," *Nature Mater.*, vol. 8, pp. 291–298, 2009.

[31] S. Marchesini, H. He, H. N. Chapman, et al., "X-ray image reconstruction from a diffraction pattern alone," *Phys. Rev. B*, vol. 68, p. 140101, 2003.

[32] J. S. Wu, U. Weierstall, J. C. H. Spence, and C. T. Koch, "Iterative phase retrieval without support," *Optics Lett.*, vol. 29, pp. 2737–2739, 2004.

[33] S. Marchesini, "A unified evaluation of iterative projection algorithms for phase retrieval," *Rev. Sci. Instrum.*, vol. 78, p. 049901, 2007.

[34] F. Pfeiffer, "X-ray ptychography," *Nature Phot.*, vol. 12, pp. 9–17, 2018.

[35] W. Hoppe, "Beugung im inhomogenen Primärstrahlwellenfeld. I. Prinzip einer Phasenmessung von Elektronenbeugungsinterferenzen," *Acta Crystallogr. A*, vol. 25, pp. 495–501, 1969.

[36]   J. M. Rodenburg, A. C. Hurst, A. G. Cullis, et al., "Hard x-ray lensless imaging of extended objects," *Phys. Rev. Lett.*, vol. 98, p. 034801, 2007.

[37]   P. Thibault, M. Dierolf, A. Menzel, et al., "High-resolution scanning x-ray diffraction microscopy," *Science*, vol. 321, pp. 379–382, 2008.

[38]   M. Dierolf, A. Menzel, P. Thibault, et al., "Ptychographic x-ray computed tomography at the nanoscale," *Nature*, vol. 467, pp. 436–440, 2010.

[39]   O. Bunk, M. Dierolf, S. Kynde, et al., "Influence of the overlap parameter on the convergence of the ptychographical iterative engine," *Ultramicroscopy*, vol. 108, pp. 481–487, 2008.

[40]   H. M. L. Faulkner and J. M. Rodenburg, "Movable aperture lensless transmission microscopy: a novel phase retrieval algorithm," *Phys. Rev. Lett.*, vol. 93, p. 023903, 2004.

[41]   P. Thibault, M. Dierolf, O. Bunk, A. Menzel, and F. Pfeiffer, "Probe retrieval in ptychographic coherent diffractive imaging," *Ultramicroscopy*, vol. 109, pp. 338–343, 2009.

[42]   M. Holler, M. Guizar-Sicairos, E. H. R. Tsai, et al., "High-resolution non-destructive three-dimensional imaging of integrated circuits," *Nature*, vol. 543, pp. 402–406, 2017.

[43]   M. Liebi, M. Georgiadis, A. Menzel, et al., "Nanostructure surveys of macroscopic specimens by small-angle scattering tensor tomography," *Nature*, vol. 527, pp. 349–353, 2015.

[44]   P. Fratzl, H. F. Jakob, S. Rinnerthaler, P. Roschger, and K. Klaushofer, "Position-resolved small-angle x-ray scattering of complex biological materials," *J. Appl. Crystallogr.*, vol. 30, pp. 765–769, 1997.

[45]   O. Bunk, M. Bech, T. H. Jensen, et al., "Multimodal x-ray scatter imaging," *New J. Phys.*, vol. 11, p. 123016, 2009.

[46]   M. Sutton, S. G. J. Mochrie, T. Greytak, et al., "Observation of speckle by diffraction with coherent x-rays," *Nature*, vol. 352, pp. 608–610, 1991.

[47]   O. G. Shpyrko, "X-ray photon correlation spectroscopy," *J. Synchrotron Rad.*, vol. 21, pp. 1057–1064, 2014.

[48]   B. Henrich, J. Becker, R. Dinapoli, et al., "The adaptive gain integrating pixel detector AGIPD: a detector for the European XFEL," *Nucl. Instrum. Methods A*, vol. 633, pp. S11–S14, 2011.

[49]   C. Donnelly, M. Guizar-Sicairos, V. Scagnoli, et al., "Three-dimensional magnetization structures revealed with X-ray vector nanotomography," *Nature*, vol. 547, pp. 328–331, 2017.

Appendix A
Cryogenic Electron Microscopy

2017 saw the Nobel Prize in Chemistry awarded to Jacques Dubochet, Joachim, Frank, and Richard Henderson 'for developing cryo-electron microscopy for the high-resolution structure determination of biomolecules in solution'.

Until well into the second decade of the twenty-first century, macromolecular x-ray crystallography (MX) enjoyed an almost complete hegemony as a tool to determine life's molecular machinery. Because hard x-rays cannot be strongly refracted, thus precluding high-magnification x-ray microscopes, the atomic-scale structure of a biological molecule could only be regained by coaxing it into forming single crystals of micron size or larger, irradiating these with x-rays, and then inverse-Fourier-transforming the resulting scattering pattern, with this latter process requiring a solution to the ubiquitous phase problem (see Chapter 6). As we have seen in the main text, MX has been one of the great successes in the natural sciences, garlanding several Nobel Prizes thanks to the insights gained using this approach.

The limitations of MX, summarized in Section 6.11, include radiation damage, problems associated with cryocooling, and the possibility that, in forcing a given protein to arrange itself in a regular crystalline array, its conformation may differ from that found *in vivo*. Indeed, many proteins that are part of a cell membrane and convey information between the confines of the cell and the external world, require their native environment to fold properly and thereby react correctly to stimuli. These membrane proteins (especially so-called 'G-protein coupled receptors', or GPCRs, and ion channels) are the targets for the majority of pharmaceutical drugs. Their generally hydrophobic surfaces and anisotropic orientation hinder crystallization in an aqueous environment, and other media such as the lipidic cubic-phase (LCP) must be brought to bear on the problem.

This brings us to perhaps the gravest bottleneck to macromolecular-structure determination using diffraction – the growth of single crystals of sufficient quality and size. The newest generation of synchrotrons, DLSRs, offer a reduction in linear dimensions close to an order of magnitude, although at present, crystals much smaller than a micron remain beyond cutting-edge technology. A technique which could break through this barrier would prove invaluable. Enter cryo-electron microscopy (cryoEM) [1–4].

Elastic scattering of x-rays is much weaker than photoabsorption, making x-ray diffraction an inefficient process. Proteins scatter electrons some ten-thousand times more strongly than they do x-rays. Because electrons are charged, they can be (a) accelerated to energies (of the order of 300 keV) with de Broglie wavelengths considerably smaller than interatomic distances (2 pm for 300 keV); and (b) efficiently and tightly focussed using electromagnetic lenses. CryoEM is, however, by no means immune to radiation damage – for each electron that contributes to the formation of an image, three others will deposit a significant fraction of their energy into the sample.

An Introduction to Synchrotron Radiation: Techniques and Applications, Second Edition. Philip Willmott.
© 2019 John Wiley & Sons Ltd. Published 2019 by John Wiley & Sons Ltd.

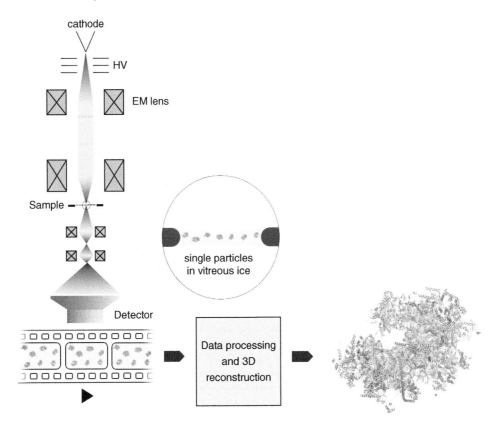

Figure A.1   A sample holder containing single particles trapped in a vitreous-ice matrix on a grid mesh is placed at the focus of an electron condenser lens. The scattered electrons are re-imaged using a series of magnifying objectives on to an electron detector which records images as quickly as possible. After data processing and 3d-reconstruction, the molecular structure is obtained.

Transmission electron microscopy (TEM) has a long and illustrious pedigree. First demonstrated in 1931 by Max Knoll and Ernst Ruska, it achieved a resolution of 5 nm by 1960 (Ruska would receive the Nobel Prize in Physics in 1986 for his pioneering work in this area, along with Gerd Binnig and Heinrich Rohrer for the development of the scanning tunnelling microscope); the best TEM resolution to date, using spherical-aberration-correcting lenses, is 0.4 Å, demonstrated on the surface of a germanium single crystal [5]. Such resolutions are only possible, however, in strongly bonded, radiation-resistant, and highly-perfect single crystals. Until approximately 2012, the resolution of images of soft matter was, with some notable exceptions, limited to approximately 10 Å. The main problems of electron microscopy of biological samples are (a) the low image contrast of biological material using high-energy electrons, which mostly pass unhindered straight through the specimens; and (b) the rapid degradation of the structural integrity caused by energy transfer from the electron beam. The former demands a high radiation dose, while the latter requires this to be as small as possible. Moreover, most traditional EM experiments are carried out in ultrahigh vacuum, unsuitable for samples that normally reside in an aqueous environment. Finally, multiple scattering of the electrons should be avoided, necessitating that the samples are very thin.

It was a series of critical breakthroughs in the last decade that heralded cryoEM as a serious alternative to MX and prompted the Nobel Prize in Chemistry in 2017.

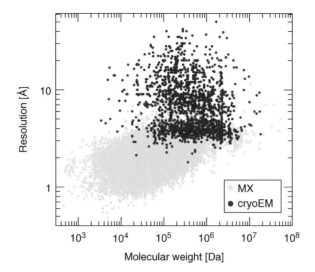

Figure A.2 A plot of resolution as a function of molecular weight for MX (121 727 structures) and cryoEM (1814 structures) for all depositions in the Protein Data Bank up to the end of 2017. Note that while the distribution for MX in this double-logarithmic plot is a diffuse cloud, that for cryoEM has an interesting cluster of data between 3 and 5 Å, and molecular weights spanning approximately 0.1 to 5 MDa, reflecting the breakthrough in this technique in the second decade of the twenty-first century. Although there are some very few notable exceptions, the lower limit of 3 Å is fairly abrupt, indicative of possibly fundamental limitations to cryoEM briefly described in the text.

The areas of expertise of the three recipients reflect these major sea changes in the field, namely sample preparation, improved microscope hardware and detectors, and sophisticated algorithms for extracting the structures from the data sets. These advances are now briefly summarized.

Jacques Dubochet and co-workers demonstrated in the early 1980s that micron-sized drops of bulk water could be vitrified (that is, frozen in a purely amorphous state). By using a grid with micron-sized apertures, individual macromolecules could be trapped in a thin film of water which was rapidly plunged into liquid ethane or propane held at liquid-nitrogen temperatures. The layers were thin enough to guarantee full vitrification but still allow a single layer of macromolecules to be trapped in their native state [6].

Improvements in hardware, including both the electron microscopes and detectors, represent the second breakthrough for cryoEM [7]. So-called 'direct-electron detectors' (DEDs), based on complementary metal oxide semiconductor (CMOS) technology, already in wide use in astronomy because of their exceptionally low noise, were made widely available in cryoEM in 2012. They also have fast readout times, making it possible to record 'movies'. This seemingly prosaic feature actually then allows one to follow specimen drift caused by energy transfer from the electron beam, which had previously caused blurring in the necessarily long-exposure images associated with the previous technologies.

Any individual cryoEM image is a noisy affair – protein scatters only about 30% more efficiently than does vitreous ice. This difference is barely above the noise for the limited acceptable dose before the particle is destroyed. Therefore, even for a single projection orientation, many images must be collected in order to obtain a sufficiently high resolution [2, 8]

A major problem in analysing weak-contrast and noisy images of thousands or even millions of randomly aligned macromolecules is obtaining a reliable handle on the particles' orientations. This is achieved by using cross-correlation functions [9]. By averaging images of many individual particles, each at a random orientation, a high-resolution three-dimensional reconstruction can be recovered via tomographic methods. Because biological structures are often nonuniform, identification of potential variations normally also requires an algorithm that

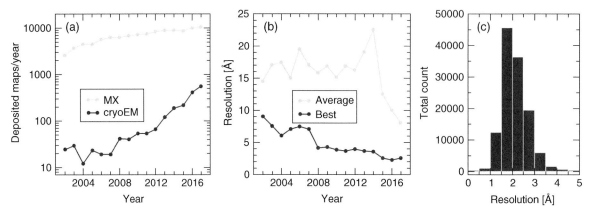

Figure A.3   Statistical comparisons between cryoEM and MX. (a) Progress in the number of deposited structures per year. Note the logarithmic ordinate. If cryoEM were to continue the exponential increase in deposition rate exhibited in the last five years (doubling every 18 months), it will catch up with MX by approximately 2025. (b) A plot of the average and best resolutions reported by cryoEM in the last decade and a half. It is the average value which really exhibits a steplike improvement after 2014, directly after the innovations described in the text were implemented. (c) The distribution of resolutions reported in MX from the 121 000 structures deposited in the PDB until the end of 2017. Note the median value between 1.5 and 2 Å.

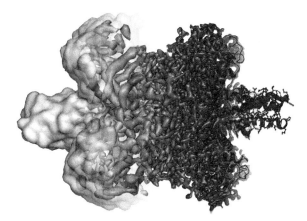

Figure A.4   The progress in resolution of cryoEM. The atomic-scale details of glutamate dehydrogenase were excluded to cryoEM prior to 2012 (left). Breakthroughs in hardware and data analysis in that year and 2013 pushed the resolution to 1.8 Å (right) [14]. Rendered from pdb file 5k12, courtesy Martin Högbom, Stockholm University.

sorts particle images not only into orientations, but also structural subclasses. This higher level of sophistication is nowadays exploited in dynamical studies.

A plot of the reported resolution as a function of molecular weight for all depositions in the Protein Data Bank up to the end of 2017 is shown in Figure A.2. The double-logarithmic plot is particularly revealing for the cryoEM data – there is a distinct cluster of structures between 3 and 5 Å, the large majority of which were deposited after 2013. Although there are indeed reports below 3 Å, these are rare. No single structure solved by cryoEM exists below $M_W = 25$ kDa. This reflects the difficulty in obtaining a sufficiently high signal-to-noise ratio in cryoEM for smaller proteins. The abruptness of this cut-off may indicate that cryoEM has fundamental limits that are fast being approached. Further statistical comparisons with MX are summarized in Figure A.3.

Nonetheless, the future of cryoEM seems very bright – it seems highly probable that, with further developments in data mining and automated analysis of images, cryoEM will be the natural method of choice for larger proteins [10, 11] (Figure A.4). Exactly how it will impact structural studies of proteins with molecular weights below 50 kDa remains to be seen, while its application in dynamical studies with resolutions better than a few milliseconds presently appears to be unlikely [12]. In particular, subtle conformational changes not involving rearrangements of large chunks of the protein may be very difficult to distinguish. Reports are emerging, at the time of writing, of cryoEM data being used as an initial low-resolution model to solve the phase problem in MX down to sub-angstrom resolution [13].

References

[1]   J. Frank, "Single-particle imaging of macromolecules by cryo-electron microscopy," *Ann. Rev. Biophys. Biomol. Struct.*, vol. 31, pp. 303–319, 2002.

[2]   X.-C. Bai, G. McMullan, and S. H. W. Scheres, "How cryo-EM is revolutionizing structural biology," *Trends Biochem. Sci.*, vol. 40, pp. 49–57, 2015.

[3]   E. Callaway, "The revolution will not be crystallized: a new method sweeps through structural biology," *Nature*, vol. 525, pp. 172–174, 2015.

[4]   S. Subramaniam, W. Kühlbrandt, and R. Henderson, "CryoEM at IUCrJ: a new era," *IUCrJ*, vol. 3, pp. 3–7, 2016.

[5]   R. Erni, M. D. Rossell, C. Kisielowski, and U. Dahmen, "Atomic resolution imaging with a sub-50-pm electron probe," *Phys. Rev. Lett.*, vol. 102, p. 096101, 2009.

[6]   J. Dubochet, M. Adrian, J. J. Chang, et al., "Cryo-electron microscopy of vitrified specimens," *Quart. Rev. Biophys.*, vol. 21, pp. 129–228, 1988.

[7]   G. McMullan, A. R. Faruqi, D. Clare, and R. Henderson, "Comparison of optimal performance at 300 keV of three direct electron detectors for use in low dose electron microscopy," *Ultramicroscopy*, vol. 147, pp. 156–163, 2014.

[8]   S. Chen, G. McMullan, A. R. Faruqi, et al., "High-resolution noise substitution to measure overfitting and validate resolution in 3D structure determination by single particle electron cryomicroscopy," *Ultramicroscopy*, vol. 135, pp. 24–35, 2013.

[9]   W. O. Saxton and J. Frank, "Motif detection in quantum noise-limited electron-micrographs by cross-correlation," *Ultramicroscopy*, vol. 2, pp. 219–227, 1977.

[10] A. Patwardhan, "Trends in the Electron Microscopy Data Bank (EMDB)," *Acta Crystallogr. D*, vol. 73, pp. 503–508, 2017.

[11] S. Subramaniam, "The cryo-EM revolution: fueling the next phase," *IUCrJ*, vol. 6, pp. 1–2, 2019.

[12] S. C. Shoemaker and N. Ando, "X-rays in the cryo-electron microscopy era: Structural biology's dynamic future," *Biochemistry*, vol. 57, pp. 277–285, 2018.

[13] L. Zeng, W. Ding, and Q. Hao, "Using cryo-electron microscopy maps for X-ray structure determination," *IUCrJ*, vol. 5, pp. 382–389, 2018.

[14] A. Merk, A. Bartesaghi, S. Banerjee, et al., "Breaking cryo-EM resolution barriers to facilitate drug discovery," *Cell*, vol. 165, pp. 1698–1707, 2016.

Appendix B

Some Helpful Mathematical Relations and Approximations

Many arguments laid out in this book, not least in the solutions to the problem sets, draw on some straightforward mathematical relations and approximations that might be unfamiliar to those readers less seasoned in mathematics. The following table lists most of those used herein. Terms given in curly parentheses in the approximations may or may not be required, depending on the nature of the argument or problem. Under this is a plot of $\sin(x)$ and $\cos(x)$, plus their two first approximations using their Taylor expansions defined in the table.

Table B.1

Relation/Approximation	Conditions
$\sin^2 x + \cos^2 x = 1$	
$\tan x = \frac{\sin x}{\cos x}$	
$\sin(x + y) = \sin(x)\cos(y) + \cos(x)\sin(y)$	
$\sin(2x) = 2\sin x \cos x$	
$\cos(x + y) = \cos(x)\cos(y) - \sin(x)\sin(y)$	
$\cos(2x) = \cos^2 x - \sin^2 x = 2\cos^2 x - 1$	
$x = \frac{-b \pm \sqrt{b^2 - 4ac}}{2a}$	$ax^2 + bx + c = 0$
$\exp(x) \approx 1 + x \{+x^2/2!\}$	$\|x\| \ll 1$
$\log_e(1 + x) \approx x \{-x^2/2\}$	$\|x\| \ll 1$
$(1 + x)^n \approx 1 + nx \{+n(n - 1)x^2/2!\}$	$\|nx\| \ll 1$
$\sin x \approx x \{-x^3/3!\}$	$\|x\| \ll 1$, x in radians
$\cos x \approx 1 - x^2/2 \{+x^4/4!\}$	$\|x\| \ll 1$, x in radians
$\tan x \approx x \{+x^3/3\}$	$\|x\| \ll 1$, x in radians
Euler's formula: $e^{ix} = \cos x + i \sin x$	x in radians

Sine rule: $\frac{\sin \alpha}{a} = \frac{\sin \beta}{b} = \frac{\sin \gamma}{c}$

Cosine rule: $a^2 = b^2 + c^2 - 2bc \cos \alpha$

$\ln n! \approx (n + 1/2)\ln n - n + \ln[(2\pi)^{1/2}] \approx (n + 1/2)\ln n - (n - 1)$

An Introduction to Synchrotron Radiation: Techniques and Applications, Second Edition. Philip Willmott.
© 2019 John Wiley & Sons Ltd. Published 2019 by John Wiley & Sons Ltd.

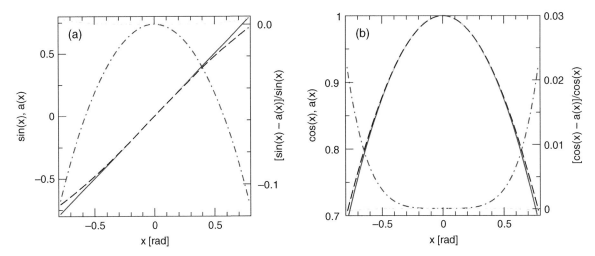

Figure B.1   The sine and cosine functions and their approximations between $x = \pm\pi/4$ (± 0.785 rad). (a) $\sin(x)$ is given by the black dashed curve. The two first approximations of this, $a(x) = x$, and $a(x) = x - x^3/3!$, are shown as the solid blue and yellow curves, respectively. The yellow curve lies almost perfectly under the true function $\sin(x)$. The deviations from the true function, $[\sin(x) - a(x)]/\sin(x)$ are given by the dot-dashed curves (corresponding colours); their values are given on the right ordinate. (b) The equivalent curves for $\cos(x)$, for which $a(x) = 1 - x^2/2$, and $a(x) = 1 - x^2/2 + x^4/4!$.

Appendix C
Fourier Series and Fourier Transforms Made Simple

C.1 Introductory Remarks

In my experience, one phrase liable to make many a young scientist freeze like a fawn in the headlights of a juggernaut is 'Fourier transform'. Fourier transforms (FTs) need not, however, be all that frightening. The goal of this appendix is to convey the quintessence of Fourier transforms and analysis with recourse to the minimum amount of mathematics.

FTs pop up all the time in everyday life. Our ears naturally perform FTs – for example, when we hear a note, we are not consciously following the longitudinal pressure variations that occur typically on a scale of hundreds to thousands of Hz, but instead we experience a sensation we call noise. In much the same way, a musician does not try to follow the voltage signal generated by a microphone that has recorded music, but instead will read the notes off sheet music (see Figure C.1). The FT is a pervasive and versatile tool used in many fields to phrase problems in terms that can be more easily solved. FTs are used in image processing, enhancements of sound recordings, and the design of stable structures in civil engineering[1], to name just three of innumerable applications.

Simply stated, a Fourier transform of an object or signal tells you what frequencies are present in it, and in what proportions. It takes a spatial (or temporal) pattern or function, it measures every possible frequency it might contain, and returns the component parts [that is, each frequency, and its amplitude and offset (or phase)].

Hence, periodic things, such as the sinusoidal electromagnetic field of monochromatic light (such as that produced by lasers), or crystals, become easier to understand when viewed from the perspective of their Fourier transforms. If this statement doesn't strike you as 'simple', consider the examples in Figure C.2.

On the left, a plane sine wave $\sin x$ is shown. This has a periodicity of $P = \lambda = 2\pi$. This is evident because the value of the function at x also assumes the same value at $(x + 2m\pi)$, where m is an integer.

The spatial frequency of this function is $f = 1/\lambda = 1/2\pi$, and a wavevector $k = 2\pi/\lambda = 1$. The Fourier transform of $\sin x$ describes it as a single-valued spike at a wavevector value equal to that of the sine wave, i.e. in this case, at $k = 1$. Note also that FTs have the same symmetry as the functions they represent, hence the FT of a sine wave has equal but opposite components at $\pm$ of the spatial frequency – that's why there are two spikes for each FT shown in Figure C.2.

---

[1] An example of the application of Fourier analysis is the suppression of the exceedingly irritating drone produced by vuvuzelas during the 2010 World Cup by Fourier-transforming the audio signal, and modifying it with two notch filters at approximately 235 and 465 Hz. Fourier analysis should have been, but wasn't, applied in the design of the Tacoma Narrows Bridge which famously collapsed in 1940, induced by resonant buffeting from not-unusual wind speeds travelling up the Tacoma Narrows Straight; it would have immediately revealed the fatal flaw associated with the construction's dynamics.

An Introduction to Synchrotron Radiation: Techniques and Applications, Second Edition. Philip Willmott.
© 2019 John Wiley & Sons Ltd. Published 2019 by John Wiley & Sons Ltd.

Figure C.1   The first few notes of the second movement 'alla Siciliana' of J.S. Bach's Concerto for three keyboards, BWV 1063, depicted both as recorded pressure variations and as sheet music. Even the most talented musician could not follow the upper trace; instead the human brain's inherent understanding of Fourier analysis in sound means that he can read the music sheet and understand that different notes represent different acoustic frequencies.

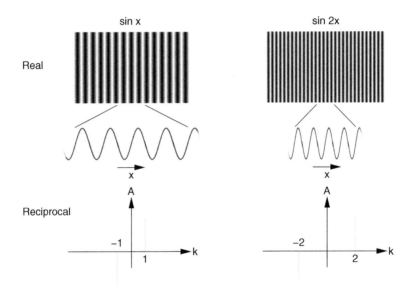

Figure C.2   Sine waves and their Fourier transforms as a function of $k = 2\pi/\lambda$.

It is worth mentioning at this juncture a small but important detail: the functions plotted in Figure C.2 are cut off – if I had wanted to plot a perfectly pure sine function, I would have had to extend the two figures to $\pm\infty$. This might seem like a pedantic point, but the action of cutting off a periodic function is to make it nonperiodic – the sine wave is suddenly no more. This causes the FT to no longer consist of a set of infinitely narrow, perfectly defined spikes, which instead assume a finite width. The FWHM of these widths is inversely proportional to the spatial extent that the periodic function covers in real space. This is important in describing real systems, such as crystals, which are not infinitely large. But let's store this confusing detail away for now …

On the right of Figure C.2, the plane-wave function $\sin 2x$ has double the frequency, in other words, $k = \pm 2$. Note that the two examples in Figure C.2 are periodic functions (ignoring the caveat above) and can be described by so-called 'Fourier series'. Fourier series are Fourier transforms of periodic objects and consist of a regular array of evenly separated discrete values. We discuss these in more detail now.

C.2   Periodic Functions

Shown in Figure C.3 are some examples of periodic functions. Any periodic function, whether a train of square pulses in one dimension, a wall of ceramic tiles (2D), or a crystal of penicillin molecules (3D), can be expressed by a Fourier series. A Fourier series in one dimension is given by the expression:

$$f(x) = \frac{a_0}{2} + \sum_{n=1}^{\infty} \left[a_n \cos \left(\frac{2\pi nx}{P} \right) + b_n \sin \left(\frac{2\pi nx}{P} \right) \right], \tag{C.1}$$

where P is the periodicity. So, each component (labelled by n) of the Fourier series has a frequency equal to an integer multiple n of the fundamental frequency and an amplitude a_n for the cosine function and b_n for the sine function. Indeed, it is logically obvious that a periodic function can only contain components that have frequencies that are integer multiples (n) of the fundamental frequency. If this were not the case, the function would not look the same as one moves from one position x to another position $x + P$ and the periodicity P would be lost.

Note that for each integer value of n, there is a cosine and a sine component.

For illustrative purposes, two further simple examples of a periodic function are shown in Figure C.4. These are described by

$$f_1(x) = \sin(2\pi x) + \frac{\sin(4\pi x)}{2},$$

$$f_2(x) = \sin(2\pi x) + \frac{\sin(4\pi x + \pi/4)}{2}.$$

$f_2(x)$ thus differs from $f_1(x)$ only in that the higher-frequency component (shown in green) is phase-shifted by an angle $\phi = \pi/4$. Note that $f_2(x)$ as given above (which includes the phase $\pi/4$ in the second component) needs to be re-expressed in a form compatible with Equation (C.1). For this, we need to determine the Fourier amplitudes a_n and b_n. The equations for these are

$$a_0 = \frac{2}{P} \int_{x_0}^{x_0+P} f(x)dx, \tag{C.2}$$

$$a_n = \frac{2}{P} \int_{x_0}^{x_0+P} f(x) \cos \left(\frac{2\pi nx}{P} \right) dx, \tag{C.3}$$

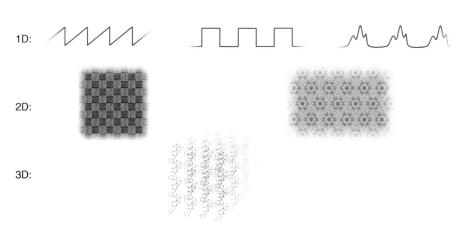

Figure C.3 Examples of one-, two-, and three-dimensional periodic functions. The faded edges imply they continue infinitely in all directions.

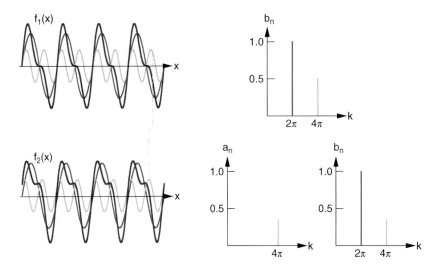

Figure C.4   Two similar periodic functions $f_1(x)$ and $f_2(x)$ (blue curves) are both composed of two sine waves, one with double the frequency of the other ($k = 2\pi$, shown in red, and $k = 4\pi$, shown in green) and their Fourier components a_n and b_n. The only difference between $f_1(x)$ and $f_2(x)$ is that the higher-frequency component in $f_2(x)$ is shifted by a phase $\pi/4$ relative to the same component in $f_1(x)$ (highlighted by the green dot-dashed dog-leg).

$$b_n = \frac{2}{P} \int_{x_0}^{x_0+P} f(x) \sin\left(\frac{2\pi n x}{P}\right) dx. \tag{C.4}$$

One need not either fully understand their origin, or indeed use them in order to understand Fourier transforms to the level required in this text. Nonetheless, for completeness, we give them here. First, it should be immediately apparent that $a_0 = 0, a_1 = 0, b_1 = 1$, and $P = 1$, as for $f_1(x)$. Using Equations (C.3) and (C.4), it turns out that $a_2 = (\sin\phi)/2$ and $b_2 = (\cos\phi)/2$ for any phase shift ϕ in the second component ($n = 2$) – the more mathematically ambitious student is encouraged to derive these coefficients for themselves. So, $f_2(x)$ can be re-expressed as

$$f_2(x) = \sin(2\pi x) + \frac{\sin\phi}{2}\cos(4\pi x) + \frac{\cos\phi}{2}\sin(4\pi x)$$

$$= \sin(2\pi x) + \frac{1}{2\sqrt{2}}\cos(4\pi x) + \frac{1}{2\sqrt{2}}\sin(4\pi x).$$

Thus, unsurprisingly, the Fourier series for $f_1(x)$ and $f_2(x)$ differ. The function $f_1(x)$ only has nonzero terms for b_n ($b_1 = 1, b_2 = 0.5$), while that for $f_2(x)$ has $b_1 = 1, b_2 = 1/2\sqrt{2}$, and $a_2 = 1/2\sqrt{2}$.

This leads us to an important property of Fourier series regarding the symmetry of the real-space object: those objects which are antisymmetric [$f(r) = -f(-r)$] contain only b_n components in their Fourier series, but none with nonzero a_n, because the cosine function is symmetric and the sine function is antisymmetric. For example, $f_1(x)$ in Figure C.4 is purely antisymmetric. Conversely, symmetric functions [$f(r) = f(-r)$] only contain cosine (a_n) terms. Note that in Figure C.4, $f_2(x)$ is neither symmetric or antisymmetric, and thus contains both types of Fourier amplitudes.

The next Fourier series we consider is that for a pulse train, such as shown in Figure C.5. This particular function has been chosen for various reasons, both didactic and because a limiting case of it is important in the description of crystals. The first thing we should recognize is that the vertical flanks of the pulses have infinite gradients (they

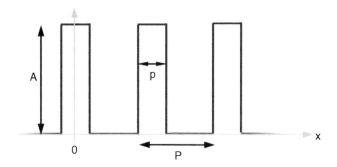

Figure C.5   A pulse train with a duty cycle of p/P and amplitude A.

increase/decrease by an amount A for zero change in x). This should, logically, lead us to conclude that we require an infinite summation ($n = 1$ to ∞), as no sine wave with a non-infinite frequency can have an infinite gradient.

Secondly, because we have set the y-axis to be in the middle of a pulse, the function is symmetric. We can therefore immediately predict that $b_n = 0$ for all n. It emerges that

$$f(x) = Ad + \sum_{n=1}^{\infty} \underbrace{\left[\frac{2A}{n\pi} \sin(n\pi d)\right]}_{a_n} \cos(2n\pi x/P), \tag{C.5}$$

where $d = p/P$ is the duty cycle. Although one needs to sum to infinity to faithfully reproduce the pulse-train function, it is well approximated after summing just a few Fourier components. The summation up to $n = 10$ for $d = 1/5$, plus the individual components, are shown in Figure C.6.

There are two interesting limiting cases to this pulse-train structure (and are the reasons I chose it as an example). The first is the case where p is made to be infinitely narrow. Note that, in three dimensions, this describes the array of points of the Bravais lattice of crystals, discussed in Section 6.3. As the limit of $d = p/P = 0$ is approached, $\sin(n\pi d) \approx n\pi d$, and the amplitudes of the Fourier components are thus all equal to $2Ad$. Importantly, for a given n, k is independent of d. In other words, the Fourier transform (or the diffraction pattern) of an infinitely long train of spikes of zero width, equally separated by P, and of the same height A (known as a 'comb function') is another comb function consisting of spikes of height $2Ad$ and periodicity $1/P$.

C.3   From Fourier Series to Fourier Transforms

The second limiting case is where we keep p constant, but increase P. If, for example, we double P, the density of Fourier components doubles, as $\Delta k = 2\pi/P$. The envelope curve of Figure C.6 (Left), however, remains unchanged when plotted as a function of k instead of n. Setting $P = \infty$ is equivalent to there being only one isolated pulse and the periodicity is lost. The Fourier series of this is now a continuous function instead of consisting of discrete values, and is given by $(4/kP) \sin(kp/2)$. This is a sinc function and is the Fourier transform of the single square pulse.

Lastly, let us take both limits together – a single, isolated, delta function, corresponding to $d = 0$ and $P = \infty$. Now, the comb-function FT of the real-space comb function is infinitely dense, and hence the FT of a single delta function is a continuum of frequencies up to infinity, all with the same amplitude (i.e. a straight horizontal line). Similarly, the FT of a constant in space or time is the frequency-independent expression a_0 given by Equation (C.2). These three limits are summarized in Figure C.7.

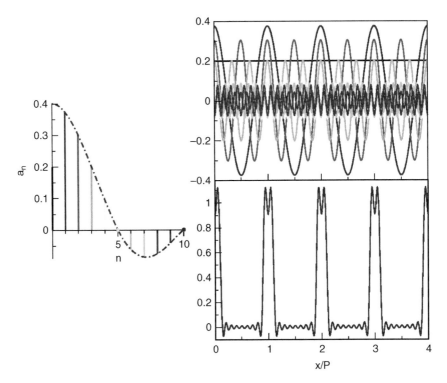

Figure C.6   Left: Values of a_n for the first 11 components of the Fourier series of a pulse train for which $A = 1$ and $d = 0.2$, as given by Equation (C.5). The dot-dashed envelope curve is given by $(2/n\pi)\sin(n\pi/5)$, which represents the limiting case of infinitely large P. Note that $a_0 = 0.2$ is only half the value of the envelope at $n = 0$; this is because we have not plotted the negative k-values, which also have their own a_0. Top right: The first 11 components plotted as a function of x/P. Bottom right: summation of the components yields a good approximation of the original pulse train.

Generally, the Fourier transform of a function $f(x)$ is given by

$$\mathscr{F}(k) = \int_{-\infty}^{\infty} f(x)e^{-2\pi ikx}dx. \tag{C.6}$$

Remembering Euler's formula, $e^x = \cos x + i\sin x$, we see in Equation (C.6) the extension of Equations (C.3) and (C.4) from a sum of discrete values to a continuum of frequencies given by the integral. Note that we need not necessarily use imaginary exponents in describing FTs; it is merely a convenient and compact notation. We could instead break up Equation (C.6) into two parts, one for the cosine and the other for the sine expressions, and indeed, for purely symmetric/antisymmetric functions, the sine/cosine terms fall away to leave

$$\mathscr{F}(k) = 2\int_{0}^{\infty} f^s(x)\cos(2\pi kx)dx$$

and

$$\mathscr{F}(k) = 2i\int_{0}^{\infty} f^a(x)\sin(2\pi kx)dx, \tag{C.7}$$

whereby $f^s(x)$ and $f^a(x)$ are, respectively, symmetric and antisymmetric functions.

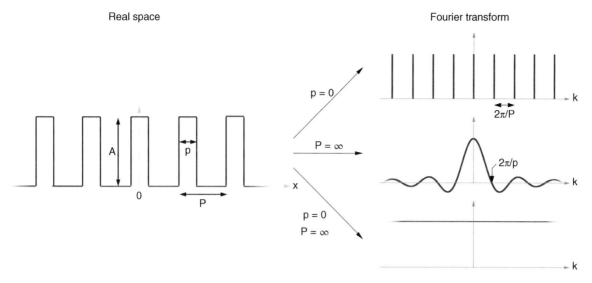

Figure C.7 Fourier transforms of a pulse train at limiting values. A pulse train (defined schematically, left) will produce a comb-function Fourier series with periodicity $k = 2\pi/P$ if p is set to zero (making the real-space function also a comb function). By increasing the periodicity P to infinity, the FT becomes a sinc function with the first crossing points at $\pm 2\pi/p$. By setting both p to zero and P to ∞, the real-space isolated delta-function has an FT composed of an infinite continuum of equally intense Fourier components.

The inverse Fourier transform is given by

$$\mathscr{F}^{-1}(k) = f(x) = \int_{-\infty}^{\infty} \mathscr{F}(k)e^{2\pi ikx}dk. \tag{C.8}$$

C.4 Mathematical Properties of Fourier Transforms

It is useful to know general mathematical properties of FTs. Table C.1 summarizes the most important of these.

Table C.1 Mathematical Relationships Between Functions and their Fourier Transforms. $k = 2\pi/x$.

Function	FT
$af(x)$	$a\mathscr{F}(k)$
$af(x) + bg(x)$	$a\mathscr{F}(k) + b\mathscr{G}(k)$
$f(x-a)$	$e^{-2\pi ika}\mathscr{F}(k)$
$\cos(k_0 x)f(x)$	$\mathscr{F}(k - k_0)$
$f(ax)$	$\frac{\mathscr{F}(k/a)}{|a|}$
$\frac{df(x)}{dx}$	$2\pi ik\mathscr{F}(k)$
$f(x) \cdot g(x)$	$\mathscr{F}(k) \otimes \mathscr{G}(k)$
$f(x) \otimes g(x)$	$\mathscr{F}(k) \cdot \mathscr{G}(k)$
$f(x)$ is real	$\mathscr{F}(k) = \mathscr{F}^*(k)$

Appendix D
Argand Diagrams and the Complex Plane

Many phenomena that we encounter in the text involve the addition of vectors – that is, properties which have both an amplitude *and* direction. Examples include a description of the difference between stochastic radiation produced by a synchrotron and coherent radiation generated by an XFEL (Figure 4.5); the addition of scattering amplitudes from individual facets in a monochromator grating to produce interference maxima (Figure 5.23); and the addition of x-ray scattering amplitudes from atoms within a unit cell to produce the structure factor (Figure 6.12).

The addition of vectors can be pleasingly and intuitively represented using Argand diagrams. These constructs also help in understanding Euler's representation of vectors using complex numbers.

Sinusoidal waves are defined by three properties: an amplitude A, a wavelength λ, and a phase ϕ. Although vectors can represent forces, electric fields, or many other properties, in the above three examples they express wavetrains, in which the length, or magnitude, of the vector indicates the amplitude A and the angle of the vector is given by its phase ϕ relative to some other property, be it another wavetrain, or some arbitrarily defined absolute value (normally zero).

We know that waves interfere with each other, as classically demonstrated by Young's double-slit experiment. Implicit in interference phenomena is that the addition of the component parts that generate the interference pattern considers their phase relationship – in other words, the relative positions of the peaks and troughs of the component waves. The four examples in Figure D.1 show that the sum of two waves of the same wavelength can be easily determined if the components are represented as vectors in the manner described above, that is, having a length proportional to the wave amplitude, and an angle determined by the phase. These are Argand diagrams.

We are taught (initially) in school that negative numbers have no square roots – any number squared will be positive, as the product of two negative numbers is always positive. Later, we learn, however, that there is a mathematical construct called i, which is defined as the square root of -1, that is, $i^2 = -1$. Any real number b (that is, $b \in \mathbb{R}$) when multiplied by i becomes a so-called *imaginary* number. A *complex* number c contains both real and imaginary components, that is, $c = a + ib$. As we will now see, this mathematical artifice is used to represent phases in Argand diagrams.

The real and imaginary components of a complex number can be plotted as x- and y-coordinates, respectively, in the so-called complex plane. In this manner, for example, the complex number $(3 + 4i)$ is at $(3, 4)$. The magnitude of a complex number $c = a + ib$ is

$$|c| = \sqrt{c \cdot c^*} = \sqrt{(a + ib) \cdot (a - ib)} = (a^2 + b^2)^{1/2},$$

where c^* is the so-called complex conjugate of c, for which the sign of the imaginary component has been reversed.

An Introduction to Synchrotron Radiation: Techniques and Applications, Second Edition. Philip Willmott.
© 2019 John Wiley & Sons Ltd. Published 2019 by John Wiley & Sons Ltd.

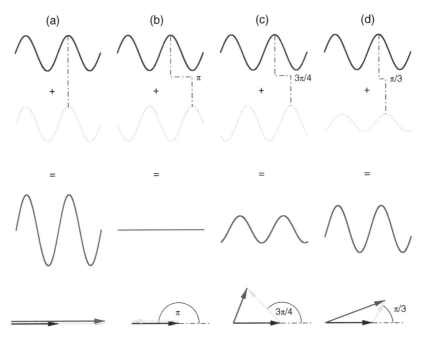

Figure D.1   Addition of waves and its representation as a vector sum. The addition of waves must be performed vectorially, whereby the wave amplitudes *and* their relative phases are considered. (a) Addition of two identical waves (blue and gold) of amplitude A, which are in-phase (zero shift in the positions of their peaks and troughs, $\phi = 0$) results in a wave of amplitude $2A$. This is the case of constructive interference. (b) Destructive interference occurs if the two component waves are exactly out-of-phase, that is, $\phi = \pi$. If the components have the same amplitude, as shown here, they will perfectly cancel one another out. (c) At angles other than 0 and π, interference of two waves of the same amplitude is *partially* destructive. In this example, $\phi = 3\pi/4$ (135°), from which it can be calculated that the summed amplitude is $2A\sin(\pi/8) = 0.765A$. (d) Summation of waves of differing amplitude. In this example, the second component has half the amplitude of the first and has a relative phase of $\pi/3$ (60°). The resultant amplitude is $\sqrt{7}A/2 = 1.323A$.

A commonly used way to represent vectors in the complex plane is not as $c = a + ib$, but as $A\exp(i\phi)$, where $A = (a^2 + b^2)^{1/2} = |c|$ and $\phi = \arctan(b/a)$ (see Figure D.2)[1]. In this description the most important variables describing the wave, A and ϕ, are used, from which it becomes apparent that the complex plane is an Argand diagram.

The equivalence of $c = a + ib$ and $A\exp(i\phi)$ is based on the so-called Euler equation and may not be obvious to the uninitiated. Its derivation is as follows.

First, it should be immediately clear that, for a vector of magnitude $A = (a^2 + b^2)^{1/2}$ at an angle ϕ to the real (x-) axis of the Argand diagram,

$$a = A\cos\phi$$
$$b = A\sin\phi.$$

---

[1] Note, however, that $\arctan\phi = \arctan(\phi - \pi)$, hence the signs of both a and b individually must be known in order to distinguish between these two possible solutions. This is often achieved in programming languages by the use of functions which explicitly require the values of both a and b, rather than a/b alone.

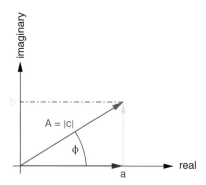

Figure D.2 Vectors in the complex-plane Argand diagram can be equivalently expressed as $c = a + ib$ or $c = A \exp(i\phi)$.

Next, the Taylor series of the exponent, cosine, and sine functions are given without derivation:

$$\exp(x) = \sum_{0}^{\infty} \frac{x^n}{n!} = 1 + x + \frac{x^2}{2!} + \frac{x^3}{3!} + \cdots \, , \tag{D.1}$$

$$\cos(x) = \sum_{0}^{\infty} \frac{-1^n}{(2n)!} x^{2n} = 1 - \frac{x^2}{2!} + \frac{x^4}{4!} - \cdots \, , \tag{D.2}$$

$$\sin(x) = \sum_{0}^{\infty} \frac{-1^n}{(2n+1)!} x^{2n+1} = x - \frac{x^3}{3!} + \frac{x^5}{5!} - \cdots \tag{D.3}$$

Now, let us consider these series for the case when x is an imaginary number $i\phi$. In this case,

$$\exp(i\phi) = 1 + i\phi - \frac{\phi^2}{2!} - \frac{i\phi^3}{3!} + \frac{\phi^4}{4!} + \frac{i\phi^5}{5!} - \cdots \, , \tag{D.4}$$

$$= 1 - \frac{\phi^2}{2!} + \frac{\phi^4}{4!} - \cdots + i\phi - \frac{i\phi^3}{3!} + \frac{i\phi^5}{5!} - \cdots \tag{D.5}$$

But from Equations (D.2) and (D.3), we see therefore that Equation (D.5) becomes

$$\exp(i\phi) = \cos\phi + i\sin\phi. \tag{D.6}$$

This is Euler's equation. From our expressions for a and b in terms of A and ϕ, the equivalence of $a + ib$ and $A \exp(i\phi)$ becomes immediately apparent.

Appendix E

Solutions to Problems

E.2 Chapter 2 – The Interaction of X-rays with Matter

1. We need to convert all the provided information into the same energy units. Easiest is to convert into eV.
   (a) 8 keV photon
   (b) A photon with $\lambda = 1.5895$ Å. Use $E = hc/\lambda = 7.8$ keV
   (c) A photon with wavevector $k = 4.0036$ Å^{-1}, $k = 2\pi/\lambda$, $E = 7.9$ keV
   (d) A photon with energy 1.2177 fJ. Conversion from J to eV, divide by 1.6022×10^{-19} eV J^{-1}. $E = 7.6$ keV
   (e) A photon with frequency $\nu = 1.8619 \times 10^{18}$ Hz. Use $E = h\nu$. $E = 7.7$ keV
   (f) A photon with angular frequency $\omega = 1.1395 \times 10^{19}$ rad. Use $\omega = 2\pi\nu$. $E = 7.5$ keV.
   The answer is therefore f - d - e - b - c - a.

2. Wien's displacement law states that the wavelength of the maximum of a blackbody spectrum, when plotted as a function of λ is given by

$$\lambda_{\text{max}} = \frac{b}{T},$$

where $b = 2.898 \times 10^{-3}$ m K. Therefore $\lambda_{\text{max}} = 1.063$ mm. This equates to an energy (in electronvolts) of $hc/e\lambda_{\text{max}} = 1.167$ meV.

3. The area of a sphere is equal to $4\pi r^2$, hence the surface area of the earth is $4 \times \pi \times 6400^2 = 5.15 \times 10^8$ km^2. But one square kilometre equals 10^6 m^2, and 1 m^2 equals 10^{28} barn. Hence the surface area of the Earth equals 5.15×10^{42} barn, or, to the nearest order of magnitude, 10^{43} barn.

4. The fractional energy loss due to Compton scattering is given by

$$f_C = \frac{h\nu_0 - h\nu}{h\nu_0} = 1 - \frac{h\nu}{h\nu_0}$$

and we know from Equation (2.4) that

$$\frac{h\nu_0}{h\nu} = 1 + \frac{\lambda_C}{\lambda_0}(1 - \cos\psi).$$

An Introduction to Synchrotron Radiation: Techniques and Applications, Second Edition. Philip Willmott.
© 2019 John Wiley & Sons Ltd. Published 2019 by John Wiley & Sons Ltd.

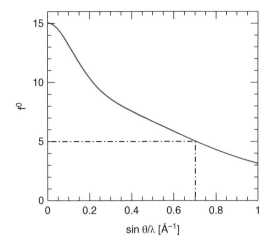

Figure E.1   The atomic form factor of phosphorus as a function of $\sin\theta/\lambda$.

Therefore,

$$f_C = 1 - \frac{1}{1 + \frac{\lambda_C}{\lambda_0}(1 - \cos\psi)}$$

$$= \frac{(\lambda_C/\lambda_0)(1 - \cos\psi)}{1 + (\lambda_C/\lambda_0)(1 - \cos\psi)}.$$

But $\lambda_0 = hc/E_0$ and $\lambda_C = h/m_e c$ and so

$$f_C = \frac{(E_0/m_e c^2)(1 - \cos\psi)}{1 + (E_0/m_e c^2)(1 - \cos\psi)} \qquad \text{Q.E.D.}$$

5. Assuming $\zeta \gg 1$, Equation (2.8) becomes

$$\sigma_C = 2Z\pi r_0^2 \left\{ \frac{1}{\zeta}\left[1 - \frac{\ln(1 + 2\zeta)}{\zeta}\right] + \frac{\ln(1 + 2\zeta)}{2\zeta} - \frac{1 + 3\zeta}{(1 + 4\zeta + 4\zeta^2)} \right\}$$

$$\approx 2Z\pi r_0^2 \left\{ \frac{1}{\zeta}\left[1 - \frac{\ln(1 + 2\zeta)}{\zeta} + \frac{\ln(1 + 2\zeta)}{2}\right] - \frac{3}{4\zeta} \right\}$$

$$\approx 2Z\pi r_0^2 \left\{ \frac{1}{\zeta}\left[1 - \frac{3}{4} + \frac{\ln(1 + 2\zeta)}{2}\right] \right\}$$

$$= \frac{Z\pi r_0^2}{2\zeta} \left\{ \frac{1}{\zeta}[1 + 2\ln(1 + 2\zeta)] \right\} \qquad \text{Q.E.D.}$$

6. Using the given coefficients, f for phosphorus is plotted in Figure E.1. From the curve, we see that $f(0.702) = 5.0$, and hence $Q = 4\pi \times 0.702 = 8.822$ Å^{-1}. The maximum wavelength that can theoretically access this form factor is when $\sin\theta = 1$ ($2\theta = \pi$), for which $\lambda = 1/0.702 = 1.425$ Å, and $E = 12.3984/1.425 = 8.704$ keV.

7. From Figure 2.29, one sees that each unit cell of GaP contains four atoms each of Ga and P. Hence, the total number of electrons in a unit cell of GaP is $4(31 + 15) = 184$. The electron density is therefore

$$\rho = 184/(5.451)^3 = 1.136 \text{ e/Å}^3.$$

8. The scattering vector $Q = (4\pi/\lambda)\sin\theta$. In this problem, $\lambda = 12.3984/20$ Å, and $\theta = 40/2 = 20°$, and therefore $Q = (80\pi/12.3984)\sin(20) = 6.933$ Å^{-1}.

9. The refractive-index decrement is given by

$$\delta = \frac{2\pi\rho r_0}{k^2},$$

whereby $k = 2\pi/\lambda$ is the wavevector of the x-rays. δ is therefore inversely proportional to the square of the photon energy.

We also know that, in practical units, the photon wavelength is given by

$$\lambda[\text{Å}] = \frac{12.3984}{E[\text{keV}]},$$

and hence a 1 Å photon equates to an energy of 12.3984 keV. The ratio of the square of the energies $(12.3984/20)^2 = 0.3843$, therefore δ at 20 keV for this material is $8 \times 10^{-6} \times 0.3843 = 3.074 \times 10^{-6}$.

10. The total atomic scattering factor, including the dispersive terms is $f = 14 - 6.1 + 1.9i = 7.9 + 1.9i$. The magnitude of f is therefore $\sqrt{7.9^2 + 1.9^2} = 8.125$. The ratio $f/f^0 = 8.125/14 = 0.580$.

11. The imaginary component f_2 varies as E^{-2}, hence is reduced by a factor of 16 accordingly. But f_2 also increases with Z^4, and the ratio of the atomic numbers of Ca to S is $20/16$, and hence this causes f_2 to increase by $1.25^4 = 2.441$. In total therefore, f_2 decreases to $1.295 \times 2.441/16 = 0.198$ e.

12. From Snell's law

$$\frac{\cos\alpha}{\cos\alpha'} = 1 - \delta.$$

Although $\alpha \geq \alpha_c$, it is stated that it remains small, and hence we can express both α and α' as the first two terms in their Taylor expansion (see Appendix B), and hence

$$\frac{1 - \alpha^2/2}{1 - \alpha'^2/2} = 1 - \delta.$$

$$\Rightarrow 1 - \alpha^2/2 = (1 - \delta)(1 - \alpha'^2/2)$$

$$= 1 - \alpha'^2/2 - \delta + \delta\alpha'^2/2.$$

The last term on the right-hand side is approximately five to six orders of magnitude smaller than the second and third terms and is therefore dropped. Thus

$$\alpha^2/2 = \alpha'^2/2 + \delta$$

$$\Rightarrow \alpha^2 - \alpha'^2 = 2\delta \qquad \text{Q.E.D.}$$

A plot of α' as a function of α is shown in Figure E.2 for silicon and platinum at 5 keV and 25 keV.

13. To obtain an expression for Δ, the perpendicular displacement of the x-ray due to refraction, we use Snell's law, that is

$$\frac{\cos\theta}{\cos\theta'} = n = 1 - \delta.$$

From Figure 2.30, we see that the length l the refracted x-ray beam travels through the material before exiting on the far side is

$$l = D/\sin\theta'.$$

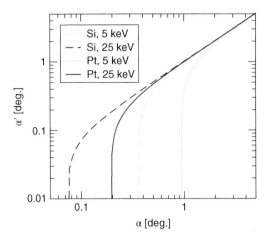

Figure E.2   Plot of α' as a function of α for Si and Pt at 5 and 25 keV and at grazing incident angles.

Also, it is immediately clear that

$$\Delta = l\sin(\theta - \theta'),$$

from which it follows that

$$\Delta = \frac{D\sin(\theta - \theta')}{\sin\theta'}.$$

But $\sin(A - B) = \sin A \cos B - \cos A \sin B$ and hence

$$
\begin{aligned}
\Delta &= \frac{D}{\sin\theta'}[\sin\theta\cos\theta' - \cos\theta\sin\theta'] \\
&= D\left[\frac{\sin\theta\cos\theta/n}{\sin\theta'} - \cos\theta\right] \\
&= D\cos\theta\left[\frac{\sin\theta}{n\sin\theta'} - 1\right] \\
&= D\cos\theta\left[\frac{\sin\theta}{n(1 - \cos^2\theta/n^2)^{1/2}} - 1\right] \\
&= D\cos\theta\left[\frac{\sin\theta}{(n^2 - \cos^2\theta)^{1/2}} - 1\right] \qquad \text{Q.E.D.} \qquad (E.1)
\end{aligned}
$$

We now attempt to simplify the above by assuming θ is small and hence $\sin\theta \approx \theta$ and $\cos\theta \approx 1 - \theta^2/2$. We also use the approximation that $(1 + x)^n \approx 1 + nx$ for $nx \ll 1$. The square of the refractive index n^2 is then accurately approximated by $n^2 = (1 - \delta)^2 \approx 1 - 2\delta$. Equation (E.1) then becomes

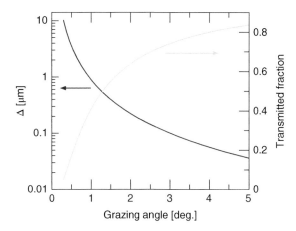

Figure E.3 Plot of the displacement Δ of a 25 keV x-ray beam due to refraction after it passes through a 0.5 mm thick parallel plate of beryllium. The refractive-index decrement is $\delta = 5.45 \times 10^{-7}$. Both the exact expression and the approximation are shown, but are indistinguishable on this scale. Also shown is the fraction of the beam that is transmitted.

$$\Delta \approx D\cos\theta \left\{ \frac{\theta}{[(1-2\delta) - (1 - \theta^2/2)^2]^{1/2}} - 1 \right\}$$

$$\approx D\cos\theta \left\{ \frac{\theta}{[1 - 2\delta - 1 + \theta^2]^{1/2}} - 1 \right\}$$

$$= D\cos\theta \left\{ \frac{\theta}{[\theta^2 - 2\delta]^{1/2}} - 1 \right\}. \tag{E.2}$$

We then make the more brutal approximation for the term before the curly brackets that $D\cos\theta \approx D$. We are also interested in the regime where $2\delta \ll \theta^2$, which holds true for $\theta > 5$ mrad (0.3°). In this case, Equation (E.2) simplifies to

$$\Delta \approx D \left\{ \frac{1}{(1 - 2\delta/\theta^2)^{1/2}} - 1 \right\}$$

$$\approx D \left\{ \frac{1}{1 - \delta/\theta^2} - 1 \right\}$$

$$= D \left\{ \frac{1}{\theta^2/\delta - 1} \right\} \qquad \text{Q.E.D} \tag{E.3}$$

We plot both Equation (E.1) and its approximation, Equation (E.3), in Figure E.3, along with the beam transmission.

14. Exactly at the critical angle, $\alpha = \alpha_c$ and hence $\alpha^2 - \alpha_c^2 = \alpha^2 - 2\delta = 0$. Equation (2.31) therefore simplifies to

$$\Lambda_c = -\frac{\lambda}{4\pi\beta^{1/2}} \qquad \text{Q.E.D.}$$

At very low angles, we set $\alpha = 0$, and Equation (2.31) becomes

$$\Lambda_0 = \frac{\lambda}{4\pi\left[\left(\sqrt{\alpha_c^4 + 4\beta^2} + \alpha_c \right)/2 \right]^{1/2}}.$$

But $\alpha_c^4 = 4\delta^2$ [Equation (2.28)], so we can ignore the $4\beta^2$ term, as it is so much smaller than $4\delta^2$. Therefore

$$\Lambda_0 = \frac{\lambda}{4\pi\sqrt{2\delta}}.$$

But $2\delta = 4\pi\rho r_0/k^2$ [Equation (2.24)] and so

$$\Lambda_0 = \frac{1}{4\sqrt{\pi\rho r_0}} \qquad \text{Q.E.D.}$$

15. K-L$_2$: allowed, as $\Delta l = 1$, $\Delta j = 0$.
    L$_1$-M$_4$: forbidden, as $\Delta l \neq 1$.
    K-M$_2$: allowed, as $\Delta l = 1$, $\Delta j = 0$.
    L$_2$-M$_4$: allowed, as $\Delta l = 1$, $\Delta j = 1$.
    L$_2$-M$_5$: forbidden, as $\Delta j = 2$.
    L$_3$-M$_3$: forbidden, as $\Delta l = 0$.

16. No helium Auger electrons are observed because helium only has two electrons and the Auger process involves three electrons – first the initial ejected photoelectron, then the relaxing electron to the core hole, and lastly, the Auger electron emitted thereafter.

E.3   Chapter 3 – Synchrotron Physics

1. We begin by squaring both sides of Equation (3.4) to obtain

$$\mathscr{E}^2 = \frac{m_e^2 c^4}{\left[1 - \left(\dfrac{v}{c}\right)^2\right]}$$

$$\Rightarrow 1 - \left(\frac{v}{c}\right)^2 = \frac{m_e^2 c^4}{\mathscr{E}^2}$$

$$\Rightarrow \frac{v}{c} = \left(1 - \frac{m_e^2 c^4}{\mathscr{E}^2}\right)^{1/2}.$$

But for relativistic particles, $m_e c^2/\mathscr{E} \ll 1$, and so

$$\frac{v}{c} \approx \left(1 - \frac{m_e^2 c^4}{2\mathscr{E}^2}\right)$$

$$\Rightarrow \frac{c}{c} - \frac{v}{c} = \frac{m_e^2 c^4}{2\mathscr{E}^2}$$

$$\Rightarrow c - v = \frac{m_e^2 c^5}{2\mathscr{E}^2} \qquad \text{Q.E.D.}$$

If we insert the values of m_e, the rest mass of an electron, c, and the storage-ring energy $\mathscr{E}$ (in appropriate units) we obtain

$$c - v = \frac{39.14}{\mathscr{E}^2[\text{GeV}^2]} \text{ m s}^{-1}.$$

This is quite an arresting result; in one second, a photon travels in vacuum a distance greater than circumnavigating the Earth seven times. Nonetheless, a 3 GeV electron starting at the same instant will, after one second, only lag behind the photon by a little over four metres!

2. We concern ourselves with the angle- and particle-velocity dependent function

$$f = \frac{\sin^2\theta}{[1 - (v/c)\cos\theta]^5}.$$

The lobe maximum occurs at an angle $\theta = \alpha_{BS}$ for which $df/d\theta = 0$. In this first part of the problem, we want to derive a general analytical expression for α_{BS} for any particle velocity v, so we cannot make any assumptions about the magnitude of θ. We begin by differentiating by parts with respect to θ:

$$\frac{df}{d\theta} = \frac{2\cos\theta\sin\theta}{(1 - \beta\cos\theta)^5} - \frac{5\beta\sin^3\theta}{(1 - \beta\cos\theta)^6}$$

$$= \frac{\sin\theta[2\cos\theta(1 - \beta\cos\theta) - 5\beta\sin^2\theta]}{(1 - \beta\cos\theta)^6}$$

$$= \frac{\sin\theta}{(1 - \beta\cos\theta)^6}(2\cos\theta - 2\beta\cos^2\theta - 2\beta\sin^2\theta - 3\beta\sin^2\theta)$$

$$= \frac{\sin\theta}{(1 - \beta\cos\theta)^6}(2\cos\theta - 2\beta - 3\beta\sin^2\theta)$$

$$= \frac{\sin\theta}{(1 - \beta\cos\theta)^6}[2\cos\theta - 2\beta - 3\beta(1 - \cos^2\theta)]$$

$$= \frac{\sin\theta}{(1 - \beta\cos\theta)^6}(2\cos\theta - 5\beta + 3\beta\cos^2\theta).$$

The term within the brackets is equal to 0 when $3\beta x^2 + 2x - 5\beta = 0$, where $x = \cos\theta = \cos\alpha_{BS}$. Solving for this quadratic equation, we obtain

$$\cos\alpha_{BS} = \frac{-2 \pm \sqrt{4 + 60\beta^2}}{6\beta}$$

$$= \frac{-1 \pm \sqrt{1 + 15\beta^2}}{3\beta}.$$

The solution where $\pm$ is negative leads to absolute magnitudes for $\cos\alpha_{BS}$ greater than unity, and is therefore to be disregarded. Because $\cos\theta = \cos(-\theta)$, we finally obtain

$$\alpha_{BS} = \pm\arccos\left[\frac{\sqrt{1 + 15\beta^2} - 1}{3\beta}\right] \qquad \text{Q.E.D.}$$

For the case of highly relativistic and forward-directed Bremsstrahlung, we begin by recognizing that for small angles $\sin\theta \approx \theta$, and $\cos\theta \approx 1 - \theta^2/2$. We then insert these approximations and the expression for β given by Equation (3.11) into Equation (3.20) to obtain

$$\frac{dP}{d\Omega} \approx \kappa a^2 \frac{\theta^2}{[1 - (1 - 1/2\gamma^2)(1 - \theta^2/2)]^5}$$

$$= \kappa a^2 \frac{\theta^2}{[1 - (1 - \theta^2/2 - 1/2\gamma^2 + \theta^2/4\gamma^2)]^5}.$$

The last term $\theta^2/4\gamma^2$ in the denominator is much smaller than all the other terms and is thus ignored. Hence

$$\frac{dP}{d\Omega} \approx \kappa a^2 \frac{\theta^2}{(\theta^2/2 + 1/2\gamma^2)^5}$$

$$= 32\kappa a^2 \frac{\theta^2}{(\theta^2 + 1/\gamma^2)^5} \qquad \text{Q.E.D.}$$

In order to determine the maximum of this expression, we differentiate by parts with respect to θ (for convenience, we ignore the prefactor $32\kappa a^2$) and calculate when this is equal to zero.

$$\frac{dP/d\Omega}{d\theta} = \frac{2\theta}{(\theta^2 + 1/\gamma^2)^5} - \frac{10\theta^3}{(\theta^2 + 1/\gamma^2)^6}$$

$$= \frac{2\theta(\theta^2 + 1/\gamma^2) - 10\theta^3}{(\theta^2 + 1/\gamma^2)^6}.$$

This is equal to zero when the factor $2(\theta^2 + 1/\gamma^2) = 10\theta^2$, which, after some further trivial algebraic manipulation, turns out to be true when $\theta = 1/2\gamma$, as we set out to demonstrate.

3. The emitted light power from the bulb is $0.02 \times 60 = 1.2$ W. The surface area of a sphere at a distance of $r = 2$ m from the light bulb is $4\pi r^2 = 5.027 \times 10^5$ cm^2. The power density at 2 m is therefore $1.2/(5.027 \times 10^5) = 2.387 \times 10^{-6}$ W cm^{-2}.

4. 13:00 British Summer Time is 12:00 midday GMT. At the equinox at midday, the sun lies directly overhead at the equator, therefore has a flux of $\cos(51.52°) = 0.622$ of this in London. One thus needs only to calculate the solar power flux and multiply by this factor.

    The power absorbed by the earth of 1.2×10^{17} W corresponds to the capture cross-section of the earth, which is simply πr^2, with $r = 6.4 \times 10^8$ cm. The solar power flux at the equator is therefore $1.2 \times 10^{17}/(1.287 \times 10^{18}) = 0.0933$ W cm^{-2}.

    The solar power flux in London is 0.0580 W cm^{-2}.

5. First, we calculate the rate of photon emission of the LED. The emitted light power is $40 \times 0.7 = 28$ mW. Each photon has an energy $hc/\lambda = 6.626 \times 10^{-34} \times 2.9979 \times 10^8/(7 \times 10^{-7}) = 2.838 \times 10^{-19}$ J ($= 1.771$ eV). The rate of photon emission is therefore $0.028/(2.838 \times 10^{-19}) = 9.866 \times 10^{16}$ photons per second.

    The LED's bandwidth is $25/700 = 3.571\%$, or 35.71 times larger than a 0.1% bandwidth.

    The solid-angle divergence of the LED is $(10 \times \pi/180)^2 = 0.0305$ rad$^2 = 3.05 \times 10^4$ mrad^2.

    Armed with the above information, the brilliance $\mathcal{B}$ of the LED is $\mathcal{B} = 9.866 \times 10^{16}/(35.71 \times 3.05 \times 10^4 \times 4) = 2.265 \times 10^{10}$ photons/s/mm^2/mrad^2/0.1% BW.

6. The approach to solve this problem is identical to that in the previous problem.

    ph/s $= 0.02/(6.626 \times 10^{-34} \times 2.9979 \times 10^8/6.33 \times 10^{-7}) = 6.373 \times 10^{16}$;

    Laser bandwidth $= 0.0158\%$, 6.33 times *smaller* than a 0.1% bandwidth;

    The solid-angle divergence is 1 mrad^2;

    The 'source size' is $\pi \times 0.5^2 = 0.785$ mm^2.

    The brilliance is therefore $6.373 \times 10^{16} \times 6.33/(1 \times 0.785) = 5.139 \times 10^{17}$ photons/s/mm^2/mrad^2/0.1% BW.

7. First we need to determine the maximum of the blackbody spectrum emitted by a human, and also what the radiant power output is. We determine the spectral peak position by using Wien's displacement law

$$\lambda_{max} = 2.898 \times 10^{-3}/T \text{ [K]}.$$

$T \approx 310.7$ K for a healthy human. From this, we obtain $\lambda_{max} = 9.3$ µm, which has a photon energy of 2.14×10^{-20} J, or 0.133 eV.

In order to determine a human's thermal-radiation power output, we assume an ambient surrounding temperature of 20 °C, or 293.2 K. In the infrared, human skin is an almost perfect blackbody (that is, it has an emissivity of 1.0), and hence, from the Stefan–Boltzmann law, we obtain an areal power output of

$$P = 5.67 \times 10^{-8}[310.7^4 - 293.2^4] \text{ W m}^{-2}$$

$$= 109.4 \text{ W m}^{-2}.$$

An adult may have a surface area of 2 m^2, hence the total power output is approximately 220 W. Note that this equates to a daily caloric requirement of 4500 kCal, more than double the normal. This is because we don't normally waltz around the place naked. Our clothes protect both our modesty and our well-being, by having a lower emissivity and poor thermal conductivity. Nevertheless, whoever set the problem stipulated 'naked', so let's press on …

We make the very rough approximation that the number of irradiated photons per second is simply the total power divided by the photon energy associated with λ_{max}, which leads to 10^{22} ph/s. The relative bandwidth is approximately 100%, or 10^3 times a bandwidth of 0.1%; the source area is 2 m^2, or 2×10^6 mm^2; and the divergence is 4π steradians, or $4\pi \times 10^6$ mrad^2. The brilliance is therefore 4×10^5 photons/s/mm^2/mrad^2/0.1% BW.

8. The product of the source size, divergence, and Si-bandwidth (relative to one of 10^{-3}) is

$$(0.005 \times 0.05)(0.015 \times 0.075)(0.14) = 3.938 \times 10^{-8} \text{ mm}^2\text{mrad}^2 0.1\% \text{ BW}.$$

The flux is therefore $2 \times 10^{19} \times 3.938 \times 10^{-8} = 7.873 \times 10^{11}$ ph/s.

The energy of each photon is $1.5 \times 10^4 \times 1.602 \times 10^{-19} = 2.403 \times 10^{-15}$ J. The power is therefore $2.403 \times 10^{-15} \times 7.873 \times 10^{11} = 1.892$ mW.

9. The time for an electron to orbit the ESRF once is simply $t = L/c$, where L is the circumference and c is the speed of light (the velocity of the relativistic electrons to a high degree of accuracy), hence here $t = 2.815$ µs. In this time, there are $m = ft = 352 \times 10^6 \times 2.815 \times 10^{-6} = 991$ klystron oscillations, which equates to the number of bunches.

The distance between bunches is $c/f = 85.2$ cm.

The amount of charge in the entire storage ring is simply the current (200 mA) multiplied by the time required for one orbit, t, which we have already determined to be 2.815 µs. The number of electrons n in an individual bunch is therefore

$$n = It/me = 3.546 \times 10^9.$$

The stored kinetic energy of the electrons within the ring is simply equal to

$$\mathscr{E}_{tot} = \mathscr{E} It/e = 6 \times 10^9 \times 0.2 \times 2.815 \times 10^{-6}$$

$$= 3378 \text{ J}.$$

The peak flux is simply the average flux divided by the duty cycle (i.e. the ratio of the pulse length τ to the pulse-to-pulse duration $T = 1/f$), that is,

$$F_{peak} = F_{av} \times \left(\frac{T}{\tau}\right),$$

which in this instance is equal to $4 \times 10^{13} \times 2.841 \times 10^{-9}/(5 \times 10^{-11}) = 2.27 \times 10^{15}$ photons per second.

The average spatial separation within any given pulse in the direction of propagation is equal to the pulse length ($c\tau = 5 \times 10^{-11} \times 3 \times 10^8 = 15$ mm) divided by the number of photons per pulse [$= F_{av}/f = 4 \times 10^{13}/(3.52 \times 10^8) = 1.136 \times 10^5$], which is thus equal to 132 nm. The temporal separation is simply this distance divided by the speed of light, that is, 0.44 fs. Note that this is approximately 100 times shorter than typical pulse durations from XFELs, which contain approximately 2×10^{12} photons per pulse. In other words, the instantaneous arrival rate of photons on a sample at an XFEL exceeds that of an undulator at a synchrotron by 10 orders of magnitude.

10. We can determine the decay-time constant τ from the provided data

$$\frac{I}{I_0} = \frac{399}{401} = \exp\left(\frac{-180}{\tau}\right)$$

$$\Rightarrow \ln\left(\frac{401}{399}\right) = \left(\frac{180}{\tau}\right)$$

$$\Rightarrow \tau = 36\,000 \text{ s} = 10 \text{ h}.$$

Note that $401/399 = 1 + 2/399 \approx 1 + 1/200$. The natural logarithm of $1 + x \approx x$ for $x \ll 1$. This accurate approximation thus simplifies the calculation.

In order to determine the Bremsstrahlung power output, we now need to calculate the number of lost electrons n_l per second in topup mode. The current loss per second $dI/dt = 0.002/180 = 1.111 \times 10^{-5}$ A/s. The number of electrons lost per second is thus $n_l = dI/dt \times L/ec = 6.663 \times 10^7$ electrons/s, whereby e is the elementary charge and c is the speed of light in vacuum. The power loss is thus

$$P = eV \times n_l$$

$$= 2.4 \times 10^9 \times 1.602 \times 10^{-19} \times 6.663 \times 10^7$$

$$= 25.6 \text{ mW}.$$

11. We use the equation describing the Lorentz factor

$$\gamma = \frac{\mathscr{E}}{m_e c^2},$$

whereby $\mathscr{E}$ is the electrons' energy in the storage ring ($= 8$ GeV $= 8000$ MeV) and $m_e c^2 = 511$ keV is the rest-mass energy of an electron. From this we obtain

$$\gamma = \frac{8000}{0.511} = 15\,656.$$

The Lorentz factor γ is related to the electron velocity v via

$$\gamma = \frac{1}{\sqrt{1 - v^2/c^2}},$$

from which it can be quickly demonstrated that

$$v/c \approx 1 - \frac{1}{2\gamma^2},$$

as long as $\gamma \gg 1$. Using the value for γ at SPring8 of 15 656, we obtain

$$v/c = 1 - 1/(4.9 \times 10^8) \approx 0.999\,999\,998.$$

The natural opening angle of synchrotron light is simply $1/\gamma$ (in radians). Therefore the natural opening angle of x-rays at SPring8 (in degrees) is

$$\theta = 180/(\pi \times 15\,656) = 3.66 \times 10^{-3} \text{ deg}.$$

The photon is faster than the electron by an amount equal to $c/2\gamma^2$, which in this instance is 0.612 m s^{-1}. After one hour, therefore, the photon will have put a distance of $3600 \times 0.612 = 2203$ m between itself and the electron, not very much when one considers they have travelled a little over one billion kilometres!

12. First, we need to convert km/hr to m/s. To do this we divide by 3.6 (1000 m/3600 s) to obtain a velocity of 12.42 m/s. The relativistic mass is γm and the increase is therefore $(\gamma - 1)m$. But $\gamma = [1 - (v/c)^2]^{-1/2} \approx 1 + 0.5(v/c)^2$. The mass increase is thus $0.5\,m(v/c)^2 = 80.7$ pg! (Note: 1 pg $= 10^{-15}$ kg.) Needless to say, this increase is negligibly small compared to the biochemical weight loss through running fast. Or, indeed, through just being alive.

13. The bending-magnet radius is related to the electron energy $\mathscr{E}$ and bending-magnet field strength B by

$$\rho = \frac{\mathscr{E}}{ceB},$$

$$\rho[\text{m}] = 3.3\frac{\mathscr{E}[\text{GeV}]}{B[\text{T}]}.$$

From the given data, therefore, $\rho = 38.82$ m, and the length of the SPring8 storage ring that is taken up with bending magnets is $2\pi \times 38.82 = 243.94$ m, or 16.99% of the storage-ring circumference.

The critical energy E_c of bending-magnet synchrotron radiation is given by

$$E_c = \hbar\omega_c = \frac{3}{2}\gamma^3\frac{c}{\rho},$$

$$E_c[\text{keV}] = 0.665\mathscr{E}^2[\text{GeV}]B[\text{T}],$$

from which we obtain $E_c = 28.94$ keV.

14. We need to be careful here in our definition of γ, which in the case of protons in the LHC is the ratio of the ring energy (6.5 TeV) to the rest-mass energy of a *proton* (*not* an electron). This latter is equal to 938.23 MeV (that is, 1836 times the rest-mass energy of an electron, as the ratio $m_p/m_e = 1836$). γ is therefore equal to $6.5 \times 10^{12}/(9.3823 \times 10^8) = 6928$.

From Equation (3.22) describing the bending radius of a magnet dipole, however, we see that this decrease in γ caused by considering the proton rest-mass is exactly cancelled by substitution of m_e with m_p. This means that in this instance, Equation (3.24) holds as true for CERN as for an electron-storage ring. This leads to $\rho = 3.3 \times 6500/5.5 = 3900$ m. The circumference of such a ring without any straight sections is $2\pi\rho = 24.5$ km, or 90.76% of the actual CERN circumference.

In order to determine the critical photon energy, we use Equation (3.25), remembering to substitute m_e with m_p:

$$\hbar\omega_c = \frac{3\hbar}{2}\left(\frac{6.5 \times 10^{12}}{9.38 \times 10^8}\right)^2\frac{e \times 5.5}{1.6726 \times 10^{-27}}$$

$$= 24.99 \text{ eV}.$$

Note that γ for the LHC is similar to that for synchrotrons – although the energies are around 2000 times larger, so is the rest mass of the particles being accelerated. But, because Equation (3.25) is also inversely proportional to the rest mass of the charged particle, the critical energy will drop accordingly. The magnetic-field strength is similar to that of superbends at synchrotrons. The ratio of the typical critical energies of the photon spectra at superbends to that at the bending magnets at LHC is therefore approximately $m_p/m_e = 2000$.

15. We begin by multiplying each term in the square brackets of Equation (3.49) by the denominator in the round brackets to obtain

$$f_{coh} = (\sigma'^p \times \sigma^p)^2 / \left\{ \left[(\sigma^p)^2 + (\sigma_x^e)^2\right]^{1/2} \left[(\sigma'^p)^2 + (\sigma_x'^e)^2\right]^{1/2} \right.$$
$$\left. \times \left[(\sigma^p)^2 + (\sigma_y^e)^2\right]^{1/2} \left[(\sigma'^p)^2 + (\sigma_y'^e)^2\right]^{1/2} \right\}.$$

Substituting in Equations (3.34), (3.35), and (3.38), we obtain

$$f_{coh} = \frac{(\epsilon^p)^2}{\epsilon_x \times \epsilon_y} \qquad \text{Q.E.D.}$$

16. We begin with the expression for constructive interference for the *m*th harmonic, given by Equation (3.70)

$$m\lambda_m = \lambda_1 = \frac{\lambda_u}{2\gamma^2}\left(1 + \frac{K^2}{2}\right)$$
$$\Rightarrow \left(1 + \frac{K^2}{2}\right) = \frac{2\lambda_1\gamma^2}{\lambda_u}.$$

Next, we insert this into Equation (3.61) to obtain

$$\sigma_x' = \frac{1}{\gamma}\left(\frac{1 + K^2/2}{2mN}\right)^{1/2}$$
$$= \left(\frac{\lambda_1}{mN\lambda_u}\right)^{1/2}$$
$$= \left(\frac{\lambda_m}{L}\right)^{1/2},$$

the expression given in Equation (3.37).

17. First, we need to determine the lower limit to the horizontal emittance of an MBA (with $M = 7$), all other parameters being equal, in particular, the swept angle per dipole θ. We obtain this from Equation (3.54) to yield a ratio of the DBA to 7-BA of $3 \times 6/8 = 9/4$.

The Lorentz factor γ for MAX-IV equals $3 \times 10^9/(5.11 \times 10^5) = 5871$. The swept angle per dipole in the twenty 7-BAs is $360/(20 \times 7)°$.

We also know from Equation (3.52) the lower limit to the value of a DBA. Hence, the lower limit to the horizontal electron emittance of the 7-BAs at MAX-IV is

$$\epsilon_{x,7BA} = \frac{4}{9}\, 2.474 \times 10^{-5}(5871)^2\, [(360/140)(\pi/180)]^3 \text{ nm rad}$$
$$= 34.3 \text{ pm rad.}$$

18. The photon energy E_m of the *m*th harmonic is simply hc/λ_m. Inserting the equation for λ_m into this, we obtain

$$E_m = \frac{2mhc\gamma^2}{\lambda_u}\frac{1}{1 + K^2/2},$$

and hence the energy difference between adjacent harmonics m and $m + 1$ is simply

$$\Delta E_m = \frac{2hc\gamma^2}{\lambda_u\,(1 + K^2/2)} \qquad \text{Q.E.D.}$$

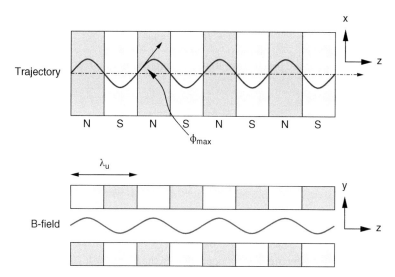

Figure E.4   Schematic of the electron trajectory through an insertion device, viewed from above showing the maximum angular deviation ϕ_{max}; and the oscillatory magnetic field, viewed from the side.

19. To solve this problem, we use the equation relating the harmonic wavelength λ_m to the undulator periodicity λ_u, the storage-ring energy $\mathscr{E}$, and the deviation parameter K, that is

$$m\lambda_m[\text{Å}] = \frac{13.056\lambda_u[\text{cm}]}{\mathscr{E}^2[\text{GeV}]}\left(1 + \frac{K^2}{2}\right).$$

Inserting the values associated with the MS beamline at the SLS, we obtain

$$m\lambda_m[\text{Å}] = 3.1733\left(1 + \frac{K^2}{2}\right).$$

For an x-ray wavelength of 1.6 Å, this becomes

$$0.5042m = 1 + \frac{K^2}{2}.$$

For the lower limit of $K - 0.65$, m (which is an integer) is rounded up to 3. At the upper limit ($K - 1.6$), m is rounded down to 4. From Figure 3.41, we see that for $K = 0.65$ and $m = 3$, $F_m(K) \approx 0.05$, while for $K = 1.6$ and $m = 4$, $F_m(K) \approx 0.13$. One should therefore access 1.6 Å radiation using the fourth harmonic.

Employing Equation (3.71), and using $\lambda_u = 1.4$ cm; $K = 1.6$; and $\mathscr{E} = 2.4$ GeV, we obtain for the first harmonic ($m = 1$) $\lambda = 7.2352$ Å, or $E = 583.6$ eV.

There are two reasons why this energy cannot be accessed at the Materials Science beamline. Firstly, the beamline is designed for hard x-rays, and cannot tolerate interactions with the strongly absorbed photons much below 5 keV. A more fundamental problem, however, is the fact that Si(111) crystals are used in the monochromator [with a lattice spacing in the (111) direction of 3.1356 Å]. The Bragg angle is $\arcsin(\lambda/2d)$, but in this instance $\lambda/2d$ is larger than unity, and hence this low photon energy cannot be accessed using the Si(111) monochromator.

20. See Figure E.4. The magnetic field $B_y(z)$ exerts a periodic Lorentz force F_L in the x-direction on the electron as it travels through the insertion device in the z-direction, given by

$$F_L(x) = B_y(z)ev = m\frac{d^2x}{dt^2},$$

whereby $m = \gamma m_e$ is the relativistic mass. If we make the approximation $v \approx c$, the electron's acceleration in the x-direction is thus given by

$$\frac{d^2x}{dt^2} = \frac{ecB_0\sin(2\pi z/\lambda_u)}{\gamma m_e}.$$

But z is simply $vt \approx ct$. We substitute this into the above equation and integrate with respect to time to obtain the electron's velocity in the x-direction.

$$\frac{dx}{dt} = \frac{ecB_0}{\gamma m_e}\int \sin\left(\frac{2\pi ct}{\lambda_u}\right)dt$$

$$= \frac{\lambda_u}{2\pi c}\frac{ecB_0}{\gamma m_e}\cos\left(\frac{2\pi ct}{\lambda_u}\right).$$

Clearly, the maximum angular deviation, ϕ_{max}, occurs when the lateral velocity in the x-direction is maximal, i.e. when the cosine term equals unity:

$$\left.\frac{dx}{dt}\right|_{max} = \frac{\lambda_u ecB_0}{2\pi c\gamma m_e}.$$

ϕ_{max} expressed in radians is simply the ratio of this velocity to the axial velocity ($\approx c$), that is

$$\phi_{max} = \frac{ecB_0\lambda_u}{2\pi\gamma\, m_ec^2} \qquad \text{Q.E.D.}$$

From the known values of the constants m_e, e, and c, we immediately obtain

$$\frac{e}{2\pi m_ec} = 93.374 \;\; [\text{T m}]^{-1}.$$

This is identical to Equation (3.57) if $\lambda_{u,w}$ is expressed in centimetres.

21. We label the on-axis and off-axis wavelengths λ_5^0, and λ_5^θ, respectively. From Equation (3.73), we directly obtain

$$\lambda_5^\theta - \lambda_5^0 = hc\left(\frac{1}{E_5^\theta} - \frac{1}{E_5^0}\right) = \frac{\lambda_u}{2m\gamma^2}(\gamma^2\theta^2)$$

$$= \frac{\lambda_u\theta^2}{2m}.$$

Inserting the given values for the two photon energies, K, and m, we obtain $\theta = 7.268 \times 10^{-5}$ rad. This is smaller than the natural opening angle $1/\gamma = 2.13 \times 10^{-4}$ rad, and also smaller than the angle subtended by the electron-beam width within the undulator, of the order of magnitude of 200 μm m^{-1} = 2×10^{-4} radians. The calculated value of θ is therefore reasonable.

E.4   Chapter 4 – Free-electron Lasers

1. The peak brilliance of the MS beamline at SLS is simply the given average photon flux divided by the number of pulses per second (5×10^8), by the pulse duration (40 ps), and by the emittance (not given, but stated to be 1000 times larger than for the LCLS). This is therefore equal to 5×10^{15} ph/s/0.1% BW divided by the emittance.

The number of photons per pulse at the LCLS is equal to the pulse energy (5 mJ) divided by the energy per photon ($8000 \times 1.6022 \times 10^{-19} = 1.282$ fJ) and is thus equal to 3.9×10^{12} photons per pulse. The relative bandwidth is $0.05/8 = 0.00625$, or 6.25 times a 0.1% relative bandwidth (the standard definition for flux and brilliance). The peak brilliance is therefore $3.9 \times 10^{12}/(6.25 \times 3 \times 10^{-14}) = 2.08 \times 10^{25}$ ph/s/0.1% BW divided by the LCLS emittance (1000 times smaller than the MS-beamline emittance). The ratio is therefore $2.08 \times 10^{28}/(5 \times 10^{15}) = 4.16 \times 10^{12}$.

2. We first determine the areal power density from the provided information, then use Equation (4.4) to determine the electric-field strength. The energy of a 1.5 Å photon is 12 398.4 eV $\times 1.602 \times 10^{-19}$ J/eV$/1.5 = 1.324 \times 10^{-15}$ J. The areal power density is thus simply this energy multiplied by 8.4×10^{11}, the number of photons per pulse, and divided by the photon-beam cross-section, 2.25×10^{-10} m^2, leading to $P/A = 7.1 \times 10^{19}$ W m^{-2}.

   From this, $E_0 = (2\mu_0 c\, P/A)^{1/2} = 2.31 \times 10^{11}$ V m^{-1}. But $B = E/c$, and hence $B_0 = 770$ T, larger than the magnetic-field strength of the undulator magnets by nearly three orders of magnitude.

3. The magnitude of the Lorentz force acting on the electrons is

$$F_L = Bec = \gamma m_e a,$$

where $\gamma = \mathscr{E}/m_e c^2 = 2.94 \times 10^4$ and a is the acceleration perpendicular to the magnetic field and the electrons' motion. From this we obtain $a = 8.43 \times 10^{10}$ m s^{-2}. This acts for a time $t = 1500/(2.9979 \times 10^8) = 5 \times 10^{-6}$ s. The distance s travelled by a particle initially at rest (which is the case in the direction of acceleration) is given by Newton's second equation of motion as $s = at^2/2$, which is therefore 1.055 m. This is a substantial distance, exceeding the undulator breadth by nearly two orders of magnitude. It is clear that the earth's magnetic field must therefore either be screened out, or compensated for in the magnet lattice.

4. As mentioned in the problem, the FWHM in the temporal domain $\Delta\tau = \sqrt{8 \ln 2}\, \sigma_t$. The Fourier transform of this is $C \exp(-\omega^2 \sigma_t^2/2) = C \exp(-2\pi^2 \sigma_t^2 \nu^2)$, which in turn has a standard deviation σ_ν equal to $1/4\pi\sigma_t$ and a FWHM of $\Delta\nu = \sqrt{8 \ln 2}/(4\pi\sigma_t)$.

   The product of these FWHMs is therefore

$$\Delta\tau\Delta\nu = \sqrt{8 \ln 2}\, \sigma_t \times \sqrt{8 \ln 2}/(4\pi\sigma_t)$$

$$= \frac{2 \ln 2}{\pi} \qquad \text{Q.E.D.}$$

5. The spectral purity of a Gaussian pulse is limited by the fact that the pulse is finite in duration and cannot, therefore, be described by a single frequency. The narrowest possible relative bandwidth in a so-called Fourier-transform limited pulse is given by

$$\frac{\Delta\nu}{\nu} = 1.472 \times 10^{-4}\, \frac{\lambda[\text{Å}]}{\Delta\tau[\text{fs}]},$$

and hence for a 1 fs pulse at 1 Å, $\Delta\nu/\nu = 1.472 \times 10^{-4}$. This is almost identical to that provided by the Darwin width of a Si(111) single-crystal, but could in principle be made to be narrower using a high-index reflection of a weakly scattering crystal such as the (333) reflection of diamond. However, this will have the effect of broadening the pulse length by suppressing the central part of the pulse.

6. First, we recognize that the attenuation length $1/\mu = 1000$ μm of x-rays in water at 8 keV is much larger than the water stream's diameter ϕ_w, which was stated as being 20 μm or smaller. Hence the product $\mu\phi_w \ll 1$ and so the transmitted fraction of the x-ray beam $e^{-\mu\phi_w}$ can be accurately approximated as being equal to $1 - \mu\phi_w$. The absorbed fraction is therefore simply $\mu\phi_w$, which is deposited in a volume of water equal to the x-ray beam cross-section multiplied by ϕ_w, the path of the x-rays through the jet.

The absorbed energy density is thus equal to

$$E/V = \frac{E_p \mu \phi_w}{(\pi \phi_x/2)^2 \phi_w} = \frac{4E_p \mu}{\pi \phi_x^2} \qquad \text{Q.E.D.}$$

Assuming $E_p = 5$ mJ, it emerges that the absorbed energy density is 6.37×10^6 J cm^{-3}. The energy required to heat one cubic centimetre of water through 1 K is 1 calorie = 4.184 J, hence the energy per unit volume required to boil water, starting at 25°C is $75 \times 4.184 = 313.8$ J cm^{-3}. This is 20 300 times smaller than the deposited energy density from the XFEL pulse – the water jet thus instantly explodes!

From our discussion about the dependence of the attenuation coefficient in Section 2.6.3, we know that it scales with the inverse-cube of the photon energy. We can therefore predict that μ at 24 keV is 27 times smaller than at 8 keV. From this, the absorbed energy density at 24 keV is approximately 2.36×10^5 J cm^{-3}, still a factor of 750 times that needed to boil water from room temperature (assuming the same energy per XFEL pulse).

7. We begin by determining the reduced mass μ_r of an iodine molecule, for which $m_1 = m_2 = m_I = 127$ a.u. Therefore

$$\mu_r = m_I/2 = \frac{127}{2 \times 6.022 \times 10^{26}} = 1.0545 \times 10^{-25} \text{ kg.}$$

From this, the moment of inertia of I_2 is $1.0545 \times 10^{-25} \times (2.666 \times 10^{-10})^2 = 7.495 \times 10^{-45}$ kg m^2. The rotational energy of the Jth quantum level is

$$E_J = \frac{J(J+1)\hbar^2}{2I}$$
$$= J(J+1) \times 4.63 \times 10^{-6} \text{ eV.}$$

The energy difference between the Jth and $(J+1)$th state is simply $2(J+1)\hbar^2/2I$. We equate this to the energy of 0.1 THz radiation, which is $h\nu = 4.136 \times 10^{-4}$ eV. From this, it emerges that the first highest rotational level with an energy difference to the next level that is larger than the THz photon energy is for $J = 43$.

8. In the harmonic approximation (that is, a quadratic relationship between energy and displacement of the bond length from its equilibrium value), the separation between vibrational quantum levels is constant and equal to

$$\Delta E_v = \hbar \sqrt{\frac{k}{\mu_r}},$$

and therefore the associated frequency ν is simply

$$\nu = \frac{1}{2\pi} \sqrt{\frac{k}{\mu_r}}.$$

In the previous problem, we calculated the reduced mass of iodine to be equal to 1.0545×10^{-25} kg. From the given value for the Hooke's force constant k, we therefore obtain a frequency $\nu = 6.428 \times 10^{12}$ Hz = 6.428 THz.

9. The areal power density P/A is $[10^{-3}/(2 \times 10^{-13})]/10^{-7} = 5 \times 10^{16}$ W m^{-2}. Using Equation (4.4), we can calculate the associated electric-field strength E_0 to be equal to $[5 \times 10^{16} \times (8\pi \times 10^{-7} \times 2.9979 \times 10^8)]^{1/2} = 6.138 \times 10^9$ V m^{-1}. From Equation (4.3), the magnetic-field strength is therefore $B_0 = 20.47$ tesla.

E.5   Chapter 5 – Beamlines

1. In the one-to-one configuration, the source-to-mirror and mirror-to-image distances $p = q = f'$, the foci of the ellipse. From the lensmaker equation, we immediately obtain a focal length $f = f'/2$. Equation (5.10) describing the relationship between the meridional radius R_m and focus f of a cylindrical mirror yields for $p = q = f'$

$$R_m = \frac{2}{\sin\theta}\frac{f'^2}{2f'} = \frac{f'}{\sin\theta},$$

where θ is the incident angle. But from Figure 5.7 it is clear that

$$\sin\theta = \frac{b}{(f'^2 + b^2)^{1/2}}.$$

Remembering that $f' = (a^2 - b^2)^{1/2}$, $\sin\theta = b/a$, and hence

$$R_m = \frac{a}{b}(a^2 - b^2)^{1/2} \approx \frac{a^2}{b} \qquad\qquad \text{Q.E.D.}$$

In general b/a is a good approximation to the incident angle and is of the order of a milliradian. Hence for a sample approximately $2f' = 40$ m downstream from the synchrotron source, a^2/b is of the order of 20 km.

2. The FWHM divergence of the beam, expressed in radians and far from the waist (that is, where the change in linear size scales approximately linearly with distance from the focus) is simply twice the gradient of the expression for w_z.

$$w_z = w_0\sqrt{1 + \left(\frac{z}{z_R}\right)^2}$$

$$\frac{dw_z}{dz} = \frac{w_0 z}{z_R^2}\left(1 + \frac{z^2}{z_R^2}\right)^{-1/2}.$$

But for $z \gg z_R$, $(1 + z^2/z_R^2)^{-1/2} \approx z_R/z$, as the term z^2/z_R^2 dominates. Thus

$$\frac{dw_z}{dz} = w_0/z_R$$

and so

$$\Theta = \frac{2w_0}{z_R}. \qquad\qquad \text{Q.E.D.}$$

3. For this problem we need to use the lensmaker's equation

$$\frac{1}{f} = \frac{1}{p} + \frac{1}{q},$$

whereby f is the FZP focal length, p is the source–FZP distance, and q is the FZP–sample distance. From the supplied information, therefore, we obtain

$$\frac{1}{0.725} = \frac{1}{40.5 - q} + \frac{1}{q}$$

$$\Rightarrow (40.5 - q)q = 0.725q + 0.725(40.5 - q)$$

$$\Rightarrow q^2 - 40.5q + 29.3625 = 0.$$

We solve this quadratic equation using the standard quadratic formula

$$q = \frac{40.5 \pm (40.5^2 - 4 \times 29.3625)^{1/2}}{2}.$$

The correct root is that using the negative sign, as we require q to be smaller than the source–sample distance of 40.5 m. We therefore obtain

$$q = 738.5 \text{ mm}.$$

4. The gradient of a parabola of the form $y = x^2/2R$ is

$$\frac{dy}{dx} = x/R.$$

The equation describing a circle in Cartesian coordinates is

$$y = \pm(R^2 - x^2)^{1/2},$$

where we can arbitrarily set the origin of the circle to $(x, y) = (0, 0)$, as we are interested in calculating the gradients only. [In Figure 5.38(a), the origin of the circle is shifted relative to the parabola by a distance R along the parabola's symmetry axis]. We consider only those rays close to the symmetry axis of the parabola, that is, when $x/R \ll 1$. In this instance,

$$y \approx \pm R[1 - (x^2/2R^2)].$$

Differentiating this with respect to x yields

$$\frac{dy}{dx} = \pm x/R.$$

The positive gradient x/R is coincident with the parabola, as we set out to prove. Note that the negative gradient $-x/R$ is associated with the opposite side of the circle furthest from the parabola.

5. The focal length f of the paraboloid refractive lens is at the position along the z-axis where the optical path difference of all rays impinging on the lens parallel to the z-axis and then being refracted to f is equal to zero, independent of r, the offset of any one incident ray to the z-axis. Comparing the central and off-centre ray in Figure E.5, we can therefore state that

$$\frac{f}{\lambda} = \frac{r^2}{2a}\frac{n}{\lambda} + \frac{1}{\lambda}\left[\left(f - \frac{r^2}{2a}\right)^2 + r^2\right]^{1/2}.$$

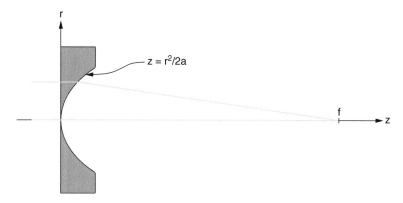

Figure E.5   Ray geometry used to determine the focal length of a paraboloid refractive lens. The optical path difference between any two paths is equal to zero.

The left-hand side of this equation is the path taken by the central ray, the right-hand side contains two terms – the first describes the path through the CRL, the second the refracted ray that meets the central ray at the focal point. Remembering that $n = 1 - \delta$, and multiplying out, we obtain

$$f = \frac{r^2(1-\delta)}{2a} + f\left[\left(1 - \frac{r^2}{2af}\right)^2 + \frac{r^2}{f^2}\right]^{1/2}$$

$$= \frac{r^2(1-\delta)}{2a} + f\left[1 + \frac{r^4}{4a^2f^2} - \frac{r^2}{af} + \frac{r^2}{f^2}\right]^{1/2}.$$

But $r/f \ll 1$, hence we can ignore the term $r^4/(4a^2f^2)$ to obtain

$$f \approx \frac{r^2(1-\delta)}{2a} + f\left[1 + r^2\left(\frac{1}{f^2} - \frac{1}{af}\right)\right]^{1/2}.$$

We can use the Taylor expansion of the right-hand term in square brackets, as $r/f \ll 1$. Therefore

$$f \approx \frac{r^2}{2a} - \frac{r^2\delta}{2a} + f + \frac{fr^2}{2}\left(\frac{1}{f^2} - \frac{1}{af}\right)$$

$$\rightarrow 0 = \frac{1}{f} - \frac{\delta}{a}$$

$$\Rightarrow f = \frac{a}{\delta} \qquad \text{Q.E.D.}$$

Note that this agrees with Equation (5.33), as for each drilled hole, *two* refracting surfaces are produced. For the single refractive paraboloid, therefore, $n_h = 0.5$.

6. In order to simplify Equation (5.52), we begin by recognizing that

$$N = \frac{1}{a^3};$$

$$\lambda = 2d_{(111)}\sin\theta;$$

$$\sin 2\theta = 2\sin\theta\cos\theta,$$

and substituting these into Equation (5.52). In addition, the (111) lattice spacing of a cubic crystal is $d_{(111)} = a/\sqrt{3}$. This yields

$$
\begin{aligned}
w_D &= \frac{2r_0|F(Q)|}{\pi} \frac{4a^2\sin^2\theta}{6a^3\sin\theta\cos\theta} \\
&= \frac{16\sqrt{2}\,r_0|f(Q)|}{3\pi a}\,\tan\theta \qquad \text{Q.E.D.} \qquad (E.4)
\end{aligned}
$$

Note that $16\sqrt{2}r_0/3\pi = 6.770\times10^{-5}$ Å. Below, we name this constant Λ.

We now calculate the Darwin widths for the three possible crystal materials. The only variables we need to calculate are $|F(Q)|$ and $\tan\theta$. $|F(Q)| = 4\sqrt{2}f(Q)$ and we can determine these from Figure 5.69, remembering that $\sin\theta/\lambda = 1/2d_{(111)} = \sqrt{3}/2a$. This is equal to 0.2428 Å^{-1}, 0.1595 Å^{-1}, and 0.1531 Å^{-1}, respectively, for diamond, Si, and Ge, from which we obtain $f(Q) = 3.007$, 10.54, and 27.37, respectively. We calculate $\tan\theta$ for the (111) reflection and 1 Å radiation to be 0.2503, 0.1616, and 0.1549, respectively (note that for these relatively small angles $\tan\theta \approx \sin\theta$). This yields

$$
\begin{aligned}
w_D^{\text{dia}} &= 14.29 \text{ μrad}, \\
w_D^{\text{Si}} &= 21.24 \text{ μrad}, \\
w_D^{\text{Ge}} &= 50.73 \text{ μrad}.
\end{aligned}
$$

Their ratios are therefore 1 : 1.486 : 3.550.

We now want to obtain an analytical expression describing the fractional bandwidth of transmitted radiation $\Delta E/E$. We begin, as suggested, by expressing E in terms of Bragg's law, that is

$$
E = \frac{hc}{\lambda} = \frac{hc}{2d\sin\theta}.
$$

We differentiate this with respect to θ:

$$
\frac{dE}{d\theta} = -\frac{hc}{2d}\frac{\cos\theta}{\sin^2\theta}
$$

and therefore

$$
\begin{aligned}
\frac{dE}{E} &= -\frac{hc}{2d}\frac{\cos\theta}{\sin^2\theta}\frac{2d\sin\theta}{hc}\,d\theta \\
&= \frac{d\theta}{\tan\theta},
\end{aligned}
$$

whereby we have dropped the sign for convenience. But $d\theta$ is equal to the Darwin width w_D, and so we substitute this with our expression for the Darwin width in Equation (E.4) to obtain

$$
\frac{dE}{E} = \frac{\Lambda f(Q)}{a}.
$$

None of the elements on the right-hand side of this equation depend on E, and hence $\Delta E/E$ is indeed independent of the photon energy. Note that for Si(111), $\Delta E/E = 1.31\times10^{-4}$, while for diamond it is equal to 5.7×10^{-5}, less than half the value for Si.

Although diamond is thermally very robust and is the best known thermal conductor, the reflected flux is limited by its small Darwin width and structure factor.

7. The range of incident angles on the monochromator $d\theta$ between the lowest measured energy 17.4733 keV at −0.9 mm and the highest, 17.4809 keV at +0.9 mm is their spatial separation (1.8 mm) divided by the distance from the undulator to the capillary (32 797 mm) = 5.488×10^{-5} rad. The range in measured energies across the beam is the difference between the two extreme values recorded, that is, $dE = 7.6$ eV. We can calculate the Bragg angle θ, and thus $\tan\theta$ using Bragg's law [most simply, by using Equation (6.12)] and our knowledge of d and E. This turns out to be $\tan\theta = 0.113\,86$. From this, we obtain a calculated range of energies equal to $17\,477 \times 1 \times 5.488 \times 10^{-5}/0.11386 = 8.42$ eV. This is very similar to the measured value, the difference being most probably attributable to the beam also bouncing off two nominally (but probably not perfectly) flat mirrors before reaching the powder station.

8. The maximum photon energy that is efficiently reflected by the Rh-coated mirror is assumed to be when the critical angle for total external reflection α_c is equal to the fixed incident angle, $0.2° = 3.491$ mrad. The relationship between α_c (in radians), the x-ray wavelength λ, and the electron density ρ is

$$\alpha_c = \lambda\sqrt{\frac{\rho r_0}{\pi}} = 2.996 \times 10^{-3}\lambda\sqrt{\rho},$$

whereby $r_0 = 2.82 \times 10^{-5}$ Å is the Thomson scattering length, λ is given in Å, and ρ is given in Å^{-3}. We therefore need to calculate the electron density of rhodium metal. From the supplied information, we know that 1 cm^3 of Rh contains $6.022 \times 10^{23} \times 12.4/102.91 = 7.256 \times 10^{22}$ atoms. Each atom has 45 electrons, hence the electron density $\rho = 7.256 \times 10^{22} \times 45/10^{24} = 3.265$ e Å^{-3}.

Hence

$$3.491 \times 10^{-3} = 2.996 \times 10^{-3}\lambda\sqrt{3.265}$$

$$\Rightarrow \lambda = 0.6448 \text{ Å},$$

corresponding to a photon energy of $12.3984/0.6448 = 19.23$ keV.

9. The meridional radius R_m required to image a source using a bendable mirror at a distance p from that source to a point a distance q downstream of the focussing mirror is given by

$$R_m = \frac{2}{\sin\theta}\frac{pq}{p+q},$$

whereby θ is the incident angle on the mirror and, in this instance, is equal to $0.16° = 2.79$ mrad. In this exercise, $p = 25$ m and $q = 15$ m. Hence

$$R_m = (2/2.79 \times 10^{-3}) \times \frac{25 \times 15}{40}$$

$$= 6714 \text{ m.}$$

10. In this example, horizontal focussing is not achieved by curved mirrors, but instead by bending the second Si(111) crystal in the double-crystal monochromator. The equation for the sagittal radius R_s is

$$R_s = 2\sin\theta\frac{pq}{p+q},$$

with $p = 23$ m, $q = 17$ m, and θ the Bragg angle for the Si(111) reflection at 1 Å. Using Bragg's law, we obtain $\sin\theta = 0.1595$ and hence $R_s = 3.117$ m.

11. The angle α_n of the nth reflection inside a conical capillary focussing tube is given by

$$\alpha_n = (2n - 1)\alpha_1.$$

Once α_n exceeds the critical angle α_c, the capillary will no longer reflect. We therefore simply need to calculate α_c for 5 keV photons impinging on platinum. In the same manner that we calculated the electron density for rhodium in a previous exercise of this chapter set (see above), we obtain $\rho = 5.165$ e/Å^3, from which we determine $\alpha_c = 0.01688$ rad $= 0.9674°$.

The ratio between α_c and α_1 is $0.9674/0.025 = 38.696 \geq (2n - 1)$. Hence $n = 19$. After the 19th reflection, the x-rays will penetrate deeply into the platinum coating and be lost.

12. First, we simplify our expression by recognizing that $\bar{\delta}^2 \ll \bar{\delta}$, as $\bar{\delta} \sim 5 \times 10^{-6}$. Remembering also that $(1 - x)^n \approx 1 - nx$ for $|x| \ll 1$, we obtain

$$m\lambda = 2\Lambda \sin\theta \left(1 - \frac{2\bar{\delta}}{2\sin^2\theta}\right).$$

But we know that the uncorrected Bragg equation $m\lambda = 2\Lambda \sin\theta$ is also a fairly good approximation, so we can substitute $\sin^2\theta$ with $m^2\lambda^2/4\Lambda^2$ to obtain

$$m\lambda = 2\Lambda \sin\theta \left(1 - \frac{4\Lambda^2\bar{\delta}}{m^2\lambda^2}\right).$$

We also know from Equation (2.24) that $\delta = \rho r_0 \lambda^2/2\pi$ (where ρ is the electron density), therefore

$$\frac{\bar{\delta}}{\lambda^2} = \frac{\bar{\rho}r_0}{2\pi},$$

and hence

$$\lambda = \frac{2\Lambda}{m} \sin\theta \left(1 - \frac{4\bar{\rho}r_0}{2\pi}\frac{\Lambda^2}{m^2}\right) \qquad \text{Q.E.D.}$$

The term $4\bar{\rho}r_0/2\pi = \kappa$ is material specific, but for most materials of interest for multilayers, ρ does not vary considerably from 1 Å^{-3}. The Thomson scattering length $r_0 = 2.82 \times 10^{-5}$ Å, and hence $\kappa \sim 2 \times 10^{-5}$ Å^{-2}, or 2×10^{-3} nm^{-2}.

13. The motivation for using multilayers with the same periodicity but with different ratios for t_A/t_B in the same monochromator is to suppress a larger number of harmonics through the phenomenon of systematic absences. In the case described here, the thickness ratio is $a/b = 1/1$, hence all reflections $m(a + b) = 2m$ are suppressed – that is, all even harmonics ($m = 2, 4, 6, \ldots$). In the second multilayer, $a/b = 1/2$, and all harmonics $m(1 + 2) = 3m$ are suppressed. Hence, by combining these, the 2nd, 3rd, 4th, and 6th harmonics are suppressed. The first harmonic above the fundamental to survive is the 5th, which in most cases will be anyway much less intense than the fundamental ($n = 1$), and can be further suppressed if necessary by, for example, one or more mirror reflections.

14. We begin by recognizing that the CRL focal length is proportional to the square of the photon energy, that is

$$f = AE^2,$$

where A is some constant. Differentiating with respect to E yields

$$\frac{df}{dE} = 2AE$$

$$\Rightarrow \frac{df}{f} = 2\frac{dE}{E}.$$

If we shift f by an amount $df = +z_R$ by changing the energy by dE, this new photon energy $E + dE$ will have a diameter at the aperture of $\sqrt{2}$ times the minimum beam waist w_0 (see Figure 5.12), meaning only half the incident intensity will be transmitted. This energy is thus the half-width at half-maximum of the transmission spectrum through the aperture, and hence the FWHM $\Delta E = 2dE$. From this

$$\frac{\Delta E}{E} = \frac{z_R}{f}. \qquad \text{Q.E.D.}$$

For a 0.2 mm diameter beam of 1 nm radiation focussed at 1 m, θ, the half angle subtended by the focussed beam, equals $0.1/1000 = 10^{-4}$ radians. But $z_R = \lambda/(\pi\theta^2) = 31.83$ mm. The beam waist at the focus is $w_0 = \lambda/(\pi\theta) = 3.183$ μm and the aperture diameter is therefore 6.37 μm. The relative bandwidth $\Delta E/E = 31.83/1000 = 3.183 \times 10^{-2}$, equating to $\Delta E = 39.5$ eV for 1 nm (1.2398 keV) radiation.

The relationship between the focal length f, the refractive-index decrement δ, the minimum radius of curvature a, and n_{ps}, the number of paraboloidal surfaces in the CRL, is given by

$$f = \frac{a}{n_{ps}\delta}.$$

From the provided information, $a = 1.363$ mm.

15. To make an estimate of the surface temperature of the insulating filter, we assume that thermal conductivity is negligible and that all the absorbed power is reradiated via blackbody radiation from an area equal to the cross-section of the incident beam.

 In thermal equilibrium, the absorbed power density is equal to the blackbody emission. The absorbed power is 2×10^{13} photons $\text{s}^{-1} \times 10^4$ eV $\times 1.6 \times 10^{-19}$ J $\text{eV}^{-1} = 32$ mW. This is absorbed over an area of 1.25×10^{-8} m^2. The areal power density is thus $J = 2.56 \times 10^6$ W m^{-2}.

 The Stefan–Boltzmann law states

$$J = \varepsilon_r \sigma(T^4 - T_a^4),$$

whereby ε_r is the relative emissivity, $\sigma = 5.67 \times 10^{-8}$ W m^{-2} K^{-4} is the Stefan–Boltzmann constant, T is the temperature of the blackbody, and T_a is the ambient surrounding temperature (in this case, approximately 300 K, i.e. room temperature). Inserting our known values for J and ε_r into the above, we obtain $T = 2834$ K.

 This value is above the melting temperature of nearly all materials. In reality, it can be expected that thermal conductivity will increase the emitting area to a certain extent, so this temperature is an upper limit. Nonetheless, the large thermal gradient induced by this very local heating will induce enormous mechanical stress on the filter material and almost certainly cause it to break. One should therefore use materials with a higher thermal conductivity (metals), or even diamond, which, although an insulator, has the highest known thermal conductivity of any material.

16. We use the Poisson distribution to calculate this – the normal-distribution approximation would be inappropriate for such a rare event. In this example, $N = 3$ and $k = 0$, and hence

$$P(0) = e^{-3} \times 3^0/0! = 0.04979.$$

There is therefore only a 5% chance that the Thames doesn't flood in a century.

17. The signal-to-noise ratio required for this experiment must exceed $20 \times 10 = 200$ in order to distinguish the signal features to better than 10%. Hence the background must have noise that is at most $1/200$ of its average value, i.e. $\Delta N/N = 1/\sqrt{N} = 1/200$, which requires $N = 200^2 = 4 \times 10^4$ counts/pixel. As the average background count rate is 100 s^{-1}, one must record for at least 400 s.

18. First, we need to calculate the photon arrival rate during the pulse. This is simply the average value multiplied by the ratio of the pulse separations $(1/f)$ to their duration (Δt), that is, $10^{13} \times 5 \times 10^{-9}/(5 \times 10^{-11}) = 10^{15}$. This is 10^8 times higher than the acceptable maximum detectable true arrival rate on the detector of $N_0 = 10^7$ per pixel. But the signal is evenly distributed across 16 pixels, hence the maximum allowable transmission of any filter used is $16/10^8 = 1.6 \times 10^{-7}$.

19. We assume that with the filter inserted, the recorded count rate N_{obs} has no 'pile-up' and the real incident count rate (that is, without filter) is $N_0 = N_{obs}/(2 \times 10^{-5})$. After removing the filter, the new observed count rate $N'_{obs} = N_{obs} \times 10^4$. Therefore

$$N_{obs} \times 10^4 = N_0 \exp(-N_0 \tau)$$

$$= \frac{N_{obs}}{2 \times 10^{-5}} \exp(-N_0 \tau)$$

$$\Rightarrow 0.2 = \exp(-N_0 \tau)$$

$$\Rightarrow N_0 = \ln 5/(1.7 \times 10^{-7}) = 9.47 \times 10^6 \text{ cts/s.}$$

The observed count rate is five times smaller, that is $N_{obs} = 1.89 \times 10^6$ cts/s.

20. First, we must calculate the attenuation length at 38 keV from the value given for 8 keV. In Section 2.6.3, we determined that, far from absorption edges, the attenuation length is proportional to the cube of the photon energy. Hence the attenuation length for silicon at 38 keV equals $69.6 \times (38/8)^3 = 7460$ μm. The transmission of a 300 μm thick sensor is therefore $\exp(-300/7460) = 0.9606$. The quantum efficiency is therefore only 3.94%. By increasing the sensor thickness to 1000 μm, the transmission is reduced to $\exp(-1000/7460) = 0.8745$ and the quantum efficiency increased to 12.55%.

21. Let the optical path difference at the recombiner gratings S_n between the direct pump beam and the beam diffracted through an angle 2θ by a grating with periodicity p_n be ΔL. After recombination again at the sample, the time delay is

$$\Delta t_n = 2\Delta L/c.$$

But

$$\Delta L = \frac{L}{2}\left(\frac{1}{\cos 2\theta} - 1\right).$$

Remember though that $\cos 2\theta = (1 - \sin^2 2\theta)^{1/2}$ and that, in this case, $\sin 2\theta \ll 1$. This leads to

$$\Delta L = \frac{L}{2}[(1 - \sin^2 2\theta)^{-1/2} - 1]$$

$$\approx \frac{L}{2}\left(1 + \frac{\sin^2 2\theta}{2} - 1\right) = \frac{L}{4}\sin^2 2\theta.$$

But because θ is small, we know that $\sin 2\theta \approx 2\sin\theta$ and, from Bragg's law, $\sin\theta = \lambda/2p_n$. Inserting this into our expression for Δt_n, we obtain

$$\Delta t_n = \frac{L\lambda^2}{2c\, p_n^2} \qquad \text{Q.E.D.}$$

E.6   Chapter 6 – Scattering Techniques

1. The interplanar spacing d_{hkl} is in general given by Equations (6.5) to (6.7). Inserting the values given for the lattice parameters of CuO, we obtain

$$X = [1 - \cos^2(99.54°)]^{1/2} = \sin(99.54°) = 0.9862$$

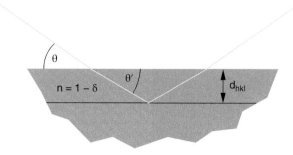

Figure E.6   The Bragg condition for constructive interference when taking refractive effects into account.

and

$$Y = \left[\left(\frac{2}{4.684} \right)^2 + \left(\frac{1}{3.423} \right)^2 \sin^2(99.54°) + \left(\frac{1}{5.129} \right)^2 \right.$$
$$\left. - \frac{4}{4.684 \times 5.129} \cos(99.54°) \right]^{1/2}$$
$$= 0.5753 \text{ Å}^{-1}$$
$$\Rightarrow d_{hkl} = \frac{X}{Y} = 1.714 \text{ Å}.$$

2. In the 'normal' derivation of Bragg's law, the optical path difference between reflections from adjacent crystal planes is an integer multiple of the x-ray wavelength. In this derivation, slight changes in the angle of the x-ray propagation due to refraction as the x-ray penetrate the material, and the very small dilation of the wavelength in the medium (of refractive index $n \approx 1 - \delta$) are ignored. Here, we take these deviations into account. Consider Figure E.6.

    The angle of the x-ray beam in the material is given by

    $$\cos \theta' = \cos \theta / n$$
    $$\Rightarrow \sin \theta' = \left(1 - \frac{\cos^2 \theta}{n^2} \right)^{1/2}. \tag{E.5}$$

In addition, the wavelength of the x-rays in the material is marginally longer:

$$\lambda' = \lambda / n. \tag{E.6}$$

The Bragg condition is therefore

$$\lambda' = 2 d_{hkl} \sin \theta'. \tag{E.7}$$

Substituting Equations (E.5) and (E.6) into Equation (E.7) and remembering $n \approx 1 - \delta$, we obtain

$$\lambda = 2 n d_{hkl} \left(1 - \frac{\cos^2 \theta}{n^2} \right)^{1/2}$$
$$= 2 d_{hkl} [(1 - \delta)^2 - \cos^2 \theta]^{1/2}$$
$$= 2 d_{hkl} [1 - 2\delta + \delta^2 - \cos^2 \theta]^{1/2}.$$

We can neglect the δ^2 term in the square brackets as this is over five orders of magnitude smaller than δ, the next smallest term. Note also the trigonometric equality $\sin^2\theta = 1 - \cos^2\theta$. From these considerations we obtain

$$\lambda \approx 2d_{hkl}[\sin^2\theta - 2\delta]^{1/2}$$

$$= 2d_{hkl}\sin\theta\left[1 - \frac{2\delta}{\sin^2\theta}\right]^{1/2}.$$

But $(1 - x)^n \approx 1 - nx$, if $x \ll 1$. This condition is met for $\delta/\sin^2\theta$, hence

$$\lambda \approx 2d_{hkl}\sin\theta\left[1 - \frac{\delta}{\sin^2\theta}\right].$$

We can to a high degree of accuracy replace $\sin\theta$ using the 'normal' Bragg equation with $\lambda/2d_{hkl}$ to obtain

$$\lambda \approx 2d_{hkl}\sin\theta\left[1 - \frac{4d_{hkl}^2\delta}{\lambda^2}\right].$$

Finally, we substitute in for δ using Equation (2.24) to obtain our final result

$$\lambda \approx 2d_{hkl}\sin\theta\left[1 - \frac{2\rho r_0 d_{hkl}^2}{\pi}\right] \qquad\qquad \text{Q.E.D.}$$

Note that this is independent of the photon energy, and produces the biggest deviations for low-index reflections in heavy materials, for which ρ and d_{hkl} are largest. If we assume $d_{hkl} = 4$ Å and $\rho = 1.5$ e Å^{-3}, the fractional change in $\sin\theta$ (and therefore also in θ) is of the order of magnitude of 5×10^{-4}.

3. The only value in Equation (6.20) not directly calculable from the information provided is $|F|$. To obtain this, we need to determine the eight scattering contributions (from the eight atoms within the unit cell of diamond) provided by Equation (6.17). The unit cell of diamond is face-centred [that is, with lattice points at $(0,0,0)$, $(1/2, 1/2, 0)$, $(1/2, 0, 1/2)$, $(0, 1/2, 1/2)$], with two atoms associated with each point, one at that point, the other displaced by $(1/4, 1/4, 1/4)$ (see Figure 6.5). When we insert these coordinates into Equation (6.17), we obtain for the (111) reflection

$$F = 4f_C(1 - i),$$

which thus has a magnitude equal to $4\sqrt{2}f_C$.

The (111) interplanar spacing is simply $a/\sqrt{3}$, and therefore

$$\Lambda'_{\text{ext}} = \frac{\sqrt{3}}{16\sqrt{2}}\frac{a^2}{r_0 f_C}$$

$$= 1.337 \text{ μm.}$$

4. Consider Figure E.7. The distance BS is simply $k_{\text{out}}\cos 2\theta$, while the distance AS is $k_{\text{out}}\cos\delta$. But AS $\cos\gamma =$ BS. Therefore

$$\cos 2\theta = \cos\gamma \,\cos\delta.$$

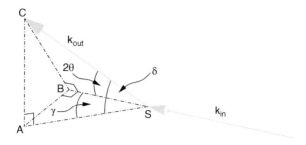

Figure E.7   The trigonometrical relationship between 2θ, γ, and δ.

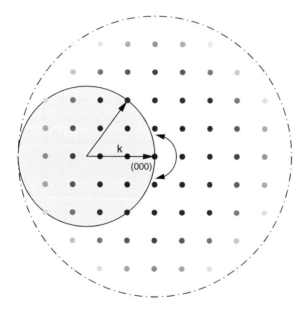

Figure E.8   Different Bragg peaks are accessed by rotating the single crystal. In reciprocal space, the diffraction pattern (the array of blue dots) rotates around the (000) forward-scattered peak. When a diffraction peak crosses the surface of the Ewald sphere (of radius $k = 2\pi/\lambda$, shown in yellow), the Bragg condition is met and the peak is detected. One can see that in principle, all Bragg peaks that lie within the volume of a sphere of radius $2k$ (shown as the dot-dashed circle) can be made to cross the Ewald sphere and thus be detected.

5. Consider Figure E.8. In general, diffraction maxima are detected when they lie on the surface of the Ewald sphere (radius $k = 2\pi/\lambda$). The consequence of rotating a single crystal is that the diffraction pattern rotates in the same manner around the (000) forward-scattered diffraction peak – this is the rotation method for recording diffraction data commonly used, for example, in protein crystallography. Hence, all diffraction spots that lie within the volume of a sphere of radius $2k$ (in reciprocal space) can in principle be recorded. The number of diffraction peaks within this volume is therefore the volume of the sphere of radius $2k$ divided by the volume of the reciprocal unit cell V^*, that is

$$N = \frac{4\pi(2k)^3/3}{V^*}$$

$$= \frac{4\pi \times 64\pi^3/3\lambda^3}{V^*}.$$

The volume of the reciprocal unit cell is

$$V^* = |a^* \cdot b^* \times c^*|,$$

whereby $a^* = 2\pi/a$, $b^* = 2\pi/b$, and $c^* = 2\pi/c$, and a, b, and c are the unit-cell lattice constants. Hence

$$V^* = \frac{8\pi^3}{|a \cdot b \times c|} = \frac{8\pi^3}{V_c},$$

whereby V_c is the volume of the unit cell in real space. From this, we obtain

$$N = \frac{32\pi V_c}{3\lambda^3} \qquad \text{Q.E.D.}$$

6. For the SrTiO$_3$ (001) reflection, we know that the phase difference between the upper and lower faces of the unit cell is 2π. We make our life simplest by setting the zero of the phase to be the central plane containing the Ti atom and two of the three oxygen atoms [note that in Figure 6.97(b), there are *two* O atoms, not *four*, as two belong to the neighbouring unit cell]. The phases of the Sr atom and remaining O atom are both π. The total structure factor is therefore $F_{(001)} = f_{Ti} + f_O - f_{Sr}$.

   Using the same approach, the phases of the Sr atom and third O atom in the (002) reflection are 2π and the total structure factor is $F_{(002)} = f_{Ti} + 3f_O + f_{Sr}$. We simply now need to determine the amplitudes of the form factors, for which we use Figure 6.97(a). The Q-value for the (001) reflection is $2\pi/3.9045 = 1.6092$ Å^{-1}, and for the (002), 3.2184 Å^{-1}. The form factors for Sr^{2+}, Ti^{4+}, and O^{2-} are 33.17, 16.81, 7.26, and 27.63, 14.02, 4.74, respectively. The ratio of the Bragg-peak intensities is therefore

$$\left(\frac{F_{(002)}}{F_{(001)}}\right)^2 = \left[\frac{27.63 + 14.02 + (3 \times 4.74)}{16.81 + 7.26 - 33.17}\right]^2 = 37.694.$$

Note that calculated powder patterns actually predict a ratio of about 10, not 38. This is because they include the so-called 'Lorentz and polarization factor', which weighs the structure factor by $1/(\sin^2\theta\cos\theta)$, which, for small angles (high Q), decreases $I_{(002)}/I_{(001)}$ approximately by a factor of 4.

7. The electron density is the Fourier transform of the set of complex structure factors, that is,

$$\rho(\mathbf{r}) = \mathscr{F}\left[\{F_{hkl}\}\right].$$

The Fourier transform of the set of measured diffraction *intensities* (the absolute squares of the structure factors) is therefore given by

$$\mathscr{F}\left[\{I_{hkl}\}\right] = \mathscr{F}\left[\{F_{hkl}\} \cdot \{F_{hkl}^*\}\right].$$

But we know from Friedel's law that

$$F_{hkl} = F_{\overline{hkl}}^*$$

and hence

$$\mathscr{F}\left[\{I_{hkl}\}\right] = \mathscr{F}\left[\{F_{hkl}\} \cdot \{F_{\overline{hkl}}\}\right].$$

But from the convolution theorem, we know that the Fourier transform of the product of two functions is equal to the Fourier transform of the first function convoluted with that of the second. Therefore

$$\mathscr{F}\left[\{I_{hkl}\}\right] = \mathscr{F}\{F_{hkl}\} \otimes \mathscr{F}\{F_{\overline{hkl}}\}$$

$$= \rho(\mathbf{r}) \otimes \rho(-\mathbf{r}) \qquad \text{Q.E.D.}$$

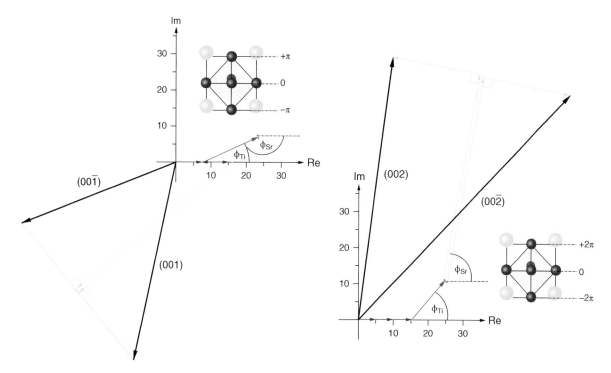

Figure E.9 The structure factors for the (001), (00$\bar{1}$), (002), and (00$\bar{2}$) reflections of PbTiO$_3$. The small blue vectors are for the three oxygen atoms, the red vectors correspond to Ti, the yellow to Sr. The axes are in units of scattering electrons.

8. (a)/(h); (b)/(g); (c)/(e); (d)/(f).
9. The Argand diagrams for the (001), (00$\bar{1}$), (002), and (00$\bar{2}$) reflections of PbTiO$_3$ are sketched in Figure E.9. We begin by setting the zero of the phase (which we can choose arbitrarily) to be in the central plane of the oxygen octahedron structure. In the case of the (001) reflection, this means that the upper (and lower) apex of the octahedron is at π ($-\pi$); for the (002) peak, they are at $\pm 2\pi$.

Two of the three form factors of the oxygen atoms for the (001) and (00$\bar{1}$) reflection therefore have positive real values (i.e. parallel to the real axis of the Argand diagram), the third one at the apex, a negative (phase $-\pi$) real value, thereby cancelling out one of the positive vectors. Next, we need to sketch the form factor of the Ti atom. This lies 0.28 Å above the central plane, so has a phase of $\phi_{Ti} = (0.28/4.1524) \times 2\pi = 24.28°$. The amplitude is given in Table 6.3. Note that this has a value for f_2 large enough to sketch (in contrast to the values for the O atoms). In this sketch, I have 'flipped' the (00$\bar{1}$) Friedel mate across the real axis so both can be sketched together, highlighting more clearly their differences (see also Figure 6.25).

Lastly, we need to draw the form factor for strontium, which begins at the arrowheads of the two small red f_2-vectors of titanium associated with the (001) and (00$\bar{1}$) reflections. The real part f_1 of this vector has an angle of $-\pi + (0.42/4.1524) \times 2\pi = -143.6°$. The imaginary components f_2 are added on to these at 90°, the (001) in an anticlockwise sense, the (00$\bar{1}$) clockwise.

We go through the same procedure for the (002) Friedel pairs, remembering that (a) now all the form factors of the O atoms are in phase (0°) and (b) the c-axis of the PbTiO$_3$ unit cell now spans $\pm 2\pi$ radians.

From the sketches, we obtain then a ratio of the (001) and (00$\bar{1}$) reflections of $56.5/47.0 = 1.20$ and that of the (002) and (00$\bar{2}$) reflections of $73.5/87.0 = 0.84$.

10. The transmitted intensity of an absorbing object of length l and attenuation coefficient μ is

$$I = I_0 \exp(-\mu l),$$

and hence the absorbed fraction is

$$I_0[1 - \exp(-\mu l)].$$

The dose D per unit mass after a time t_D is therefore

$$D = \frac{nh\nu \times t_D \times [1 - \exp(-\mu l)]}{Al\rho}.$$

We assume that $\mu l \ll 1$ (i.e. the beam is only marginally attenuated by the crystal). Remembering that $\exp(-x) \approx 1 - x$ for $|x| \ll 1$, we obtain

$$D = \frac{nh\nu \times t_D \times \mu l}{Al\rho},$$

$$\Rightarrow t_D = \frac{DA\rho}{nh\nu\mu} \qquad \text{Q.E.D.}$$

In the case of reaching the Garman limit of $D = 3 \times 10^7$ J kg^{-1}, $t_D = t_{\text{Gm}}$.

Inserting the given values for $D = 10^4$ J g^{-1}, $A = 2.5 \times 10^{-7}$ cm^2, $\rho = 1.37$ g cm^{-3}, $h\nu = 1.99 \times 10^{-15}$ J, $n = 5 \times 10^{12}$ s^{-1}, and $\mu = 2.0$ cm^{-1} into the above equation (note the unit conversions!), we calculate that the exposed volume therefore accumulates a dose of 10^7 Gy within 0.171 s.

11. The cubic protein crystal has a mass of $(0.02)^3 \times 1.35 = 1.08 \times 10^{-5}$ g. The temperature increase after one second is

$$\Delta T = \frac{E}{mC_p},$$

where E is the absorbed photon energy, which can be accurately approximated to be $nh\nu \times (200/5000)$ for this case in which the attenuation length is much larger than the crystal dimensions. Therefore $E = 794\,\mu$J. From this, we obtain $\Delta T = 147$ K s^{-1}! In reality, the surrounding cryocooled medium and the holder will remove a good fraction of this heat, underlining the importance of cooling.

12. The half-angle subtended by the detector is $\arctan(220/140) = 57.53°$. This is twice the maximum Bragg angle that can be detected, that is, $\theta_{\text{max}} = 28.76°$ and $\sin\theta_{\text{max}} = 0.481$. But the maximum scattering vector Q_{max} is given by

$$Q_{\text{max}} = \frac{4\pi}{\lambda}\sin\theta_{\text{max}}.$$

But $\lambda = 12.3984/12.66 = 0.979$ Å, and hence $Q_{\text{max}} = 6.172$ Å^{-1}. The smallest resolvable feature in the electron-density map emerging from the diffraction-data analysis is $2\pi/Q_{\text{max}} = \lambda/2\sin\theta_{\text{max}} = 1.018$ Å.

The photon energy of 12.66 Å corresponds to the K-edge of selenium. This relatively heavy element ($Z = 34$) is often substitutionally introduced (isomorphously replacing sulfur) in MIR/MAD and SAD experiments.

13. We begin with the Bragg equation

$$\lambda = \frac{2d}{m}\sin\theta,$$

$$\Rightarrow m = \frac{2d}{\lambda}\sin\theta.$$

We differentiate this to obtain

$$\frac{dm}{d\theta} = \frac{2d}{\lambda} \cos \theta.$$

The smallest change in m is $dm = 1$, resulting in a change in θ; the change in 2θ (called δ in the problem) is therefore

$$\delta = \frac{\lambda}{d \cos \theta} \qquad \text{Q.E.D.}$$

The angular separation in radians between the two Bragg peaks is accurately approximated by $\delta = 0.91/90 = 0.010\,11$, from which we conclude that $d = (12.3984/12.66)/(\delta \cos \theta)$. But $\cos \theta \approx 1$, as we have been told that the Bragg peaks are close to the direct beam. Therefore $d = 96.87$ Å.

14. Figure 6.14 must now be modified to include the imaginary components f_2 of the form factors. These always point in the imaginary direction, as f_1 is always along the real axis. Hence the Friedel pairs have form factors with the same magnitude, that is $|F_{hkl}| = |F_{\overline{hkl}}|$, but now $\phi_{hkl} = \phi_{\overline{hkl}} = \arctan(f_1/f_2)$.

15. We need to calculate $N_j^{1/2} f_{2j}$ for C, N, O, and S. We take the values for f_{2j} from Table 6.1 for 6 keV, although, because their ratios remain constant far above any absorption edges, the result we are calculating is valid for any energy above approximately 3 keV (see Figure 6.56). We obtain $N_j^{1/2} f_{2j} = 0.2588, 0.2567, 0.5078$, and 2.328 electrons, respectively. The statistical sum of the C, N, and O contributions, assuming no correlation between them, is $(0.2588^2 + 0.2567^2 + 0.5078^2)^{1/2} = 0.6251$, almost four times smaller than the sulfur signal. Hence sulfur does dominate in native-SAD studies of crambin.

16. From Figure 6.100, we can determine $k_{out,\|}$, the in-plane component of k_{out} using the cosine rule, that is

$$k_{out,\|}^2 = k_{in}^2 + Q_\|^2 - 2k_{in}Q_\| \cos \phi.$$

But from Pythagoras' theorem,

$$k_{out}^2 = l^2 + k_{out,\|}^2.$$

Remembering that the magnitudes of k_{in} and k_{out} are equal, we therefore obtain

$$l = (k^2 - k^2 + 2kQ_\| \cos \phi - Q_\|^2)^{1/2}$$
$$= (2kQ_\| \cos \phi - Q_\|^2)^{1/2} \qquad \text{Q.E.D.}$$

In the given example, the in-plane component of the (21)-CTR has a magnitude of $Q_\| = 2\pi/d_{210} = 2\pi\sqrt{5}/4 = 3.512$ Å^{-1}; while $|k| = 2\pi \times 16/12.3984 = 8.108$ Å^{-1}. Inserting these values into our expression for l and setting $\phi = 0°$, we obtain $l_{max} = 6.680$ Å^{-1}. This, in terms of $2\pi/c$, the out-of-plane reciprocal lattice unit (with $c = 4$ Å in the case of this cubic system), means that one can probe as high as $l = 4.25$ r.l.u. along this rod.

17. The critical angle α_c for an x-ray wavelength λ is given by

$$\alpha_c = \lambda \sqrt{\frac{\rho r_0}{\pi}},$$

whereby $r_0 = 2.82 \times 10^{-5}$ Å is the Thomson scattering length and ρ is the electron density. From our knowledge of the size and atomic structure of the Si-unit cell, $\rho = (14 \times 8)/5.431^3 = 0.6992$ e Å^{-3}, from which we obtain

$$\alpha_c = 1.2398 \sqrt{\frac{2.82 \times 10^{-5} \times 0.6992}{\pi}}$$
$$= 3.106 \text{ mrad.}$$

The reflectivity of an atomically smooth surface for incident angles α significantly larger than α_c is given by

$$R = \left(\frac{\alpha_c}{2\alpha}\right)^4.$$

Hence for atomically flat silicon at 2° (0.034 91 rad), we obtain

$$R = \left(\frac{3.106}{2 \times 34.91}\right)^4 = 3.91 \times 10^{-6}.$$

Germanium has an electron density $\rho = 1.422$ e Å^{-3}. As the critical angle α_c is proportional to the square root of ρ and R is proportional to the fourth power of α_c, we would expect the reflectivity of perfectly flat germanium to be $(1.422/0.6992)^2$ times larger than that for Si, i.e. $R = 1.62 \times 10^{-5}$. However, we need to take the surface roughness into account. This suppresses the reflectivity by a factor $\mathscr{R}$ given by

$$\mathscr{R} = \exp[-(2k\alpha\sigma)^2],$$

whereby k is the x-ray wavevector and σ is the root-mean-square roughness. In this example, $k = 2\pi/1.2398 = 5.068$ Å^{-1}, and $\sigma = 8$ Å, leading to $\mathscr{R} = 3.31 \times 10^{-4}$. The reflectivity is therefore 5.36×10^{-9}.

18. From Snell's law, we have

$$\frac{\cos\alpha}{\cos\alpha'} = 1 - \delta,$$

whereby α' is the angle of the transmitted ray relative to the surface. From this, we obtain $\alpha' = 0.48938°$. The reflectivity amplitude is $r = (\alpha - \alpha')/(\alpha + \alpha') = 0.010\,74$. The reflected intensity $R = r^2 = 1.152 \times 10^{-4}$.

19. The ratio $\mathscr{R}$ of the reflectivity of a surface with root-mean-square roughness σ to that of an ideally flat surface of the same material is given by

$$\mathscr{R} = \exp[-(2k\alpha\sigma)^2],$$

whereby $k = 2\pi/\lambda$ is the wavevector and α is the incident and reflected angle. From Figure 6.101, we know that $\mathscr{R}$ drops to 50% at $\alpha = 0.5882° = 10.266$ mrad. From this we therefore obtain

$$\sigma = \frac{1.23984 \times 10^{-10} \sqrt{\ln 2}}{4\pi \times 1.0266 \times 10^{-2}}$$

$$= 8 \text{ Å}.$$

20. Let $Qr_2 = X$ and $Qr_1 = x$. We begin by inserting the given expression for the first-order spherical Bessel function into our expression for F to obtain

$$\begin{aligned}F(Q, X, x) &= \frac{3}{r_2^3 - r_1^3}\left\{\frac{r_2^3}{X}\left[\frac{\sin X}{X^2} - \frac{\cos X}{X}\right]\right.\\[1mm] &\quad\left. - \frac{r_1^3}{x}\left[\frac{\sin x}{x^2} - \frac{\cos x}{x}\right]\right\}\\[2mm] &= \frac{3}{r_2^3 - r_1^3}\left\{\frac{r_2^3}{X^3}[\sin X - X\cos X]\right.\\[2mm] &\quad\left. - \frac{r_1^3}{x^3}[\sin x - x\cos x]\right\}.\end{aligned} \qquad (E.8)$$

We are concerned in this problem with the Guinier regime, for which X and $x \ll 1$. We can therefore approximate the sine and cosine terms by their Taylor expansions. This begs the question as to the number of terms that should be included. Because we want to obtain some sort of dependence on the physical dimensions (r_1 and r_2) and because both terms bounded by square brackets in Equation (E.8) are divided by x^3 (or X^3), our expansions should exceed the third power. Thus, we substitute $\sin x$ with $x - x^3/6 + x^5/120$ and $\cos x$ with $1 - x^2/2 + x^4/24$ (and similarly for X), to obtain

$$
\begin{aligned}
F(Q, X, x) &= \frac{3}{r_2^3 - r_1^3} \left\{ \frac{r_2^3}{X^3} \left[X - \frac{X^3}{6} + \frac{X^5}{120} - X + \frac{X^3}{2} - \frac{X^5}{24} \right] \right. \\
&\qquad \left. - \frac{r_1^3}{x^3} \left[x - \frac{x^3}{6} + \frac{x^5}{120} - x + \frac{x^3}{2} - \frac{x^5}{24} \right] \right\} \\
&= \frac{3}{r_2^3 - r_1^3} \left\{ \frac{r_2^3}{X^3}(X^3/3 - X^5/30) - \frac{r_1^3}{x^3}(x^3/3 - x^5/30) \right\} \\
&= \frac{3}{r_2^3 - r_1^3} \left\{ \frac{r_2^3 - r_1^3}{3} - \frac{r_2^3 X^2}{30} + \frac{r_1^3 x^2}{30} \right\} \\
&= 1 - \left(\frac{r_2^3 X^2 - r_1^3 x^2}{10(r_2^3 - r_1^3)} \right) \\
&= 1 - \frac{Q^2}{10} \left(\frac{r_2^5 - r_1^5}{r_2^3 - r_1^3} \right).
\end{aligned}
\tag{E.9}
$$

We insert the square of our expression for F in Equation (E.9) into Equation (6.76), to obtain

$$
\left[1 - \frac{Q^2}{10} \left(\frac{r_2^5 - r_1^5}{r_2^3 - r_1^3} \right) \right]^2 = \exp \left(\frac{-Q^2 r_g^2}{3} \right).
$$

Because $Qr \ll 1$, we can expand both sides, so that

$$
1 - \frac{Q^2}{5} \left(\frac{r_2^5 - r_1^5}{r_2^3 - r_1^3} \right) = 1 - \frac{Q^2 r_g^2}{3}
$$

$$
\Rightarrow r_g^2 = \frac{3}{5} \left(\frac{r_2^5 - r_1^5}{r_2^3 - r_1^3} \right) \qquad \text{Q.E.D.}
$$

21. Consider Figure E.10. In cylindrical coordinates, the volume element $dV = r\, d\phi\, dl\, dr$. The moment of that element around the centre of the rod is $\rho\, dV s^2$, where $\rho = M/(\pi R^2 L)$ and $s^2 = r^2 + l^2$. We integrate this expression with respect to the three cylindrical coordinates ϕ, l, and r to obtain the moment of inertia of the rod, I_{rod}:

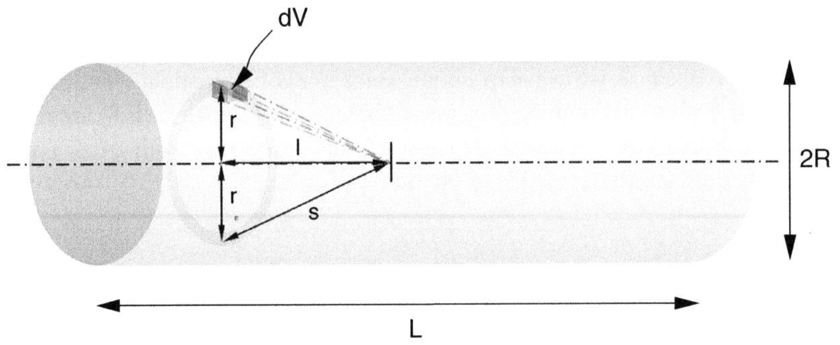

Figure E.10 The geometry and volume element, dV, in cylindrical coordinates, of a solid cylinder of radius R and length L.

$$I_{\text{rod}} = \rho \int_{-L/2}^{+L/2} \int_0^R \int_0^{2\pi} r(r^2 + l^2) d\phi \, dr \, dl$$

$$= 2\pi\rho \int_{-L/2}^{+L/2} \int_0^R r^3 + l^2 r \, dr \, dl$$

$$= 2\pi\rho \int_{-L/2}^{+L/2} \left(\frac{R^4}{4} + \frac{l^2 R^2}{2} \right) dl$$

$$= 2\pi\rho \left(\frac{R^4 L}{4} + \frac{L^3 R^2}{24} \right).$$

But $\rho = M/(\pi R^2 L)$, and so

$$I_{\text{rod}} = M \left(\frac{R^2}{2} + \frac{L^2}{12} \right).$$

Also $r_g^2 = I_{\text{rod}}/M$, hence

$$r_g^2 = \left(\frac{R^2}{2} + \frac{L^2}{12} \right) \qquad \text{Q.E.D.}$$

22. Equation (6.72) contains two terms. We handle each separately. Both, however, contain the expression $(\sin x)/x$. We begin by simplifying this:

$$\frac{\sin x}{x} \approx \frac{x - x^3/6}{x} = 1 - x^2/6.$$

We insert this into our first term of Equation (6.72):

$$\frac{2 \int_0^{QL} \frac{\sin u}{u} du}{QL} \approx \frac{2 \int_0^{QL} 1 - (u^2/6) \, du}{QL} = 2 - \frac{(QL)^2}{9}.$$

The second term is

$$\left[\frac{\sin(QL/2)}{(QL/2)} \right]^2 \approx \left[1 - \frac{(QL)^2}{24} \right]^2$$

$$= 1 - \frac{(QL)^2}{12} + \frac{(QL)^4}{576}.$$

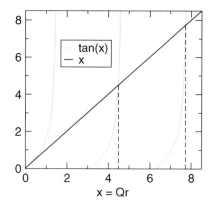

Figure E.11   The solutions for $\tan(x - n\pi) = x$ (integer $n \geq 1$) can be determined where the curves $y = x$ and $y = \tan(x)$ cross one another. The first two solutions $x = 4.493$ and $x = 7.725$ are shown.

We ignore the fourth-power term in (QL) as $QL \ll 1$. Now, bringing both these simplified terms together, we obtain

$$I(QL \ll 1) \approx 2 - \frac{(QL)^2}{9} - 1 + \frac{(QL)^2}{12} = 1 - \left(\frac{QL}{6}\right)^2$$

$$\approx \exp\left[-\left(\frac{QL}{6}\right)^2\right].$$

We equate this to $\exp(-Q^2 r_g^2/3)$, from which it emerges that $r_g^2 = L^2/12$, as we set out to demonstrate.

23. From the hint supplied, all we need to do is determine at which values of $x = Qr$ the numerator within the curly brackets is equal to zero, that is, when

$$\sin x - x\cos x = 0.$$

If we divide this by $\cos x$, one might assume that the first minimum occurs when $\tan x = x$. This is incorrect, as this leads to $x = 0$ and we are dividing zero by itself. But $\tan(x - n\pi) = \tan x$ (n being an integer), therefore our first minimum occurs when

$$\tan(x - \pi) = x,$$

from which we can numerically determine that $x \approx 4.4935$ (see Figure E.11). As n increases, x asymptotically approaches $(n + 1/2)\pi$.

In order to determine the value of $I(Q)/I_0$ at $x = 0$, we need to substitute the cosine and sine terms in the numerator with their Taylor expansions so that the latter is expressed up to powers of x^3, i.e. that of the denominator. The Taylor expansions of $\sin x$ and $\cos x$ are, respectively,

$$\sin x \approx x - \frac{x^3}{6},$$

$$\cos x \approx 1 - \frac{x^2}{2},$$

and hence

$$\frac{\sin x - x \cos x}{x^3} \approx \frac{x - x^3/6 - x(1 - x^2/2)}{x^3}$$
$$= \frac{x - x^3/6 - x + x^3/2}{x^3}$$
$$= \frac{1}{3}.$$

Substituting this back into our original equation, we immediately see that indeed $I(Q = 0)/I_0 = 1$.

24. The beating due to interference between reflections from the outer and inner shell walls of apoferritin produces envelope minima in the SAXS plot at multiples of $Q = 2\pi/\Delta r$. This is estimated to be at $Q = 0.23$ Å^{-1}, from which we obtain $\Delta r = 27$ Å, meaning apoferritin has an inner diameter of $130 - 54 = 76$ Å.

25. The equation describing the radius of gyration r_g of a hollow ellipsoid is

$$r_g^2 = \frac{(1 - \alpha^3\beta\gamma)a^2 + (1 - \alpha\beta^3\gamma)b^2 + (1 - \alpha\beta\gamma^3)c^2}{5(1 - \alpha\beta\gamma)}, \tag{E.10}$$

whereby a, b, and c are the outer semi-axes, and αa, βb, and γc are the inner semi-axes. In this example, $a = b = 250$ nm, $c = 1000$ nm, and $\alpha = \beta = (250 - 40)/250 = 0.84$ and $\gamma = (1000 - 40)/1000 = 0.96$. Inserting these into Equation (E.10), we obtain

$$r_g^2 = \frac{(2 \times 0.52204 \times 250^2) + (0.37573 \times 1000^2)}{5 \times 0.32262} [\text{nm}^2]$$
$$= \frac{65255 + 375730}{1.61312} [\text{nm}^2]$$
$$= 273374 \text{ nm}^2$$
$$\Rightarrow r_g = 523 \text{ nm}.$$

In order to measure this in a SAXS experiment, we must record the scattering curve in the regime where $Q \ll 2\pi/r_g$ (the Guinier regime). In the problem, it was given that this is satisfied if $Q \le \pi/2r_g$. But $Q = 4\pi \sin\theta/\lambda$ and therefore

$$Q \le \frac{\pi}{2r_g} = \frac{4\pi}{\lambda} \sin\theta$$
$$\Rightarrow \sin\theta_{max} = \frac{1}{4r_g} = \frac{1}{20\,920},$$

whereby R was expressed in Å. The maximum angle of the Guinier regime on the detector is at $2\theta = 1/10\,460$ and this angle should subtend at least 10 pixels, or 750 μm. For this to be satisfied, the sample–detector distance L is given by

$$\frac{0.75}{L[\text{mm}]} = \frac{1}{10\,460}$$
$$\Rightarrow L = 7845 \text{ mm}.$$

26. The scattering intensity of a Gaussian chain is given by Equation (6.73), that is,

$$I(Q) = \frac{2I_0}{\phi}(e^{-\phi} + \phi - 1).$$

At large Q, $\exp(-\phi)$ tends to zero, and $\phi \gg 1$. Therefore

$$\frac{I(Q)}{I_0} = \frac{2}{\phi^2}\phi = \frac{2}{\phi}.$$

But $\phi = Q^2 a^2 n/6$. A plot of $Q^2 I(Q)$ against Q therefore tends asymptotically to $12/(a^2 n)$.

E.7   Chapter 7 – Spectroscopic Techniques

1. Equation (2.16) describes the relationship between the imaginary component of the dispersion correction, $f'' = f_2$, the photon wavelength λ, and the absorption cross-section σ_a, namely

$$f''(\hbar\omega) = \frac{\sigma_a}{2 r_0 \lambda},$$

where $r_0 = 2.82 \times 10^{-5}$ Å is the Thomson scattering length. But the photon energy $E = hc/\lambda$, and hence

$$f''(\hbar\omega) = \frac{E \sigma_a}{2 h c r_0}.$$

All we need to do now is determine the constant $1/2hcr_0$, remembering we want to express E and σ_a in eV and barn, respectively. This is therefore

$$\frac{1.6022 \times 10^{-19} \times 10^{-28}}{2 \times 6.626 \times 10^{-34} \times 2.9979 \times 10^8 \times 2.82 \times 10^{-15}} = 1.431 \times 10^{-8} \quad \text{Q.E.D.}$$

2. From Figure 5.30, we immediately see that

$$\frac{D}{l} = \sin\theta,$$

whereby l is the distance AB between the two reflections. Similarly,

$$\frac{\Delta y}{l} = \sin 2\theta$$

$$= 2\sin\theta\cos\theta.$$

Substituting for l using our first equation, we obtain

$$\Delta y = 2D\cos\theta \qquad \text{Q.E.D.}$$

Using Bragg's law, we obtain $\sin\theta = 0.2471$ at 8 keV, and $\sin\theta = 0.1797$ at 11 keV, from which we can determine the corresponding values of $\cos\theta$ to be 0.9690 and 0.9837, respectively. The difference between these, multiplied by 20 mm, yields a shift in beam height upwards of 0.2947 mm as the monochromator energy is changed from 8 to 11 keV.

3. To solve this problem, we need only to determine the thickness of a 100%-dense pellet of $SrTiO_3$ that would produce a tenfold decrease in transmission as one scans across the Sr K-edge. In general, the fractional transmitted intensity of x-rays passing through a material having an absorption coefficient μ and thickness z is

$$\frac{I}{I_0} = \exp(-\mu z).$$

Therefore the ratio of the transmitted intensity just below the Sr K-edge and that just above it is

$$\frac{I_{\text{below}}}{I_{\text{above}}} = \frac{\exp(-\mu_- z)}{\exp(-\mu_+ z)}$$

$$= \exp[(\mu_+ - \mu_-)z]$$

$$\Rightarrow \ln(10) = z(0.03146 - 0.007999)$$

$$\Rightarrow z = 98.15 \; \mu\text{m}.$$

Note that such a thin pellet (less than a tenth of a mm) is unlikely to be mechanically robust, and is the reason why a transparent polymer-powder filler is added. The volume of a pure $SrTiO_3$ pellet is $\pi r^2 h = \pi \times 0.65^2 \times 0.009816 = 0.01303 \; \text{cm}^3$. The mass of $SrTiO_3$ to be added is therefore 67 mg.

4. To solve this problem of two unknowns (the thicknesses x and y of the Si and Al, respectively), we are provided with two pieces of information, namely the transmissions, Tr, of the coated diode at two x-ray energies. From the Beer–Lambert equation, we can therefore state

$$\text{Tr}(6 \text{ keV}) = \exp[-(\mu_{\text{Si}}^{(6)} x + \mu_{\text{Al}}^{(6)} y)],$$

$$\text{Tr}(10 \text{ keV}) = \exp[-(\mu_{\text{Si}}^{(10)} x + \mu_{\text{Al}}^{(10)} y)]. \tag{E.11}$$

We rearrange the first of these two equations to obtain

$$\ln\left(\frac{1}{\text{Tr}(6 \text{ keV})}\right) = \mu_{\text{Si}}^{(6)} x + \mu_{\text{Al}}^{(6)} y$$

$$\Rightarrow x = \frac{\ln\left[\frac{1}{\text{Tr}(6 \text{ keV})}\right] - \mu_{\text{Al}}^{(6)} y}{\mu_{\text{Si}}^{(6)}}.$$

Inserting this expression for x into Equation (E.11) leads to

$$\ln\left(\frac{1}{\text{Tr}(10 \text{ keV})}\right) = \mu_{\text{Al}}^{(10)} y + \mu_{\text{Si}}^{(10)} \left\{ \frac{\ln[1/\text{Tr}(6 \text{ keV})] - \mu_{\text{Al}}^{(6)} y}{\mu_{\text{Si}}^{(6)}} \right\}$$

$$= \left[\mu_{\text{Al}}^{(10)} - \mu_{\text{Al}}^{(6)} \frac{\mu_{\text{Si}}^{(10)}}{\mu_{\text{Si}}^{(6)}} \right] y$$

$$+ \frac{\mu_{\text{Si}}^{(10)}}{\mu_{\text{Si}}^{(6)}} \ln[1/\text{Tr}(6 \text{ keV})]. \tag{E.12}$$

We determine the attenuation coefficients for Si and Al from the cxro website. These are $\mu_{\text{Si}}^{(6)} = 0.033\,005$ μm^{-1}, $\mu_{\text{Al}}^{(6)} = 0.029\,76 \; \mu\text{m}^{-1}$, $\mu_{\text{Si}}^{(10)} = 0.007\,479 \; \mu\text{m}^{-1}$, and $\mu_{\text{Al}}^{(10)} = 0.006\,684 \; \mu\text{m}^{-1}$. The only remaining unknown in Equation (E.12) is y, the thickness of the aluminium layer, which we can therefore directly calculate from the provided information:

$$2.444\,17 = \left[0.006\,684 - 0.029\,76 \times \frac{0.007\,479}{0.033\,005} \right] y + \frac{0.007\,479}{0.033\,005} \times 10.793\,92$$

$$\Rightarrow 2.444\,17 = -6.080\,94 \times 10^{-5} y + 2.445\,92$$

$$\Rightarrow y = 28.85 \; \mu\text{m}.$$

Now it is a simple exercise to determine x, the Si thickness, from either of the total transmissions [see e.g. Equation (E.11)], which yields

$$x = 301 \ \mu\text{m}.$$

5. The energy of the helium atom is 100 meV, or 1.6022×10^{-20} J. The velocity is $\sqrt{2E_{\text{He}}/m_{\text{He}}} = 2196.4$ m s^{-1}. The de Broglie wavelength is therefore $\lambda_{\text{He}} = h/(m_{\text{He}} v_{\text{He}}) = 0.454$ Å.

6. The kinetic energy of a free electron moving at a velocity $c/10$ is approximately

$$\mathscr{E}_e = \frac{m_e c^2}{200} = 2553 \ \text{eV},$$

or approximately 2500 eV.

In the low-energy regime, that is, $eV \ll m_e c^2$, we ignore the eV in the round brackets of Equation (7.38), and obtain

$$\lambda_e \approx \frac{h}{\sqrt{2m_e eV}}.$$

But $eV = m_e v^2/2$ equals the nonrelativistic kinetic energy of the electron, and hence

$$\lambda_e \approx \frac{h}{\sqrt{2m_e \times m_e v^2/2}} = \frac{h}{p},$$

the classical expression for the de Broglie wavelength of nonrelativistic electrons.

In the other limit, we now ignore the rest-mass energy, leading to

$$\lambda_e \approx \frac{hc}{eV}.$$

This expression is simply that for the wavelength of a *photon*. In other words, as particles with mass become highly relativistic, they assume increasingly the character of a photon. Note also that for a given energy, photons *always* have longer wavelengths than those of particles with mass. Figure E.12 shows this asymptotic behaviour for electrons, muons (with a mass approximately 200 times that of an electron), and protons.

The turning point in energy $\mathscr{E}_e$ (that is, the position where there is a change in gradient in the double-logarithmic plot of Figure E.12) midway between the two limiting cases for a particle with mass (here, an electron) is when

$$\frac{h}{\sqrt{2m_e \mathscr{E}_e}} = \frac{hc}{\mathscr{E}_e}$$

$$\Rightarrow \frac{1}{2m_e \mathscr{E}_e} = \frac{c^2}{\mathscr{E}_e^2}$$

$$\Rightarrow \mathscr{E}_e = 2m_e c^2 = 1.022 \ \text{MeV}.$$

7. The electron wavelength is given by the de Broglie law

$$\lambda_e = h/p = \frac{h}{m_e v}$$

$$= \frac{h}{\sqrt{2m_e E}},$$

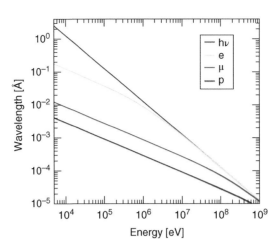

Figure E.12   The change of the de Broglie wavelengths of electrons, muons, and protons compared to that of photons, as a function of total energy. In this double-logarithmic plot, the gradient changes from being $-1/2$ in the nonrelativistic regime well below mc^2 to -1 at highly relativistic energies.

whereby E is the common energy of the photon and electron. This is equal to hc/λ_p for the photon. Therefore

$$\lambda_e^2 = \frac{h^2 \lambda_p}{2m_e hc}$$

$$\Rightarrow \frac{\lambda_e}{\lambda_p} = \frac{h^2}{2m_e hc \lambda_e} = \frac{h^2}{2m_e hc \times h/(m_e v)}$$

$$= \frac{v}{2c} \qquad \text{Q.E.D.}$$

8. The energy-dependent equation describing the universal curve follows the form

$$y = \frac{A}{x^2} + Bx^{1/2},$$

where y is the IMFP and x is the electron energy. To find the minimum of this, we need to determine where $dy/dx = 0$. But

$$\frac{dy}{dx} = \frac{-2A}{x^3} + \frac{B}{2x^{1/2}} = 0$$

$$\Rightarrow \frac{2x^{1/2}}{B} = \frac{x^3}{2A}$$

$$\Rightarrow x = \left(\frac{4A}{B}\right)^{2/5} = 40.74 \text{ eV}.$$

9. The binding energy of selenium $3d$ electrons is 55.1 eV, hence the kinetic energy of the ARPES and SX-ARPES electrons are 39.5 and 1939.5 eV, respectively. Inserting these into the best-fit equation for the universal curve, we obtain $\Lambda_e^{\text{SX-ARPES}} = 23.78$ Å, and $\Lambda_e^{\text{ARPES}} = 4.31$ Å. Their ratio is therefore 5.52.

10. The energy separation between adjacent Rydberg states n and $n + 1$ is

$$\Delta E = R_H \left[\frac{1}{n^2} - \frac{1}{(n + 1)^2} \right]$$

$$= R_H \left[\frac{2n + 1}{n^2(n + 1)^2} \right].$$

We assume that $n \gg 1$ and make the approximation that $n^2(n + 1)^2 \approx n^4$ and $2n + 1 \approx 2n$, from which it follows that

$$\Delta E \approx \frac{2R_H}{n^3}.$$

Because these states are very close to the ionization threshold, the photon energy will be approximately $R_H = 13.6$ eV. The fractional bandwidth of the synchrotron radiation is $\Delta E/E = \Delta E/R_H = 10^{-4}$, from which we obtain

$$\frac{1}{10\,000} = \frac{2}{n^3},$$

which is most closely approximated for $n = 27$.

    Side note: We could also have arrived at our expression for ΔE by differentiating $E_n = -R_H/n^2$ with respect to n – that is, $dE_n = 2R_H/n^3 \, dn$, which, for the smallest increment in n of $dn = 1$, yields $\Delta E = 2R_H/n^3$.

11. Beginning with Bragg's law $\lambda = 2d \sin\theta$, and substituting for $\lambda = hc/E$, we have

$$E = \frac{hc}{2d \sin\theta}.$$

We differentiate with respect to θ to obtain

$$\frac{dE}{d\theta} = \frac{\Delta E}{\Delta\theta} = \frac{-hc\cos\theta}{2d\sin^2\theta} = \frac{-E}{\tan\theta}$$

$$\Rightarrow \frac{\Delta E}{E} = \frac{-\Delta\theta}{\tan\theta}.$$

But for shallow Bragg angles, $\tan\theta \approx \sin\theta = hc/(2Ed)$, and so (ignoring the sign), we obtain

$$\Delta\theta = \frac{hc}{2d}\frac{\Delta E}{E^2} \qquad \text{Q.E.D.}$$

The K-edge of palladium is at 24.35 keV. Using our practical expression for $\Delta\theta$ given in Equation (7.22), the angular stroke need to cover ± 500 eV is

$$\Delta\theta = \frac{355.188}{3.1356 \times 24.35^2} \times 1 = 0.191°.$$

This stroke needs to occur within 0.02 s, meaning an angular rotation rate of 9.552 degrees/second. The spectral resolution needs to be 1 eV or better, meaning that the intensity monitors need to read out at a rate of at least $1000/0.02 = 5 \times 10^4$ Hz.

12. The number of photons/s in a 2 mW beam of 1.24 angstrom (10 keV) photons is $0.002/(1.6022 \times 10^{-15}) = 1.248 \times 10^{12}$. The beam is incident at an angle of $0.5°$, and thus travels a distance of $2/\sin(0.5°) = 229$ μm. This is much larger than the attenuation length, and hence we can assume that all the photons are absorbed. The fluorescence efficiency is 0.8, hence the number of fluorescent photons/second is 10^{12}.

13. The absorption length $1/\mu$ of ZnO at 9900 eV is much larger than the film thickness z, hence the transmission $\exp(-\mu z)$ can be approximated by $(1 - \mu z)$, and the fraction of absorbed photons is therefore $\mu z = 0.09/9 = 0.01$. The number of absorbed photons is simply $0.01 \times 10^{13} = 10^{11}$. But from Figure 2.28, we determine that only half of these produce fluorescence; the other half result in Auger electrons. Therefore the total rate of fluorescent photons is 5×10^{10} s^{-1} emitted isotropically (that is, over 4π steradians).

The detector captures $\pi 2^2/(4\pi 20^2) = 0.01$ of these. The recorded count rate is thus 5×10^8 s^{-1}.

14. Here, we use the energy-balance equation

$$h\nu = E_B + E_e + e\phi,$$

whereby $h\nu$ is the photon energy, E_B is the binding energy, E_e is the photoelectron's kinetic energy, and $e\phi$ is the work function. But

$$E_e[\text{eV}] = \left(\frac{12.28}{\lambda_e[\text{Å}]}\right)^2,$$

which in this example leads to $E_e = 35.88$ eV. Therefore, we have

$$500 = 460 + 35.88 + e\phi,$$

and hence $e\phi = 4.12$ eV.

15. From lowest binding energy to highest: Ti, TiO, TiN, Ti_2O_3, TiO_2, TiF_3.

16. From the EDM, we know that $E_k/k_\parallel^2 = \hbar^2/2m_{\text{eff}} = 0.025/0.01 = 2.5$ eV Å^2. Converting this to J m^2 (i.e. multiplying by 1.6022×10^{-39}), we obtain $m_{\text{eff}} = 1.39 \times 10^{-30}$ kg $= 1.53\ m_e$.

17. The photoelectrons emerging from the Fermi surface of the (conducting) TiC layer have a kinetic energy $\mathscr{E}_e = 460 - 5 = 455$ eV. The associated wavevector has a magnitude

$$k\ [\text{Å}^{-1}] = 0.512\ \sqrt{\mathscr{E}_e[\text{eV}]}$$

$$= 10.92\ \text{Å}^{-1}.$$

The maximum detectable in-plane component of k is $k_\parallel = k\ \sin(8°) = 0.1392\ k = 1.520$ Å^{-1}. But the nth Brillouin zone has its boundaries at $\pm n\pi/a$, whereby here $a = 4.4$ Å, and hence $\pi/a = 0.714$ Å^{-1}. In the above configuration, therefore, the first two Brillouin zones ($n = 1, 2$) can be accessed. Their boundaries are at the angles $\pm \arcsin(0.714/10.92) = \pm 3.75°$ and $\pm \arcsin(2 \times 0.714/10.92) = \pm 7.51°$, respectively.

E.8   Chapter 8 – Imaging Techniques

1. From the definition of HU, we require

$$11.5 = \frac{\mu_X - \mu_w}{\mu_w},$$

whereby μ_X and μ_w are the reciprocal values of the data provided in Figure 8.35. This is shown in Figure E.13, from which we can glean that the tomography experiment was carried out at a photon energy of approximately 26.8 keV. Note that a second solution would be at around 4 keV (outside the range of photon energies shown in Figure 8.35), though here, the attenuation length of bone is only approximately 10 μm, rendering the experiment impractical for any sample much larger that a few tens of micron. Also, the question states that the experiment was carried out at a hard x-ray beamline.

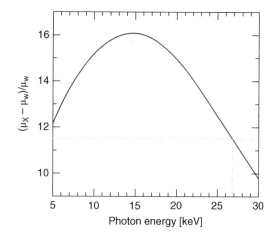

Figure E.13   The function $(\mu_x - \mu_w)/\mu_w$ between 5 and 30 keV.

2. Let $f(x, y)$ be a two-dimensional function. We project this on to the x-axis by integrating in the y-direction to obtain the projection $p(x)$:

$$p(x) = \int_{-\infty}^{\infty} f(x, y) dy. \tag{E.13}$$

The two-dimensional Fourier transform $\mathscr{F}(k_x, k_y)$ of $f(x, y)$ is

$$\mathscr{F}(k_x, k_y) = \int_{-\infty}^{\infty} \int_{-\infty}^{\infty} f(x, y) \exp[-2\pi i(xk_x + yk_y)] dx\, dy.$$

The one-dimensional slice $s(k_x)$ through the origin of this (i.e. at $k_y = 0$) is

$$s(k_x) = \mathscr{F}(k_x, 0) = \int_{-\infty}^{\infty} \int_{-\infty}^{\infty} f(x, y) \exp(-2\pi i xk_x) dx\, dy.$$

$$= \int_{-\infty}^{\infty} \left[\int_{-\infty}^{\infty} f(x, y) dy \right] \exp(-2\pi i xk_x) dx.$$

But we have already identified the term within the square parentheses in Equation (E.13) as being $p(x)$, and therefore

$$s(k_x) = \int_{-\infty}^{\infty} p(x) \exp(-2\pi i xk_x) dx,$$

which is simply the one-dimensional Fourier transform of the projection $p(x)$. Q.E.D.

3. The optimal regime for edge enhancement is when the sample–detector distance Z_0 is approximately a^2/λ, where a is the characteristic length to be resolved. In this experiment, $\lambda \approx 0.5$ Å. This yields $Z_0 \approx 500$ mm for the study of the fly wing, and $Z_0 \approx 1.25$ mm for the retinal cells.

4. μ varies with the inverse third power of the photon energy, hence we can expect the absorption coefficient μ at 50 keV to be $6.48 \times 10^{-5}/(2.5)^3 = 4.147 \times 10^{-6}$ μm^{-1}. The transmission is therefore $\exp(-4.147 \times 10^{-6} \times 2 \times 10^3) = 0.992$.

5. The phase shift is equal to $2\pi L\delta/\lambda$. But δ is inversely proportional to the square of the photon energy, and λ is inversely proportional to the photon energy, hence ϕ is itself inversely proportional to the photon energy. As in this problem, the energy is doubled, ϕ must be halved, that is, equal to $20°$.

6. We see that the phase ring is situated at the focal length of the objective lens. But from the lensmaker equation, we know $1/f = 1/p + 1/q$ and hence $f = pq/(p+q)$.

   For $M = 50$ and $p = 50$ mm, $q = 2500$ mm, the position of the image plane downstream of the objective. Hence $f = (50 \times 2500)/2550 = 49.02$ mm.

7. To determine I_g/I_b in absorption-contrast mode (i.e. without phase ring), we need simply to add the vectors for each case. Let $A_g = A_b = A = A_r/40$, where A_r is the amplitude of the reference wave. For the green scattering vector, the vector sum is

$$A\{[40 + \cos(15)]^2 + \sin^2(15)\}^{1/2} = A(1678.274)^{1/2}.$$

For the blue scattering vector, it is

$$A\{[40 + \cos(45)]^2 + \sin^2(45)\}^{1/2} = A(1657.569)^{1/2}.$$

From this, we obtain $I_g/I_b = 1678.274/1657.569 = 1.0125$.

Before we can determine I_g/I_b when the phase ring is in place, we need to determine the thickness to obtain a phase shift of $\pi/2$ and through this, also its degree of attenuation. We use Equation (8.7) to obtain

$$\pi/2 = \frac{2\pi \times 7 \times 10^{-7} L}{5 \times 10^{-11}}$$

$$\Rightarrow L = 17.857 \; \mu\text{m}.$$

Thus the transmission of the ring is $\exp(-17.857/11) = 0.1972$. This means that the amplitude $A_r = 0.1972 \times 40A = 7.889\,A$.

Now, the vector sum of A_r and A_g is

$$A_r + A_g = A\{[7.889 - \sin(15)]^2 + \cos^2(15)\}^{1/2} = A(59.158)^{1/2}.$$

For the blue scattering vector, it is

$$A_r + A_b = A\{[7.889 - \sin(45)]^2 + \cos^2(45)\}^{1/2} = A(52.079)^{1/2}.$$

From this, we obtain $I_g/I_b = 59.158/52.079 = 1.136$. The contrast using the phase ring is thus over 10 times stronger than in absorption mode.

8. The number of pixels that provide a useable signal is equal to $\pi 70^2/0.172^2 = 5.2 \times 10^5$. The object has a volume in terms of voxels of $500/0.2^3 = 62\,500$. The redundancy is therefore $5.2 \times 10^5/(6.25 \times 10^4) = 8.32$.

9. The symmetry of the setup shown in Figure 8.36 demands that the incident beam be parallel to the *l*-axis of the reciprocal lattice. In this configuration, the plane containing the eight equivalent (311) reflections thus lies perpendicular to the incident beam and crosses it at $l = 1$ [i.e. at the (001) Bragg spot]. The circle in reciprocal space that intercepts all these equivalent reflections has a radius $(h^2 + k^2)^{1/2} = (3^2 + 1^2)^{1/2} = \sqrt{10}$. This is equal to $Q_{\parallel}$, the component of the scattering vector Q that lies parallel to the *hk*-plane (see Figure E.14).

   We need to determine the photon energy that produces an Ewald sphere that intersects the (000) direct beam (which any Ewald sphere will do) *and* simultaneously the eight equivalent reflections. The radius of the Ewald sphere is $|k|$, which has the same magnitude as k_{in} and k_{out}. In terms of the reciprocal lattice vector $2\pi/a$, where $a = 5.431$ Å is the cubic unit cell size of silicon, this is simply $k_{\text{in}} = (2\pi/\lambda)/(2\pi/a) = a/\lambda$. Therefore, using Pythagoras' theorem, we can state

$$\left(\frac{a}{\lambda} - 1\right)^2 + Q_{\parallel}^2 = \left(\frac{a}{\lambda}\right)^2 \Rightarrow \lambda = \frac{2a}{11},$$

from which we obtain $\lambda = 2 \times 5.431/11 = 0.98745$ Å, or $E = 12.556$ keV.

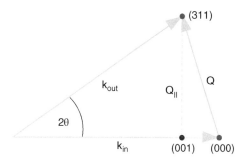

Figure E.14   The geometry of the incident and scattered beams and the scattering vector Q relative to pertinent points in reciprocal space, in reference to the setup sketched in Figure 8.36.

Also from Figure E.14, we see that $\sin 2\theta = \sqrt{10}/(a/\lambda) = 2\sqrt{10}/11 = 0.57496$, from which we obtain $\theta = 17.548°$.

10. From the definition of brilliance as being the number of photons per unit time, generated per unit area source size, per unit solid angle divergence, and per fractional bandwidth (BW) $\Delta\lambda/\lambda$, the photon flux F_{coh} of coherent radiation is therefore

$$F_{\text{coh}} = \mathscr{B} \times (w_h \times w_v) \times \left(\frac{\xi_{t,h}}{R} \times \frac{\xi_{t,v}}{R} \right) \times \left(\frac{\Delta\lambda}{\lambda} \right)$$

$$= \mathscr{B} \times (w_h \times w_v) \times \left(\frac{\lambda R}{w_h R} \times \frac{\lambda R}{w_v R} \right) \times \left(\frac{\Delta\lambda}{\lambda} \right)$$

$$= \mathscr{B} \times \lambda^2 \times \frac{\Delta\lambda}{\lambda} \qquad \text{Q.E.D.} \qquad (E.14)$$

The transverse coherent lengths $\xi_{t,h}$ and $\xi_{t,v}$ are given by

$$\xi_{t,h} = \frac{\lambda R}{w_h},$$

$$\xi_{t,v} = \frac{\lambda R}{w_v},$$

whereby $w_h = 200\,\mu\text{m}$ and $w_v = 50\,\mu\text{m}$ are the horizontal and vertical source size, respectively. At a distance $R = 40$ m and a wavelength of 1.77 Å (7-keV energy), $\xi_{t,h} = 35.4\,\mu\text{m}$ and $\xi_{t,v} = 142\,\mu\text{m}$.

In order to determine the coherent flux F_{coh}, we must multiply the brilliance $\mathscr{B}$ by the source size, the source divergence, and the bandwidth. We have already derived the appropriate expression in Equation (E.14). However, one must proceed very carefully to ensure that the same units have been used in defining all the above properties. In the 'normal' definition of $\mathscr{B}$ the source size is given in mm², so we must express λ also in mm. This turns out to be 1.77×10^{-7} mm. Also, in our derivation above, we expressed the solid angle in radians², but in the definition of $\mathscr{B}$, it is given in mrad², hence we need to take into account a factor of 10^6. Finally, the monochromator bandwidth is $\Delta\lambda/\lambda = 1.3 \times 10^{-4}$, or 0.13 compared to a 0.1% bandwidth. From this we obtain

$$F_{\text{coh}} = 10^{21} \times (1.77 \times 10^{-7})^2 \times 10^6 \times 0.13 = 4.07 \times 10^{12} \text{ photons per second.}$$

Appendix F

Glossary

Table F.1 Glossary of abbreviations

Abbreviation	Definition
AES	Auger-electron spectroscopy
APPES	Ambient-pressure photoemission spectroscopy
APPLE	Advanced planar polarized light emitter
ARPES	Angle-resolved photoemission spectroscopy
BC	Bunch compression
BM	Bending magnet
BPM	Beam-position monitor
BSB	Bremsstrahlung blocker
BZ	Brillouin zone
CCD	Charge-coupled device
CCM	Channel-cut monochromator
CMOS	Complementary metal oxide semiconductor
CHA	Concentric hemispherical analyser
CRL	Compound refractive lens
cryoEM	Cryogenic electron microscopy
CTR	Crystal truncation rod
CVD	Chemical vapour deposition
CXDI	Coherent x-ray diffractive imaging
DBA	Double-bend achromat
DCD	Double-crystal deflector
DCM	Double-crystal monochromator
DED	Direct electron detector
DLSR	Diffraction-limited storage-ring
DMM	Double-multilayer monochromator
EDC	Energy-dispersion curve
EDS/EDX	Energy-dispersive x-ray spectrometer

(Continued)

An Introduction to Synchrotron Radiation: Techniques and Applications, Second Edition. Philip Willmott.
© 2019 John Wiley & Sons Ltd. Published 2019 by John Wiley & Sons Ltd.

Table F.1   (*Continued*)

Abbreviation	Definition
EDXAS	Energy-dispersive x-ray absorption spectroscopy
EEHG	Echo-enabled harmonic generation
EGA	Entirely gratuitous abbreviation
EXAFS	Extended x-ray absorption fine structure
FS	Fermi surface
FSM	Fermi-surface map
FT	Fourier transform
FWHM	Full-width at half-maximum
FZP	Fresnel zone plate
GIC	Gridded ionization chamber
GISAXS	Grazing-incidence small-angle x-ray scattering
HAXPES	Hard x-ray photoemission spectroscopy
HB-SASE	High-brightness self-amplified spontaneous emission
HEROS	High energy-resolution off-resonant spectroscopy
HOMO	Highest occupied molecular orbital
HPAD	Hybrid pixel-array detector
HPXPS	High-pressure x-ray photoemission spectroscopy
ID	Insertion device
IFT	Inverse Fourier transform
IMFP	Inelastic mean free path
LCP	Left circularly polarized
LCP	Lipidic cubic phase
LEED	Low-energy electron-diffraction
LINAC	Linear accelerator
LUMO	Lowest unoccupied molecular orbital
MAD	Multi-wavelength anomalous dispersion
MBA	Multibend achromat
MCP	Multichannel plate
MDC	Momentum distribution curve
MIR	Multiple isomorphous replacement
MLL	Multilayer Laue lens
MR	Molecular replacement
MX	Macromolecular crystallography
NEG	Nonevaporable getter
NEXAFS	Near-edge x-ray absorption fine-structure (see XANES)
OPD	Optical path difference
PDF	Pair-distribution function
PDOS	Partial density of states
PEEM	Photoemission electron microscopy
PES	Photo(emission) electron spectroscopy
PM	Patterson map
PMT	Photomultiplier tube
PSB	Pseudo single-bunch
PSF	Point-spread function
PX	Protein crystallography (same as MX)
PZP	Phase zone plate
QEXAFS	Quick extended x-ray absorption fine-structure
RCP	Right circularly polarized

(*Continued*)

Table F.1 (*Continued*)

Abbreviation	Definition
RESOXS	Resonant soft x-ray scattering
RESPES	Resonant photoemission spectroscopy
RF	Radio frequency
RHEED	Reflection high-energy electron-diffraction
RIXS	Resonant inelastic x-ray scattering
r.l.u.	Reciprocal lattice unit
RT	Room-temperature
RXS	Resonant x-ray scattering
SAD	Single-wavelength anomalous diffraction
SASE	Self-amplified spontaneous emission
SAXS	Small-angle x-ray scattering
SB	Superbend
SEXAFS	Surface-sensitive extended x-ray absorption fine-structure
SF	Structure factor
SFX	Serial femtosecond crystallography
SIRAS	Single isomorphous replacement with anomalous scattering
SR	Synchrotron radiation
SSR	Superstructure rod
SSX	Serial synchrotron crystallography
STXM	Scanning transmission x-ray microscopy
SX-ARPES	Soft x-ray angle-resolved photoemission spectroscopy
SXRD	Surface x-ray diffraction
TBA	Triple-bend achromat
UPS	Ultraviolet photoemission spectroscopy
WDS/WDX	Wavelength-dispersive x-ray spectrometer
XANES	X-ray absorption near-edge structure
XAS	X-ray absorption spectroscopy
XFEL	X-ray free-electron laser
XMCD	X-ray magnetic circular dichroism
XMLD	X-ray magnetic linear dichroism
XPCS	X-ray photon correlation spectroscopy
XPD	X-ray photoelectron diffraction
XPS	X-ray photoemission spectroscopy
XRD	X-ray diffraction
XRF	X-ray fluorescence
XRR	X-ray reflectometry
XSW	X-ray standing waves
XTM	X-ray tomography

Appendix G

Physical Constants Relevant to Synchrotron Radiation

Table G.1

Constant	Symbol	Value
Speed of light in vacuum	c	$2.997\ 924\ 58 \times 10^8$ m s^{-1}
Planck's constant	h	$6.626\ 068\ 76 \times 10^{-34}$ J s
Reduced Planck's constant	$\hbar = h/2\pi$	$1.054\ 571\ 596 \times 10^{-34}$ J s
Rest mass of an electron	m_e	$9.109\ 381\ 88 \times 10^{-31}$ kg
Rest mass energy of an electron	$m_e c^2$	$81.871\ 041$ fJ $= 510.998\ 88$ keV
Elementary charge	e	$1.602\ 176\ 53 \times 10^{-19}$ C
Permeability of free space	$\mu_0 = 4\pi \times 10^{-7}$	$1.256\ 637\ 1 \times 10^{-6}$ V s/(A m)
Permittivity of free space	$\epsilon_0 = 1/\mu_0 c^2$	$8.854\ 187\ 82 \times 10^{-12}$ A s/(V m)
Classical electron radius (Thomson length)	$r_0 = e^2/4\pi\epsilon_0 m_e c^2$	$2.817\ 9 \times 10^{-15}$ m
Thomson cross-section	$\sigma_T = \frac{8}{3}\pi r_0^2$	$0.665\ 245$ barn
Fine structure constant	$\alpha = \mu_0 c e^2/2h$	$1/137.035\ 999\ 76$
Boltzmann's constant	k_B	$1.380\ 650\ 3 \times 10^{-23}$ J K^{-1}
Avogadro's number	N_A	$6.022\ 141\ 99 \times 10^{23}$ mol^{-1}
Absolute zero	θ_0	-273.15 °C
Gas constant	$R = k_B N_A$	$8.314\ 472$ J K^{-1} mol^{-1}
Normal pressure	p_n	$101\ 325$ Pa

An Introduction to Synchrotron Radiation: Techniques and Applications, Second Edition. Philip Willmott.
© 2019 John Wiley & Sons Ltd. Published 2019 by John Wiley & Sons Ltd.

Index

An Introduction to Synchrotron Radiation: Techniques and Applications, Second Edition. Philip Willmott.
© 2019 John Wiley & Sons Ltd. Published 2019 by John Wiley & Sons Ltd.